DEVELOPMENTAL SOCIAL NEUROSCIENCE AND CHILDHOOD BRAIN INSULT

Developmental Social Neuroscience and Childhood Brain Insult

THEORY AND PRACTICE

Edited by

Vicki Anderson
Miriam H. Beauchamp

THE GUILFORD PRESS
New York   London

© 2012 The Guilford Press
A Division of Guilford Publications, Inc.
72 Spring Street, New York, NY 10012
www.guilford.com

Printed in the United States of America

This book is printed on acid-free paper.

Last digit is print number: 9   8   7   6   5   4   3   2   1

The authors have checked with sources believed to be reliable in their efforts to provide information that is complete and generally in accord with the standards of practice that are accepted at the time of publication. However, in view of the possibility of human error or changes in behavioral, mental health, or medical sciences, neither the authors, nor the editor and publisher, nor any other party who has been involved in the preparation or publication of this work warrants that the information contained herein is in every respect accurate or complete, and they are not responsible for any errors or omissions or the results obtained from the use of such information. Readers are encouraged to confirm the information contained in this book with other sources.

Library of Congress Cataloging-in-Publication Data

Developmental social neuroscience and childhood brain insult : theory and practice / edited by Vicki Anderson, Miriam H. Beauchamp.
    p. cm.
  Includes bibliographical references and index.
  ISBN 978-1-4625-0429-9 (hardcover : alk. paper)
  1. Brain—Wounds and injuries—Complications.   2. Pediatric neurology.
3. Developmental neurobiology.   4. Social psychology.   5. Neurosciences.   I. Anderson, Vicki.
II. Beauchamp, Miriam H.
  RD594.D52 2012
  618.92′8—dc23

                              2012012537

About the Editors

Vicki Anderson, PhD, is Director of Psychology at the Royal Children's Hospital in Melbourne, Australia; Director of Critical Care and Neuroscience Research at the Murdoch Childrens Research Institute; and Professor of Pediatrics and Psychology at the University of Melbourne. Dr. Anderson's work focuses on the outcomes of developmental and acquired brain disorders in children, particularly traumatic brain injury. She has served on the Board of Governors of the International Neuropsychological Society and is past president of the Australian Society for the Study of Brain Impairment.

Miriam H. Beauchamp, PhD, is Assistant Professor in the Department of Psychology, University of Montréal, Québec, Canada, where she leads the ABCs Developmental Neuropsychology Laboratory. She is also a researcher at the Sainte-Justine Hospital Research Center and Adjunct Professor in the Department of Neurology and Neurosurgery at McGill University. Dr. Beauchamp's work focuses on investigating the environmental, cognitive, and neural substrates of social functioning in children and adolescents using both behavioral and neuroimaging methodologies.

Contributors

Amélie M. Achim, PhD, School of Psychology, Laval University Robert-Giffard Research Center, and Department of Psychiatry and Neurosciences, Faculty of Medicine, Laval University, Québec, Québec, Canada

Vicki Anderson, PhD, Department of Psychology, Royal Children's Hospital, Child Neuropsychology, Murdoch Childrens Research Institute, Paediatrics and Psychology, University of Melbourne, Parkville, Victoria, Australia

Miriam H. Beauchamp, PhD, University of Montréal, Sainte-Justine Hospital Research Center, Montréal, Québec, Canada

Gary Bedell, PhD, OTR, FAOTA, Department of Occupational Therapy, Tufts University, Medford, Massachusetts

Annie Schulz Begle, PhD, Department of Human Development, University of Maryland, College Park, Maryland

Erin D. Bigler, PhD, Department of Psychology, Brigham Young University, Provo, Utah; Department of Psychiatry, University of Utah, Salt Lake City, Utah

Rosée Bruneau-Bhérer, PhD, School of Psychology, Center for Interdisciplinary Research in Rehabilitation and Social Integration and Laval University Robert-Giffard Research Center, Québec, Québec, Canada

Cathy Catroppa, PhD, Department of Psychology, Royal Children's Hospital, Child Neuropsychology, Murdoch Childrens Research Institute, Paediatrics and Psychology, University of Melbourne, Parkville, Victoria, Australia

Kim Cornish, PhD, Centre for Developmental Psychiatry and Psychology, School of Psychology and Psychiatry, and Monash Institute for Brain Development and Repair, Monash University, Clayton, Victoria, Australia

Louise Crowe, PhD, Child Neuropsychology, Murdoch Childrens Research Institute, Parkville, Victoria, Australia

Julian J. Dooley, PhD, Sellenger Centre for Research in Law, Justice and Social Change, School of Law and Justice, Edith Cowan University, Joondalup, Western Australia, Australia

Baudouin Forgeot d'Arc, MD, PhD, Fernand Seguin Research Center, Rivière-des-Prairies Hospital, Montréal, Québec, Canada

Cynthia A. Gerhardt, PhD, Department of Pediatrics, The Ohio State University, and Center for Biobehavioral Health, The Research Institute at Nationwide Children's Hospital, Columbus, Ohio

Alison Gomes, BA(Hons), School of Psychology and Psychiatry, Monash University, Clayton, and Waiora Clinic, South Yarra, Victoria, Australia

Kylie M. Gray, PhD, Centre for Developmental Psychiatry and Psychology, School of Psychology and Psychiatry, and Monash Institute for Brain Development and Repair, Monash Medical Centre, Clayton, Victoria, Australia

Gerri Hanten, PhD, Department of Physical Medicine and Rehabilitation, Baylor College of Medicine, and Department of Psychology, Rice University, Houston, Texas

Paul D. Hastings, PhD, Department of Psychology, University of California–Davis, Davis, California

Stephanie Burnett Heyes, PhD, Department of Brain Repair and Rehabilitation, Institute of Neurology, University College London, London, United Kingdom; Department of Experimental Psychology, University of Oxford, Oxford, United Kingdom

Philip L. Jackson, PhD, School of Psychology, Center for Interdisciplinary Research in Rehabilitation and Social Integration and Laval University Robert-Giffard Research Center, Québec, Québec, Canada

Kathrin Cohen Kadosh, PhD, Department of Experimental Psychology, University of Oxford, Oxford, United Kingdom

Harvey S. Levin, PhD, Departments of Physical Medicine and Rehabilitation, Neurology, and Neurosurgery, Baylor College of Medicine, Houston, Texas

Jeffrey E. Max, MBBCh, Department of Psychiatry, University of California–San Diego, La Jolla, California; Rady Children's Hospital, San Diego, California

Kari L. Maxwell, MA, Department of Technology, Learning and Culture, West Virginia University, Morgantown, West Virginia

Kristina L. McDonald, PhD, Department of Psychology, University of Alabama, Tuscaloosa, Alabama

Skye McDonald, PhD, School of Psychology, University of New South Wales, Sydney, New South Wales, Australia

Laurent Mottron, MD, PhD, Fernand Seguin Research Center, Rivière-des-Prairies Hospital, Montréal, Québec, Canada

Frank Muscara, DPsych, Child Neuropsychology, Murdoch Childrens Research Institute, Parkville, Victoria, Australia

Mary R. Newsome, PhD, Department of Physical Medicine and Rehabilitation, Baylor College of Medicine, Houston, Texas

Ronald M. Rapee, PhD, Department of Psychology, Centre for Emotional Health, Macquarie University, Sydney, New South Wales, Australia

Amy E. Root, PhD, Department of Technology, Learning and Culture, West Virginia University, Morgantown, West Virginia

Stefanie Rosema, MSc, Child Neuropsychology, Murdoch Childrens Research Institute, Parkville, Victoria, Australia

Kenneth H. Rubin, PhD, Department of Human Development, University of Maryland, College Park, Maryland

Randall S. Scheibel, PhD, Department of Physical Medicine and Rehabilitation, Baylor College of Medicine, Houston, Texas

Catherine L. Sebastian, PhD, Department of Clinical, Educational and Health Psychology and Institute of Cognitive Neuroscience, University College London, London, United Kingdom

Cheryl Soo, PhD, Child Neuropsychology, Murdoch Childrens Research Institute, Parkville, Victoria, Australia

Terry Stancin, PhD, Departments of Pediatrics and Psychiatry, Case Western Reserve University and MetroHealth Medical Center, Cleveland, Ohio

Bradley C. Taber-Thomas, MA, Department of Neurology and Neuroscience Graduate Program, University of Iowa, Iowa City, Iowa

Robyn L. Tate, PhD, Rehabilitation Studies Unit, Northern Clinical School, University of Sydney, Sydney, New South Wales, Australia

H. Gerry Taylor, PhD, Department of Pediatrics, Case Western Reserve University and Rainbow Babies and Children's Hospital, Cleveland, Ohio

Leanne Togher, PhD, Discipline of Speech Pathology, Faculty of Health Sciences, University of Sydney, Lidcombe, New South Wales, Australia

Daniel Tranel, PhD, Departments of Neurology and Psychology, University of Iowa, Iowa City, Iowa

Lyn S. Turkstra, PhD, Department of Communicative Disorders, University of Wisconsin–Madison, Madison, Wisconsin

Kathryn Vannatta, PhD, Department of Pediatrics, The Ohio State University, and Center for Biobehavioral Health, The Research Institute at Nationwide Children's Hospital, Columbus, Ohio

Damith T. Woods, PhD, Child Neuropsychology, Murdoch Childrens Research Institute, Parkville, Victoria, Australia

Keith Owen Yeates, PhD, Department of Pediatrics, The Ohio State University, and Center for Biobehavioral Health, The Research Institute at Nationwide Children's Hospital, Columbus, Ohio

Preface

For several decades now, those working with individuals suffering from early brain insult and developmental disabilities have focused much effort on improving our understanding of the cognitive impairments experienced by these young people. As a result, an extensive body of literature now describes the cognitive consequences of childhood brain disorders, as well as the factors that might predict their presence and severity. For example, we know that early brain insult is commonly linked to deficits in attention and memory, executive function, and speed of processing, as well as to academic failure. Furthermore, the literature has established that greater severity of insult, younger age at injury, and dysfunctional social and family environments are associated with poor outcomes. In contrast, there is surprisingly little information regarding the impact of early brain insults on social skills or the factors that might predict them. This neglect of the social domain is surprising, given that (1) young people and their families frequently complain that social problems are the most devastating of all consequences following early brain insult, and (2) social dysfunction has been found to lead to significant reductions in quality of life and increased risk of mental health problems.

Social skills and interactions form the foundation of human consciousness. They emerge gradually through childhood and adolescence, through a dynamic interplay between the individual and his or her environment. They are central to a child's capacity to develop and sustain lasting relationships, and to participate and function within the community. The recent burgeoning of the field of social neurosciences has produced evidence that a number of social domains are affected by brain disruptions, and that these affected domains in turn can influence broader spheres of functioning—such as children's capacity to adapt to their environment, establish rewarding friendships and relationships, and perform in school settings. Yet many aspects of social function and social cognition remain poorly understood, whether in the context of pathology or in typical development. Prevalence, predictors, psychological and biological bases, and developmental pathways are all largely unknown. The impact of disruption as a result of brain insult or environmental influences is even less well understood, but it is likely to be dramatic as social skills are developing and emerging during childhood and adolescence, resulting in psychological distress, social isolation, and reduced self-esteem.

There are several potential explanations for the lack of progress in the social domain in the context of brain insult, and early brain insult in particular. A major factor has been the failure to integrate theoretical and empirical knowledge across disciplines. For example, social psychology provides a number of well-respected theoretical models and significant empirical data with which to contextualize social skills. Despite this, the scanty literature examining social outcomes from early brain insult has mostly ignored these models, and has taken a largely atheoretical approach. Similarly, findings from social development research (e.g., social-developmental milestones, impact of family and parenting practices) are often neglected by those considering the consequences of brain disorders. A second significant limitation for the field has been the lack of a coherent model describing the brain bases of social skills. Finally, the absence of age-appropriate, psychometrically sound assessment tools has made diagnosis and systematic description of social impairments problematic in all children. To date, there has been a heavy emphasis on parent and teacher ratings of children's social skills, with little reference to direct, objective child measures or to self-report tools.

With the emergence of the social neurosciences, the biological bases of social function are gradually becoming better understood, providing a framework for exploring the neural underpinning of social skills. Although this field has gained increasing attention in recent years, to date there has been little focus on (1) relating its findings to children and adolescents, whose brains are still developing rapidly; (2) drawing on established developmental principles, which could provide insight into typical and atypical patterns of maturation and the environmental factors that might influence these; and (3) determining the potential clinical and treatment implications. We believe that the emergence of multidimensional theoretical paradigms is essential for gaining a fuller understanding of social function in the context of child development, and in particular for contributing to clinical practices—specifically, identification, diagnosis, and intervention.

The primary aim of this book is to address these current gaps in the field by providing the reader with perspectives from a range of different but complementary paradigms, by encouraging communication across relevant disciplines, and by employing a developmental focus. Our rationale is that to gain a full understanding of the basis of disrupted social function in the context of developmental or acquired brain-related disorders, it is essential to draw on (1) a coherent theoretical framework, supported by an evidence base; (2) sound knowledge of typical brain development and social development across childhood and adolescence; and (3) appropriate assessment tools covering social function and brain function, and incorporating direct child measures. Diagnostic models and interventions that take into account the three key dimensions of *brain*, *function*, and *development* can then be developed and implemented.

VICKI ANDERSON
MIRIAM H. BEAUCHAMP

Acknowledgments

This book would not have been possible without the valuable contributions of many people. We would first like to thank our authors, all of whom are experts in their fields. Together their work has resulted in a book that we hope will add significantly to the emerging domain of developmental neuroscience. Their contributions made our task as editors both stimulating and rewarding. More important, we have learned an enormous amount from them along the way. We are also grateful for the guidance of Rochelle Serwator, Senior Editor at The Guilford Press, for her input and assistance at all stages of conceptualization, editing, and publication, as well as for the opportunity to publish under Guilford's banner. We greatly appreciate the technical assistance of Nicholas Anderson and Michael Anderson and their help with editing and references.

VICKI ANDERSON
MIRIAM H. BEAUCHAMP

Contents

PART I
INTRODUCTION

CHAPTER 1

SOCIAL

A Theoretical Model of Developmental Social Neuroscience

Vicki Anderson and Miriam H. Beauchamp

S ocial skills and interactions form the foundation of human consciousness, and much human thought and activity are dictated by this domain. Social skills emerge gradually through childhood and adolescence, through a dynamic interplay between an individual and his or her environment. These skills are critical for the individual's capacity to develop and sustain lasting relationships and participate and function within the community. Disruptions to social skills can result in psychological distress, social isolation, and reduced self-esteem, which have major implications for quality of life. Although still in its infancy, social neuroscience, which focuses on the mental operations and neural substrates thought to underpin social function (i.e., social cognition), offers an opportunity to establish a conceptual foundation for better understanding the social domain in general and social cognition in particular. Despite these advances, many aspects of social function and social cognition remain poorly understood; prevalence, predictors, psychological and biological bases, and developmental pathways are largely unknown. The impact of disruption as a result of either brain insult or environmental influences is even less well understood, but it is likely to have dramatic effects as these skills are developing and emerging during childhood and adolescence.

Until recently, the majority of research investigating social difficulties in "at-risk" childhood populations has lacked a theoretical foundation, and this deficiency has impeded progress. Existing theoretical models, such as the social information-processing (SIP) framework, though well accepted in the adult literature, have generally not incorporated either developmental or neuroscience dimensions. Recently, however, the SIP framework has been adapted to include developmental aspects of social function (Crick & Dodge, 1994; Rubin, Begle, & McDonald, Chapter 2, this volume). Briefly, the SIP model proposes that several sequential problem-solving stages occur in response to a social situation: interpreting cues, clarifying goals, generating alternative responses, selecting and implementing a specific response and outcome evaluation. Effective processing of social information across each stage is necessary for socially appropriate interactions (for further details, see Rubin et al., Chapter 2, this volume). From a clinical perspective, when social problems arise, the SIP model provides the opportunity to isolate deficient processing stages and to target the specific SIP skills that may require intervention. To date, however, the SIP framework has rarely been implemented in a context of disruption, either biological or environmental. To address this gap, Yeates and colleagues (2007 and Chapter 10, this volume) propose a heuristic targeted specifically at social skills in the context of brain insult acquired during childhood, and incorporating theoretical and empirical evidence from neuroscience, psychology, and neuropsychology. These authors suggest that social skills are mediated by social/affective and cognitive/executive processes and subsumed by the social brain network, with some integration of both developmental issues and the potential impact of early disruption, due to brain insult.

The emergence of multidimensional theoretical paradigms is essential for gaining a fuller understanding of social function in the context of child development, and in particular for contributing to clinical practices—specifically, identification, diagnosis, and intervention. The aim of this chapter is to propose a theoretical *developmental social neuroscience* framework that will contribute to our understanding of social function in typical children, as well as those at either environmental or biological risk. The socio-cognitive integration of abilities model (SOCIAL; Beauchamp & Anderson, 2010) defines the core dimensions of social skills (biological–psychological–social) and their interactions within a developmental framework founded on empirical research and clinical principles. SOCIAL acknowledges the importance of both typical and atypical biological and environmental influences on the development of social skills.

Social Skills: Definitions and Epidemiology

The term *social skills* incorporates a range of components. For the purposes of this discussion, we have chosen to employ three specific elements: social competence, social interaction, and social adjustment. *Social competence* we define as the ability to achieve personal goals in social interaction while simultaneously maintaining positive relationships with others over time and across situations. *Social interaction* refers to the social actions and reactions between individuals or groups modified to their interaction partners, while *social adjustment* represents the capacity of individuals to adapt to the demands of their social environment (Beauchamp & Anderson, 2010; Bedell, Chapter 9, this volume).

The development of social skills is precarious, as illustrated by the presence of social problems in many neurological and developmental disorders; in several chronic medical and psychiatric conditions; and in the context of social disadvantage, parental psychopathology, or environmental deprivation. Social problems occur either directly through disruption of a particular social or cognitive skill, or as a secondary consequence of stigma, restrictions in social participation, or paucity of social opportunities. Despite their frequency, health care professionals have tended to overlook the importance of social problems for their patients, possibly because they are difficult to observe and assess in clinical contexts. Parents also tend to underrate this domain, seeing it as of secondary importance compared to their children's health and educational status. In contrast, children rate their social skills as being of primary importance (Bohnert, Parker, & Warschausky, 1997).

To date, there are no global estimates of the prevalence of social problems in children or adults, although Asher (1990) reports that 10% of typically developing children are affected by social problems. Figures are much higher in *at-risk* groups, with some studies reporting rates of up to 50% (e.g., in children with autism, traumatic brain injury, or epilepsy) (Catroppa, Anderson, Morse, Haritou, & Rosenfeld, 2008; Rutter, 1983). The lack of epidemiological data is hardly surprising, given that social problems exist not only as independent entities, but also as symptoms or secondary consequences of many other conditions. Even when they are identified as potential difficulties, identification and diagnosis of social problems are hindered by a lack of appropriate assessment tools and of clearly defined diagnostic criteria.

The burden of social problems is undisputed. Children with poor social skills are at higher risk for delinquent or criminal behaviors in adolescence and adulthood (Hawkins, Kosterman, Catalano, Hill, & Abbott, 2005). Reduced social function is also linked with aggression and violence (Boxer, Goldstein, Musher-Eizenman, Dubow, & Heretick, 2005), sexual offenses (Righthand & Welch, 2004), alcohol and drug use (Botvin & Kantor, 2000; Henry & Slater, 2007), conduct disorder (Hill, 2002), and bullying (Camodeca & Goossens, 2005). Poor social skills are frequently associated with neurological and psychiatric conditions, and are overrepresented in incarcerated populations (Butler et al., 2006; Slaughter, Fann, & Ehde, 2003).

Development: The Emergence of Social Functions

Research examining infant behavior has generally focused on social aspects of perception and conduct, demonstrating the innate bias of infants for social interaction. Newborns are sensitive to the correct configuration of facial stimuli (i.e., direct gaze, upright faces, straight heads), reflecting the early foundations of social development. By 2–3 months of age, infants display social initiatives and preferences in processing and recognition of visual and auditory stimuli (Kelly et al., 2005; Rochat & Striano, 2002). By 7 months, they can integrate emotional information across modalities to recognize emotions in faces and voice (Grossmann, Striano, & Friederici, 2006). These milestones are paralleled by a gradual functional specialization throughout the brain, as measured by physiological techniques such as event-related potentials, heart rate, and brain metabolism (Halit, Csibra, Volein, & Johnson, 2004; Purhonen, Kilpelainen-Lees, Valkonen-Korhonen, Karhu, & Lehtonen, 2004; Tzourio-Mazoyer et al., 2002). Joint attention then emerges at about 9 months of age, as evidenced by sharing of attention between a child and an adult in reference to some third event, person, or object (Carpenter, Nagell, & Tomasello, 1998). This phase of infant social development is considered to be a precursor to complex social cognition (e.g., theory of mind or ToM), as it underpins the infant's ability to understand the thoughts and intentions of others, which tends to emerge at about 4–5 years (Tomasello, Carpenter, Call, Behne, & Moll, 2005). The progression from the toddler's egocentric considerations of the environment to the preschooler's capacity for cooperative play and perspective taking, as well as a gradual increase in social communication, are mirrored in the neuroanatomical and functional maturation of ToM and language abilities. Essentially, the emergence of early childhood social abilities that underpin ToM, such as understanding of intentions and false-beliefs, provides a basis for the specialization of more complex mentalizing abilities (Flavell, 1999; Russell, 2005), which continue to mature throughout childhood and into adolescence.

Adolescence is a developmental stage of particular interest for social skills, being characterized by substantial neurostructural change as well as environmental and biological changes that increase teenagers' exposure to social situations (Burnett, Sebastian, & Cohen Kadosh, Chapter 3, this volume; Choudhury, Blakemore, & Charman, 2006). From an environmental perspective, social maturation is a function of increasing personal independence and peer group interaction when the primary goal is focused on the importance of friendships and relationships. Concurrently, the adolescent brain undergoes marked changes in areas that have been shown to underlie social cognition, including a decrease in gray matter volume and an increase in white matter density in frontal and parietal cortices (Casey, Galvan, & Hare, 2005; Giedd et al., 1999; Gogtay et al., 2004).

The Socio-Cognitive Integration of Abilites Model

The emergence and maturation of social skills clearly constitute a protracted process, with many opportunities for derailment. SOCIAL (Beauchamp & Anderson, 2010) assumes that the development of intact social skills is dependent on the typical maturation of the brain, cognition, and behavior, within a supportive environmental context (i.e., a biopsychosocial approach). Specifically, social skills are subsumed by the *social brain network* and are vulnerable to environmental influences. SOCIAL extends current perspectives to provide an integrated representation of the cognitive and affective subskills that contribute to social function, and to explore how these may be influenced by both internal and external forces (e.g., neural, environmental). We focus on social development and its central place in the overall maturation process, beginning in early infancy and continuing throughout adulthood.

The first component of the model (Figure 1.1) represents factors that will shape the emergence of social function. *External factors* are environmental factors that may contribute to the quality and nature of social interactions, such as family environment, socioeconomic status, or culture. *Internal factors* are components of an individual's self that affect the way in which the person interacts with others in social situations, such as temperament, personality, or physical attributes. *Brain development and integrity* are the

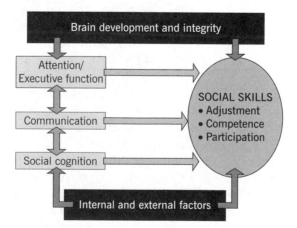

FIGURE 1.1. The socio-cognitive integration of abilities model. Adapted from Beauchamp and Anderson (2010). Copyright 2010 by the American Psychological Association. Adapted by permission.

neural underpinnings of social skills. It may be argued that almost every brain function and structure is involved in social skills (Beer & Ochsner, 2006). The second element of the model refers to the abilities of the child, specifically those that we see as essential for intact social skills: *attention and executive functions, communication,* and *social cognition* (or *socio-emotional skills*). We propose that these processes are interrelated at both the behavioral and neural levels, forming a functional social system.

Internal and External Influences

External Factors

A child's social experience during early development is largely constrained to the social environment of the family. Socioeconomic status (SES) plays an established role in social development for typically developing children (McLoyd, 1998). Social disadvantage has been associated with poor adjustment and problematic peer interactions in school settings (Ackerman & Brown, 2006; Bulotsky-Shearer, Fantuzzo, & McDermott, 2008). SES may also be relevant for the frequency and quality of social interactions and opportunities: Children from socially disadvantaged backgrounds may be at risk for limited exposure to stimulating social environments (Ellaway, Kirk, Macintyre, & Mutrie, 2007). In the context of developmental and neurological disorders, SES has been found to contribute to long-term social outcomes (Anderson, Morse, Catroppa, Haritou, & Rosenfeld, 2004; Yeates et al., 2004), with a combination of greater social disadvantage and more serious disruption resulting in severe social problems (Breslau, 1990).

Family factors are also important. Research confirms that maternal attachment greatly influences the development of social skills in infancy (Bowlby, 1962; Root, Hastings, & Maxwell, Chapter 5, this volume). In adolescence, parent–child interactions predict and moderate social behavior, with autonomy and attachment differentially influencing the emergence of delinquency in teenagers (Allen et al., 2002). This illustrates how parental involvement and preexisting social risk factors may positively or negatively affect children's social competence, even in typically developing children who have the cognitive and emotional skills to establish satisfying relationships. In the context of childhood brain disease, SES and family environment both contribute to long-term outcomes, with lower SES and family dysfunction exacerbating negative social outcomes (Anderson et al., 2004; Yeates et al., 2004).

Cultural influences also play a fundamental role in social development. They permeate child-rearing practices, educational systems, and customs, and dictate social norms. For example, differences in infant facial and emotional expressions have been noted in research comparing Chinese and American babies (Camras et al., 1998). In studying cultural differences in social impressions, Crystal, Watanabe, Weinfurt, and Wu (1998) found that American students notice differences in appearance, attractiveness, and material resources, whereas Japanese students focus on physical features and Chinese students on the behaviors of their peers. Cultural factors also shape other social processes, such as moral reasoning and decision making (Kirmayer, Rousseau, & Lashley, 2007). Such cultural divergences are highlighted by the differences in prevalence rates for social problems across cultures, suggesting ethnic influences on the existence, expression, and reporting of social problems (Rapee & Spence, 2004).

Internal Factors

Factors intrinsic to the individual, such as temperament or personality, are often neglected in the consideration of social outcomes, despite their potential to influence a person's social skills.

Several aspects of personality have been linked to social function. Extraversion and openness are associated with proactive socialization (Wanberg & Kammeyer-Mueller, 2000), while higher levels of self-esteem predict better social and interpersonal relations (Delugach, Bracken, Bracken, & Schicke, 1992). Ozer and Benet-Martinez (2006) note the consequences of personality for the quality of relationships with peers, family members, and romantic others, as well as for community involvement, criminal activity, and political ideology. Shyness and embarassability have obvious consequences for social interaction. Shyness is related to anxious and inhibited social interactions, characterized by poor social skills, reduced communication, reduced eye contact, and a tendency to sit away from other people (Cheek & Buss, 1981; Greco & Morris, 2001; Leary, 1983). At a more debilitating level, social withdrawal and social phobia are associated with poor social development, and therefore constitute a risk factor with the potential to limit a child's social experience (Rubin et al., Chapter 2, this volume).

Physical differences provide a source of stigmatization, which negatively influences social interactions by challenging interpersonal relationships and limiting opportunities to engage in social interactions (Major & O'Brien, 2005). For example, research demonstrates an association between obesity and social contact (Doll, Paccaud, Bovet, Burnier, & Wietlisbach, 2002). Craniofacial anomalies (Kapp-Simon, Simon, & Kristovich, 1992), oral clefts (Slifer et al., 2006), and motor coordination problems (Cummins, Piek, & Dyck, 2005) have also been linked to social stigmatization, subsequent emotional difficulties, and poor social skills. Conversely, it has been suggested that physical attractiveness has a positive impact on social interactions (e.g., Thornhill & Gangestad, 1999); interestingly, it has been found to elicit activity in the medial orbitofrontal cortex, a brain area involved in a number of social skills (O'Doherty et al., 2003).

Neural Bases of Social Cognition and Function

The recent burgeoning of the social neurosciences reflects a growing interest in the biological bases of social function, and has established that social functions involve complex behaviors that require the efficient working of an intricate neural system. Early localizationist literature identified a number of isolated brain regions as potentially important for mediating aspects of social skills, whereas contemporary views have incorporated these isolated areas into functional neural systems.

Lesion studies provide a basis for our knowledge of the social brain network by demonstrating that damage to the prefrontal cortex, and the orbitofrontal cortex in particular, leads to significant changes in behavior, emotion regulation, personality, and social functioning (see Eslinger, Flaherty-Craig, & Benton, 2004, for a review). Similarly, information about the social brain comes from studies of traumatic brain injury because of its potential for causing focal lesions to the frontal and temporal areas of the brain (Anderson, Rosema, Gomes, & Catroppa, Chapter 11, this volume; Hanten, Levin, Newsome, & Scheibel, Chapter 12, this volume; Levin et al., 1989; Max, Chapter 14, this volume), thought to be of importance for social cognition. Such studies offer a link among deficits in social cognition, poor social outcomes and damage to particular areas of the frontal and temporal cortices.

Neuroimaging research also contributes substantially to the growing literature on social skills and their biological bases. Most of this work to date has focused on determining the capacities of the social brain by studying brain activity during particular tasks of social cognition. Rapidly evolving technology, such as functional magnetic resonance imaging (fMRI) and diffusor tensor imaging (DTI), has enabled the investigation of these neural substrates (Adolphs, 2003; Blakemore, 2008). In parallel, animal studies have contributed to our understanding of social neural networks (Cunningham & Janson, 2007; Goursaud & Bachevalier, 2007). It is likely that the social brain network develops and becomes refined through childhood and adolescence, although there is still much work to do to track these developmental trajectories. The protracted development of social skills, and their neural substrates, reflects the multifaceted nature of underlying processes and their interactions; it also highlights the system's vulnerability to dysfunction should any of the many regions involved or their connections be disrupted by biological or environmental elements. Damage or disruption involving the neural substrates of social cognition may lead to socially inappropriate behaviors (or functioning), such as those observed in many clinical populations and discussed in later sections. For a detailed review of this literature, we refer readers to several chapters in this volume (Burnett et al., Chapter 3; Thomas & Tranel, Chapter 4; Dooley, Rosema, & Beauchamp, Chapter 8; and Hanten et al., Chapter 12), which all address aspects of the social brain.

Development, Biology, and Environment

Social skills provide an opportunity to examine the integration of development, biology, and environment. Animal work, genetics research, and neuroimaging studies highlight the bidirectional influences of biology and environment on brain structure and development. Kolb and Whishaw (1998) have documented the brain's capacity to adapt both structurally and functionally to changes in experience and environment, and to translate these changes into social behavior. In particular, enriched environments and social interactions can induce experience-dependent neural changes in the rat's brain, including brain size, cortical thickness, neuron size, dendritic branching, spine density, synapses per neuron, and glial number (Silasi, Hamilton, & Kolb, 2008). Genetic influences and environmental experience may also affect the structure of the brain, and influence the resultant social "phenotype" variably at different ages (Gray & Cornish, Chapter 13, this volume; Kolb, Forgie, Gibb, Gorny, & Rowntree, 1998). All this suggests a complex interplay among biological and environmental dimensions, development, and their implications for the emergence of social function.

Social development is also susceptible to environmental exposure and experience. Cases of maternal separation, parental loss, and social isolation highlight the increased risk in such children of developing indiscriminate social behaviors and social dysfunction more generally (Colvert et al., 2008). These findings suggest that, like other sensory and cognitive systems, social development may be susceptible to adequate social stimulation during decisive developmental periods, and that early nurturing is critical to socioemotional development.

Cognitive Abilities

Although it could be argued that all mental functions, from basic perceptual processes to complex cognitive functions, are necessary for social interactions to occur adequately, three higher-order cognitive domains are critical for social functioning: attention and executive skills, communication, and social cognition. These domains include multiple subskills, which interact selectively and dynamically to determine social competence in varying situations.

Attention and Executive Skills

Attention and executive skills, generally considered to be subsumed by frontal brain structures, are critical for efficient functioning in everyday life and are therefore central to models of social function. *Executive function* is an umbrella term for a broad range of abilities, including (1) *attentional control,* (2) *cognitive flexibility,* and (3) *goal setting* (Anderson, 2008). Attentional control (i.e., self-monitoring, response inhibition, and self-regulation) is especially critical for social skills and development. For example, a child who cannot wait his or her turn while playing a game attracts negative peer responses, which in turn affect social interactions. Goal setting and cognitive flexibility also play important roles in facilitating social interactions (Jacobs & Anderson, 2002). For example, persistent lateness and disorganization, or an inability to adapt to changes in a routine or conversation, have social implications for friendships and work relationships.

The attention/executive dimension has been linked to social outcomes in the context of behaviors, including emotional dyscontrol, aggression, delinquency, and antisocial behavior (Hill, 2002; Jorge et al., 2004). Furthermore, in the context of long-term consequences of childhood acquired brain injury, Muscara, Catroppa, and Anderson (2008) describe the mediating influences of executive skills for social problem solving. Interventions focusing on attentional control have been shown to improve social competence (Greenberg, 2006; McDonald, Turkstra, & Togher, Chapter 16, this volume; Riggs, Greenberg, Kusche, & Pentz, 2006).

Communication

Communication provides the basis for our experience of thought, intentions, and information, and thus underpins the quality of our social relationships. The evolution of communication marks important milestones in the acquisition of social skills—evident first in the appearance of the social smile in the young baby, and then in the emergence of intentional imitative behavior. The emergence of joint attention marks the transition between engaging in dyadic interactions to participating in triadic exchanges (Grossmann & Johnson, 2007). Social skills then progress rapidly following the emergence of expressive language. Children with good linguistic abilities are more able to communicate emotional information, and this facilitates better-quality social interactions (Mostow, Izard, Fine, & Trentacosta, 2002).

Expressive and receptive language skills have clear implications for communication, and thus for social interactions, as they have a direct impact on the expression or comprehension of the message being passed between individuals. Subtle aspects of language processing are also essential. Pragmatics (conveying meaning beyond words used and without ambiguity) are fundamental for creating logical sequences and determining the burden of conversation (as in turn taking), whereas monitoring the appropriateness of utterances provides insights into aspects of social functioning (Turkstra, McDonald, & Kaufman, 1996). The ability to grasp nonlinguistic social signals is also significant. Children who are unable to decode subtle differences in prosody (pitch, loudness/intensity, intonation, rate, stress, rhythm) are at a social disadvantage (McCann, Peppé, Gibbon, O'Hare, & Rutherford, 2007), given the importance of prosody for communicating emphasis, clarification, and contradiction of word meanings. A child who is unable to detect the underlying meaning of ironic or deceptive messages does not receive the social cues necessary to respond appropriately and may consequently breach social rules. Interventions to minimize these difficulties have been successfully implemented in adults and adolescents, as described by McDonald et al. (Chapter 16, this volume).

Social Cognition

Social cognition begins with basic aspects of face and emotion perception, and extends to complex cognitive processes involving understanding mental states, which modulate appropriate behavior within social contexts.

Face/Emotion Perception

Faces provide a wealth of information important for social functioning, including identity, personality, intent, emotion, beliefs, and gaze direction (Calder & Young, 2005; Vuilleumier & Pourtois, 2007). Perceiving, recognizing, and identifying facial expressions and emotions are equally crucial to social reciprocity (McClure, 2000), and are sometimes impaired in socio-emotional processing disorders such as autism (d'Arc & Mottron, Chapter 15, this volume). Face processing and perception of emotional expressions involves a widespread neural network including the fusiform gyrus, amygdala, inferior occipital gyrus, and parts of the superior temporal sulcus and anterior temporal pole, as well as the orbitofrontal, cingulate, insular, and somatosensory cortices (Haxby, Hoffman, & Gobbini, 2000; Vuilleumier & Pourtois, 2007). Body language also has a role in social cognition, connecting both communication and emotion processing domains; moreover, as in face perception, misperception of such signs can lead to poor peer relationships (de Gelder, 2006; Fujiki, Spackman, Brinton, & Illig, 2007).

Attribution

Attribution may be seen as a mediator between basic face and emotion perception and higher-order cognitive domains such as theory of mind. It refers to the way in which people ascribe causes or intent to the behavior of others (*intent attribution*) or to the way they ascribe lasting personality characteristics to others (*trait attribution*). Intentions are inferred and influence the way a person is perceived, shaping the course of social interaction (Harris, Todorov, & Fiske, 2005). Attribution of traits is involved in the social phenomenon of stereotyping, where mental associations are frequently made between particular traits and groups of people (Krueger, Hasman, Acevedo, & Villano, 2003). They are also involved in social judgments of truthfulness and deception (O'Sullivan,

2003). The ability to ascribe social meaning (mental states, emotions) to external stimuli (social attribution) (see Klin, 2006 & Jones), is correlated with more basic social processing skills such as face processing, as well as with high-level mentalizing skills such as ToM (Abell, Happé, & Frith, 2000; Campbell et al., 2006).

Developmentally, a sense of intended actions or goals is apparent from early infancy (Frith & Frith, 1999), but children only begin to interpret social events in terms of actual personality traits by school age (Miller & Aloise, 1989). Van Overwalle (2009) reports that the temporoparietal junction and medial prefrontal cortex, including the anterior cingulate region, are implicated in this aspect of social cognition. Biases in attribution, such as a tendency to perceive intentions as hostile, can explain aggressive and antisocial behavior (Orobio de Castro, Veerman, Koops, Bosch, & Monshouwer, 2002), with obvious implications for social function.

Theory of Mind

Theory of mind is the ability to attribute mental states (beliefs, intents, desires, pretending, knowledge) to oneself and others, and to understand that others have mental states different from one's own (Frith & Frith, 2006). ToM is closely related to empathy, which may be defined as the emotional reaction in the observer to the affective state of another (Blair, 2005). ToM can be seen as a form of "cognitive empathy," requiring an understanding of the mental state of another individual. Unlike cognitive perspective taking, empathy requires the observer to be in an affective state, highlighting the added emotional involvement needed for empathy to occur. These two processes appear to engage different neural networks. Cognitive empathy elicits activity in some of the same regions as intent attribution (medial prefrontal cortex, temporoparietal junction, superior temporal sulcus), whereas affective empathy involves the additional recruitment of areas involved in emotional processing (anterior cingulate, insula, somatosensory cortex) (de Vignemont & Singer, 2006; Vollm et al., 2006).

ToM has a strong developmental basis. It begins to emerge in the preschool period (Brune & Brune-Cohrs, 2006). By about 8 years of age, children typically have a sense that other individuals may have feelings or thoughts differing from their own, and this continues to develop in adolescence (Damon & Hart, 1982). A further milestone is attained when children can recognize that beliefs can be false (Birch & Bloom, 2004)—an understanding that is positively correlated with improvements in social skills such as social problem solving (Baird & Astington, 2004). Not surprisingly, the consequences of impairments in ToM and empathy are profound for prosocial behaviors and other complex social behaviors (Walker, 2005), as is evident from the study of social disorders such as autism (Hill & Frith, 2003).

Moral Reasoning

Moral reasoning allows an individual to make decisions about right and wrong. It is closely linked to ToM, since the ability to understand and represent another person's perspective is necessary for moral behaviors. From a social cognition perspective, the moral domain is concerned with how individuals develop and understand knowledge about social events by relying on moral judgments (Arsenio & Lemerise, 2004). According to Kohlberg (1984), moral development has six identifiable stages progressing from immature reasoning with an egocentric focus, based on obedience and punishment, to more adequate, mature, and society-oriented reasoning. A number of specific neural regions have been associated with moral reasoning—in particular, the medial prefrontal, orbitofrontal, and cingulate cortices, as well as the superior temporal sulcus and amygdala (Moll et al., 2002). Impairments in moral reasoning often result in socially unacceptable actions, characteristic of antisocial, rule-breaking behaviors (Arsenio & Lemerise, 2004).

Social Outcomes

The components of SOCIAL interact dynamically to determine an individual's level of social competence. Internal and external factors, shaped by biology and the environment, interact bidirectionally with the ongoing development of the brain to influence the emergence of cognitive function. For example, internal factors such as personality exert a direct effect on cognition and may together predispose the individual to social dysfunction—for example, social or personality disorders (Kimbrel, 2008; Raine & Yang, 2006). External factors also influence cognitive factors directly; for example, attachment styles and family function are known to affect both cognitive and social development (Green & Goldwyn, 2002). Favorable SES is linked to better cognitive and social outcomes (Conger & Donnellan, 2007; Hackman & Farah, 2009). Cognitive and affective processes dictate an individual's ability to navigate social interactions and environments (i.e., social function and competence), and these processes exert an effect on each other. Changes in cognitive abilities can also have an impact back on mediators of social function (brain structure/function, biology, and the environment).

Developing an Assessment Framework for SOCIAL

SOCIAL provides a conceptual framework for understanding the development of social skills in both typical and atypical development. It proposes that social competence requires cognitive and affective capacities, with the potential for other factors (both extrinsic and intrinsic) to exert influence on social development. In order to have clinical utility, the various components of SOCIAL need to be accessible via standard psychological assessment and treatments.

Many of the subskills in the cognitive domains (attention and executive skills, communication, social cognition) of SOCIAL can be evaluated by using standardized measures developed to explore specific aspects of such skills. There currently exist a number of well-recognized, standardized tools for this purpose within the child and adolescent age ranges. For example, both the Test of Everyday Attention for Children (Manly, Robertson, Anderson, & Nimmo-Smith, 1998) and the Delis–Kaplan Executive Function System (Delis, Kaplan, & Kramer, 2001) are widely used to evaluate attention and executive functions across a wide age range. In contrast, while there exists a wide range of experimental tools for the evaluation of the socio-cognitive domain (see Bruneau-Bhérer, Achim, & Jackson, Chapter 7, this volume), there are few reliable, valid, and standardized tools available for assessing social cognition within a developmental context. Existing tools tend to lack norms; to be restricted to older age groups; to be limited to use in research contexts;

to be available only as indirect, parent-based opinions of child functioning; or to be too general to identify the specific nature of socio-emotional problems. Given the presence of specific social neural networks in the brain that underlie social skills, both structural and functional brain imaging techniques may play an important role in the identification of potential social problems, particularly following brain injury or altered neural development (Bedell, Chapter 9, this volume; Dooley et al., Chapter 7, this volume).

The assessment of general social outcomes is an area that appears to have received more attention, and there are numerous questionnaires and rating scales available to tap into social interactions, social adjustment, and social participation (Crowe, Beauchamp, Catroppa, & Anderson, 2011; Muscara & Crowe, Chapter 6, this volume). Most of these tools are broad-based and employ only a limited number of items tapping specific social skills (e.g., the Adaptive Behavior Assessment Scale—Second Edition; Harrison & Oakland, 2003), and of those that are specific to social outcomes, few have been tested in children with developmental or acquired brain insults (Crowe et al., 2011).

Thus there exists a significant need to develop age-appropriate, psychometrically sound, and ecologically valid assessment tools that can measure the skills contributing to social outcomes (Dooley et al., Chapter 7, this volume), as well as the various dimensions of these outcomes. Such tools are essential for guiding evidence-based interventions for children within the social domain (Soo, Tate, & Rapee, Chapter 18, this volume; Woods, Catroppa, & Anderson, Chapter 17, this volume).

Conclusions

SOCIAL offers a potential framework for integrating the dynamic and multifaceted inter-actions among cognitive, socio-emotional, communicative, biological and environmental dimensions; their relationships; and the ways they play out to determine social competence. Environmental and biological factors, including brain development and injury, are argued to have a direct impact on the development and integrity of an individual's cognitive and affective abilities. We believe that these factors combine to determine the presence of social skills and dictate the individual's level of social competence. Disruption during development can lead to social dysfunction, such as that seen in autism spectrum disorders, genetic conditions, and acquired brain injury. From a clinical perspective, SOCIAL provides a basis for understanding the biopsychosocial components that may contribute to social problems in childhood and later life. The development of valid and reliable measures tapping into each aspect of SOCIAL would provide a way of obtaining a systemic view of social function in both typical and atypical development. Future avenues of research should seek to operationalize components of the model within the domain of social neuroscience.

References

Abell, F., Happé, F., & Frith, U. (2000). Do triangles play tricks?: Attribution of mental states to animated shapes in normal and abnormal development. *Cognitive Development, 15*, 1–20.
Ackerman, B. P., & Brown, E. D. (2006). Income poverty, poverty co-factors, and the adjust-

ment of children in elementary school. *Advances in Child Development and Behavior, 34*, 91–129.

Adolphs, R. (2003). Investigating the cognitive neuroscience of social behavior. *Neuropsychologia, 41*(2), 119–126.

Allen, J. P., Marsh, P., McFarland, C., McElhaney, K. B., Land, D. J., Jodl, K. M., et al. (2002). Attachment and autonomy as predictors of the development of social skills and delinquency during mid-adolescence. *Journal of Consulting and Clinical Psychology, 70*(1), 56–66.

Anderson, P. (2008). Towards a developmental model of executive function. In V. A. Anderson, R. Jacobs, & P. J. Anderson (Eds.), *Executive functions and the frontal lobes: A lifespan perspective.* Hove, UK: Psychology Press.

Anderson, V. A., Morse, S. A., Catroppa, C., Haritou, F., & Rosenfeld, J. V. (2004). Thirty month outcome from early childhood head injury: A prospective analysis of neurobehavioural recovery. *Brain, 127*(Pt. 12), 2608–2620.

Arsenio, W. F., & Lemerise, E. A. (2004). Aggression and moral development: Integrating social information processing and moral domain models. *Child Development, 75*(4), 987–1002.

Asher, S. R. (1990). Recent advances in the study of peer rejection. In S. R. Asher & J. D. Coie (Eds.), *Peer rejection in childhood.* Cambridge, UK: Cambridge University Press.

Baird, J. A., & Astington, J. W. (2004). The role of mental state understanding in the development of moral cognition and moral action. *New Directions for Child and Adolescent Development, 103*, 37–49.

Beauchamp, M., & Anderson, V. (2010). SOCIAL: An integrative framework for the development of social skills. *Psychological Bulletin, 136*(1), 39–64.

Beer, J. S., & Ochsner, K. N. (2006). Social cognition: A multi-level analysis. *Brain Research, 1079*(1), 98–105.

Birch, S. A., & Bloom, P. (2004). Understanding children's and adults' limitations in mental state reasoning. *Trends in Cognitive Sciences, 8*(6), 255–260.

Blair, R. J. (2005). Responding to the emotions of others: Dissociating forms of empathy through the study of typical and psychiatric populations. *Consciousness and Cognition, 14*(4), 698–718.

Blakemore, S. J. (2008). The social brain in adolescence. *Nature Reviews Neuroscience, 9*(4), 267–277.

Bohnert, A. M., Parker, J. G., & Warschausky, S. A. (1997). Friendship and social adjustment of children following a traumatic brain injury: An exploratory investigation. *Developmental Neuropsychology, 13*, 477–486.

Botvin, G. J., & Kantor, L. W. (2000). Preventing alcohol and tobacco use through life skills training. *Alcohol Research and Health, 24*(4), 250–257.

Bowlby, J. (1962). *Deprivation of maternal care.* Geneva: World Health Organization.

Boxer, P., Goldstein, S. E., Musher-Eizenman, D., Dubow, E. F., & Heretick, D. (2005). Developmental issues in school-based aggression prevention from a social-cognitive perspective. *Journal of Primary Prevention, 26*(5), 383–400.

Breslau, N. (1990). Does brain dysfunction increase children's vulnerability to environmental stress. *Archives of General Psychiatry, 47*, 15–20.

Brune, M., & Brune-Cohrs, U. (2006). Theory of mind—evolution, ontogeny, brain mechanisms and psychopathology. *Neuroscience and Biobehavioral Reviews, 30*(4), 437–455.

Bulotsky-Shearer, R. J., Fantuzzo, J. W., & McDermott, P. A. (2008). An investigation of classroom situational dimensions of emotional and behavioral adjustment and cognitive and social outcomes for Head Start children. *Developmental Psychology, 44*(1), 139–154.

Butler, T., Andrews, G., Allnutt, S., Sakashita, C., Smith, N. E., & Basson, J. (2006). Mental disorders in Australian prisoners: A comparison with a community sample. *Australian and New Zealand Journal of Psychiatry, 40*(3), 272–276.

Calder, A. J., & Young, A. W. (2005). Understanding the recognition of facial identity and facial expression. *Nature Reviews Neuroscience, 6*(8), 641–651.

Camodeca, M., & Goossens, F. A. (2005). Aggression, social cognitions, anger and sadness in bullies and victims. *Journal of Child Psychology and Psychiatry, 46*(2), 186–197.

Campbell, R., Lawrence, K., Mandy, W., Mitra, C., Jeyakuma, L., & Skuse, D. (2006). Meanings in motion and faces: developmental associations between the processing of intention from geometrical animations and gaze detection accuracy. *Development and Psychopathology, 18*(1), 99–118.

Camras, L. A., Oster, H., Campos, J., Campos, R., Ujiie, T., Miyake, K., et al. (1998). Production of emotional facial expressions in European American, Japanese, and Chinese infants. *Developmental Psychology, 34*(4, Serial No. 255), 616–628.

Carpenter, M., Nagell, K., & Tomasello, M. (1998). Social cognition, joint attention, and communicative competence from 9 to 15 months of age. *Monographs of the Society for Research in Child Development, 63*(4, Serial No. 255), 1–143.

Casey, B. J., Galvan, A., & Hare, T. A. (2005). Changes in cerebral functional organization during cognitive development. *Current Opinion in Neurobiology, 15*(2), 239–244.

Catroppa, C., Anderson, V. A., Morse, S. A., Haritou, F., & Rosenfeld, J. V. (2008). Outcome and predictors of functional recovery 5 years following pediatric traumatic brain injury (TBI). *Journal of Pediatric Psychology, 33*(7), 707–718.

Cheek, J. M., & Buss, A. H. (1981). Shyness and sociability. *Journal of Personality and Social Psychology, 41,* 330–339.

Choudhury, S., Blakemore, S. J., & Charman, T. (2006). Social cognitive development during adolescence. *Social Cognitive and Affective Neuroscience, 1*(3), 165–174.

Colvert, E., Rutter, M., Beckett, C., Castle, J., Groothues, C., Hawkins, A., et al. (2008). Emotional difficulties in early adolescence following severe early deprivation: Findings from the English and Romanian adoptees study. *Development and Psychopathology, 20*(2), 547–567.

Conger, R. D., & Donnellan, M. B. (2007). An interactionist perspective on the socioeconomic context of human development. *Annual Review of Psychology, 58,* 175–199.

Crick, N., & Dodge, K. (1994). A review and reformulation of social information-processing mechanisms in children's social adjustment. *Psychological Bulletin, 115,* 74–101.

Crowe, L. M., Beauchamp, M. H., Catroppa, C., & Anderson, V. (2011). Social function assessment tools for children and adolescents: A systematic review from 1988 to 2010. *Clinical Psychology Review, 31,* 767–785.

Crystal, D. S., Watanabe, H., Weinfurt, K., & Wu, C. (1998). Concepts of human differences: A comparison of American, Japanese, and Chinese children and adolescents. *Developmental Psychology, 34*(4), 714–722.

Cummins, A., Piek, J. P., & Dyck, M. J. (2005). Motor coordination, empathy, and social behaviour in school-aged children. *Developmental Medicine and Child Neurology, 47*(7), 437–442.

Cunningham, E., & Janson, C. (2007). A socioecological perspective on primate cognition, past and present. *Animal Cognition, 10*(3), 273–281.

Damon, W., & Hart, D. (1982). The development of self-understanding from infancy through adolescence. *Child Development, 53*(4), 841–864.

de Gelder, B. (2006). Towards the neurobiology of emotional body language. *Nature Reviews Neuroscience, 7*(3), 242–249.

de Vignemont, F., & Singer, T. (2006). The empathic brain: How, when and why? *Trends in Cognitive Sciences, 10*(10), 435–441.

Delis, D., Kaplan, E., & Kramer, J. H. (2001). *Delis–Kaplan Executive Function System.* San Antonio, TX: Psychological Corporation.

Delugach, R. R., Bracken, B. A., Bracken, M. J., & Schicke, M. C. (1992). Self concept: Multidimensional construct validation. *Psychology in the Schools, 29,* 213–223.

Doll, S., Paccaud, F., Bovet, P., Burnier, M., & Wietlisbach, V. (2002). Body mass index, abdominal adiposity and blood pressure: consistency of their association across developing and developed countries. *International Journal of Obesity and Related Metabolic Disorders, 26*(1), 48–57.

Ellaway, A., Kirk, A., Macintyre, S., & Mutrie, N. (2007). Nowhere to play?: The relationship between the location of outdoor play areas and deprivation in Glasgow. *Health and Place, 13*(2), 557–561.

Eslinger, P. J., Flaherty-Craig, C. V., & Benton, A. L. (2004). Developmental outcomes after early prefrontal cortex damage. *Brain and Cognition, 55*(1), 84–103.

Flavell, J. H. (1999). Cognitive development: Children's knowledge about the mind. *Annual Review of Psychology, 50,* 21–45.

Frith, C. D., & Frith, U. (1999). Interacting minds: A biological basis. *Science, 286*(5445), 1692–1695.

Frith, C. D., & Frith, U. (2006). The neural basis of mentalizing. *Neuron, 50*(4), 531–534.

Fujiki, M., Spackman, M. P., Brinton, B., & Illig, T. (2007). Ability of children with language impairment to understand emotion conveyed by prosody in a narrative passage. *International Journal of Language and Communication Disorders, 43*(3), 330–345.

Giedd, J. N., Blumenthal, J., Jeffries, N. O., Castellanos, F. X., Liu, H., Zijdenbos, A., et al. (1999). Brain development during childhood and adolescence: A longitudinal MRI study. *Nature Neuroscience, 2*(10), 861–863.

Gogtay, N., Giedd, J. N., Lusk, L., Hayashi, K. M., Greenstein, D., Vaituzis, A. C., et al. (2004). Dynamic mapping of human cortical development during childhood through early adulthood. *Proceedings of the National Academy of Sciences USA, 101*(21), 8174–8179.

Goursaud, A. P., & Bachevalier, J. (2007). Social attachment in juvenile monkeys with neonatal lesion of the hippocampus, amygdala and orbital frontal cortex. *Behavioural Brain Research, 176*(1), 75–93.

Greco, L. A., & Morris, T. L. (2001). Treating childhood shyness and related behavior: empirically evaluated approaches to promote positive social interactions. *Clinical Child and Family Psychology Review, 4*(4), 299–318.

Green, J., & Goldwyn, R. (2002). Annotation: Attachment disorganisation and psychopathology: New findings in attachment research and their potential implications for developmental psychopathology in childhood. *Journal of Child Psychology and Psychiatry, 43*(7), 835–846.

Greenberg, M. T. (2006). Promoting resilience in children and youth: Preventive interventions and their interface with neuroscience. *Annals of the New York Academy of Sciences, 1094,* 139–150.

Grossmann, T., & Johnson, M. H. (2007). The development of the social brain in human infancy. *European Journal of Neuroscience, 25*(4), 909–919.

Grossmann, T., Striano, T., & Friederici, A. D. (2006). Cross-modal integration of emotional information from face and voice in the infant brain. *Developmental Science, 9*(3), 309–315.

Hackman, D. A., & Farah, M. J. (2009). Socioeconomic status and the developing brain. *Trends in Cognitive Sciences, 13*(2), 65–73.

Halit, H., Csibra, G., Volein, A., & Johnson, M. H. (2004). Face-sensitive cortical processing in early infancy. *Journal of Child Psychology and Psychiatry, 45*(7), 1228–1234.

Harris, L. T., Todorov, A., & Fiske, S. T. (2005). Attributions on the brain: Neuro-imaging dispositional inferences, beyond theory of mind. *NeuroImage, 28*(4), 763–769.

Harrison, P. L., & Oakland, T. (2003). *Adaptive Behavior Assessment System—Second Edition.* San Antonio, TX: Psychological Corporation.

Hawkins, J. D., Kosterman, R., Catalano, R. F., Hill, K. G., & Abbott, R. D. (2005). Promoting positive adult functioning through social development intervention in childhood: Long-term effects from the Seattle Social Development Project. *Archives of Pediatrics and Adolescent Medicine, 159*(1), 25–31.

Haxby, J. V., Hoffman, E. A., & Gobbini, M. I. (2000). The distributed human neural system for face perception. *Trends in Cognitive Sciences, 4*(6), 223–233.

Henry, K. L., & Slater, M. D. (2007). The contextual effect of school attachment on young adolescents' alcohol use. *Journal of School Health, 77*(2), 67–74.

Hill, E. L., & Frith, U. (2003). Understanding autism: insights from mind and brain. *Philosophical Transactions of the Royal Society of London. Series B: Biological Sciences, 358*(1430), 281–289.

Hill, J. (2002). Biological, psychological and social processes in the conduct disorders. *Journal of Child Psychology and Psychiatry, 43*(1), 133–164.

Jacobs, R., & Anderson, V. (2002). Planning and problem solving skills following focal frontal brain lesions in childhood: Analysis using the Tower of London. *Child Neuropsychology, 8*(2), 93–106.

Jorge, R. E., Robinson, R. G., Moser, D., Tateno, A., Crespo-Facorro, B., & Arndt, S. (2004). Major depression following traumatic brain injury. *Archives of General Psychiatry, 61*(1), 42–50.

Kapp-Simon, K. A., Simon, D. J., & Kristovich, S. (1992). Self-perception, social skills, adjustment, and inhibition in young adolescents with craniofacial anomalies. *Cleft Palate–Craniofacial Journal, 29*(4), 352–356.

Kelly, D. J., Quinn, P. C., Slater, A. M., Lee, K., Gibson, A., Smith, M., et al. (2005). Three-month-olds, but not newborns, prefer own-race faces. *Developmental Science, 8*(6), F31–F36.

Kimbrel, N. A. (2008). A model of the development and maintenance of generalized social phobia. *Clinical Psychology Review, 28*(4), 592–612.

Kirmayer, L. J., Rousseau, C., & Lashley, M. (2007). The place of culture in forensic psychiatry. *Journal of the American Academy of Psychiatry and the Law, 35*(1), 98–102.

Klin, A., & Jones, W. (2006). Attributing social and physical meaning to ambiguous visual displays in individuals with higher-functioning autism spectrum disorders. *Brain and Cognition, 61*(1), 40–53.

Kohlberg, L. (1984). *Essays on moral development* (Vol. 2). San Francisco: Harper.

Kolb, B., Forgie, M., Gibb, R., Gorny, G., & Rowntree, S. (1998). Age, experience and the changing brain. *Neuroscience and Biobehavioral Reviews, 22*(2), 143–159.

Kolb, B., & Whishaw, I. Q. (1998). Brain plasticity and behavior. *Annual Review of Psychology, 49*, 43–64.

Krueger, J. I., Hasman, J. F., Acevedo, M., & Villano, P. (2003). Perceptions of trait typicality in gender stereotypes: Examining the role of attribution and categorization processes. *Personality and Social Psychology Bulletin, 29*(1), 108–116.

Leary, M. R. (1983). Social anxiousness: The construct and its measurement. *Journal of Personality Assessment, 47*(1), 66–75.

Levin, H., Amparo, E., Eisenberg, H., Miner, M., High, W., Jr., Ewing-Cobbs, L., et al. (1989). MRI after closed head injury in children. *Neurosurgery, 24*, 223–227.

Major, B., & O'Brien, L. T. (2005). The social psychology of stigma. *Annual Review of Psychology, 56*, 393–421.

Manly, T., Robertson, I. H., Anderson, V., & Nimmo-Smith, I. (1998). *The Test of Everyday Attention for Children (TEA-Ch).* Bury St. Edmunds, UK: Thames Valley Test Company.

McCann, J., Peppé, S., Gibbon, F. E., O'Hare, A., & Rutherford, M. (2007). Prosody and its relationship to language in school-aged children with high-functioning autism. *International Journal of Language and Communication Disorders, 42*(6), 682–702.

McClure, E. B. (2000). A meta-analytic review of sex differences in facial expression processing and their development in infants, children, and adolescents. *Psychological Bulletin, 126*(3), 424–453.

McLoyd, V. C. (1998). Socioeconomic disadvantage and child development. *American Psychologist, 53*(2), 185–204.

Miller, P. J., & Aloise, R. R. (1989). Young children's understanding of the psychological causes of behavior. *Child Development, 60,* 257–285.

Moll, J., de Oliveira-Souza, R., Eslinger, P. J., Bramati, I. E., Mourao-Miranda, J., Andreiuolo, P. A., et al. (2002). The neural correlates of moral sensitivity: A functional magnetic resonance imaging investigation of basic and moral emotions. *Journal of Neuroscience, 22*(7), 2730–2736.

Mostow, A. J., Izard, C. E., Fine, S., & Trentacosta, C. J. (2002). Modeling emotional, cognitive, and behavioral predictors of peer acceptance. *Child Development, 73*(6), 1775–1787.

Muscara, F., Catroppa, C., & Anderson, V. (2008). Social problem solving skills as a mediator between executive function and long-term social outcome following pediatric traumatic brain injury. *Journal of Neuropsychology, 2,* 445–461.

O'Doherty, J., Winston, J., Critchley, H., Perrett, D., Burt, D. M., & Dolan, R. J. (2003). Beauty in a smile: The role of medial orbitofrontal cortex in facial attractiveness. *Neuropsychologia, 41*(2), 147–155.

O'Sullivan, M. (2003). The fundamental attribution error in detecting deception: The boy-who-cried-wolf effect. *Personality and Social Psychology Bulletin, 29*(10), 1316–1327.

Orobio de Castro, B., Veerman, J. W., Koops, W., Bosch, J. D., & Monshouwer, H. J. (2002). Hostile attribution of intent and aggressive behavior: A meta-analysis. *Child Development, 73*(3), 916–934.

Ozer, D. J., & Benet-Martinez, V. (2006). Personality and the prediction of consequential outcomes. *Annual Review of Psychology, 57,* 401–421.

Purhonen, M., Kilpelainen-Lees, R., Valkonen-Korhonen, M., Karhu, J., & Lehtonen, J. (2004). Cerebral processing of mother's voice compared to unfamiliar voice in 4-month-old infants. *International Journal of Psychophysiology, 52*(3), 257–266.

Raine, A., & Yang, Y. (2006). Neural foundations to moral reasoning and antisocial behavior. *Social Cognitive and Affective Neuroscience, 1*(3), 203–213.

Rapee, R. M., & Spence, S. H. (2004). The etiology of social phobia: Empirical evidence and an initial model. *Clinical Psychology Review, 24*(7), 737–767.

Riggs, N. R., Greenberg, M. T., Kusche, C. A., & Pentz, M. A. (2006). The mediational role of neurocognition in the behavioral outcomes of a social-emotional prevention program in elementary school students: Effects of the PATHS Curriculum. *Prevention Science, 7*(1), 91–102.

Righthand, S., & Welch, C. (2004). Characteristics of youth who sexually offend. *Journal of Child Sexual Abuse, 13*(3–4), 15–32.

Rochat, P., & Striano, T. (2002). Who's in the mirror?: Self–other discrimination in specular images by four- and nine-month-old infants. *Child Development, 73*(1), 35–46.

Russell, J. (2005). Justifying all the fuss about false belief. *Trends in Cognitive Sciences, 9*(7), 307–308.

Rutter, M. (Ed.). (1983). *Developmental neuropsychiatry.* New York: Guilford Press.

Silasi, G., Hamilton, D. A., & Kolb, B. (2008). Social instability blocks functional restitution following motor cortex stroke in rats. *Behavioural Brain Research, 188*(1), 219–226.

Slaughter, B., Fann, J. R., & Ehde, D. (2003). Traumatic brain injury in a county jail population: Prevalence, neuropsychological functioning and psychiatric disorders. *Brain Injury, 17*(9), 731–741.

Slifer, K. J., Pulbrook, V., Amari, A., Vona-Messersmith, N., Cohn, J. F., Ambadar, Z., et al. (2006). Social acceptance and facial behavior in children with oral clefts. *Cleft Palate–Craniofacial Journal, 43,* 226–236.

Thornhill, R., & Gangestad, S. W. (1999). Facial attractiveness. *Trends in Cognitive Sciences, 3*(12), 452–460.

Tomasello, M., Carpenter, M., Call, J., Behne, T., & Moll, H. (2005). Understanding and sharing intentions: The origins of cultural cognition. *Behavioral and Brain Sciences, 28*(5), 675–691; discussion 691–735.

Turkstra, L. S., McDonald, S., & Kaufmann, P. M. (1996). Assessment of pragmatic communication skills in adolescents after traumatic brain injury. *Brain Injury, 10*(5), 329–345.

Tzourio-Mazoyer, N., De Schonen, S., Crivello, F., Reutter, B., Aujard, Y., & Mazoyer, B. (2002). Neural correlates of woman face processing by 2–month-old infants. *NeuroImage, 15*(2), 454–461.

Van Overwalle, F. (2009). Social cognition and the brain: A meta-analysis. *Human Brain Mapping, 30*(3), 829–858.

Vollm, B. A., Taylor, A. N., Richardson, P., Corcoran, R., Stirling, J., McKie, S., et al. (2006). Neuronal correlates of theory of mind and empathy: A functional magnetic resonance imaging study in a nonverbal task. *NeuroImage, 29*(1), 90–98.

Vuilleumier, P., & Pourtois, G. (2007). Distributed and interactive brain mechanisms during emotion face perception: Evidence from functional neuroimaging. *Neuropsychologia, 45*(1), 174–194.

Walker, S. (2005). Gender differences in the relationship between young children's peer-related social competence and individual differences in theory of mind. *Journal of Genetic Psychology, 166*(3), 297–312.

Wanberg, C. R., & Kammeyer-Mueller, J. D. (2000). Predictors and outcomes of proactivity in the socialization process. *Journal of Applied Psychology, 85*(3), 373–385.

Yeates, K. O., Bigler, E. D., Dennis, M., Gerhardt, C. A., Rubin, K. H., Stancin, T., et al. (2007). Social outcomes in childhood brain disorder: A heuristic integration of social neuroscience and developmental psychology. *Psychological Bulletin, 133*(3), 535–556.

Yeates, K. O., Swift, E., Taylor, H. G., Wade, S. L., Drotar, D., Stancin, T., et al. (2004). Short- and long-term social outcomes following pediatric traumatic brain injury. *Journal of the International Neuropsychological Society, 10*(3), 412–426.

PART II
THEORETICAL CONTRIBUTIONS

CHAPTER 2

Peer Relations and Social Competence in Childhood

Kenneth H. Rubin, Annie Schulz Begle, and Kristina L. McDonald

Historically, it has been commonplace to assume that adaptive and maladaptive social development during childhood and adolescence emanates from the parenting and parent–child relationship experiences that children have had from the earliest years of their lives. The roles of parenting and parent–child relationship experiences were described in the psychoanalytic writings of Freud (1933); the social learning theory and research of such scholars as Sears, Maccoby, and Levin (1957) and Bandura and Walters (1963); and the seminal perspectives of attachment theory developed by Bowlby (1958). Clearly, parents do play a role in the development of adaptive and maladaptive development. So too do genetic and biological factors. Nevertheless, it is the case that from the very earliest years of life, children come into social contact with other familial (siblings, grandparents) and extrafamilial (caregivers, teachers, coaches, peers) sources of developmental influence. And with increasing age, the amount of time spent in the company of these nonparental influences increases significantly. Thus, in the present chapter, our primary focus is on the interactions and relationships that children and young adolescents experience with their peers.

The chapter is organized in the following manner. We begin with a conceptual schema for the study of child and adolescent social development. Thereafter, we briefly review various theoretical perspectives pertaining to child and adolescent peer interactions and relationships. We next describe the construct of *social competence*; we propose that this phenomenon derives in large part from a child's ability to think about the social world and to contemplate ways and means of navigating the social milieu. In so doing, we focus not only on typical children, but also on children who deviate from the norm in meaningful ways. Given much of the material found within the present book, we link the content of our discussion with the developmental social neuroscience field by addressing the ways in which children who have suffered a traumatic brain injury (TBI) might think about and experience the world of peers. Subsequently, we describe factors that predict,

23

correlate with, and result from peer acceptance and rejection. Following this, we describe the nature of child and early adolescent friendship; the features of interpersonal attraction and friendship formation; and the correlates and consequences of supportive and dysfunctional friendships.

A Conceptual Framework

The conceptual framework for this chapter is drawn largely from the writings and research of Hinde (e.g., Hinde, 1987; Hinde & Stevenson-Hinde, 1976). To begin with, it is argued that each individual brings into the world numerous defining characteristics, including such physical phenomena as age, sex, ethnicity, race, height, weight, physical attractiveness, and so on. Other characteristics that are part and parcel of the child's individual makeup include such biologically based phenomena as temperament (e.g., reactivity/regulation, fearfulness, sociability), as well as physical, physiological, and psychological distinctiveness that may have been incurred through illness, injury, or heredity. We include, among the child's individual characteristics, the child's intellect, ability to process information about the social and non-social world, and his or her ability to use such information to solve problems in social and nonsocial worlds. Again, given the focus of this book, we would also include, among a child's individual characteristics, those features that may result from TBI.

Taken together, these individual characteristics can affect the child's *interactions*—not only with peers, but also with parents, siblings, cousins, aunts and uncles, teachers, mentors, coaches, and so on. Given that the focus of this chapter has much to do with peers, we suggest simply that over the short term, social interactions with other children may vary in form and function in response to fluctuations in the parameters of the social situation, such as the partner's individual characteristics, overtures, and responses. One consequence of regularly finding oneself in the company of another person and either interacting (or not interacting) in particular ways with that person is the development of identifiable social *relationships*. If interactions do occur between peers, the type of relationship that is formed by the members of the dyad may be influenced by the *quality* of these interactions. Thus, positive interactions (e.g., helping, caring, sharing interactions) may lead one to think and feel positively about the individual with whom such exchanges have taken place; be attracted to that person; and seek to develop a meaningful, supportive, and constructive *friendship* with the individual. Negative, agonistic interactions may result in unenthusiastic thoughts, feelings of repulsion, and the development of enmity or bully–victim relationships. In this regard, relationships are influenced by both past and anticipated future interactions. Significantly, choosing not to interact with others may lead to fewer or qualitatively poor social relationships. Taken together, the nature of children's dyadic social relationships is defined partly by the individual characteristics of its members *and* its constituent interactions (Hinde, 1987; Hinde & Stevenson-Hinde, 1976).

The concept of a *group* introduces an even higher order of complexity into the study of children's peer experiences. A group is a collection of interacting individuals with some degree of reciprocal influence over one another. Groups often form spontaneously, out of common interests or circumstances (*homophily*), but they are also established formally, the most ubiquitous example being the school class (Rubin, Bukowski, & Parker, 2006).

Hinde (1987) suggested that a group is the structure that emerges from the features and patterning of the relationships and interactions present within a population of children, such as a classroom. Accordingly, groups possess properties that arise from the manner in which the relationships are patterned but are not present in the individual, dyadic relationships themselves. Rubin et al. (2006) have noted that examples of such group properties include *cohesiveness*, or the degree of unity and inclusiveness exhibited by the children; *hierarchy*, or the extent of intransitivity in the ordering of the individual relationships along such group-defining dimensions as dominance (e.g., if *A* dominates *B* and *B* dominates *C*, does *A* dominate *C*?); and *heterogeneity*, or consistency across members in the ascribed or achieved personal characteristics (e.g., sex, race, age, intelligence, attitudes toward school). Finally, every group has *norms*, or distinctive patterns of behaviors and attitudes that characterize group members and differentiate them from members of other groups (see Rubin, Bukowski, & Laursen, 2009, for relevant reviews). Significantly, the emergent properties of groups shape the experiences of the constituent members. For example, crowd labels (e.g., "jocks" or "artsies") constrain adolescents' freedom to explore new identities; status hierarchies influence the formation of new friendships; segregation influences the diversity of children's experiences with others; and cohesiveness influences children's sense of belonging. As such, the group can influence the individual.

In summary, the schema that provides the foundation for the following review suggests that children's social universes and development are influenced by several interconnected factors. For instance, a child's social behaviors, whether adaptive or maladaptive, depend on a wide variety of sources: (1) sociodemographic or environmental characteristics, such as gender, age, and cultural environment; (2) dispositional characteristics, such as temperament, personality, and behavioral traits; (3) the relationships between the parents and the child, as well as the parents' child-rearing practices and styles; (4) the ways in which the child's own behavior is interpreted by the child and others; (5) the social-cognitive or social information-processing (SIP) patterns endorsed by the child; (6) the child's social reputation; (7) the nature of the relationships within which the child displays the behavior; and (8) the groups within which the child is a member.

Lastly, it must be noted that all of the above must be interpreted culturally. Cultural beliefs and norms help interpret the acceptability of individual characteristics and the types and ranges of interactions and relationships that are likely or permissible (Rubin et al., 2006). Yet most of what we know about children's social competence and peer relationships derives from research that has been conducted in Western European and North American samples. Unfortunately, space constraints preclude a discussion here of cultural "meanings" of social behaviors and relationships. Instead, we refer the reader to a recently published review by Rubin, Cheah, and Menzer (2010).

Theoretical Perspectives

Theoretical frameworks for contemporary research on children's peer relationships and social competence can be traced to the early 20th century. One of the earliest influences was the theory of Jean Piaget (e.g., 1932), who, to some extent, moved the etiological and epistemological study of children's ways of thinking about and behaving in the social world from one involving the impact of parents to one examining the potential effect of peers on child development. Piaget proposed that peer interactions and relationships can be clearly distinguished from those of children with their parents. He portrayed the latter as vertical, asymmetrical, and involving power assertion, thereby leading to young children's unquestioned obedience and acceptance of parental or caregivers' values, beliefs, and norms even if they do not completely understand them. In contrast, he characterized peer interactions and relationships as more balanced and egalitarian. According to Piaget, it is in the peer context that children are more likely to experience conflicting ideas and explanations, negotiation and discussion of multiple perspectives, and the resolution of conflict and disagreement with compromise. In short, Piaget's influence provided an early-20th-century perspective on how children come to think about things social and how these thoughts influence their interactions with peers. As the reader will note in the remainder of this chapter, Piaget's theoretical influence remains strong in contemporary research on such topics as perspective taking, theory of mind, SIP, and social competence.

Sullivan (1953) shared with Piaget the notion that concepts of mutual respect, egalitarianism, and reciprocity derive in large part from peer experiences. Sullivan's distinct contribution was his focus on the role of best friendships in the emergence of these concepts. Sullivan believed that children are relatively unaffected by their peers in early childhood; he thought that by middle childhood, however, individuals are able to identify and recognize each other's qualities, becoming more relevant as personality-shaping agents. Sullivan's theory has been shown to be influential in the contemporary research on children's friendships, as well as in the study of the consequences of not having close dyadic relationships with friends (Asher & Paquette, 2003).

Mead (1934) argued that over the early years of life, the ability to reflect on the *self* develops gradually and primarily as a function of peer interaction. This theoretical perspective has been highly influential in the contemporary study of the association between the self-system and the quality of children's peer relationships (Boivin & Hymel, 1997).

According to social learning theorists (e.g., Bandura & Walters, 1963), children learn about the social domain and how to behave in it by observing their peers and being guided directly by them. Thus peers become agents of behavior control and change, reinforcing or punishing those children whose social behaviors support or defy social norms and values within the particular context or culture (Shortt, Capaldi, Dishion, Bank, & Owen, 2003).

The ethological perspective has had a significant influence on the study of peer relationships and groups. Ethologists have studied the biological origins of particular forms of social behavior (e.g., aggression, prosocial behavior), as well as of the abilities to initiate, maintain, or disengage from particular relationships (Vaughn & Santos, 2009). Another contribution of human ethological research has been the provision of detailed observations of the organization and structure of social groups (e.g., Vaughn & Santos, 2009). In this regard, ethological theory has influenced the development of observational methods by which children's peer interactions, relationships, and groups have been studied in natural settings.

Lastly, Harris's (2009) group socialization theory has challenged the perspective that a child's personality is shaped primarily by parents, granting the peer group a much stronger role in the child's personality and social development. Harris argued that children must adapt to and follow the prevailing norms of those groups within which they spend most of their time, such as the peer group. Drawing from several social-psychological concepts regarding the significance of group norms, ingroup biases and outgroup hostilities, and social-cognitive perspectives of group processes, Harris suggested that a child's identity develops primarily from experiences with peers. Harris's perspective gives researchers an opportunity to address central questions about the causal roles of genes, biology, family, and peers in child and adolescent adjustment and maladjustment.

Social Competence and SIP

We begin with a discussion of social competence and SIP, two *individual* factors that are conceptually and empirically associated with the types of interactions and relationships children have with their peers (Lemerise, Gregory, & Fredstrom, 2005).

Defining Social Competence within an Interpersonal Problem-Solving Framework

A quarter of a century ago, Dodge (1985) wrote that there are probably as many definitions of social competence as there are researchers currently gathering data on the topic. Dodge's observation was less "tongue in cheekish" than one might have hoped. We offer below, a smattering of long-standing definitions culled from the many that have been suggested by investigators of social competence: "an organism's capacity to interact effectively with its environment" (White, 1959, p. 297); "the effectiveness or adequacy with which an individual is capable of responding to various problematic situations which confront him" (Goldfried & D'Zurilla, 1969, p. 161); "a judgment by another that an individual has behaved effectively" (McFall, 1982, p.1); the ability "to make use of environmental and personal resources to achieve a good developmental outcome" (Waters & Sroufe, 1983; p. 81); and "the ability to engage effectively in complex interpersonal interaction and to use and understand people effectively" (Oppenheimer, 1989, p. 45).

The definitions noted above share several properties. First, most of the authors, in defining social competence, refer to "effectiveness." Second, competence seems to involve the manipulation of others to meet one's own needs or goals. It is important, however, to distinguish between the effective manipulation of others in the Machiavellian sense and the effective manipulation of others by using conventionally accepted means in accord with "common sense." Thus McFall's notion that social competence refers to a "judgment" based on the display of skilled behavior is an important one.

Drawing from the definitions noted above, Rubin and Rose-Krasnor (1992, p. 285) defined social competence as "the ability to achieve personal goals in social interaction while simultaneously maintaining positive relationships with others over time and across situations." This definition has, over the years, served as a useful heuristic in guiding our own research program. It has also served us well in understanding and mentally "filing" the burgeoning numbers of published papers, chapters, and books concerning social skills, social competence, and social relationships in children.

The definition encapsulates the functional properties of most social behaviors. Most social behaviors are goal-oriented; most involve the demonstration of appropriate and acceptable *strategies* to achieve these goals; and most strategies can be judged as successful or not. Thus we turn now to a discussion of the significance of social competence for typical development and to a description of several traditional and contemporary models of social competence, SIP, and interpersonal problem solving.

Models of SIP

Children regularly face social dilemmas ranging in magnitude from miniscule to over-bearing. Take, for example, the following questions that a child may ask him- or herself during any given day: "How can I get that computer game from my older sister?" "Should I try to get to know the new kid next door, and if I should, how can I do it?" "How can I get the kids in my class to help out with the recycling project?" "How can I convince my parents that one half hour of playing computer games per day won't kill me?" For children who can think through their social dilemmas and eventually solve them effectively, the social world is a welcome and reinforcing place. For children whose SIP and interpersonal problem-solving skills are lacking, there may be considerable developmental risk.

In the past 25 years, researchers have reported that children whose attempts to meet their social goals result in repeated failure eventually develop negative thoughts and feelings about themselves (Nelson, Rubin, & Fox, 2005). And when some children fail in achieving their social goals, they react with anger (e.g., Lemerise et al., 2005) or with anxiety and fear (e.g., Burgess, Wojslawowicz, Rubin, Rose-Krasnor, & Booth-LaForce, 2006). These negative emotions are destructive, in that they often lead to aggression on the one hand or passive withdrawal on the other (Crick & Dodge, 1994). The production of aggressive or socially withdrawn behavior often leads to rejection, not only by peers, but also by adults; in turn, rejection leads to an inability to form and maintain supportive social relationships (Rubin et al., 2006). Given that these social relationships and the interpersonal exchanges of ideas, perspectives, and actions experienced within them supposedly *promote* the development of social competence, one can begin to see that a destructive cycle of negativity may eventuate. If it is the case that social relationships and social exchange promote the development of social skills, and if the lack of mature and competent social thinking and behavior negates opportunities for the development and maintenance of productive social relationships, one comes quickly to realize what the costs of poor social skills may be, both interpersonally and intrapersonally.

Rubin and Rose-Krasnor (1992) outlined steps in *SIP* that may affect how children form and maintain relationships with peers. When children encounter difficult social situations, they first select a *social goal*. These goals may include gaining attention or acknowledgment, obtaining information, giving and/or receiving help, acquiring desired objects, initiating social play, and avoiding anger and/or loss of face. The goal, by definition, is a mental representation of the *end state* of the problem-solving process—"This is what I want to accomplish." Second, they *examine the task environment,* which involves attending to and interpreting relevant social cues. Social cues are extensive, including the status of peers, their familiarity, and their age and gender (Krasnor & Rubin, 1983). Furthermore, Rubin and Rose-Krasnor (1992) acknowledge the significance of context. For example, how familiar are the social partners? Is the interaction occurring in private or in the company of others? Is the interaction taking place on "neutral ground" or in one of the partners' homes? Next, they *access and select strategies* for achieving the social goal. Fourth, they *implement the chosen strategy,* and afterward, they *evaluate the outcome of the strategy,* assessing the situation to determine the relative success of the chosen course of action in achieving the social goal.

Crick and Dodge (1994) proposed a similar model, which was developed primarily to account for aggressive behavior in children. These researchers suggested that when children find themselves in social company, and when a dilemma confronts them, they first encode and interpret social cues and information; next, they access their cognitive repertoires, decide upon possible responses to the given situation, and evaluate these; lastly, they select and enact the chosen response.

More recently, Lemerise and Arsenio (2000) have integrated emotional experiences within Crick and Dodge's (1994) SIP model. These emotions in turn may influence the information attended to, the information recalled, and the ways children respond to negative events befalling them. Nonaggressive children, particularly those who may be described as socially wary and withdrawn, may view interpersonal situations such as peer group entry as stressful and anxiety-producing; in their case, avoidance evoked by fear or wariness may be the social consequence. Thus an inability to regulate emotional arousal under certain circumstances may influence several steps of the information-processing and behavioral enactment process.

We turn now to the literature on the relations between SIP and children's social behaviors and peer relationships. This review is necessarily brief; we suggest that the reader refer to Dodge, Coie, and Lyman (2006) for a relevant review.

Aggressive Children

There are consistent findings in the literature that aggressive children have social-cognitive schemas that lead them to expect others to treat them harshly. For example, aggressive behavior has been linked to *rejection sensitivity*, or the tendency to defensively expect, readily perceive, and overreact to rejection (Downey, Lebolt, Rincon, & Freitas, 1998). Such sensitivity has been thought to develop from rejection experiences with parents or other significant adults (teachers), as well as peers. Downey and colleagues hypothesized and found that expectations of rejection may evoke anger, which in turn leads to the demonstration of aggressive behavior in the peer group and to a relative lack of social competence.

Relative to typical children, aggressive children also believe that aggression is more legitimate or acceptable in social situations, and they feel more efficacious enacting verbally and physically aggressive acts than do their typical peers (Erdley & Asher, 1996). There is also evidence that aggressive children favor these behaviors because they believe it will yield positive outcomes—either through tangible rewards or the termination of aversive behavior by others (Lochman & Dodge, 1994). Aggressive children are also more likely than their typical age-mates to believe that aggression will increase self-esteem and respect from peers, and less likely to recognize that a victim of their aggression will suffer (Slaby & Guerra, 1988).

As for *SIP*, it has been shown consistently that aggressive children are more likely than their nonaggressive peers to pay attention to and recall hostile cues in social situations (Gouze, 1987). Aggressive children are also more likely than their nonaggressive counterparts to make negative interpretations of their peers' actions. For example, aggressive children demonstrate a hostile attribution bias: When they are confronted with a situation in which it is unclear whether another person has intended to harm them, these children assume that the other person had malevolent intent (Dodge et al., 2003). These interpretational biases appear to lead to the generation and selection of hostile goals, rather than to goals that may establish or enhance their social relationships. Thus aggressive children are more likely than typical children to endorse revenge and dominance goals in response to ambiguous harm or in situations involving minor conflict (Rose &

Asher, 1999). Finally, when choosing behavioral responses, aggressive children have more limited strategic repertoires. They are also more likely than their nonaggressive peers to generate hostile, aggressive, or dominant strategies and to assess these strategies more positively (Dodge et al., 2003). Taken together, these schema and *SIP* patterns support continued aggressive behavior and problematic relationships with peers.

Withdrawn Children

Compared to the preponderance of aggressiveness in research regarding *SIP*, knowledge of how socially withdrawn children process social information is somewhat limited. However, there is evidence that socially anxious/withdrawn children (Burgess et al., 2006) differ from socially competent youth in some or all steps or aspects of *SIP*. To begin with, it has been found that the social goals of anxious/withdrawn children are less "costly" to their social partners than those of their typical age-mates. Moreover, they are more likely to use indirect, unassertive means to meet their social goals than typical children. Yet, given the lower "cost" of their social goals and the lesser use of direct means to meet their goals, anxious/withdrawn children are less likely than typical children to actually meet or obtain their goals (e.g., Rubin, 1982).

Furthermore, young withdrawn children have a less flexible interpersonal problem-solving repertoire, and they both express and demonstrate a greater reliance on adult intervention than on using independent strategies to meet their social goals. In addition to suggesting adult-dependent strategies to resolve their interpersonal dilemmas, withdrawn children also use less productive and assertive social strategies to deal with peer conflict, endorsing interpersonal avoidance (Rubin, 1982). In a more recent study, withdrawn children differed from typical children by internalizing blame, internalizing their emotional reactions, and coping through avoidance (Burgess et al., 2006).

When anxious/withdrawn children are asked to explain why the strategies that they suggest to solve interpersonal problems may fail, they are more likely than typical children to assign blame to themselves and to attribute their failures to internal, stable causes (Wichmann, Coplan, & Daniels, 2004). In addition to self-blame, these children are more likely to react to negative events by demonstrating (or suggesting) that they would feel upset and fearful. It may well be that these emotions are largely responsible for the choice of unassertive or submissive strategies to meet solve their interpersonal dilemmas.

In summary, the ways in which children think and feel about the interpersonal dilemmas they face predict, in part, the ways in which they go about attempting to solve these dilemmas. Thus the social-cognitive profiles of typical, anxious/withdrawn, and aggressive children are quite distinct.

SIP and Interpersonal Problem Solving in Children with TBI

The *SIP* skills and social competence of children who have suffered a TBI have not been well studied. Much of the extant literature has focused on issues pertaining to theory of mind and perspective taking. In this regard, researchers have demonstrated that children with TBI may have difficulty understanding the emotions and intentions of others during complicated social interactions. For example, Tonks, Williams, Frampton, Yates, and Slater (2007) demonstrated that children with TBI had more difficulty in reading emotions (as conveyed by facial expressions, eyes, and voices) than did children without TBI. Children with TBI may also have difficulty interacting with others or solving interpersonal dilemmas because they have difficulty recognizing deceptive emotions, or expressed emotions that are incongruent with felt emotions (Dennis, Barnes, Wilkinson, & Humphreys, 1998). This is the sort of difficulty that may preclude the ability to distinguish between literal and nonliteral social interchanges. For example, by preschool age, typical children are quite competent in distinguishing rough-and-tumble play from aggression.

Children with TBI also have difficulty recognizing the intentions of others; although they do not appear to have difficulties understanding literal statements made by others, they perform more poorly when attempting to interpret others' statements that involve sarcasm or are deceptive (this is precisely the issue raised above vis-à-vis rough-and-tumble play). In short, nonliterality presents a problem for children with TBI. In this regard, researchers would do well to examine whether behavioral demonstrations that involve nonliterality are problematic for children with TBI. The lack of an understanding of such nonliteral behaviors (or statements) may "mark" children with TBI as prime candidates for peer rejection.

Significantly, deficits in theory of mind or perspective taking have been found in adults and adolescents with TBI (Bibby & McDonald, 2005). Given that research with typically developing children has found that misunderstanding peers' thoughts, feelings, and literal visual perspectives has been associated with behavior problems and peer rejection, children with TBI who evidence these social-cognitive deficits are also likely to have trouble getting along with peers.

Research also indicates that children with TBI may have trouble generating sophisticated and competent strategies in challenging social situations. Warschausky, Cohen, Parker, Levendosky, and Okun (1997) found that children with TBI generated fewer ideas for hypothetical situations involving peer group entry. The solutions that these children generated were also more indirect, less positive, and less assertive than the solutions generated by children who were uninjured. In addition, Yeates, Schultz, and Selman (1991) compared children and adolescents with severe and moderate TBI and youth with orthopedic injuries. The participants with TBI were more likely to endorse less sophisticated strategies to resolve social dilemmas, and they employed lower-level social-cognitive reasoning to explain the effectiveness of the strategies they had selected. It was also suggested that the children with TBI showed long-term deficits in the social problem-solving skills, producing further social difficulties or impoverished social competence.

The significant prediction of social competence from *SIP* and social problem-solving skills in the group of children and adolescents with TBI has been confirmed in other studies (Janusz, Kirkwood, Yeates & Taylor, 2002; Walz, Yeates, Wade, & Mark, 2009). It is important to note, however, that studies of interpersonal problem solving *in vivo* are few and far between. Nevertheless, given the SIP and social-cognitive differences noted above, one would predict that children with TBI would demonstrate deficits in observed social competence and difficulties with their peer relationships. In the one study we could identify that bears on this issue, Ellison, Rubin, and Yeates (2011) found that children with TBI were less likely than age-mates who had been orthopedically injured to meet success when they attempted to initiate social interchanges with peers.

We turn now to the literature on children's peer relationships. In our review, we describe how children's individual characteristics and social interactions are associated with the extent to which they are accepted or rejected by the peer group.

Peer Relationships

Defining Peer Acceptance and Rejection

Peer acceptance refers to the experience of being liked and accepted by the peer group, whereas *peer rejection* represents the experience of being disliked by peers. In the following sections, we briefly examine the methods used by researchers to assess acceptance and rejection. We also present findings concerning the behavioral correlates and consequences of peer acceptance and rejection.

Measuring Peer Acceptance and Rejection

The procedures typically used to assess the quality of children's peer relationships address two central issues: "Is the child liked?" and "What is the child like?" (Asher & McDonald, 2009). Children spend a great deal of time with each other in school, and so they are privy to low-frequency but significant social *interactions* (e.g., the demonstration of an aggressive act, providing help to someone in need), which can lead to the establishment and maintenance of particular social reputations and *relationships*. In addition, *peer group* assessments of children's behaviors and relationships provide multiple perspectives pertaining to the two central questions posed above (Rubin et al., 2006). Thus peer reports of behavior and relationships have become the dominant methodology in contemporary research concerning the quality, determinants, correlates, and consequences of children's peer relationships (for a review, see Rubin et al., 2009).

Of course, one may obtain teacher, parent, and self-reports of children's behavioral characteristics and peer relationships. However, teachers and parents bring with them a perspective that may paint a picture of adult values about social behaviors; these values may vary considerably from those of children (Ladd & Profilet, 1996). And self-assessments may be particularly biased or colored in many ways by a child's own relationships with peers. With this in mind, we only review methods derived from peer assessments.

Sociometric Nominations

Based on Moreno's (1934) work on the topic of interpersonal attraction and repulsion with groups, Coie, Dodge, and Coppotelli (1983) created a method of assessment in which children nominate three to five peers whom they "like" and "dislike." From these nominations, children are categorized into status groups based on the number of positive and negative nominations they receive from peers. To control for class size, nominations are standardized within each classroom or grade (sometimes scores are standardized within gender, depending on the study goals).

On the basis of these nominations, children have been classified into one of five sociometric categories. Sociometrically *popular* children are those who receive many "like" and few "dislike" nominations, whereas children who receive many "dislike" and fewer "liked" nominations are considered *rejected*. Children who receive many "like" and many "dislike" nominations are labeled as *controversial*; *neglected* children receive few nominations in both categories; and *average* children are those who are near the mid-range for both "like" and "dislike" nominations (for a detailed description of sociometric categorization schemes, refer to Cillessen, 2009).

An alternative to sociometric nominations is a rating scale measure of peer acceptance. With this method, children are provided with a class roster and asked to rate each of their classmates (or a randomly selected group of grade-mates for children in middle or high school) on a scale (e.g., 1 = "not at all" to 5 "very much") of how much they like to play or work each person. This too yields a continuous indicator of peer acceptance. An advantage of rating scale measures is that each child in a class or grade receives an equal number of ratings, rather than just the few who are prominent, as may be the case when nominations are used.

Perceived Popularity

Perceived popularity is different from *sociometric* popularity, as the former pertains to those children who are perceived by the group as popular, and the latter is a personal choice regarding whom an individual child likes. To measure perceived popularity, researchers typically ask children to nominate peers they perceive as "popular" (e.g., Cillessen & Mayeux, 2004). The number of nominations a child receives is then divided by the number of possible nominations he or she could have received, and this fraction is standardized within class or grade. The relation between perceived and sociometric popularity ranges from moderate to strong, but in adolescence this association seems to decline and does so more for girls than for boys.

Correlates of Peer Acceptance and Rejection

Not surprisingly, much of children's social behavior is related to their popularity and/or rejection by peers. Well-liked children are often prosocial, appropriately assertive, and seen as leaders by their peers (Asher & McDonald, 2009). They are also skilled at initiating peer interactions and entering ongoing group activities; they consider the frame of reference common to the ongoing play group, and establish themselves as sharing in the group's frame of reference without drawing unwarranted attention to themselves (Putallaz & Wasserman, 1990). In addition, popular or well-liked children typically negotiate, compromise, and deal with peer conflict in competent ways (e.g., Troop-Gordon & Asher, 2005). In short, sociometrically popular children appear to be socially competent.

Cillessen and Mayeux (2004) have found perceived popularity to be associated with both physical and relational aggression (the latter of which is intended to harm relationships and friendships, especially for older children), as well as with prosocial behavior. Perceived popularity has also been associated with having a good sense of humor, and academic competence, and athletic ability, as well as with being attractive, stylish, and wealthy (Vaillancourt & Hymel, 2006).

Rejected children are more heterogeneous in their behavioral characteristics. Childhood aggression, in its varied forms, is the strongest behavioral predictor of peer dislike (see Rubin et al., 2006, for a review). Some of the most significant studies in which causal links between behavior and rejection have been established have involved making laboratory-based observations of unfamiliar peers in play groups and obtaining sociometric nominations over time (e.g., Dodge, 1983). This procedure has allowed the examination of behaviors that predict peer acceptance and rejection within newly formed groups. Findings have indicated that rejected boys in these groups display both more instrumental and reactive aggression, less cooperation, and more time off task than do the popular boys. Significantly, Cillessen and Mayeaux (2004) have reported that the negative associations between physical aggression and peer acceptance decrease in strength as children progress through junior high school, perhaps as children form groups with similar peers (Cairns, Cairns, Neckerman, Gest, & Gariepy, 1988). In addition, although aggressive behavior is generally associated with peer rejection, there are important exceptions. Aggressive children who have other valued individual characteristics (e.g., athletic ability) may not be rejected by the peer group, but instead may be identified as sociometrically "controversial" (Parkhurst & Hopmeyer, 1998) or perceived to be popular (Cillessen & Mayeux, 2004).

Other individual characteristics are associated with peer rejection. For example, disruptive, hyperactive, or inattentive and immature behaviors are associated with rejection (Miller-Johnson, Coie, Maumary-Gremaud, Bierman, & the Conduct Problems Prevention Research Group, 2002). Socially anxious, timid, and withdrawn behaviors are also associated with peer rejection. It appears likely that withdrawn children are rejected because they are viewed by peers as "easy marks" or "whipping boys" (Olweus, 1993) who produce few social overtures, appear overly compliant to peers' requests and demands, and shy away from the demonstration of assertive, confident behavior in the company of peers (see Rubin, Bowker, & Gazelle, 2010, for a review). These children exhibit a lack of social competence, in that the ways in which they attempt to meet their social goals often result in peer noncompliance and outright behavioral rejection (e.g., Stewart & Rubin, 1995).

Gender appears to influence the relation between social withdrawal and peer rejection. For instance, socially withdrawn boys are more likely to be rejected than similarly behaved girls (Gazelle, 2008), probably because withdrawn behavior violates male gender norms (Caspi, Elder, & Bem, 1988). In general, examinations of how behavior is associated with rejection support the idea that any form of social behavior considered deviant from the norm is likely to be associated with peer rejection.

Predictive Outcomes of Peer Relationship Difficulties

Several conceptual models exist that explain the possible relations between peer rejection and negative developmental outcomes. For example, in a causal model, those who interact with peers in a non-normative fashion are viewed, in a negative light by those peers (Parker & Asher, 1987). These negative perceptions result in rejection and exclusion by peers, which in turn limits the child's positive socialization experiences with peers and promotes negative social experiences, leading to poorer psychological outcomes. An *incidental* model (Parker & Asher, 1987) suggests that the same underlying behavioral tendencies that account for children's peer rejection also serve as primary predictors of later negative developmental outcomes. Finally, Ladd and Burgess (2001) suggested the possibility of an *additive* model, in which relational risk factors (e.g., peer rejection) increase the likelihood of later dysfunction beyond the risks associated with behavioral characteristics. These researchers have also offered a *moderated risk adjustment* model, in which relational risks (e.g., peer rejection) exacerbate maladjustment among children who are behaviorally at risk.

Results from longitudinal studies support both of these latter models, providing evidence that peer rejection both adds to and exacerbates the risks associated with such behaviors as aggression and social withdrawal. For example, longitudinal studies have indicated that peer rejection in childhood predicts a wide range of externalizing problems in adolescence, including delinquency, conduct disorder, attentional difficulties, and substance abuse (Kupersmidt & Coie, 1990). More specifically, *early* peer rejection provides a unique increment in the prediction of later antisocial outcomes, even when the researchers control for previous levels of aggression and externalizing problems (Ladd & Burgess, 2001). In addition, Dodge et al. (2003) reported that peer rejection predicted "growth" in aggression over time, after controlling for original levels of aggression from early childhood through to adolescence. Relatedly, Prinstein and La Greca (2004) found that girls' childhood aggression predicted later substance use and sexual risk behavior, but only for those girls who were disliked or rejected in junior high school.

Researchers have also reported that being anxious/withdrawn is contemporaneously and predictively associated with internalizing problems across the lifespan, including low self-esteem, anxiety problems, loneliness, and depressive symptoms (Coplan, Arbeau, & Armer, 2008; Rubin, Chen, McDougall, Bowker, & McKinnon, 1995). Significantly, the full impact of social withdrawal requires that peer exclusion be considered not only as a consequence of anxious/withdrawn behavior, but also as a factor that may change the course of withdrawal itself, as well as withdrawn children's social and emotional adjustment more broadly. Gazelle and Ladd (2003) found that only those anxious/withdrawn children who were excluded by peers in early grade school displayed greater stability in anxious solitude and elevated levels of depression over the course of middle childhood. Similarly, Gazelle and Rudolph (2004) have shown that over a 2-year period, high exclusion by middle school peers led anxious/solitary youth to maintain or exacerbate the extent of their social avoidance and depression, whereas the experience of low exclusion predicted increased social approach and less depression.

Although research indicates that children with disabilities or other neurological dysfunction are less well liked by peers (e.g., Noll et al., 2007), to date few researchers have used classroom peer nominations to investigate the social acceptance of children with TBI (Yeates et al., 2007). We hypothesize that children with TBI may be rejected by peers because of behavioral changes that accompany head injury. For example, children with TBI show marked changes in their personalities (Max et al., 2006) that may make them less enjoyable play partners or companions.

In addition, children with TBI are rated by parents as less socially competent and as having more externalizing problems than children without TBI (Boehnert, Parker, & Warschausky, 1997; Levin et al., 2004; Walz et al., 2009). Children with severe TBI also self-report lower social competence, compared to those with less severe injuries (Walz et al., 2009). Taken together with the social-cognitive deficits accompanying TBI discussed above, it follows that children with TBI should also be prime candidates for peer rejection. In following the moderated risk adjustment model, we suggest that peer rejection may exacerbate the social-behavioral problems of children with TBI because they may miss out on positive peer interactions, which may improve social-cognitive skills and socially competent behavior.

Children's Friendships

Functions of Friendship

Friendship has been defined as a close, mutual, and voluntary dyadic relationship (Rubin et al., 2006). This definition allows friendship to be differentiated (1) from parent–child and other familial relationships, since a friendship is *voluntary* (both parties choose to become involved in the relationship); and (2) from the construct of popularity at the peer group level, which refers to the experience of being accepted or liked by one's peers. Some of the defining attributes of friendship include the feeling of *equality* between the members of the relationship, and *reciprocity,* which refers to the return of corresponding affection and behaviors between partners. Furthermore, friendship is characterized by reciprocal *affection,* in the sense that both members share positive affect and a liking for each other. Although homophily or similarities along different domains (surface characteristics, behaviors, interests, etc.) may help draw individuals together and attract them into developing friendships, mutual affection constitutes the basis of friendship (Rubin et al., 2006).

Children's friendships have been recognized as serving such functions as the provision or promotion of (1) support, self-esteem, and positive self-evaluation; (2) emotional security; (3) affection; (4) intimacy and opportunities for intimate disclosure; (5) consensual validation of interests, hopes, and fears; (6) instrumental and informational assistance; (7) the promotion and growth of interpersonal sensitivity; and (8) prototypes for later romantic, marital, and parental relationships (Newcomb & Bagwell, 1995). However, the most salient function of friendship is to provide its constituent members with an extrafamilial base of security, from which it is possible for them to explore the effects of their behaviors on their environment, their peers, and themselves.

The functions of friendship vary according to age (Rubin et al., 2006). During early childhood, friendship serves to maximize excitement, provides opportunities for amusement and enjoyable play, and helps children regulate their behavior. By middle childhood, friendships serve the function of helping children learn about behavioral and social norms as well as the social skills necessary for successful self-presentation. These skills are especially vital in middle childhood, when anxiety in regard to peer relationships starts to develop. Friendship also confers emotional and social support and provides instrumental aid and assistance during this developmental stage (Newcomb & Bagwell, 1995). Finally, during late childhood and adolescence, friendships provide opportunities for the expression and regulation of affect (Denton & Zarbatany, 1996) and assist individuals in their quest for successful identity development (Sullivan, 1953).

Children's Understanding of Friendship

We have suggested what friendship provides, but from the child's perspective, the understanding of what friendship entails changes with age. By the ages of 7–8 years, children describe friends as companions who are enjoyable to be with, live nearby, and share their own expectations about play activities (Selman & Schultz, 1990). In late childhood (10–11 years), shared values become increasingly significant, and children begin expecting their friends to be loyal and to help them deal with their inter- and intrapersonal difficulties. During early adolescence, sharing similar interests, engaging in intimate disclosure, and actively attempting to understand each other are primary features identified in friend-ship (Schneider & Tessier, 2007). By adolescence, commitment is believed to be a central defining feature of friendship (Gummerum & Keller, 2008).

Berndt (1981) has suggested that the understanding of friendship represents a cumu-lative assimilation of essentially unrelated themes, such as shared play interests and intimate disclosure. According to this viewpoint, when children begin recognizing the importance of intimacy and loyalty, they do not necessarily abandon their initial notions that play, enjoying each other's company, and mutuality are defining properties of friend-ship. Furthermore, Berndt notes that with increasing age, children become better able to distinguish between friendships and other dyadic relationships. In this regard, it has been noted that children's individual inclinations, their thoughts about varying social situations (e.g., how to resolve an interpersonal dilemma), and the behaviors they choose to meet their social goals vary in accord with the relationship they have with a partner in a given social situation. Put another way, children think about social situations and behave differently in these situations when they are in the company of friends rather than acquaintances or unfamiliar age-mates (Burgess et al., 2006).

Implications for Adjustment

Friendship can serve as a buffering agent insofar as psychological well-being and positive adjustment are concerned (Hartup & Stevens, 1997). Children who have a best friend are less likely to experience loneliness and more likely to report higher self-esteem than those who are friendless (Brendgen, Vitaro, & Bukowski, 2000). Indeed, by early adolescence, a best friendship can protect children from negative outcomes when they lack a support-ive parent–child relationship (Rubin et al., 2004) or when they are repeatedly targets of peer victimization and rejection (Laursen, Bukowski, Aunola, & Nurmi, 2007). In this regard, friendship can serve as a primary source of social and emotional support and can provide children with a sense of felt security. Further, although there exist very few lon-gitudinal studies on the outcomes of having friends during childhood, results from one investigation revealed that friendship involvement in preadolescence (age 10 years) posi-tively predicted self-esteem during early adulthood (age 23 years; Bagwell, Newcomb, & Bukowski, 1998), even after the researchers controlled for peer acceptance/rejection.

Friendship Quality

Considerable variability exists in the *quality* of children's friendships. Researchers have reported consistently that some friendships are supportive, whereas others are conflict-ridden. Those friendships characterized by *positive* relationship qualities, such as inti-macy, caring, and support, have been shown to be associated with positive thoughts about oneself and with general psychological well-being during childhood and early adolescence (Hartup & Stevens, 1997). Children and adolescents with positive or high-quality friendships report fewer psychological difficulties than children with less positive friendships do (e.g., Schmidt & Bagwell, 2007).

Significantly, some friendships can prove debilitating (Rose, Carlson, & Waller, 2007). For example, there appears to be a positive relation between friendship quality and young adolescent girls' tendency to *co-ruminate*, or repeatedly discuss problems and negative events with a friend. These discussions may promote intimacy and closeness, but they may also reinforce or promote feelings of anxiety and depression, especially when the co-ruminative discussion is dominated by shared negativity about the self.

The individual characteristics of the child, the characteristics of the best friend, and whether the friendship is stable also appear to have an impact on psychological adjustment. Children who have an aggressive best friend seem to be at risk for increases in aggressive behavior (Vitaro, Pedersen, & Brendgen, 2007), behavior problems (Berndt, Hawkins, & Jiao, 1999), emotional maladjustment (Brendgen et al., 2000), and academic difficulties (Véronneau, Vitaro, Pedersen, & Tremblay, 2008). Similarly, in a recent growth curve analysis of longitudinal data, Oh et al. (2008) found that those anxious/withdrawn young adolescents whose best friends were similarly withdrawn were at risk for being placed on a trajectory of increased anxious/withdrawn behavior throughout the years of middle school (ages 10–13 years).

In summary, research clearly indicates that being friends with peers who themselves have behavioral difficulties may lead to increased adjustment difficulty, especially for children with behavioral difficulties of their own. Significantly, however, there is also evidence that having prosocial and sociable friends can promote school adjustment and social competence (Wentzel, Barry, & Caldwell, 2004).

TBI and Friendship

The specific implications of friendship for the adjustment of children who have suffered a TBI may differ from those of the typical population. Social dysfunction is not only commonly reported by survivors of TBI, but is also very often rated as the most debilitating of all sequelae, with negative effects on various areas of daily life and overall quality of life (Greenham, Spencer-Smith, Anderson, Coleman, & Anderson, 2010). However, very little is known about the social and emotional concomitants of TBI (Yeates et al., 2007 and Chapter 10, this volume).

In the integrative model of social outcomes of children with TBI proposed by Yeates et al. (2007), not only is social adjustment moderated by injury-related and environmental factors, but it may be mediated or bidirectionally related to social-cognitive and SIP factors. As noted above, these factors can predict various forms of social interaction, which in turn help to determine the quality of relationships that children form in the peer group and with friends. Again, these individual, interaction, and relationship factors are likely to affect the children's social and emotional adjustment.

One of the most effective ways of fostering the development of friendships in children is through participating with peers in enjoyable activities. In the case of children with disabilities, opportunities for engaging in recreational and pleasurable social activities are fewer; when such participation occurs, it is often in the company of adults rather than peers (Solish, Perry, & Minnes, 2010). Thus it is not surprising that friendships of children with severe TBI have been characterized as involving difficulties in managing conflict, coordinating play, and developing intimacy (Boehnert et al., 1997). Nevertheless, the extant studies are few and far between. Much remains to be discovered about the prevalence, stability, quality, and correlates of friendships for children with TBI. It will also be important to examine whom they befriend and how they conceptualize friendship. Clearly, as in almost all topics reviewed in this chapter, there are many remaining questions with very few extant answers.

Conclusions

In this chapter, we have reviewed literature concerning children's social competence and peer relationships. It should be clear that experiences garnered by children in the peer group and with their friends represent significant developmental phenomena. Thus children who are socially competent, are accepted by peers, and have qualitatively rich friendships appear to fare better throughout childhood than children who are socially incompetent, are rejected and excluded by the peer group, or are lacking in friendships.

Although we have learned a great deal about the significance of nondisordered children's individual characteristics, social interactions, relationships, and groups, there remain many important questions to address vis-à-vis the social lives of children with TBI. For example, although researchers are only now beginning to examine *SIP* skills in children with TBI, the extant research has not focused on the severity of injury, the locale of injury, or the age at which the children experienced injury. The same issues apply regarding the types of social interaction biases that children with TBI may have (are the children less socially skilled, more aggressive, more withdrawn?), or the extent to which they are rejected or victimized in the peer group. Frankly, virtually nothing is known about the friendships of children with TBI.

Children's peer interactions, relationships, and groups are not only relevant insofar as psychological and emotional adjustment is concerned; they are clearly important entities as children attempt to make their ways through their everyday lives at home, at school, or in their neighborhoods. Clearly, it is time to develop an extensive program of research on these topics for children who have suffered a TBI.

References

Asher, S. R., & McDonald, K. L. (2009). The behavioral basis of acceptance, rejection, and perceived popularity. In K. H. Rubin, W. M. Bukowski, & B. Laursen (Eds.), *Handbook of peer interactions, relationships, and groups* (pp. 232–248). New York: Guilford Press.

Asher, S. R., & Paquette, J. (2003). Loneliness and peer relations in childhood. *Current Directions in Psychological Science, 12,* 75–78.

Bagwell, C. L., Newcomb, A. F., & Bukowski, W. M. (1998). Preadolescent friendship and peer rejection as predictors of adult adjustment. *Child Development, 69,* 140–153.

Bandura, A., & Walters, R. H. (1963). *Social learning and personality development.* New York: Holt, Rinehart & Winston.

Berndt, T. J. (1981). Relations between social cognition, nonsocial cognition, and social behavior: The case of friendship. In J. Flavell & L. Ross (Eds.), *Social cognitive development* (pp. 176–199). Cambridge, UK: Cambridge University Press.

Berndt, T. J., Hawkins, J. A., & Jiao, Z. (1999). Influences of friends and friendships on adjustment to junior high school. *Merrill–Palmer Quarterly, 45,* 13–41.

Bibby, H., & McDonald, S. (2005). Theory of mind after traumatic brain injury. *Neuropsychologia, 43,* 99–114.

Boehnert, A. M., Parker, J. G., & Warschausky, S. A. (1997). Friendship and social adjustment of children following traumatic brain injury: An exploratory investigation. *Developmental Neuropscyhology, 13,* 477–486.

Boivin, M., & Hymel, S. (1997). Peer experiences and social self-perceptions: A sequential model. *Developmental Psychology, 33*(1), 135–145.

Bowlby, J. (1958). The nature of the child's tie to his mother. *International Journal of Psycho-Analysis, 39*, 350–373.

Brendgen, M., Vitaro, F., & Bukowski, W. M. (2000). Deviant friends and early adolescents' emotional and behavioral adjustment. *Journal of Research on Adolescence, 10*(2), 173–189.

Burgess, K. B., Wojslawowicz, J. C., Rubin, K. H., Rose-Krasnor, L., & Booth-LaForce, C. (2006). Social information processing and coping styles of shy/withdrawn and aggressive children: Does friendship matter? *Child Development, 77*, 371–383.

Cairns, R. B., Cairns, B. D., Neckerman, H. J., Gest, S., & Gariepy, J. L. (1988). Peer networks and aggressive behavior: Peer support or peer rejection? *Developmental Psychology, 24*, 815–823.

Caspi, A., Elder, G. H., Jr., & Bem, D. J. (1988). Moving away from the world: Life-course patterns of shy children. *Developmental Psychology, 24*, 824–831.

Cillessen, A. H. N. (2009). Sociometric methods. In K. H. Rubin, W. M. Bukowski, & B. Laursen (Eds.), *Handbook of peer interactions, relationships, and groups* (pp. 82–99). New York: Guilford Press.

Cillessen, A. H. N., & Mayeux, L. (2004). From censure to reinforcement: Developmental changes in the association between aggression and social status. *Child Development, 75*, 147–163.

Coie, J. D., Dodge, K. A., & Coppotelli, H. (1983). Continuities and changes in children's social status: A five-year longitudinal study. *Merrill–Palmer Quarterly, 29*, 261–282.

Coplan, R. J., Arbeau, K. A., & Armer, M. (2008). Don't fret, be supportive!: Maternal characteristics linking child shyness to psychosocial and school adjustment in kindergarten. *Journal of Abnormal Child Psychology, 36*, 359–371.

Crick, N. R., & Dodge, K. A. (1994). A review and reformulation of social information-processing mechanisms in children's social adjustment. *Psychological Bulletin, 115*, 74–101.

Dennis, M., Barnes, M. A., Wilkinson, M., & Humphreys, R. P. (1998). How children with brain injury represent real and deceptive emotions in short narratives. *Brain and Language, 61*, 450–483.

Denton, K., & Zarbatany, L. (1996). Age differences in support processes in conversations between friends. *Child Development, 67*, 1360–1373.

Dodge, K. A. (1983). Behavioral antecedents of peer social status. *Child Development, 54*, 1386–1399.

Dodge, K. A. (1985). Attributional bias in aggressive children. In P. C. Kendall (Ed.), *Advances in cognitive-behavioral research and therapy* (Vol. 4, pp. 73–110). San Diego, CA: Academic Press.

Dodge, K. A., Coie, J. D., & Lynam, D. (2006). Aggression and antisocial behavior in youth. In W. Damon & R. M. Lerner (Series Eds.) & N. Eisenberg (Vol. Ed.), *Handbook of child psychology: Vol. 3. Social, emotional, and personality development* (6th ed., pp. 719–788). Hoboken, NJ: Wiley.

Dodge, K. A., Lansford, J. E., Burks, V. S., Bates, J. E., Pettit, G. S., Fontaine, R., et al. (2003). Peer rejection and social information-processing factors in the development of aggressive behavior problems in children. *Child Development, 74*, 374–393.

Downey, G., Lebolt, A., Rincon, C., & Freitas, A. L. (1998). Rejection sensitivity and children's interpersonal difficulties. *Child Development, 69*, 1074–1091.

Ellison, K., Rubin, K. H., & Yeates, K. O. (2011, April). *Peer rejection experiences of children who have sustained a traumatic brain injury.* Poster presented at the Biennial Meeting of the Society for Research in Child Development, Montréal.

Erdley, C. A., & Asher, S. R. (1996). Children's social goals and self-efficacy perceptions as influences on their responses to ambiguous provocation. *Child Development, 67*, 1329–1344.

Freud, S. (1933). *New introductory lectures on psycho-analysis* (W. J. H. Sprott, Trans.). New York: Norton.

Gazelle, H. (2008). Behavioral profiles of anxious solitary children and heterogeneity in peer relations. *Developmental Psychology*, *44*, 1604–1624.

Gazelle, H., & Ladd, G. W. (2003). Anxious solitude and peer exclusion: A diathesis–stress model of internalizing trajectories in childhood. *Child Development*, *74*, 257–278.

Gazelle, H., & Rudolph, K. D. (2004). Moving toward and away from the world: Social approach and avoidance trajectories in anxious solitary youth. *Child Development*, *75*(3), 829–849.

Goldfried, M. R., & D'Zurilla, T. J. (1969). A behavioral-analytic model for assessing competence. In C. Spielberger (Ed.), *Current topics in clinical and community psychology* (Vol. 1, pp. 151–196). New York: Academic Press.

Gouze, K. R. (1987). Attention and social problem solving as correlates of aggression in preschool males. *Journal of Abnormal Child Psychology*, *15*, 181–197.

Greenham, M., Spencer-Smith, M. M., Anderson, P. J., Coleman, L., & Anderson, V. A. (2010). Social functioning in children with brain insult. *Frontiers in Human Neuroscience*, *4*, 1–10.

Gummerum, M., & Keller, M. (2008). Affection, virtue, pleasure, and profit: Developing an understanding of friendship closeness and intimacy in Western and Asian cultures. *International Journal of Behavioral Development*, *32*, 130–143.

Harris, J. R. (2009). *The nurture assumption: Why children turn out the way they do (Revised and updated)*. New York: Free Press.

Hartup, W. H., & Stevens, N. (1997). Friendships and adaptation across the life span. *Current Directions in Psychological Science*, *8*, 76–79.

Hinde, R. A. (1987). *Individuals, relationships and culture*. Cambridge, UK: Cambridge University Press.

Hinde, R. A., & Stevenson-Hinde, J. (1976). Toward understanding relationships: Dynamic stability. In P. Bateson & R. A. Hinde (Eds.), *Growing points in ethology* (pp. 451–479). Cambridge, UK: Cambridge University Press.

Janusz, J. A., Kirkwood, M. W., Yeates, K. O., & Taylor, H. G. (2002). Social problem-solving skills in children with traumatic brain injury: Long-term outcomes and prediction of social competence. *Child Neuropsychology*, *8*, 179–194.

Krasnor, L., & Rubin, K. H. (1983). Preschool social problem solving: Attempts and outcomes in naturalistic interaction. *Child Development*, *54*, 1545–1558.

Kupersmidt, J. B., & Coie, J. D. (1990). Preadolescent peer status, aggression, and school adjustment as predictors of externalizing problems in adolescence. *Child Development*, *61*, 1350–1362.

Ladd, G. W., & Burgess, K. B. (2001). Do relational risks and protective factors moderate the linkages between childhood aggression and early psychological and school adjustment? *Child Development*, *72*, 1579–1601.

Ladd, G. W., & Profilet, S. (1996). The Child Behavior Scale: A teacher report measure of young children's aggressive, withdrawn, and prosocial behaviors. *Developmental Psychology*, *32*, 1008–1024.

Laursen, B., Bukowski, W. M., Aunola, K., & Nurmi, J. (2007). Friendship moderates prospective associations between social isolation and adjustment problems in young children. *Child Development*, *78*(4), 1395–1404.

Lemerise, E. A., & Arsenio, W. F. (2000). An integrated model of emotion processes and cognition in social information processing. *Child Development*, *71*, 107–118.

Lemerise, E. A., Gregory, D. S., & Fredstrom, B. K. (2005). The influence of provocateurs' emotion displays on the social information processing of children varying in social adjustment and age. *Journal of Experimental Child Psychology*, *90*, 344–366.

Levin, H. S., Zhang, L., Dennis, M., Ewing-Cobbs, L., Schachar, R., Max, J., et al. (2004). Psychosocial outcome of TBI in children with unilateral frontal lesions. *Journal of the International Neuropsychological Society*, *10*, 305–316.

Lochman, J. E., & Dodge, K. A. (1994). Social-cognitive processes of severely violent, moder-ately aggressive, and nonaggressive boys. *Journal of Consulting and Clinical Psychology, 62,* 366–374.

Max, J. E., Levin, H. S., Schachar, R. J., Landis, J., Saunders, A. E., & Ewing-Cobbs, L. (2006). Predictors of personality change due to traumatic brain injury in children and adolescents six to twenty-four months after injury. *Journal of Neuropsychiatry and Clinical Neurosciences, 18,* 21–32.

McFall, R. M. (1982). A review and reformulation of the concept of social skills. *Behavioral Assessment, 4*(1), 1–33.

Mead, G. H. (1934). *Mind, self, and society.* Chicago: University of Chicago Press.

Miller-Johnson, S., Coie, J. D., Maumary-Gremaud, A., Bierman, K., & the Conduct Problems Prevention Research Group. (2002). Peer rejection and aggression and early starter models of conduct disorder. *Journal of Abnormal Child Psychology, 30,* 217–230.

Moreno, J. L. (1934). *Who shall survive? A new approach to the problem of human inter-relations.* Washington, DC: Nervous and Mental Disease Publishing.

Nelson, L. J., Rubin, K. H., & Fox, N. A. (2005). Social and nonsocial behaviors and peer accep-tance: A longitudinal model of the development of self-perceptions in children ages 4 to 7 years. *Early Education and Development, 20,* 185–200.

Newcomb, A. F., & Bagwell, C. L. (1995). Children's friendship relations: A meta-analytic review. *Psychological Bulletin, 117,* 306–347.

Noll, R. B., Reiter-Purtill, J., Moore, B. D., Schorry, E. K., Lovell, A. M., Vannatta, K., et al. (2007). Social, emotional, and behavioral functioning of children with NF1. *American Jour-nal of Medical Genetics, 143A,* 2261–2273.

Oh, W., Rubin, K. H., Bowker, J. C., Booth-LaForce, C., Rose-Krasnor, L., & Laursen, B. (2008). Trajectories of social withdrawal from middle childhood to early adolescence. *Journal of Abnormal Child Psychology, 36*(4), 553–566.

Olweus, D. (1993). Victimization by peers: Antecedents and long-term outcomes. In K. H. Rubin & J. B. Asendorpf (Eds.), *Social withdrawal, inhibition, and shyness in childhood* (pp. 315–341). Hillsdale, NJ: Erlbaum.

Oppenheimer, L. (1989). The nature of social action: Social competence versus social conformism. In B. H. Schneider, G. Attili, J. Nadel, & R. P. Weissberg (Eds.), *Social competence in devel-opmental perspective* (pp. 41–69). New York: Kluwer Academic/Plenum.

Parker, J. G., & Asher, S. R. (1987). Peer relations and later personal adjustment: Are low-accepted children at risk? *Psychological Bulletin, 102*(3), 357–389.

Parkhurst, J. T., & Hopmeyer, A. (1998). Sociometric popularity and peer-perceived popularity: Two distinct dimensions of peer status. *Journal of Early Adolescence, 18*(2), 125–144.

Piaget, J. (1932). *The moral judgment of the child.* Glencoe, IL: Free Press.

Prinstein, M. J., & La Greca, A. M. (2004). Childhood rejection, aggression, and depression as predictors of adolescent girls' externalizing and health risk behaviors: A six-year longitudinal study. *Journal of Consulting and Clinical Psychology, 72,* 103–112.

Putallaz, M., & Wasserman, A. (1990). Children's entry behaviors. In S. R. Asher & J. D. Coie (Eds.), *Peer rejection in childhood* (pp. 60–89). New York: Cambridge University Press.

Rose, A. J., & Asher, S. R. (1999). Children's goals and strategies in response to conflicts within a friendship. *Developmental Psychology, 35,* 69–79.

Rose, A. J., Carlson, W., & Waller, E. M. (2007). Prospective associations of co-rumination in friendship and emotional adjustment: Considering the socioemotional trade-offs of co-rumination. *Developmental Psychology, 43*(4), 1019–1031.

Rubin, K. H. (1982). Social and social-cognitive developmental characteristics of young isolate, normal and sociable children. In K. H. Rubin & H. S. Ross (Eds.), *Peer relationships and social skills in childhood* (pp. 353–374). New York: Springer-Verlag.

Rubin, K. H., Bowker, J. C., & Gazelle, H. (2010). Social withdrawal in childhood and adolescence: Peer relationships and social competence. In K. H. Rubin & R. J. Coplan (Eds.), *The development of shyness and social withdrawal* (pp. 131–156). New York: Guilford Press.

Rubin, K. H., Bukowski, W. M., & Laursen, B. (Eds.). (2009). *Handbook of peer interactions, relationships, and groups.* New York: Guilford Press.

Rubin, K. H., Bukowski, W. M., & Parker, J. G. (2006). Peer interactions, relationships, and groups. In W. Damon & R. Lerner (Series Eds.) & N. Eisenberg (Vol. Ed.), *Handbook of child psychology: Vol. 3. Social, emotional, and personality development* (6th ed., pp. 571–645). Hoboken, NJ: Wiley.

Rubin, K. H., Cheah, C., & Menzer, M. M. (2010). Peers. In M. H. Bornstein (Ed.), *Handbook of cultural developmental science* (pp. 223–237). New York: Psychology Press.

Rubin, K. H., Chen, X., McDougall, P., Bowker, A., & McKinnon, J. (1995). The Waterloo Longitudinal Project: Predicting adolescent internalizing and externalizing problems from early and mid-childhood. *Development and Psychopathology, 7,* 751–764.

Rubin, K. H., Dwyer, K. M., Booth-LaForce, C., Kim, A. H., Burgess, K. B., & Rose-Krasnor, L. (2004). Attachment, friendship, and psychosocial functioning in early adolescence. *Journal of Early Adolescence, 24,* 326–356.

Rubin, K. H., & Rose-Krasnor, L. (1992). Interpersonal problem solving. In V. B. Van Hasselt & M. Hersen (Eds.), *Handbook of social development* (pp. 283–323). New York: Plenum Press.

Schmidt, M. E., & Bagwell, C. L. (2007). The protective role of friendships in overtly and relationally victimized boys and girls. *Merrill–Palmer Quarterly, 53,* 439–460.

Schneider, B. H., & Tessier, N. G. (2007). Close friendship as understood by socially withdrawn, anxious early adolescents. *Child Psychiatry and Human Development, 38*(4), 339–351.

Sears, R. R., Maccoby, E. E., & Levin, H. (1957). *Patterns of child rearing.* Evanston, IL: Row, Peterson.

Selman, R. L., & Schultz, L. H. (1990). *Making a friend in youth: Developmental theory and pair therapy.* Chicago: University of Chicago Press.

Shortt, J. W., Capaldi, D. M., Dishion, T. J., Bank, L., & Owen, L. D. (2003). The role of adolescent peers, romantic partners, and siblings in the emergence of the adult antisocial lifestyle. *Journal of Family Psychology, 17,* 521–533.

Slaby, R. G., & Guerra, N. G. (1988). Cognitive mediators of aggression in adolescent offenders: 1. Assessment. *Developmental Psychology, 24,* 580–588.

Solish, A., Perry, A., & Minnes, P. (2010). Participation of children with and without disabilities in social, recreational and leisure activities. *Journal of Applied Research in Intellectual Disabilities, 23,* 226–236.

Stewart, S. L., & Rubin, K. H. (1995). The social problem solving skills of anxious-withdrawn children. *Development and Psychopathology, 7,* 323–336.

Sullivan, H. S. (1953). *The interpersonal theory of psychiatry.* New York: Norton.

Tonks, J., Williams, W. H., Frampton, I., Yates, P., & Slater, A. (2007). Reading emotions after child brain injury: A comparison between children with brain injury and non-injured controls. *Brain Injury, 21*(7), 731–739.

Troop-Gordon, W., & Asher, S. R. (2005). Modifications in children's goals when encountering obstacles to conflict resolution. *Child Development, 76*(3), 568–582.

Vaillancourt, T., & Hymel, S. (2006). Aggression and social status: The moderating roles of sex and peer-valued characteristics. *Aggressive Behavior, 32*(4), 396–408.

Vaughn, B. E., & Santos, A. J. (2009). Structural descriptions of social transactions among young children: Affiliation and dominance in preschool groups. In K. H. Rubin, W. M. Bukowski, & B. Laursen (Eds.), *Handbook of peer interactions, relationships, and groups* (pp. 195–214). New York: Guilford Press.

Véronneau, M., Vitaro, F., Pedersen, S., & Tremblay, R. E. (2008). Do peers contribute to the likelihood of secondary school graduation among disadvantaged boys? *Journal of Educational Psychology, 100*(2), 429–442.

Vitaro, F., Pedersen, S., & Brendgen, M. (2007). Children's disruptiveness, peer rejection, friends' deviancy, and delinquent behaviors: A process-oriented approach. *Development and Psychopathology, 19*, 433–453.

Walz, N. C., Yeates, K. O., Wade, S. L., & Mark, E. (2009). Social information processing skills in adolescents with traumatic brain injury: Relationships with social competence and behavior problems. *Journal of Pediatric Rehabilitation Medicine, 2*, 285–295.

Warschausky, S., Cohen, E. H., Parker, J. G., Levendosky, A. A., & Okun, A. (1997). Social problem-solving skills of children with traumatic brain injury. *Pediatric Rehabilitation, 1*, 77–81.

Waters, E., & Sroufe, L. A. (1983). Social competence as a developmental construct. *Developmental Review, 3*(1), 79–97.

Wentzel, K. R., Barry, C., & Caldwell, K. A. (2004). Friendships in middle school: Influences on motivation and school adjustment. *Journal of Educational Psychology, 96*(2), 195–203.

White, R. W. (1959). Motivation reconsidered: The concept of competence. *Psychological Review, 66*(5), 297–333.

Wichmann, C., Coplan, R. J., & Daniels, T. (2004). The social cognitions of socially withdrawn children. *Social Development, 13*, 377–392.

Yeates, K. O., Bigler, E. D., Dennis, M., Gerhardt, C. A., Rubin, K. H., Stancin, T., et al. (2007). Social outcomes in childhood brain disorder: A heuristic integration of social neuroscience and developmental psychology. *Psychological Bulletin, 133*, 535–556.

Yeates, K. O., Schultz, L. H., & Selman, K. R. (1991). The development of interpersonal negotiation strategies in thought and action: A social-cognitive link to behavioral adjustment and social status. *Merrill–Palmer Quarterly, 37*, 369–406.

CHAPTER 3

Brain Development and the Emergence of Social Function

Stephanie Burnett Heyes, Catherine L. Sebastian,
and Kathrin Cohen Kadosh

We humans are an intensely social species. We show a repertoire of social abilities—from rapidly and automatically detecting the presence of another human in our environment; to making inferences about their emotions, beliefs, and enduring character traits; and finally to using this knowledge to guide interactions (Frith & Frith, 2008, 2010). The last two decades have seen significant progress in understanding the neural underpinnings of human social abilities. The noninvasive *in vivo* neuroimaging technique functional magnetic resonance imaging (fMRI) has played an important role in this research. fMRI studies have begun to reveal how the functional brain correlates of social-cognitive processes change during development (see "Functional Neuroimaging and Behavioral Studies of Social Cognition," below).

The collection of brain regions that subserves social cognition is referred to as the *social brain* (Brothers, 1990; Frith, 2007). The social brain includes the fusiform face area (FFA), the posterior superior temporal sulcus (pSTS), the amygdala, the temporoparietal junction (TPJ), the anterior rostral medial prefrontal cortex (MPFC), the anterior cingulate cortex (ACC), the anterior temporal cortex (ATC), and the inferior frontal gyrus. fMRI studies show developmental differences in the patterns of activity within a number of these regions, and more recently in their patterns of functional connectivity. Furthermore, anatomical magnetic resonance imaging (MRI) studies indicate continuing structural development during adolescence within certain regions of the social brain.

In fMRI, participants perform experimental (e.g., social cognition) and control tasks, and neural activity during task performance is inferred by measuring local changes in blood flow, which occur in response to the metabolic demands of synaptic activity: the blood-oxygenation-level-dependent (BOLD) response (Attwell & Iadecola, 2002; Logothetis, Pauls, Augath, Trinath, & Oeltermann, 2001). fMRI has relatively high spatial resolution relative to other noninvasive *in vivo* methods, but its temporal resolution for describing neural activity is limited by the hemodynamic response, which occurs over 1–5 seconds. A practical consequence of this is that the technique is somewhat more straightforward to implement in adolescent participants, than young children since the latter find it difficult to remain stationary during scanning. Partially for this reason, much of the developmental fMRI material covered in this chapter focuses on adolescence.

Empirical behavioral studies are vital to interpret and qualify developmental neuroimaging findings in terms of maintained versus changing cognitive abilities. It is likely that neuroanatomical reorganization within social brain regions alters their functionality, causing changes at a cognitive-behavioral level. Therefore, adolescence is a particularly suitable developmental period for investigating how developmental trajectories of cortical specialization for social information processing emerge. In addition to providing a better understanding of developmental trajectories, studying developmental cognitive neuroscience can also inform our understanding of the mature functional and structural architecture of human social cognition.

In this chapter, we first summarize selected topics of research on social-cognitive development and its functional neural correlates—beginning with research on face processing, and proceeding to mentalizing, peer influence, and then social evaluation. Subsequently, we summarize current theoretical neurocognitive models accounting for developmental, and in particular adolescent, behavioral and functional neuroimaging changes. Finally, we evaluate evidence relating to the interpretation of developmental fMRI findings in the context of structural MRI findings, with consideration of potential neurophysiological mechanisms.

Functional Neuroimaging and Behavioral Studies of Social Cognition

Face Processing

Basic Face Processing

A fundamental requirement for social interaction is the ability to note the presence of another human being rapidly, from visual, auditory, and other cues. A particularly salient source of person information is the presence of visual cues indicating a face. Behavioral work with newborn infants has shown a preference for face-like objects within hours of being born, with newborns turning preferentially toward photographs and cartoons of faces in comparison to inverted faces or nonface objects (Farroni, Csibra, Simion, & Johnson, 2002; Morton & Johnson, 1991). It has also been shown that infants prefer looking at faces that engage them in mutual eye gaze, rather than those showing averted gaze (Cassia, Turati, & Simion, 2004). In view of these early preferences, it has therefore been a surprising discovery that both the behavioral and neural bases of face-processing abilities continue to develop throughout the first two decades of life.

An early study by Carey, Diamond, and Woods (1980) showed improvement in facial identity recognition across the first decade of life, followed by a brief dip in performance at age 12. In a more recent study (Mondloch, Geldart, Maurer, & Grand, 2003), a matching task was administered to 6-, 8-, and 10-year-old children and adults, which required participants to compare faces on the basis of identity (with facial expression and head orientation varying), facial expression, gaze direction, and sound being spoken. Results showed that in comparison to adults, the 6-year-olds made more errors on every task, and the 8-year-olds made more errors on three of the five tasks (namely, on matching the direction of gaze and on the two identity tasks). The 10-year-olds made more errors than did adults on the identity task in which head orientation varied. This suggests that basic face-processing abilities—here, the ability to recognize identity in a context-invariant manner—continue to develop until at least the end of the first decade of life.

Recent cross-sectional developmental fMRI studies suggest that the prolonged acquisition of face-processing abilities is mirrored by protracted functional specialization within the cortical face-processing network. In one fMRI study, children (n = 23, 7–11 years, 13 females), adolescents (n = 10, 12–16 years, 5 females) and adults (n = 17, 18–35 years, 8 females) passively viewed photographic images of faces versus objects, places, or abstract patterns (Golarai et al., 2007). Results showed an age-related increase in the spatial extent of face-selective suprathreshold activation within right fusiform cortex (the FFA): FFA was significantly larger in adults than in children, and the adolescent group showed an intermediate pattern. The expansion of FFA into surrounding cortex was correlated with a behavioral improvement in recognition memory for facial identity.

In another fMRI study in which children (n = 10, 5–8 years, 4 females), adolescents (n = 10, 11–14 years, 4 females), and adults (n = 10, 20–23 years, 4 females) freely viewed dynamic displays of faces, places, and objects, an age-related increase in size of face-selective FFA was observed between childhood and adolescence, as well as an increase in face-selective superior temporal sulcus (STS) (Scherf, Behrmann, Humphreys, & Luna, 2007). In this study, no further changes in functional specificity for face processing were noted between adolescence and adulthood, but activity became more bilateral with age.

More recent evidence from two developmental fMRI studies contrast sharply with the studies reviewed above (Cantlon, Pinel, Dehaene, & Pelphrey, 2011; Pelphrey, Lopez, & Morris, 2009). In the first study, Cantlon and colleagues observed a robust FFA response in 4-year-old children (n = 15, 4–6 years, 7 females) for faces in comparison to other categories, such as shoes, letters, numbers, or scrambled images. Most importantly, with regard to the question of cortical specialization, they observed that responses for the nonpreferred stimulus categories decreased with age. This finding led the authors to suggest that cortical specialization might reflect category-selective "pruning" in a given brain region, which in turn might be due to the ongoing synaptic pruning during the first two decades of life (Huttenlocher, De Courten, Garey, & Van der Loos, 1982).

In the second study, Pelphrey et al. compared neural responses to faces, flowers, objects, and bodies in the ventral–temporal stream in adults (N = 10, 19–29 years, 7 females) and children (n = 22, 7–12 years, 14 females). No developmental changes in face selectivity were found in the FFA from midchildhood when face-specific responses were contrasted with those to flower stimuli. It is less clear, however, whether similar results would have been obtained in this study if face responses had been contrasted with neural responses to the other stimulus categories (i.e., objects, bodies)—a question that is particularly relevant, as these categories are preferentially processed in adjacent cortical areas in the mature brain.

Based on the review of these often conflicting findings, an outstanding empirical issue is whether childhood and adolescence are marked by increasing functional neural specialization for faces. To resolve this issue, future studies should adopt a multimethod approach, which combines longitudinal measures of functional and structural brain development, together with models of changing network connectivity.

A recent developmental fMRI study implemented dynamic causal modeling analysis to examine task-dependent causal interactions among cortical face-processing regions during a match-to-sample task. This connectivity analysis enabled investigation of age group differences in the impact of differing task demands (matching based on identity, emotion, or gaze) on effective connectivity between regions, in children ($n = 16$, 7–8 years years, 8 females), preadolescents ($n = 8$, 10–11 years, 4 females), and adults ($n = 13$, 19–37 years, 7 females) (Cohen Kadosh, Cohen Kadosh, Dick, & Johnson, 2011). The same basic cortical network, comprising FFA, STS, and the inferior occipital gyrus (the occipital face area) was present in all age groups. However, there was an age-related increase in extent of differential top-down modulation of specific intranetwork connections depending on task context. This finding was interpreted as a cumulative effect of exposure and training, such that that the cortical network for face processing becomes increasingly fine-tuned with age (Plate 3.1 on color insert).

The functional neuroimaging and behavioral data summarized above can be interpreted within the *interactive-specialization* theoretical framework (Cohen Kadosh, 2011) (Plate 3.1 on color insert). According to this account, discrete cognitive functions (e.g., facial identity recognition) are an emergent product of interactions between spatially and functionally discrete brain regions, and between the whole brain and its external environment. Age-associated changes in functional brain activity, such as those described in this section for faces, are hypothezised to reflect emerging task selectivity or fine-tuning within localized neural components, as a consequence of interactions between brain regions and with the environment. This fine-tuning leads to improvement in behavioral performance.

Facial Emotion Processing

A secondary aspect of face processing, which is particularly important for social interaction, is the ability to interpret facial displays of emotion. This cognitive function recruits a number of brain regions in addition to the basic face-processing regions described above. For example, fMRI studies in which participants view emotional face stimuli often report activity in the amygdala, which is involved in automatic emotion processing (e.g., fear and avoidance), and in parts of the prefrontal cortex (PFC), including those implicated in emotion evaluation and high-level social processing (e.g., OFC, ACC, and MPFC).

There is some behavioral evidence showing continuing development during adolescence in the ability to recognize facial emotions (for a review, see Blakemore, 2008). An early behavioral study in which male and female participants ages 10–17 years matched emotional face pictures to emotion words showed evidence for a brief developmental regression at about the start of adolescence (McGivern, Andersen, Byrd, Mutter, & Reilly, 2002). One recent study tested facial emotion recognition accuracy by using morphed faces that varied along continua from neutral to fear, neutral to anger, and fear to anger (Thomas, De Bellis, Graham, & LaBar, 2007). Participants were children ($n = 31$, 7–13 years, 18 females), adolescents ($n = 23$, 14–18 years, 9 females), and adults ($n = 48$, 25–57 years, 41 females). Across all expression morphs, adults were more accurate at identifying the emotion shown than were children and adolescents. However, whereas recognition accuracy for fear showed a linear improvement across the three age groups, anger showed a quadratic trend, with sharp improvement between adolescence and adulthood. This suggests that adolescence is characterized by continuing improvement in facial emotion recognition, but that the shape of the developmental trajectory may differ between emotions. Thomas et al. suggest that this finding may reflect discrete neural underpinnings (each with a distinct developmental trajectory) for the detection of anger relative to fear. This interpretation is consistent with evidence from fMRI studies in adults, which show distinct neural components for facial expressions of anger and fear. It is conceivable that these mature functionally at different rates during adolescence.

At the neural level, there is some evidence for continuing development during adolescence of functional activity within brain regions subserving facial emotion processing. It remains a challenge to dissociate (both theoretically and empirically) the fMRI substrates of facial emotion identification from those that subserve downstream processing (e.g., social inference and emotional self-regulation). Perhaps for this reason, inconsistencies are reported in the adult fMRI literature, and these pose a challenge to interpretation of developmental findings (see Guyer et al., 2008, for a discussion). Bearing in mind these challenges, we now describe three noteworthy developmental fMRI studies of emotional face processing. In the first (Monk et al., 2003), functional brain activity was compared between fearful and neutral face viewing in adolescents ($n = 17$, 9–17 years, 8 females) and adults ($n = 17$, 25–36 years, 8 females). On some trials participants passively viewed the faces, whereas on other trials participants were instructed to rate their emotional response to the faces, or to pay attention to a nonemotional feature (nose width). Adolescents showed greater activity than did adults within the right amygdala, ACC, and OFC bilaterally during passive viewing of fearful relative to neutral faces—a result that may correspond to the trajectory of emerging behavioral competence in fearful face recognition (Thomas et al., 2007). Relative to adults, adolescents showed greater modulation by task context (attention to nose width vs. passive viewing) of ACC activity during fear versus neutral face processing, although this result was not shown in a follow-up study.

In the follow-up study (Guyer et al., 2008), participants were adolescents ($n = 31$, 9–17 years, 15 females) and adults ($n = 30$, 21–40 years, 13 females). The study confirmed the previous finding showing an age-related decrease in amygdala engagement during passive viewing of fearful faces, and also showed greater activity in the FFA in adolescents relative to adults during this contrast. The latter result extends findings in adults that have shown greater activity in FFA during emotional relative to neutral face processing.

Last, a recent study used a novel approach to pinpoint emerging cortical networks supporting emotional expression processing across the complete developmental trajectory. This study tested children and adults ($n = 48$, 7–37 years, 25 females) in three face-processing tasks (identity task, expression task, gaze task) (Cohen Kadosh, Johnson, Dick, Cohen Kadosh, & Blakemore, 2011). Cohen Kadosh and colleagues were able to show for the first time that the three face aspects exhibited distinct developmental trajectories across a wide age range. In particular, they found that developmental trajectories for emotional expression processing exhibited substantial development across the age range. Moreover, they were able to dissociate widespread functional frontotemporoparietal networks, whose activity varied with age, from anterior frontolimbic networks, whose activity varied according to proficiency level. The majority of the frontoparietal age effects, but only the limbic proficiency effects, were due to underlying gray and white matter development in the same regions. Instead, the results support the notion of a continuous strengthening and segregation of core and extended cortical networks for basic face-processing abilities throughout the first three decades of life.

This section has reviewed behavioral and fMRI studies of emotional face processing. Behavioral studies suggest continuing maturation in facial emotion recognition accuracy during adolescence, and early evidence from fMRI studies suggests that the neural substrates of emotional face processing show concurrent development during this time. More evidence is needed to establish whether the observed developmental changes, such as the decrease in FFA and amygdala reactivity to emotional faces from adolescence into adulthood, are the result of emerging behavioral proficiency, brain development, or both.

Mentalizing

Mentalizing, or *theory of mind,* is the ability to infer mental states such as intentions, beliefs, and desires (Premack & Woodruff, 1978). This social ability enables one to understand and predict other agents' behaviors that arise as a direct consequence of their mental states. For example, in an interpersonal context, an agent's action toward a coffee pot is more parsimoniously understood in terms of a *desire* for coffee than in terms of mechanical forces. A substantial functional neuroimaging literature indicates that mentalizing in adults recruits a circumscribed set of brain regions, comprising the ATC, pSTS, TPJ, and the anterior rostral portion of MPFC (Frith & Frith, 2003). Briefly, the ATC is thought to represent semantic social information, and the pSTS is important for decoding social gestures and signals to form predictions of action or intent. Much evidence implicates both TPJ and anterior rostral MPFC in representing or attending to mental states, although their respective roles are not yet clear (Amodio & Frith, 2006).

Mentalizing Tasks

A number of fMRI studies have investigated the neural correlates of mentalizing in adolescence (for a review, see Blakemore, 2008). These adolescent studies have used a variety of mentalizing tasks: reflecting on one's intentions to carry out particular actions (Blakemore, den Ouden, Choudhury, & Frith, 2007); thinking about the preferences and dispositions of oneself or a fictitious story character (Pfeifer, Lieberman, & Dapretto, 2007; Pfeifer et al., 2009); judging someone's sincerity (Wang, Lee, Sigman, & Dapretto, 2006); reading someone's facial expression (Gunther Moor et al., 2012), reflecting on one's own and another's emotional response during social situations (Figure 3.1) (Burnett, Bird, Moll, Frith, & Blakemore, 2009); and understanding how individuals are likely to act, based on their emotional state (Sebastian et al., 2012). Despite the variety of tasks used, these studies have consistently shown that activity within anterior rostral MPFC, during mentalizing relative to control tasks, correlates negatively with age between adolescence and adulthood. Some of these studies have shown that activity within posterior and temporal components of the mentalizing system, including pSTS, TPJ, and/or ATC, shows the opposite developmental pattern. It has been shown that the shift in activity within the mentalizing system is accompanied by a change in task-dependent interactions (effective connectivity) between anterior rostral MPFC and pSTS/TPJ (Burnett & Blakemore, 2009).

   Given the purported role of anterior rostral MPFC in representing mental states, one hypothesis regarding the age-related decrease in anterior rostral MPFC activity, and changes in its effective connectivity profile, is that it may reflect or underlie a change in mentalizing proficiency or strategy. Alternatively, or in addition, the shift in activity may represent increasing regional specialization, or efficiency within integrated neural circuits (Brown, Petersen, & Schlaggar, 2006; Durston et al., 2006). The development

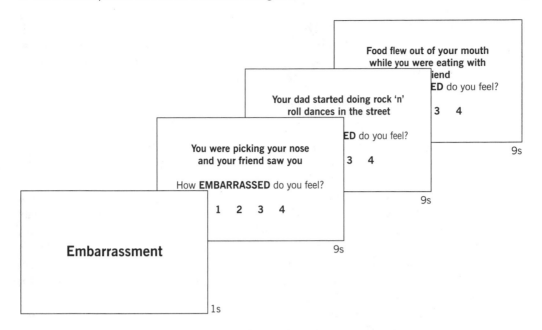

FIGURE 3.1. A mentalizing task used in developmental fMRI. Social emotions, such as guilt and embarrassment, require processing the mental states of real or projected others. For example, in order to feel embarrassed, an individual must represent the notion that an observer would think he or she looks ridiculous or undignified. Basic emotions, such as disgust and fear, carry a lesser requirement for mentalizing. Studies have shown that fMRI signal is greater in the mentalizing network during social (vs. basic) emotion processing, and that activity within this network shifts between adolescence and adulthood.

of mentalizing proficiency up to the age of 5 has been studied extensively and is well characterized (Frith & Frith, 2003), but very little is known about the development of mentalizing beyond early childhood. This could be due to a lack of suitable paradigms: In order to create a mentalizing task that does not elicit ceiling performance in children age 5 or older, the linguistic and executive demands of the task must be increased. This renders any age-associated improvement in performance difficult to attribute to improved mentalizing ability per se. However, in view of the increasing complexity of social relationships in adolescence, and reports of age-related increases in tolerance for diversity in others' beliefs (Wainryb, Shaw, Laupa, & Smith, 2001), mentalizing proficiency or its interaction with other cognitive functions is expected to show continuing development across adolescence.

A recent study used a novel mentalizing task in which children (*n* = 71, 7–11 years, all females) adolescents (*n* = 70, 11–18 years, all females) and adults (*n* = 36, 19-28 years, all female) were instructed to sequentially move objects between a set of shelves as instructed by a "Director" character (Dumontheil, Apperly, & Blakemore, 2010). The Director could see the contents of only some of the shelves; therefore, correct interpretation of the Director's instructions required participants to take into account the Director's visual perspective, and to use this mental state information online in a communicative situation. Results showed continuing development during late adolescence (between age 17 and early adulthood) in performance on visual perspective (mentalizing) trials, relative to rule-based control trials relying on executive functions only. Advanced mentalizing paradigms are now needed that are less heavily reliant on visual perspective taking. The next section discusses behavioral economic paradigms, which test different aspects of mentalizing ability.

Behavioral Economic Games

Behavioral economic paradigms have been used to investigate the use of mental state inferences in strategic social decision making. Games such as the Ultimatum Game and the Trust Game, which engage participants in structured competitive or cooperative interactions, reveal subtle differences in the degree of mental perspective taking (among other social variables). This is quantified as the amount of money or number of tokens exchanged. A number of fMRI studies in adults have shown task-related activity within the brain's reward system (e.g., the nucleus accumbens) during economic games, consistent with the desire to win monetary rewards; and also within the mentalizing system (e.g., pSTS, TPJ, and anterior rostral MPFC), consistent with the processing of one's own and the other player's actions and intentions (Montague, 2007).

Behavioral studies have shown evidence that the tendency to use mental state information strategically to win money in such games continues to develop during adolescence, and functional neuroimaging studies have shown evidence that this shift is accompanied by a change in neural processing. For example, a study using a modified Ultimatum Game showed that the tendency to make a generous offer of money was increasingly modulated by the perceived power of a participant's co-player to punish a selfish offer between ages 9 and 18 (Experiment 2: $n = 56$, 9–18 years, 26 females; Güroğlu, van den Bos, & Crone, 2009). In contrast, the tendency to act upon basic, inflexible social principles such as strict reciprocity is present from a relatively early age. A study using a modified Dictator Game showed that an individual's conception of fairness becomes increasingly egalitarian during adolescence. Whereas early adolescent children are likely to view only equal splits of a pot of tokens between two players as fair, older adolescents are more likely to view splits based on the amount of effort each player has devoted to the task as fair (Almås, Cappelen, Sørensen, & Tungodden, 2010).

fMRI has been used to explore the neural bases of these behavioral effects. A study that used the Trust Game showed that an increase during adolescence in the tendency to reciprocate fair offers of money was accompanied by an increase in activity within the TPJ and dorsolateral PFC during reciprocal versus subreciprocal social exchanges (van den Bos, van Dijk, Westenberg, Rombouts, & Crone, 2011). Young adolescents, who reciprocated less, showed greater activity in the anterior MPFC. The authors suggest that these findings are consistent with an account whereby younger adolescents adopt an egocentric, MPFC-based mentalizing strategy. In contrast, older adolescents and adults may be more effective at taking the perspective of their partner (via TPJ) and employing cognitive control to override self-oriented impulses (via dorsolateral PFC).

Peer Influence

Peer influence is a nebulous construct in cognitive terms, but it is one that has been studied with particular interest with regard to adolescent social cognition and behavior. Peer influence is conceptualized as acting on multiple levels—from implicit influences effects on bodily gestures and mood, to broader influences on an individual's social attitudes and activities. In this section, we review developmental studies that investigate the influence of peers on self-reported attitudes, as well as the influence of peers on risk taking in an experimental context. Finally, we touch upon structural MRI and fMRI correlates.

Much evidence indicates that peer influence in adolescence can promote engagement in beneficial and prosocial behaviors. However, for practical (e.g., public health) reasons, empirical studies tend to focus on peer influence on potentially harmful or risky behavior. One study investigated differences among adolescents ($n = 106$, 13–16 years, 54 females), young adults ($n = 105$, 18–22 years, 53 females), and older adults ($n = 95$, 24+ years, 48 females) in the number of risky decisions made in a driving simulation game—for example, speeding through a yellow traffic light (Gardner & Steinberg, 2005). The game was played alone or in the presence of two peers, and adolescents took many more risks in the presence of peers than when they were alone. In contrast, older adults' risk taking did not differ between the social and solitary conditions, and young adults showed an intermediate effect. It would be valuable to identify which aspect of the peers-present condition was most important in inducing high risk taking in adolescence: distinct adolescent mechanisms mediating group influence, age differences in norms for risk seeking, or some other factor.

Another study used a self-report questionnaire to chart development during adolescence in the tendency to resist peer influence (Steinberg & Monahan, 2007). More than 3600 male and female participants ages 10–23+ completed a Resistance to Peer Influence (RPI) questionnaire consisting of items assessing applicability to self of both morally valenced and neutral (personal preference) statements such as "Some people go along with their friends just to keep their friends happy." The results showed a linear increase in RPI scores between 14 and 18 years.

Neuroimaging studies are beginning to investigate the neural bases of peer influence. In one fMRI study, preadolescent (10-year-old) children were divided into two groups on the basis of a median split on scores on the Steinberg and Monahan (2007) RPI questionnaire described above (Grosbras et al., 2007). Children then underwent fMRI while passively viewing clips of angry hand and face gestures. The children who scored more highly on the RPI measure showed stronger functional connectivity between brain regions underlying action perception (e.g., STS) and decision making (e.g., lateral PFC, premotor cortex) than did the group who scored low on the measure. These results are consistent with an account whereby increasing functional integration within task-related networks underpins age-related development in cognitive abilities (e.g., resistance to peer influence) and is associated with development of certain brain structures, although more direct studies are needed. In particular, one outstanding question for theoretical and empirical work is the extent to which these behavioral and neurobiological findings regarding adolescent peer influence should be interpreted within a common conceptual framework.

One of the first longitudinal studies to address neurocognitive development in adolescence explored the relationship between the development of resistance to peer influence and the brain's response to emotional facial expressions (Pfeifer et al., 2011). Participants underwent fMRI while passively viewing emotional and neutral faces at two time points between the ages of 10 and 13. During each testing session, they also completed the RPI questionnaire and a measure of risky behavior. Neural responses to emotional (relative to neutral) faces increased with age in the ventral striatum (VS). Of relevance to understanding peer influence, this increase was positively correlated with the ability to resist peer influence and refrain from risky behavior. Although the causal and conceptal links between increasing VS activity and RPI scores are not fully known, the authors suggest that the longitudinal design may be capturing the maturation of circuitry underpinning successful emotion regulation (in which VS is hypothesized to play a role). This, in turn, is important for the ability to resist peer influence successfully.

Social Evaluation: Acceptance and Rejection

Social psychology studies have shown that adolescents, and by some accounts particularly female adolescents, are more sensitive to being excluded from a social interaction by peers than either adults or younger children are. In cases where risky behavior is a group norm, this effect could contribute to the impact of peer influence on risky behavior. A recent study (Sebastian, Viding, Williams, & Blakemore, 2010) investigated social rejection experimentally, using a computerized ball-passing paradigm known as Cyberball. In Cyberball, participants are told that they are playing a ball-passing game over the Internet with two other players, represented by cartoon drawings. In reality, the "players" are preprogrammed computer algorithms that systematically include (by passing the ball to) or exclude participants. In this study, young female adolescents (n = 26, 11–13 years) and female midadolescents (n = 25, 14–16 years, females) showed significantly reduced self-reported positive mood following episodes of exclusion (social rejection), compared to female adults (n = 26, 22–47 years). In addition, levels of anxiety were disproportionately increased following social rejection in the younger adolescents (11–13 years) relative to the adults, while anxiety was sustainedly high in the mid adolescents (14–16 years). Thus female adolescents show heightened sensitivity to social rejection in an experimental context.

A number of neuroimaging studies have investigated the neural basis of this effect. One study used the Cyberball paradigm combined with fMRI in a group of adolescents (n = 23, 12–13 years, 14 females) (Masten et al., 2009). The results showed patterns of brain activity that were similar to those in a previous study in adults: Positive correlations were found between self-reported distress on the one hand, and activity within visceral-pain-related and negative-affect-related regions (e.g., insula) on the other, during social exclusion versus inclusion; negative correlations were found between self-reported distress and activity within emotion regulation regions (e.g., ventrolateral PFC). There were some differences between the adult and adolescent findings. Most notably, the study in adults found a positive correlation between self-reported distress and activity in dorsal ACC, whereas the study in adolescents found no effect in dorsal ACC, but did find a positive relationship with activity in subgenual ACC.

Another study (Sebastian et al., 2011) directly compared female adolescents (n = 19, 14–16 years) and female adults (n = 16, 23–28 years) during a modified version of the Cyberball task and found that the adults activated ventrolateral PFC to a greater extent during exclusion than inclusion conditions, while the adolescents exhibited the reverse pattern. Since right ventrolateral PFC has previously been associated with the regulation of distress during social exclusion, it is possible that a reduced engagement of this region in adolescents in response to rejection-related stimuli underlies the increased affective response seen in adolescents in behavioral studies.

A similar result was found in an fMRI study exploring neural responses to the automatic processing of rejection-related information (Sebastian, Roiser, et al., 2010). This study compared female adolescents (*n* = 19, 14–16 years) and female adults (*n* = 16, 23–28 years) on a rejection-themed emotional Stroop task in which participants were asked to indicate the ink color in which rejection, acceptance, and neutral words were written. In thwe adults, rejection words activated the right ventrolateral PFC to a greater extent than acceptance or neutral words did. However, in the adolescent group, this regulatory region did not discriminate between rejection and neutral words, and responded more to acceptance words than to rejection. These studies are consistent with theories suggesting that prefrontal regulatory mechanisms continue to develop between midadolescence and adulthood (Nelson, Leibenluft, McClure, & Pine, 2005), and this may be one factor underlying observed adolescent hypersensitivity to rejection.

Rejection by peers is an extreme form of peer evaluation. This was the subject of a recent fMRI study that used an Internet chat room paradigm with male and female participants ages 9–17 years (*n* = 34, 16 females; Guyer, McClure-Tone, Shiffrin, Pine, & Nelson, 2009). Results showed that in females only, there was an age-related increase in activity during expectation of peer evaluation within brain regions involved in affective processing (nucleus accumbens, hypothalamus, hippocampus, and insula), but no differences within the ACC or other social brain regions. The finding of gender differences in the neural response to social evaluation is in line with reports of greater social anxiety in female than in male adolescents, in response to negative social evaluations in everyday life. However, the possibility that female adolescents are more sensitive to episodes of experimentally induced social rejection than males are has yet to be tested empirically. Guyer et al. (2009), for example, did not find gender differences in behavior on their task.

An fMRI study investigated peer evaluation and rejection across age, in groups of preadolescents (*n* = 12, 8–10 years, 7 females), young adolescents (*n* = 14, 12–14 years, 8 females), midadolescents (*n* = 15, 16–17 years, 7 females), and adults (*n* = 16, 19–25 years, 8 females) (Gunther Moor, van Leijenhorst, Rombouts, Crone & Van der Molen, 2010). In this experiment, participants were rated by a panel of peers. Results showed an age-related increase in activity within ventral MPFC, ACC, and striatum during evaluation and predicted social feedback, whereas predicted social rejection resulted in activity within affect regulation regions (such as OFC and lateral PFC) that increased linearly across age. In addition, activity within parts of OFC during social rejection was positively correlated with self-rated social anxiety in 8- to 17-year-olds. An interesting point to note in this study is the age-related increase in ventral MPFC activity, while mentalizing studies show age-related decreases in the more dorsally situated anterior rostral MPFC. This difference could relate to possible functional subdivisions within MPFC, a large and incompletely functionally characterized brain area.

The above-mentioned fMRI studies of social evaluation show age differences in neural activity. Often these neural changes correlate with behaviorally assessed changes in the impact of social rejection on mood. For example, fMRI findings correlate with the ability to regulate emotional responses to peer evaluation and rejection. This echoes findings from the studies of peer influence reviewed above. However, new behavioral measures of emotional self-regulation are needed to investigate these relationships. In addition, it is likely that an adolescent's wider social-cognitive skill set has an impact on the response to (and risk of) social rejection. Finally, although the studies reported above hint at possible gender differences in the affective and neural response to peer evaluation, the hypothesis that females are more sensitive to social evaluation should be tested empirically.

This review of the developmental fMRI and behavioral literature on social cognition is not exhaustive. However, the evidence reviewed above suggests continuing development across adolescence in the neural correlates of social-cognitive processes, including face processing, mentalizing, peer influence, and the emotional response to social evaluation and rejection. Concurrently, behavioral changes have been demonstrated in many of these domains. In the following section, these behavioral and functional neuroimaging findings are placed in the context of theoretical accounts of neural and cognitive development in adolescence.

Theoretical Models of Adolescent Neurocognitive Development

Several models have been proposed in which key behavioral and cognitive characteristics of adolescence, as well as the corresponding patterns of fMRI activity, are accounted for as consequences of neuroanatomical development. These models include the *social information-processing network* (SIPN) model (Nelson et al., 2005), the *triadic* model (Ernst & Fudge, 2009), and other *developmental-mismatch* models (Casey, Jones, & Hare, 2008; Steinberg, 2008). The models are broadly compatible, differing in their degree of focus on social cognition and in the level of detailed description at each level of explanation.

The SIPN model (Plate 3.2 on color insert), which is the most explicitly social model, proposes that a process of "social reorienting" takes place during adolescence, in partial consequence of neuroanatomical remodeling within "affective" and "cognitive-regulatory" brain nodes. This results in key behavioral characteristics of adolescence, such as risk taking in the presence of peers, as well as the increasing importance of peer relationships and peer approval. Neuroanatomical remodeling is proposed to result in part from the effects of pubertal gonadal steroids on limbic regions (the affective node), which are densely innervated by gonadal steroid receptors, and partly from the gradual maturation of PFC (the cognitive-regulatory node), enabling more sophisticated cognitive processing and top-down control (see Blakemore, Burnett, & Dahl, 2010, for a discussion of hormones). It is hypothesized that greater top-down control results from developing connectivity between these components, as well as from local development within the affective and cognitive nodes. This review has discussed some recent data on connectivity development, and it is likely that this will continue to be an important direction for research.

In common with the SIPN model, the triadic model of Ernst and Fudge (2009) distinguishes between affective-motivational and cognitive-regulatory neural systems, which develop anatomically during adolescence. The triadic model subdivides the affective-motivational system into "approach" and "avoidance" components, centered on the striatum and amygdala, respectively. According to the model, an imbalance favoring the approach node over the avoidance node in adolescence (relative to adulthood) contributes to adolescent risk-taking and social-affiliative characteristics. Greater influence during adolescence of the striatal approach system relative to the amygdalar avoidance system is proposed to be due to anatomical development within each node, as well as to protracted development in their regulation by the regulatory node (MPFC/OFC). The imbalance is hypothesised to result in distinct patterns of activity shown in fMRI studies, and in high risk taking in adolescence.

The triadic model shares some key features with developmental-mismatch models. A number of researchers have drawn attention to the pattern of heightened risk taking and emotional sensitivity in adolescence, relative to both childhood *and* adulthood. This nonlinearity across age suggests heterochronous maturation within neural systems that subserve these processes. It is hypothesized that the limbic system, including the amygdala and striatum, attains functional maturity earlier in development than does the PFC, and that the greatest mismatch in development of these systems occurs during adolescence. Consequently, during the time lag in functional maturity between PFC and limbic regions, individuals are more greatly affected by emotional context (e.g., reward immediacy, anticipated social rejection) when making decisions. Developmental patterns of activity during fMRI studies that employ emotional and reward-based tasks are also thought to reflect this functional mismatch. In a variant of this model, Steinberg (2008) suggests that remodeling of the dopamine system during adolescence (e.g., reduced limbic and prefrontal dopamine receptor density) increases the salience of social rewards such as peer approval, while gonadal steroid hormone release leads to an increase in sensitivity to social stimuli via effects on oxytocin receptors. The impact of increased pubertal levels of circulating gonadal (as well as adrenal) hormones on the brain and behavioral responsiveness in humans is not fully understood, but studies in nonhuman animal models are consistent with this theory (Spear, 2000).

Implicit in these models is the reasonable assumption that adolescent behavioral and cognitive development is causally related to changes in functional brain activity measured in fMRI, and that the changes in functional brain activity are related to neuroanatomical development. However, there are a number of potentially bidirectional relationships within this scheme. In the following section, we review evidence for relationships between structural and functional (including fMRI) brain development, and effects on behavior, as this is of central importance for interpreting the social brain findings reviewed here.

Structure–Function Relationships in the Adolescent Social Brain

Structural MRI Findings

MRI studies show continuing neuroanatomical development during adolescence. Two main age-associated changes have been described in volumetric MRI studies. First, cortical gray matter measures (volume, density, thickness) decrease across adolescence in a region-specific and commonly nonlinear manner (Paus, 2005). Second, white matter volume and density increase across the brain. By many accounts, this increase is linear during the second decade of life, decelerating into adulthood (Giedd et al., 1999). Increases in white matter volume are accompanied by progressive changes in MRI measures of white matter integrity, such as the magnetization–transfer ratio (MTR) in MRI, and fractional anisotropy (FA) in diffusion tensor MRI (Giorgio et al., 2010).

Research has not yet systematically characterized regional gray and white matter development in specific social brain components (e.g., functionally defined FFA) within longitudinal samples. However, the region specificity in patterns of cortical gray matter development can be interpreted as indicating that many social brain regions continue to develop during adolescence. In general, gray matter development is completed prior to adolescence within primary sensory processing regions (e.g., the occipital lobe), and during adolescence in association regions (e.g., the frontal and temporal lobes). Gray matter density in the frontal and temporal lobes has been reported to follow an inverted-U-shaped pattern of development, peaking at about puberty onset in the frontal lobe (age 11 in girls and 12 in boys), and at about 16–17 years in the temporal lobe. These peaks in gray matter density are followed by an extended profile of decline throughout the remainder of adolescence and early adulthood.

There is evidence for distinct trajectories of gray matter development within subregions of each cortical lobe (Shaw et al., 2008). In the frontal lobe, precentral (motor) gray matter density peaks prior to adolescence, whereas dorsolateral PFC and parts of MPFC attain peak gray matter volume later, at about puberty onset or beyond. Thus subregions of the frontal lobe implicated in social cognition and executive functions show protracted gray matter change in adolescence. In the temporal lobe, regions implicated in social cognition, such as the superior temporal lobe, attain peak gray matter density later (~14 years) than more middle and inferior temporal lobe regions involved in object and perceptual functions (~11–12 years). Trajectories of adolescent white matter development in specific tracts related to social-cognitive functioning have yet to be outlined.

Microstructural Mechanisms of Developmental MRI Findings

It has been suggested that adolescent changes in gray matter density, shown in MRI studies, may reflect regional alterations in synaptic density. Histological studies have shown evidence for synaptic proliferation in certain brain regions at about the start of adolescence, with protracted synaptic pruning occurring across the remainder of adolescence (e.g., human frontal cortex: Huttenlocher & Dabholkar, 1998; primate frontal cortex: Bourgeois & Rakic, 1993; Rakic, Bourgeois, Eckenhoff, Zecevic, & Goldman-Rakic, 1986). However, whether changes in synaptic density would be visible as volumetric changes in MRI scans is debated (see Paus, Keshavan, & Giedd, 2008). It is reasonable to hypothesize that adolescent changes in synaptic density might be accompanied by yoked changes in glial and other cellular components. Elsewhere, it has been suggested that adolescent decreases in gray matter volume shown in MRI studies are predominantly due to intracortical myelination, resulting in an increase in the volume of tissue that is classified as white matter in MRI scans (Paus et al., 2008).

A possible consequence of early adolescent prefrontal synaptogenesis could be a temporary dip in signal-to-noise ratio within these neural circuits (Blakemore, 2008; Rolls & Deco, 2011). Subsequent synaptic pruning, perhaps occurring via experience-dependent mechanisms, would then result in more finely tuned, robust, and efficient neural circuits. More direct neurophysiological evidence is needed in support of this hypothesis with regard to adolescent development specifically. However, results from a recent primate electrophysiology study suggest that the pruning of excitatory synapses in dorsolateral PFC during adolescence may have a functional impact. Unlike during early childhood, the synapses that undergo pruning during adolescence are already functionally mature (Gonzalez-Burgos et al., 2008).

Adolescent changes in white matter shown in structural and diffusion MRI studies (i.e., increasing volume, density, MTR, and FA) are hypothesized to reflect processes including myelination and increasing axonal caliber (Giorgio et al., 2010; Paus et al., 2008; Perrin et al., 2009). As mentioned above, it has been suggested that these cellular processes would also contribute to volumetric gray matter changes. The probable result of both myelination and increasing axonal caliber would be an increase in axonal conduction speed (Waxman, 1980). This would not only "speed up" neural processing, but allow for greater temporal precision (e.g., interregional synchronization) (Fornari, Knyazeva, Meuli, & Maeder, 2007; Paus et al., 1999).

Mechanistic Relationships between MRI Findings and fMRI Signal

A crucial question for adolescent fMRI studies is whether there is a functionally meaningful relationship between structural brain development, reviewed above, and the BOLD signal in fMRI. This would suggest that age differences in task-elicited BOLD signal index neuronal properties that contribute to age differences in cognition and behavior. As yet there is no direct evidence focusing on the developmental relationship across adolescence between microstructural neuronal properties (e.g., synapse density and myelination) and the BOLD signal, although suggestive relationships have been shown (e.g., Fornari et al., 2007). Therefore, at this point, structure–function correlations in fMRI should be interpreted with caution.

It has been observed that there is a "diffuse-to-focal" shift in BOLD signal between childhood and adulthood (Durston et al., 2006; Thomason, Burrows, Gabrieli, & Glover, 2005). The challenge now is to quantify this emerging focality. A good starting point would be to investigate the point spread function (PSF), which in the context of fMRI is the hemodynamic response to a minimal stimulus—for example, the striatal BOLD response to a faint spot of light (Shmuel, Yacoub, Chaimow, Logothetis, & Ugurbil, 2007; Sirotin, Hillman, Bordier, & Das, 2009) It would be challenging to establish an equivalent "minimal stimulus" for association cortex, such as the TPJ or temporal poles.

The BOLD response is thought to be generated by local excitatory synaptic activity (Logothetis et al., 2001), leading to precisely localized changes in blood flow in response to the release of glutamate (Attwell & Iadecola, 2002). Therefore, changes in BOLD signal with age could result from a change in local excitatory synaptic drive, due either to progressive myelination (resulting in increased synchronization of afferent inputs to a region), to increased excitatory synaptic density, or to some other mechanism. In addition, increasing synapse density may cause local hypoxia, triggering an increase in capillary density with associated changes in neurovascular coupling. This could also give rise to more "focal" BOLD signals.

It is not known whether the relationship between excitatory synaptic activity and local changes in blood flow (neurovascular coupling) remains constant across age. At the low field strengths suitable for scanning children and adolescents (<7 Tesla), the PSF in a region of cortex is thought to be largely determined by proximity to local blood vessels, rather than by local gray matter density, horizontal connections, synapses, and so on (Shmuel et al., 2007; Sirotin et al., 2009) There is evidence that local blood supply alters adaptively in response to local excitatory activity, but it is not known whether the nature of this responsiveness changes across development, or indeed whether there are developmental periods of "neurovascular mismatch." Further studies are needed to determine the extent to which adolescent BOLD signal changes are attributable to spatial "sharpening" of broad axonal or dendritic arbouurs, or to increasing density of brain tissue perfusion by capillaries (D'Esposito, Deouell, & Gazzaley, 2003; Harris, Reynell, & Attwell, 2011).

Conclusions

Neuroimaging studies have shown that the social brain—the complex network of brain regions that participates in understanding and interacting with social agents—continues to develop during adolescence. Using a number of social-cognition tasks, fMRI studies have shown changes in functional brain activity occurring alongside emerging social cognitive proficiency and neuroanatomical development.

Evidence is awaited that will shed light on the causal links between adolescent social-cognitive and social brain development via underlying neurophysiological mechanisms. Neuroanatomical development, leading to enhanced local and interregional neural processing capacity, may be necessary for transitions between each stage of cognitive development. It is also reasonable to predict that experience-dependent behavioral and cognitive progress, leading to differential recruitment of the neural substrates for social cognition, leads to modification of structural brain properties. A conceptual framework that accommodates these potentially bidirectional causal relationships is provided within the interactive-specialization account (Johnson, Grossmann, & Cohen Kadosh, 2009) (Plate 3.1 on color insert). Within this framework, discrete cognitive functions are conceptualized as emergent products of interactions between brain regions, and between the brain and its external environment, via a process of yoked neural and behavioral fine-tuning.

Understanding developmental fMRI findings on the social brain, and elucidating how these are related to concurrent changes in social cognition and structural brain development, will increase our understanding of adolescence. Delineating typical brain development, and the sequence of emerging social-cognitive abilities, will contribute to a better understanding of the rise in vulnerability to certain psychiatric illnesses in adolescence, including affective illness anxiety, schizophrenia, and addiction. It is also important to consider the role of individual differences in, for example, genetic variation or early life experiences. Development of the social brain may expose an adolescent to certain vulnerabilities presented in an adverse social environmental, but at the same time it presents a unique window of opportunity for fostering resilience.

References

Almås, I., Cappelen, A. W., Sørensen, E. Ø., & Tungodden, B. (2010). Fairness and the development of inequality acceptance. *Science, 328*(5982), 1176–1178.

Amodio, D. M., & Frith, C. D. (2006). Meeting of minds: The medial frontal cortex and social cognition. *Nature Reviews Neuroscience, 7*(4), 268–277.

Attwell, D., & Iadecola, C. (2002). The neural basis of functional brain imaging signals. *Trends in Neurosciences, 25*(12), 621–625.

Blakemore, S.-J. (2008). The social brain in adolescence. *Nature Reviews Neuroscience, 9*(4), 267–277.

Blakemore, S.-J., Burnett, S., & Dahl, R. E. (2010). The role of puberty in the developing adolescent brain. *Human Brain Mapping, 31*(6), 926–933.

Blakemore, S.-J., den Ouden, H., Choudhury, S., & Frith, C. (2007). Adolescent development of the neural circuitry for thinking about intentions. *Social Cognitive and Affective Neuroscience, 2*(2), 130–139.

Bourgeois, J. P., & Rakic, P. (1993). Changes of synaptic density in the primary visual cortex of the macaque monkey from fetal to adult stage. *Journal of Neuroscience, 13*(7), 2801–2820.

Brothers, L. (1990). The neural basis of primate social communication. *Motivation and Emotion, 14*(2), 81–91.

Brown, T. T., Petersen, S. E., & Schlaggar, B. L. (2006). Does human functional brain organization shift from diffuse to focal with development? *Developmental Science, 9*(1), 9–11.

Burnett, S., Bird, G., Moll, J., Frith, C., & Blakemore, S.-J. (2009). Development during adolescence of the neural processing of social emotion. *Journal of Cognitive Neuroscience, 21*(9), 1736–1750.

Burnett, S., & Blakemore, S.-J. (2009). Functional connectivity during a social emotion task in adolescents and in adults. *European Journal of Neuroscience, 29*(6), 1294–1301.

Cantlon, J. F., Pinel, P., Dehaene, S., & Pelphrey, K. A. (2011). Cortical representations of symbols, objects, and faces are pruned back during early childhood. *Cerebral Cortex, 21*(1), 191–199.

Carey, S., Diamond, R., & Woods, B. (1980). Development of face recognition: A maturational component? *Developmental Psychology, 16*(4), 257–269.

Casey, B. J., Jones, R. M., & Hare, T. A. (2008). The adolescent brain. *Annals of the New York Academy of Sciences, 1124,* 111–126.

Cassia, V. M., Turati, C., & Simion, F. (2004). Can a nonspecific bias toward top-heavy patterns explain newborns' face preference? *Psychological Science, 15*(6), 379–383.

Cohen Kadosh, K. (2011). What can emerging cortical face networks tell us about mature brain organisation? *Developmental Cognitive Neuroscience, 1*(3), 246–255.

Cohen Kadosh, K., Cohen Kadosh, R., Dick, F., & Johnson, M. H. (2011). Developmental changes in effective connectivity in the emerging core face network. *Cerebral Cortex, 21*(6), 1389–1394.

Cohen Kadosh, K., Johnson, M. H., Dick, F., Cohen Kadosh, R., & Blakemore, S. J. (2011). *Effects of age, task performance, and structural brain development on face processing.* Manuscript submitted for publication.

D'Esposito, M., Deouell, L. Y., & Gazzaley, A. (2003). Alterations in the BOLD fMRI signal with ageing and disease: a challenge for neuroimaging. *Nature Reviews Neuroscience, 4*(11), 863–872.

Dumontheil, I., Apperly, I. A., & Blakemore, S.-J. (2010). Online usage of theory of mind continues to develop in late adolescence. *Developmental Science, 13*(2), 331–338.

Durston, S., Davidson, M. C., Tottenham, N., Galvan, A., Spicer, J., Fossella, J. A., et al. (2006). A shift from diffuse to focal cortical activity with development. *Developmental Science, 9*(1), 1–8.

Ernst, M., & Fudge, J. L. (2009). A developmental neurobiological model of motivated behavior: Anatomy, connectivity and ontogeny of the triadic nodes. *Neuroscience and Biobehavioral Reviews, 33*(3), 367–382.

Farroni, T., Csibra, G., Simion, F., & Johnson, M. H. (2002). Eye contact detection in humans from birth. *Proceedings of the National Academy of Sciences USA, 99*(14), 9602–9605.

Fornari, E., Knyazeva, M. G., Meuli, R., & Maeder, P. (2007). Myelination shapes functional activity in the developing brain. *NeuroImage, 38*(3), 511–518.

Frith, C. D. (2007). The social brain? *Philosophical Transactions of the Royal Society: Series B. Biological Sciences, 362*(1480), 671–678.

Frith, C. D., & Frith, U. (2008). Implicit and explicit processes in social cognition. *Neuron, 60*(3), 503–510.

Frith, U., & Frith, C. D. (2003). Development and neurophysiology of mentalizing. *Philosophical Transactions of the Royal Society of London: Series B. Biological Sciences, 358*(1431), 459–473.

Frith, U., & Frith, C. D. (2010). The social brain: Allowing humans to boldly go where no other species has been. *Philosophical Transactions of the Royal Society of London: Series B. Biological Sciences, 365*(1537), 165–176.

Gardner, M., & Steinberg, L. (2005). Peer influence on risk taking, risk preference, and risky decision making in adolescence and adulthood: An experimental study. *Developmental Psychology, 41*(4), 625–635.

Giedd, J. N., Blumenthal, J., Jeffries, N. O., Castellanos, F. X., Liu, H., Zijdenbos, A., et al. (1999). Brain development during childhood and adolescence: A longitudinal MRI study. *Nature Neuroscience, 2*, 861–862.

Giorgio, A., Watkins, K. E., Chadwick, M., James, S., Winmill, L., Douaud, G., et al. (2010). Longitudinal changes in gray and white matter during adolescence. *NeuroImage, 49*(1), 94–103.

Golarai, G., Ghahremani, D. G., Whitfield-Gabrieli, S., Reiss, A., Eberhardt, J. L., Gabrieli, J. D. E., et al. (2007). Differential development of high-level visual cortex correlates with category-specific recognition memory. *Nature Neuroscience, 10*(4), 512–522.

Gonzalez-Burgos, G., Kroener, S., Zaitsev, A. V., Povysheva, N. V., Krimer, L. S., Barrionuevo, G., et al. (2008). Functional maturation of excitatory synapses in layer 3 pyramidal neurons during postnatal development of the primate prefrontal cortex. *Cerebral Cortex, 18*(3), 626–637.

Grosbras, M.-H., Jansen, M., Leonard, G., McIntosh, A., Osswald, K., Poulsen, C., et al. (2007). Neural mechanisms of resistance to peer influence in early adolescence. *Journal of Neuroscience, 27*(30), 8040–8045.

Gunther Moor, B., Op de Macks, Z. A., Güroglu, B., Rombouts, S. A. R. B., Van der Molen, M. W., & Crone, E. A. (2012). Neurodevelopmental changes of reading the mind in the eyes. *Social Cognitive and Affective Neuroscience, 7*(1), 44–52.

Gunter Moor, B., van Leijenhorst, L., Rombouts, S. A., Crone, E. A., & Van der Molen, M. W. (2010). Do you like me? Neural correlates of social evaluation and developmental trajectories. *Social Neuroscience, 5*(5–6), 461–482.

Güroğlu, B., van den Bos, W., & Crone, E. A. (2009). Fairness considerations: Increasing understanding of intentionality during adolescence. *Journal of Experimental Child Psychology, 104*(4), 398–409.

Guyer, A. E., McClure-Tone, E. B., Shiffrin, N. D., Pine, D. S., & Nelson, E. E. (2009). Probing the neural correlates of anticipated peer evaluation in adolescence. *Child Development, 80*(4), 1000–1015.

Guyer, A. E., Monk, C. S., McClure-Tone, E. B., Nelson, E. E., Roberson-Nay, R., Adler, A. D., et al. (2008). A developmental examination of amygdala response to facial expressions. *Journal of Cognitive Neuroscience, 20*(9), 1565–1582.

Harris, J. J., Reynell, C., & Attwell, D. (2011). The physiology of developmental changes in BOLD functional imaging signals. *Developmental Cognitive Neuroscience, 1*(3), 199–216.

Huttenlocher, P. R., & Dabholkar, A. S. (1998). Regional differences in synaptogenesis in human cerebral cortex. *Journal of Comparative Neurology, 387*(2), 167–178.

Huttenlocher, P. R., De Courten, C., Garey, L. J., & Van der Loos, H. (1982). Synaptic development in human cerebral cortex. *International Journal of Neurology, 16–17*, 144–154.

Johnson, M. H., Grossmann, T., & Cohen Kadosh, K. (2009). Mapping functional brain development: Building a social brain through interactive specialization. *Developmental Psychology, 45*(1), 151–159.

Logothetis, N. K., Pauls, J., Augath, M., Trinath, T., & Oeltermann, A. (2001). Neurophysiological investigation of the basis of the fMRI signal. *Nature, 412*(6843), 150–157.

Masten, C. L., Eisenberger, N. I., Borofsky, L. A., Pfeifer, J. H., McNealy, K., Mazziotta, J. C., et al. (2009). Neural correlates of social exclusion during adolescence: Understanding the distress of peer rejection. *Social Cognitive and Affective Neuroscience, 4*(2), 143–157.

McGivern, R. F., Andersen, J., Byrd, D., Mutter, K. L., & Reilly, J. (2002). Cognitive efficiency on a match to sample task decreases at the onset of puberty in children. *Brain and Cognition, 50*(1), 73–89.

Mondloch, C. J., Geldart, S., Maurer, D., & Grand, R. L. (2003). Developmental changes in face processing skills. *Journal of Experimental Child Psychology, 86*(1), 67–84.

Monk, C. S., McClure, E. B., Nelson, E. E., Zarahn, E., Bilder, R. M., Leibenluft, E., et al. (2003). Adolescent immaturity in attention-related brain engagement to emotional facial expressions. *NeuroImage, 20*(1), 420–428.

Montague, P. R. (2007). Neuroeconomics: A view from neuroscience. *Functional Neurology, 22*(4), 219–234.

Morton, J., & Johnson, M H. (1991). CONSPEC and CONLERN: A two-process theory of infant face recognition. *Psychological Review, 98*(2), 164–181.

Nelson, E. E., Leibenluft, E., McClure, E. B., & Pine, D. S. (2005). The social re-orientation of adolescence: A neuroscience perspective on the process and its relation to psychopathology. *Psychological Medicine, 35*(2), 163–174.

Paus, T. (2005). Mapping brain maturation and cognitive development during adolescence. *Trends in Cognitive Sciences, 9*(2), 60–68.

Paus, T., Keshavan, M., & Giedd, J. N. (2008). Why do many psychiatric disorders emerge during adolescence? *Nature Reviews Neuroscience, 9*(12), 947–957.

Paus, T., Zijdenbos, A., Worsley, K., Collins, D. L., Blumenthal, J., Giedd, J. N., et al. (1999). Structural maturation of neural pathways in children and adolescents: *In vivo* study. *Science, 283*(5409), 1908.

Pelphrey, K. A., Lopez, J., & Morris, J. P. (2009). Developmental continuity and change in responses to social and nonsocial categories in human extrastriate visual cortex. *Frontiers in Human Neuroscience, 3*, 25.

Perrin, J. S., Leonard, G., Perron, M., Pike, G. B., Pitiot, A., Richer, L., et al. (2009). Sex differences in the growth of white matter during adolescence. *NeuroImage, 45*(4), 1055–1066.

Pfeifer, J. H., Lieberman, M. D., & Dapretto, M. (2007). "I know you are but what am I?!": Neural bases of self- and social knowledge retrieval in children and adults. *Journal of Cognitive Neuroscience, 19*(8), 1323–1337.

Pfeifer, J. H., Masten, C. L., Borofsky, L. A., Dapretto, M., Fuligni, A. J., & Lieberman, M. D. (2009). Neural correlates of direct and reflected self-appraisals in adolescents and adults: When social perspective-taking informs self-perception. *Child Development, 80*(4), 1016–1038.

Pfeifer, J. H., Masten, C. L., Moore, W. E., 3rd, Oswald, T. M., Mazziotta, J. C., Iacoboni, M., et al. (2011). Entering adolescence: Resistance to peer influence, risky behavior, and neural changes in emotion reactivity. *Neuron, 69*(5), 1029–1036.

Premack, D., & Woodruff, G. (1978). Does the chimpanzee have a theory of mind? *Behavioral and Brain Sciences, 1*(4), 515–526.

Rakic, P., Bourgeois, J. P., Eckenhoff, M. F., Zecevic, N., & Goldman-Rakic, P. S. (1986). Concurrent overproduction of synapses in diverse regions of the primate cerebral cortex. *Science, 232*(4747), 232–235.

Rolls, E. T., & Deco, G. (2011). A computational neuroscience approach to schizophrenia and its onset. *Neuroscience and Biobehavioral Reviews, 35*(8), 1644–1653.

Scherf, K. S., Behrmann, M., Humphreys, K., & Luna, B. (2007). Visual category-selectivity for faces, places and objects emerges along different developmental trajectories. *Developmental Science, 10*(4), F15–F30.

Sebastian, C. L., Fontaine, N. M. G., Bird, G., Blakemore, S. J., De Brito, S. A., McCrory, E. J. P., et al. (2012). Neural processing associated with cognitive and affective theory of mind in adolescents and adults. *Social Cognitive and Affective Neuroscience, 7*(1), 53–63.

Sebastian, C. L., Roiser, J. P., Tan, G. C. Y., Viding, E., Wood, N. W., & Blakemore, S-J. (2010). Effects of age and MAOA genotype on the neural processing of social rejection. *Genes, Brain, and Behavior, 9*(6), 628–637.

Sebastian, C. L., Tan, G. C. Y., Roiser, J. P., Viding, E., Dumontheil, I., & Blakemore, S.-J. (2011). Developmental influences on the neural bases of responses to social rejection: Implications of social neuroscience for education. *NeuroImage, 57*(3), 686–694.

Sebastian, C., Viding, E., Williams, K. D., & Blakemore, S. J. (2010). Social brain development and the affective consequences of ostracism in adolescence. *Brain and Cognition, 72*(1), 134–145.

Shaw, P., Kabani, N. J., Lerch, J. P., Eckstrand, K., Lenroot, R., Gogtay, N., et al. (2008). Neurodevelopmental trajectories of the human cerebral cortex. *Journal of Neuroscience, 28*(14), 3586–3594.

Shmuel, A., Yacoub, E., Chaimow, D., Logothetis, N. K., & Ugurbil, K. (2007). Spatiotemporal point-spread function of fMRI signal in human gray matter at 7 Tesla. *NeuroImage, 35*(2), 539–552.

Sirotin, Y. B., Hillman, E., Bordier, C., & Das, A. (2009). Spatiotemporal precision and hemodynamic mechanism of optical point spreads in alert primates. *Proceedings of the National Academy of Sciences USA, 106*(43), 18390.

Spear, L. P. (2000). The adolescent brain and age-related behavioral manifestations. *Neuroscience and Biobehavioral Reviews, 24*(4), 417–463.

Steinberg, L. (2008). A social neuroscience perspective on adolescent risk-taking. *Developmental Review, 28*(1), 78–106.

Steinberg, L., & Monahan, K. C. (2007). Age differences in resistance to peer influence. *Developmental Psychology, 43*(6), 1531–1543.

Thomas, L. A., De Bellis, M. D., Graham, R., & LaBar, K. S. (2007). Development of emotional facial recognition in late childhood and adolescence. *Developmental Science, 10*(5), 547–558.

Thomason, M. E., Burrows, B. E., Gabrieli, J. D., & Glover, G. H. (2005). Breath holding reveals differences in fMRI: BOLD signals in children and adults. *NeuroImage, 25*(3), 824–837.

van den Bos, W., van Dijk, E., Westenberg, M., Rombouts, S. A. R. B., & Crone, E. A. (2011). Changing brains, changing perspectives: The neurocognitive development of reciprocity. *Psychological Science, 22*(1), 60–70.

Wainryb, C., Shaw, L. A., Laupa, M., & Smith, K. R. (2001). Children's, adolescents', and young adults' thinking about different types of disagreements. *Developmental Psychology, 37*(3), 373–386.

Wang, A. T., Lee, S. S., Sigman, M., & Dapretto, M. (2006). Developmental changes in the neural basis of interpreting communicative intent. *Social Cognitive and Affective Neuroscience, 1*(2), 107–121.

Waxman, S. G. (1980). Determinants of conduction velocity in myelinated nerve fibers. *Muscle and Nerve, 3*(2), 141–150.

CHAPTER 4

Social and Moral Functioning
A Cognitive Neuroscience Perspective

Bradley C. Taber-Thomas and Daniel Tranel

The social and moral faculties are among the most celebrated capacities of living beings. It is becoming increasingly clear that affective processes are particularly important for economizing the guidance of effective and appropriate behavior in real-world social contexts—highly complex, and often ambiguous and risky, places. Accordingly, and contrary to some of the well-known traditional views such as those of Kohlberg (1969), contemporary research in cognitive neuroscience and psychology has strongly supported a central role for emotions in moral cognition (Damasio, 1994; Greene, 2009; Haidt, 2001). In this chapter, we discuss cognitive neuroscience research on the neural basis of social and moral functioning in humans. This work has demonstrated that socio-moral cognitive processes rely on a frontolimbic neural network that integrates basic emotional and motivational information into the organization of behavior (Beauchamp & Anderson, 2010; Damasio, Tranel, & Damasio, 1998).

The chapter is organized around themes that have come to the fore in the recently expanding field of the cognitive neuroscience of morality. Although there is inconsistency in how the term *moral* is defined in this literature (Killen & Smetana, 2008), these studies have in common the goal of understanding the neural basis of cognitive processing related to social norms (Moll, R. Zahn, de Oliveira-Souza, Krueger, & Grafman, 2005). It is in this sense that we use the notion of *moral*. We begin by reviewing a framework for understanding the neural basis of moral functioning—namely, the *somatic marker hypothesis* (SMH; Damasio, 1994)—as well as research on the neural structures involved specifically in moral cognition and emotion-related behavior. We then discuss cognitive neuroscience research on more basic social functions (e.g., the perception of social stimuli and theory of mind). In our view, there is a functional processing hierarchy underlying socio-moral functioning, from more basic functions (including basic drives, processing of signals from faces, and theory of mind) to higher cognitive processes such as moral cognition. The SMH describes how these processes work together to generate adaptive socio-moral behavior.

The Role of Emotion in Social and Moral Functioning

Important Cases of Brain Injury

The lesion–deficit approach has led to many fundamental discoveries in cognitive neuroscience, and the domain of socio-moral functions is no exception. Early insight into the neural bases of these functions was provided by the 19th-century case of Phineas Gage. Gage, who worked as the foreman of a rock excavation group at a railroad company, was a "great favorite" among his men and was known as "a shrewd, smart business man, very energetic and persistent in executing all his plans of operation" (Harlow, 1868, p. 340). In the fall of 1848, an unfortunate accident occurred. Gage was tamping explosives into a hole when his iron tamping rod struck a spark and the explosives were set off prematurely, firing the rod up and out the top of his skull. A modern-day study of Gage's skull revealed that the bilateral ventromedial prefrontal cortex (vmPFC) was the probable locus of his brain injury (Damasio, Grabowski, Frank, Galaburda, & Damasio, 1994; for other perspectives, see Ratiu, Talos, Haker, Lieberman, & Everett, 2004). The vmPFC encompasses the medial orbitofrontal cortex and the ventral portion of the medial prefrontal cortex (see Plate 4.1 on color insert). The vmPFC is extensively connected to limbic structures—the amygdala, insula, and hippocampus—as well as the hypothalamus and brainstem (Öngür & Price, 2000; Rolls, 2000). Through these connections, it maintains higher-order control over the execution of bodily components of emotions, as well as the encoding of emotional values and the values of the consequences of actions (Öngür & Price, 2000; Rangel, Camerer, & Montague, 2008; Rolls, 2000).

Remarkably, Gage made a full medical recovery within less than a year. He was walking and talking, and would have appeared to the casual observer to have remarkably escaped the incident without complication. However, although largely invisible to superficial observation, changes in his personality and cognitive functioning were profound. He reapplied for his previous position as foreman, but, as his physician John Harlow noted in one of the most well-known passages in the field of social-cognitive neuroscience,

> His contractors, who regarded him as the most efficient and capable foreman in their employ previous to his injury, considered the change in his mind so marked that they could not give him his place again. The equilibrium or balance, so to speak, between his intellectual faculties and animal propensities, seems to have been destroyed. (Harlow, 1868, p. 339)

Gage's brain injury resulted in severe behavioral and personality disturbances. He seemed to have become irresponsible, whereas he was once a paragon of responsibility. He lost his once natural respect for social conventions. His new behavioral patterns were ill advised, and it seemed that he was unable to learn from the negative consequences of his mistakes. Harlow accordingly concluded that "his mind was radically changed, so decidedly that his friends and acquaintances said he was 'no longer Gage' " (Harlow, 1868, p. 340).

Changes similar to those seen in Gage have been observed in a modern-era patient studied extensively in our laboratory at the University of Iowa, patient E. V. R. (Eslinger

& Damasio, 1985). E. V. R. suffered from a large bilateral orbitofrontal meningioma that was successfully removed, but left him with a disruptive behavioral syndrome that has been likened to "acquired sociopathy" (Eslinger & Damasio, 1985). Like Gage, prior to his brain insult E. V. R. was a well-respected individual and employee, epitomizing social deftness and interpersonal savvy. After his brain damage, however, he developed profound changes to his personality and his ability to execute advantageous plans and decisions. He became unreliable and unable to maintain employment, and his behavioral pattern became one of misguided business and social decisions (Damasio et al., 1998). And, like Gage, he seems unable to learn from the negative consequences of his disadvantageous behavior (Damasio, 1994).

Not surprisingly, the striking changes in social and moral functioning observed in the cases of Gage and E. V. R. tend to attract most of the attention from neuroscientists. However, it is important to contrast these changes with the many domains of cognition and behavior that remain generally intact after vmPFC injury, including sensory perception; memory, language, and general intellectual abilities; the ability to express basic emotions like anger, disgust, and happiness (although often at inappropriate times); and basic social knowledge and processing (e.g., face recognition, cognitive theory of mind) (Damasio, Anderson, & Tranel, 2011). E. V. R., for example, has demonstrated postmorbid Verbal and Performance IQ scores of 129 and 135, respectively, on the Wechsler Adult Intelligence Scale—Revised; Minnesota Multiphasic Personality Inventory scores in the normal range; and normal performances or better on all manner of neuropsychological tests (Eslinger & Damasio, 1985). Moreover, despite his deficits in real-world social behavior, he is able to give appropriate responses to hypothetical social and ethical scenarios (e.g., about stealing; Saver & Damasio, 1991). In sum, patients with vmPFC injury show circumscribed postmorbid changes specific to real-world decision making, particularly in the social and moral domains.

The Somatic Marker Hypothesis

The SMH is a systems-level neurological theory of the relationship between basic motivational, affective processes and the coordination of complex reasoning and action, as in the social and moral domains (Damasio, 1994; Damasio et al., 1998). The concomitant impairments in emotion and decision making in cases of vmPFC injury highlight the link between these two seemingly distinct functions (Damasio, 1994). However, despite the striking behavioral profile common in people with vmPFC injury, their generally intact functioning makes it difficult to capture and empirically document that profile in the laboratory (Damasio et al., 2011). Tasks commonly employed in both clinical and research applications of neuropsychology are designed (purposefully in most cases) to be highly specified and with limited open-ended complexities, presenting well-structured problems with clearly defined rules and goals (Lezak, Howieson, Loring, & Tranel, 2012). This is in contrast to the complex, ambiguous, poorly specified problems frequently encountered in the real world, where patients with vmPFC injury manifest the greatest impairments.

To address the inadequacies of extant laboratory tasks, Bechara, Damasio, Damasio, and Anderson (1994) devised the Iowa Gambling Task (IGT), which uses unpredictable combinations of risk and ambiguity to simulate real-world decision making in an experimental neuropsychological instrument. In the IGT, participants are given an initial loan of (facsimile) money, with a goal of winning as much money as possible during the task.

Play consists of drawing cards one at a time from any of four decks of face-down cards. Each draw provides an immediate reward; in addition, punishments of various magnitudes occur at unpredictable intervals. Overall, the decks differ in their reward and punishment schedules, so that there are two "good" (advantageous in the long run) and two "bad" (disadvantageous in the long run) decks. Drawing from good decks will result in a net profit over the course of the task, as they provide modest payoffs with each card (e.g., $50), and the occasional punishments are modest too (e.g., totaling only $250 after 10 card draws). The bad decks, however, will result in a net loss, as they provide larger payoffs with each card (e.g., $100), but the occasional punishments are even larger (e.g., totaling $1250 after 10 draws). Thus, in one version of the IGT, after 10 draws from a good deck the net profit would be $250 ($500 gained and $250 lost), while there would be a net loss of $250 after 10 draws from a bad deck ($1000 gained and $1250 lost).

Over time, neurologically intact individuals demonstrate a strong preference for the advantageous strategy of drawing more from the good decks than from the bad ones (Bechara et al., 1994). Participants also develop anticipatory skin conductance responses (SCRs) prior to drawing cards from bad decks, even before they report conscious awareness of the decks' being disadvantageous, suggesting that covert autonomic information is informing their decision-making process (Bechara, Damasio, Tranel, & Damasio, 1997). In contrast, patients with vmPFC damage do not demonstrate a preference for selecting from good decks, even after they report knowing which decks are good (Bechara et al., 1994). Moreover, although both patients with vmPFC damage and normal comparison participants generate SCRs in immediate response to gain or loss after drawing a card on the IGT, patients with vmPFC damage fail to develop anticipatory SCRs prior to choosing from disadvantageous decks (Bechara et al., 1994; Bechara, Tranel, Damasio, & Damasio, 1996). Thus it seems that patients with vmPFC injury continue to choose from bad decks in part due to the missing anticipatory SCRs, demonstrating an impairment in using the emotional consequences of past experience to guide behavior.

Patients with vmPFC damage also show abnormal SCRs in response to socially meaningful stimuli, but normal responses to other, basic unconditioned stimuli such as loud noises (Damasio, Tranel, & Damasio, 1990). As Damasio et al. (2011, p. 441) put it, "[social] stimuli failed to activate somatic states previously associated with specific social situations," and this deficit can lead to disadvantageous real-world behavior because these somatic states "marked the anticipated outcomes of response options as advantageous or not."

This idea is the core of the SMH, which holds that the vmPFC is critical for reactivating emotional states that influence decision making (Damasio, 1994, 1995). During an initial learning phase, emotions arise in the body as somatic, bioregulatory states associated with the rewarding or punishing consequences of an option that has been chosen, and this emotion–option relationship is encoded in the vmPFC. When similar options are considered in the future, the emotional states associated with them are reactivated by the vmPFC through its interactions with subcortical and cortical structures, producing "somatic markers" that bias behavior. In this way, *a somatic marker is the learned anticipation of future emotional experience that may occur if one makes a certain choice* (Bechara & Damasio, 2005). When a neurologically healthy individual plays the IGT, somatic markers are activated when the person is considering drawing from bad decks, as indicated by anticipatory SCRs. The markers anticipate the negative emotional experience of losing money, which has been learned from previous draws from the bad decks.

According to the SMH, a preference for good decks is developed because somatic markers bias choice away from disadvantageous decks. Patients with vmPFC injuries, then, do not develop a preference for good decks (and, analogously, have deficits in social behavior) because they are unable to reactivate the somatic markers that guide advantageous decision making.

Tying this back to the case of E. V. R., Eslinger and Damasio (1985) note that "E. V. R. had a defect of analysis and integration of stimuli pertaining to real-life situations that may be due, in part, to ineffectual access to previously learned strategies of action" (p. 1739). The emphasis on a deficit in strategy selection is important. It is not that E. V. R. cannot experience emotion; rather, he has an impairment in engaging decision-making strategies that involve the reactivation of somatic markers. When one is making decisions in the real world, a deficit in the implementation of somatic-marker-based strategies can lead to the selection of disadvantageous options, friends, business decisions, and so on.

In support of the SMH, psychophysiological and functional neuroimaging studies have demonstrated a role for the vmPFC in social emotions, predicting the future rewarding consequences of behavior, and encoding the value of behavioral consequences (Naqvi, Shiv, & Bechara, 2006). The vmPFC is known to be involved in reward- and punishment-related behavior, encoding the primary reinforcing values of stimuli, and encoding the values of conditioned stimuli that predict reward or punishment (Rolls, 2000, 2004). Recent work by Wallis and colleagues in primates has shown that orbital and medial prefrontal neurons encode abstract value signals, which is consistent with a value-marking function (Kennerley, Dahmubed, Lara, & Wallis, 2009; Wallis & Kennerley, 2010). Neuroimaging studies have extended these findings to humans at the level of brain regions; a meta-analysis has shown that the vmPFC is involved in the representation and monitoring of reward (Kringelbach & Rolls, 2004).

A functional magnetic resonance imaging (fMRI) study found that vmPFC activation in response to conditioned stimuli is reduced by reducing the value of the unconditioned stimulus (Gottfried, O'Doherty, & Dolan, 2003). This suggests, in accordance with the SMH, that the vmPFC may predict the future rewards associated with a behavior based on previous experience with its consequences (Naqvi et al., 2006). Another study showed that the vmPFC is involved in encoding the subjective values of future rewards, rather than purely objective values like reward magnitude or temporal delay (Kable & Glimcher, 2009).

Also, body state representations have been shown to have a role in vmPFC processing, as the SMH predicts. Using magnetoencephalography, Rudrauf and colleagues (2009) found that body responses to emotionally arousing stimuli are represented in the vmPFC at the earliest stages of emotion processing. Another study used false physiological feedback to show that information about somatic states plays a causal role in the formation of preferences (Batson, Engel, & Fridell, 1999). In sum, the SMH is an empirically supported theory of the neural mechanisms critical for engaging decision-making strategies that integrate emotional information into the decision-making process, which is crucial in the domain of social and moral functioning.

As a historical aside, it is worth noting that the SMH builds on earlier ideas about relationships among emotion, the body, and complex behavior (Dalgleish, 2004), such as the James–Lange theory, which understands emotion as the perception of physiological changes in the body (James, 1884). This is the basis for the notion of a somatic marker represented in terms of the internal milieu.

In psychology and philosophy there is a long tradition of separating reason on the one hand from emotion and the body on the other, which persisted throughout most of the 20th century (Edwards, 1954). Following popular perception, emotions were often viewed as impediments to cool, rational behavior, rather than as critical participants in such behavior. Nonetheless, over the years building up to Damasio's theorizing, an increasing number of researchers came to appreciate the role of emotion in the organization of adaptive complex behavior. Echoed in Damasio's theory is Wundt's notion of affective primacy, which held that "the clear apperception of ideas in acts of cognition and recognition is always preceded by special feelings" (Wundt, 1902, p. 237). In Wundt's view, much like the SMH, the bringing to mind of ideas (or behavioral options in the context of decision making) is inherently accompanied by affective information. Later, Robert Zajonc (1980) built on the idea of affective primacy by arguing for a tight link between emotion and reason. Zajonc gave emotion an essential role in the organization of complex behavior like decision making.

Perhaps the theory that most closely presaged Damasio's SMH came from the neuroanatomist and systems neuroscientist Walle Nauta (1971). Nauta (and later Damasio) understood the prefrontal cortex in terms of its extensive bidirectional connections with the limbic system, hypothalamus, and brain stem—nervous system structures closely linked to the internal milieu. Given this neuroanatomical arrangement and the deficits in behavioral organization and emotion observed after frontal cortex lesions, Nauta noted that "the behavioral effects of frontal-lobe destruction could be seen as the consequence of an 'interoceptive agnosia', i.e., an impairment of the subject's ability to integrate certain informations from his internal milieu with the environmental reports provided by his neocortical processing mechanisms" (1971, p. 182). Foreshadowing Damasio's notion of somatic markers, Nauta went on to suggest that interoceptive information acts as a "navigational marker" to guide "complex goal-directed forms of behavior" (Nauta, 1971, p. 184). The SMH builds these early ideas into a neurologically and cognitively specified theory of the role of emotion in the organization of complex behavior.

The Neural Basis of Moral Functions

Moral Cognition

The SMH holds that the vmPFC is critical for using emotions to guide decision making toward the selection of adaptive behaviors—a function that is particularly important in the social and moral domains (Damasio et al., 2011). From an evolutionary perspective, a fundamental aspect of adaptive behavior is the ability to follow social norms (engaging in socially and morally appropriate behavior), which helps maintain membership in social groups and fosters cooperation in order to promote survival (Silk, Alberts, & Altmann, 2003; de Waal, 2006). But the social world is a tricky place, fraught with subtle and variable information. Accordingly, socio-moral cognition has been shown to rely on affective mechanisms specialized for such circumstances.

Over the past decade, building on the early work leading to the development of the SMH, findings from functional imaging studies (Greene, Sommerville, Nystrom, Darley, & Cohen, 2001; Heekeren, Wartenburger, Schmidt, Schwintowski, & Villringer, 2003; Moll, de Oliveira-Souza, Bramati, & Grafman, 2002; Schaich Borg, Hynes, Van Horn,

Grafton, & Sinnott-Armstrong, 2006), behavioral studies (Schnall, Haidt, Clore, & Jordan, 2008; Valdesolo & DeSteno, 2006; Wheatley & Haidt, 2005), and lesion studies (Ciaramelli, Muccioli, Làdavas, & di Pellegrino, 2007; Koenigs et al., 2007; Thomas, Croft, & Tranel, 2011) have converged to demonstrate the role of emotion in moral cognition, and the vmPFC as a crucial neural structure for integrating emotions into socio-moral processing (Casebeer, 2003; Greene, 2009; Young & Koenigs, 2007). As described below, the vmPFC has been shown to be critical for normal moral cognition under conditions of ambiguity, uncertainty, and conflict, when it is important to integrate somatic information into the reasoning process (Thomas et al., 2011). Under such conditions, the "cognitive–emotional" representation of options involves the vmPFC and its limbic connections (Moll, de Oliveira-Souza, & Zahn, 2008, p. 167).

In one prominent line of research, the vmPFC has been implicated in making moral judgments about high-conflict moral dilemmas that pit strong emotional aversion to causing harm against utilitarian considerations (Greene et al., 2001; Koenigs et al., 2007; Thomas et al., 2011). For example, consider the classic "footbridge" dilemma: You are standing on a footbridge over some trolley tracks and see a run-away trolley heading toward five workers on the tracks (Thomson, 1985). To save the workers, you must push a large man off the footbridge onto the tracks, where his weight will stop the trolley. This dilemma pits the utilitarian consideration of maximizing the number of lives saved against an emotionally evocative harm, and a large majority of people tend to reject the utilitarian option to avoid pushing the man (Hauser, Cushman, Young, Kang-Xing Jin, & Mikhail, 2007). High-conflict dilemmas can involve emotionally aversive acts that are either direct harms (e.g., pushing someone off a bridge) or indirect harms with high personal involvement (as in flipping a switch that results in the death of your daughter) (Thomas et al., 2011). Although more work is needed on this point, the key seems to be whether a high degree of conflict is elicited between the options, rather than the actual content of the options.

In contrast to a high-conflict dilemma, a low-conflict dilemma limits the competition between the choice options—for example, by reducing the socio-emotional aversion to performing the action. Consider the "switch" dilemma, in which saving the workers requires that you flip a switch, redirecting the trolley onto a side track where it will kill one stranger (rather than your daughter, as in the high-conflict version of this dilemma). The majority of people tend to endorse utilitarian options in such a low-conflict dilemma (Hauser et al., 2007).

When a person considers the appropriateness of causing emotionally aversive harm in order to secure a utilitarian outcome, as in a high-conflict moral dilemma, an anticipatory emotional response (or somatic marker, cf. Damasio, 1994) is activated by the vmPFC and biases the decision away from that choice (Greene, 2007; Koenigs et al., 2007; Thomas et al., 2011). For example, a somatic marker might encode the anticipation of the socio-emotional consequences of the action (e.g., living with the knowledge that you pushed a person off a footbridge; Thomas et al., 2011). In support of this role for the vmPFC in moral cognition, Greene et al. (2001) found increased vmPFC activation when participants responded to high-conflict versus low-conflict dilemmas. Moreover, patients with vmPFC injury, who lack a prepotent emotional response (Bechara et al., 1997; Damasio et al., 1990), are more likely to judge the utilitarian action as appropriate on high-conflict moral dilemmas (Figure 4.1; Ciaramelli et al., 2007; Koenigs et al., 2007; Thomas et al., 2011).

(a)

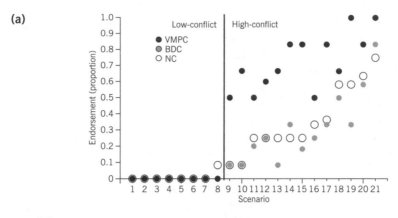

(b)

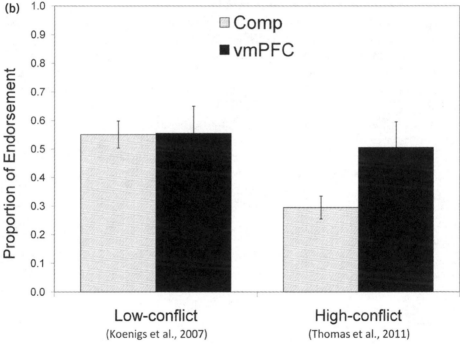

FIGURE 4.1. Moral judgments for low- versus high-conflict moral dilemmas that pit utilitarian outcomes against social-emotionally aversive direct (a) or indirect (b) harms. The *y* axis represents mean proportion of endorsement of the utilitarian outcome (i.e., accepting causing the aversive harm to maximize aggregate welfare). The direct-harm graph (a) breaks the data down by individual scenario (vmPFC, group with vmPFC damage; BDC, brain-damaged comparison group; NC, normal comparison group). The indirect-harm graph (b) depicts mean proportions of endorsement for low- versus high-conflict scenarios (error bars represent standard error of the mean; Comp, comparison group of brain-damaged and normal comparison participants). In both the direct and indirect cases, patients with vmPFC damage showed significantly increased endorsement of utilitarian outcomes in the high-conflict, but not the low-conflict, condition. Figure 4.1a from Koenigs et al. (2007). Copyright 2007 by the Nature Publishing Group. Reprinted by permission from Macmillan Publishers. Figure 4.1b adapted from Koenigs et al. (2007) and Thomas, Croft, and Tranel (2011). Copyright 2007 by the Nature Publishing Group and 2011 by The MIT Press. Adapted by permission.

Furthermore, patients with vmPFC injury made atypically utilitarian judgments both when the agent in the dilemma was the self (e.g., "Is it appropriate for *you* to push the person?") or when the agent was someone else (e.g., "Bob") (Thomas et al., 2011). All participants, though, were more likely to endorse utilitarian outcomes in the "other" as compared to the "self" conditions, suggesting an aversion to endorsing the performance of an aversive act by oneself. This self–other asymmetry in moral judgment was observed for both the vmPFC and comparison groups, suggesting that the vmPFC is not critical for generating the effect. The self–other asymmetry may be due to a simple heightened aversion to imagining oneself, as compared to another person, causing a social-emotion-arousing harm. An fMRI study of moral judgments of one's own actions versus another person's actions suggests that this process may be underpinned by the amygdala: Greater amygdala activation was observed in the self as compared to the other condition (Berthoz, Grèzes, Armony, Passingham, & Dolan, 2006).

Consistent with the SMH, it seems that the vmPFC is critical for typical reasoning in ambiguous, uncertain contexts that place demands on the anticipation of the future consequences of an action, as in high-conflict dilemmas in the moral domain, or the IGT in the domain of economic decision making (Bechara et al., 1997; Naqvi et al., 2006). The notion of "anticipating the future consequences of an action" does not entail that the emotion is not presently experienced. Instead, a somatic marker is understood as an emotion that represents how one would feel in the future if an action were to be performed; it is an affective forewarning of future emotional experience. In fact, we have come to think that this affective anticipation of future emotion may be the defining feature of the moral emotions (see below).

It is important to recognize that the affective reasoning mechanisms organized by the vmPFC are not always critical for typical moral cognition. In contrast to moral judgment about high-conflict dilemmas, the vmPFC does not appear to be critical for moral cognition about low-conflict dilemmas (Greene et al., 2001; Koenigs et al., 2007) or reasoning about ordinary hypothetical moral scenarios (e.g., stealing; Anderson, Bechara, Damasio, Tranel, & Damasio, 1999; Damasio et al., 2011; Saver & Damasio, 1991). In such contexts, moral cognition appears to rely on more cognitive rule-oriented processes underpinned by the dorsolateral prefrontal cortex (dlPFC) and temporal lobe structures (Greene et al., 2001), such as semantic memory, rule processing, and cognitive executive processes (Bunge, 2004). Here emotional information is not weighed as heavily during moral cognition. Thus, as with reasoning and cognitive processing in general (Lieberman, 2003), there seem to be both semantic (cognitive/executive processing) and somatic (affective processing) strategies for moral cognition, which are adaptively engaged under different circumstances. More research is needed to clarify further the extent to which these two distinct modes of moral cognition depend on separable neural underpinnings.

One prominent theory has proposed a dual-process view of the neural basis of moral cognition (Greene, 2007). According to Greene, during moral cognition cognitive and affective processes compete with the final moral judgment corresponding to the outcome of this competition. In this view, the vmPFC is responsible for promoting prepotent emotion-based moral judgments that conform to deontological principles such as "Do no harm." In contrast, the more "rational," less affective processing of the dlPFC is thought to promote utilitarian moral judgments. However, patients with vmPFC damage are only more likely to endorse utilitarian outcomes on high-conflict moral dilemmas, rather than showing greater utilitarian judgments on all types of stimuli, as Greene's theory would predict. Furthermore, it is not clear that the vmPFC and dlPFC perform *competing* processes, rather than working in concert to produce efficient socio-moral cognitive functioning across a variety of contexts (Moll & de Oliveira-Souza, 2007; Ochsner et al., 2004).

In our view, the vmPFC is not specialized for producing moral judgments that are consistent with any particular moral theory; rather, the vmPFC is critical for selecting an option based on its previous history with punishment and reinforcement. That is, the vmPFC's role in moral cognition is not directly tied to a moral theory, but only to the process of emotion–reasoning integration. As in other domains of reasoning and rule processing, the neurocognitive systems engaged during moral cognition are likely to be determined by whether there is an emotion-eliciting stimulus that can be associated with past reward and punishment learning, and the specific circumstances at hand (e.g., is it an ambiguous and uncertain situation?). It may turn out that in certain contexts this type of affective reasoning process is more likely to generate judgments consistent with certain moral theories (e.g., deontological judgments of rejecting causing a harm that would maximize welfare) than others (e.g., utilitarian judgments of accepting causing such a harm). But we believe that it is not the normative content of the rules in play that distinguishes the brain systems involved. The details of the stimuli and context, along with individual-person factors, will determine whether a more or less "affective," vmPFC-based strategy is likely to be engaged.

Moral Emotions

Moral emotions, underpinned by the medial prefrontal cortex and its limbic and subcortical connections, are essential for normal social and moral functioning. Several frameworks have been suggested for the philosophical and psychological structure of the moral emotions (Moll, de Oliveira-Souza, Zahn, & Grafman, 2008; Parrott, 2001). How the notion of *moral emotions* fits with other similar notions, such as *secondary emotions* (Kemper, 1987) and *complex emotions* (Solomon, 2002), is also an issue that has attracted considerable discussion and debate. For our present purposes, we focus on moral emotions, which can also be called social emotions, including guilt, shame, indignation, and pride (Eisenberg, 2000; Moll, de Oliveira-Souza, Zahn, et al., 2008). Moral emotions have been shown to engage a frontolimbic network involving the vmPFC, cingulate cortex, amygdala, insula, hypothalamus, basal forebrain, and brainstem (Moll, de Oliveira-Souza, & Zahn, 2008).

A subclass of the moral emotions, the *self-conscious* moral emotions, such as guilt, regret, and embarrassment, provide immediate affective feedback (punishment or reward) about one's behavior during self-reflection (Tangney, Stuewig, & Mashek, 2007). Impairments in self-conscious moral emotions are associated with vmPFC damage (Anderson et al., 1999; Beer, Heerey, Keltner, Scabini, & Knight, 2003; Camille et al., 2004; Eslinger & Damasio, 1985; Stuss & Benson, 1984). Patients with vmPFC damage are less influenced by the socio-emotional factor of guilt when making choices in economic games (e.g., the Dictator, Ultimatum, and Trust Games) that involve the social component of determining whether to emphasize payoffs for oneself or for others (Krajbich, Adolphs, Tranel, Denburg, & Camerer, 2009). Passive viewing of emotional scenes both with and without moral implications evokes amygdala, limbic, and brainstem activation (Moll, de Oliveira-Souza, Eslinger, et al., 2002). However, viewing scenes that evoke moral emotions, as compared to neutral, nonmoral unpleasant, nonmoral pleasant, or interesting scenes, is associated with greater activation of the vmPFC. This suggests a highly specialized role for the vmPFC in *moral* emotional processing.

One key moral emotion that has received increasing attention is moral disgust. For example, in one particularly intriguing study, Wheatley and Haidt (2005) showed that people made harsher moral judgments when hypnotized to experience disgust. Moll, de Oliveira-Souzza, et al. (2005) also found greater orbitofrontal cortex activation for moral disgust (indignation) than for pure disgust (e.g., in response to unpleasant statements about disgusting animals or bodily excretions). Another study (Schaich Borg, Lieberman, & Kiehl, (2008) had participants read statements designed to evoke moral disgust (incest), nonsexual moral emotion (murder), and pathogen-related disgust (eating a scab) while undergoing fMRI scanning. Contrasting the moral conditions (disgust and nonsexual moral) with the nonmoral pathogen condition revealed greater activation in the medial prefrontal cortex (crossing the ventral, anterior, and dorsal sectors), further supporting the role of this region in processing moral information. Moreover, comparing the moral disgust and nonsexual moral conditions showed increased activation in the medial prefrontal cortex (even more broadly than in the previous contrast) and the left amygdala. The evidence on moral disgust thus suggests that the medial prefrontal cortex, including the ventral and dorsal sectors, plays an important role in this emotion, and moral disgust may involve heightened medial prefrontal cortex processing above and beyond that involved in other moral emotions. It remains to be explored precisely how the socio-moral functions of the dorsomedial prefrontal cortex (dmPFC) and vmPFC might differ.

In discussing moral cognition and the SMH for integrating emotion into decision making, an important point of emphasis is the *anticipation* of emotion (e.g., anticipating the emotional consequences of selecting a certain option). This affective anticipation has also been recognized as important in the context of moral emotions. For example, *expected guilt*[1] is experienced prior to choices where selecting an option is likely to engender the feeling of guilt; that is, there is an expectation of the possible future consequences of one's behavior, one of which is the experience of guilt, which can dissuade one from choosing an option (Moll, de Oliveira-Souza, & Zahn, 2008; Tangney et al., 2007). The expectation of future moral emotional experience can be a behaviorally motivating factor (Tangney et al., 2007)—a point also emphasized by the SMH (Damasio, 1996).

In our view, the anticipation of future emotional experience plays a quite fundamental role in the moral emotions, even beyond what has been alluded to above in the notion of *expected guilt*. There is good reason to think that the common nature of the various moral emotions may lie in the *anticipation* of emotion. For one, the vmPFC is a critical neural structure both for moral emotional processes and for the anticipation of the emotional consequences of behavior, suggesting an underlying neurological link between these two functions. Moreover, in considering the psychological structure of moral emotions, it seems that they involve a forecast of emotional experience. For instance, guilt, at

---

[1] The notion of *expected guilt* is frequently referred to as *anticipatory guilt* (Tangney et al., 2007). We use the term *expected guilt* here to distinguish it from the "anticipatory theory of moral emotions."

its most basic level, might be an anticipation of negative emotions due to social punishment for one's behavior. But this is not the expected guilt discussed above; rather, it is very much the basic guilt that one feels *after* doing wrong. The point is that such ordinary guilt may be grounded in the anticipation of negative emotional experience. This "anticipatory theory of moral emotions" provides a parsimonious explanation of the vmPFC's dual roles in decision making and moral emotion, by explaining both in terms of the underlying vmPFC function of anticipating the emotional consequences of behavior. Needless to say, this idea requires further empirical investigation.

Moral Behavior

Moral behavior refers to the implementation of moral-cognitive processes in contexts in which one's actions have meaningful consequences in the external world. Laboratory tasks of moral judgment frequently do not focus on moral behavior; although one's actions (judgments) in such tasks may have internal consequences (e.g., feelings of guilt for having made a certain judgment), they typically lack meaningful external consequences (e.g., making or losing a friend). Studies that focus directly on moral functions in contexts involving clear external consequences have shown that moral behavior critically depends on a network of brain regions similar to those involved in moral cognition and socio-moral emotion as discussed above, although there is not a thorough understanding of the potential differences in neural mechanisms involved in those processes (which are mostly not behavioral) in contrast to moral behavior.

Patients with vmPFC injury have impairments in social emotions and moral behavior in their everyday lives, where external consequences are obvious and almost always at hand (Anderson, Barrash, Bechara, & Tranel, 2006). The patients display a lack of concern for others, socially inappropriate or callous behavior (e.g., inappropriate or rude comments), a reduction in the display of guilt and shame, and increased aggressive behavior (Anderson et al., 1999, 2006; Beer, John, Scabini, & Knight, 2006; Damasio et al., 1990; Dimitrov, Phipps, Zahn, & Grafman, 1999; Eslinger & Damasio, 1985; Fellows, 2007; Rolls, Hornak, Wade, & McGrath, 1994; Stuss & Benson, 1984). The behavior of the patient E. V. R., for example, was described as "sociopathic" (Eslinger & Damasio, 1985). In a case study of a patient with childhood-onset vmPFC injury, Anderson et al. (1999) note that the patient lied frequently and unnecessarily; maintained no lasting friendships; and displayed little empathy, guilt, or remorse for his behavior.

Research on the neural basis of moral behavior has been largely limited to studies of clinical populations and case studies. The functional imaging literature has begun to touch on issues related to moral behavior, such as the detection of prevarication and dishonest behavior (Greene & Paxton, 2009; Spence et al., 2001). Although controversy exists, this literature suggests that deceptive behavior involves an executive process and is typically associated with increased activity in the lateral prefrontal cortex (Spence et al., 2004). Although controlled experiments designed to examine moral behavior other than lying are lacking, some evidence can be gleaned from research in neuroeconomics, which employs economic games that involve social factors such as fairness and altruism. In the Ultimatum Game, for example, two players are given the chance to split a sum of money. The money is first given to the proposer, who must make an offer to the responder of how to split the money. The responder can accept the offer and the players receive the money as proposed, or the responder can reject the offer and neither player gets anything.

Whereas a purely economical decision would be to accept all offers, responders are sensitive to the social dynamics of the game and typically reject low offers, presumably to punish unfair proposers (Bolton & Zwick, 1995; Sanfey, Rilling, Aronson, Nystrom, & Cohen, 2003).

The Ultimatum Game has been investigated in research using fMRI, where it has been shown that there is greater anterior insula activation in response to unfair offers than fair offers, and in response to unfair offers that are later rejected than those later accepted (Sanfey et al., 2003). Furthermore, right anterior insula activation is inversely correlated with the acceptance rates of unfair offers. Thus the insula, a cortical limbic structure known to be involved in processing emotion, is involved in generating socio-moral behavior in the context of an economic game with a social component.

It has further been shown that patients with vmPFC damage are more likely than comparison participants to reject unfair offers in the Ultimatum Game (Koenigs & Tranel, 2007). This is yet another illustration of the vmPFC's critical role in the ability to integrate emotion strategically and advantageously into the behavior selection process. In this case, the patients' behavior appears to be guided by unchecked negative affective input that drives the punishment of unfair proposers, while failing to account for the potential negative emotional consequences of doing so (e.g., the person punished might continue to make unfair offers in retaliation, or the punisher might feel guilty about his or her behavior in retrospect).

Basic Social and Emotional Processing

Social and moral functioning critically depend on a core frontolimbic network centered on the vmPFC that uses emotion to guide the selection of advantageous behaviors. This network, however, does not function in isolation. It relies on many other, more basic social functions, ranging from theory of mind to face processing to basic affective and motivational mechanisms (Adolphs, 2003; Casebeer, 2003).

Theory of Mind and Empathy

Theory of mind, or *mentalizing*—the ability to understand the mental states of others—is an important component of social functions (Frith & Frith, 1999). In the socio-moral domain, understanding what other people are thinking, what their plans and intentions are, and how they are feeling are relevant factors for understanding their social and moral behavior, and for adjusting one's behavior accordingly (as in empathetic behavior; Astington, 2004; Sokol, Chandler, & Jones, 2004). Several brain regions have been implicated in theory of mind, including the dmPFC and temporoparietal junction (TPJ, including the posterior superior temporal gyrus and sulcus and portions of the inferior parietal lobule; see Plate 4.1; Saxe & Powell, 2006; Young, Cushman, Hauser, & Saxe, 2007). The dmPFC is important for social-cognitive processes such as theory of mind (Mitchell, Banaji, & Macrae, 2005), self-referential processing (Northoff & Bermpohl, 2004), and the resting-state default network of the brain (Buckner, Andrews-Hanna, & Schacter, 2008). The TPJ is involved in theory of mind; default mode processing; and the detection of biological motion, such as processing movements of the eyes and mouth (Allison, Puce, & McCarthy, 2000).

Young and colleagues have pursued the question of how theory of mind and its neural substrates are involved in moral judgment (Young & Saxe, 2011; Young, Camprodon, Hauser, Pascual-Leone, & Saxe, 2010). In a recent fMRI study, these researchers had participants read about a person's action and beliefs about that action, and then read either moral or nonmoral facts about the action (Young & Saxe, 2011). The dmPFC and right TPJ, as well as the precuneus, were more active when participants read the moral as compared to nonmoral facts, even though the amount of belief information was the same in both conditions. These results support the notion that the neural underpinnings of theory of mind play an important role in moral information processing.

In another study, Young and colleagues (2007) examined brain activation when participants made moral judgments about vignettes describing actions that were or were not intended to cause harm (negative vs. neutral intent), and either did or did not actually cause harm (negative vs. neutral outcome; see Figure 4.2 for examples). The right TPJ showed increased activation in response to these moral judgment conditions as compared to a control condition requiring judgments of physical attributes. Moreover, the dmPFC and right TPJ were significantly more active in response to neutral outcomes (no harm caused) when the action was intended to harm than when it was not. A similar effect of intention on dmPFC and right TPJ activation was not observed when the outcome was negative. Thus, when mental states are more important for moral judgment, such as when an outcome is neutral and guilt can only be inferred by inferring bad intentions, the dmPFC and TPJ are more engaged. This again shows that neural structures involved in theory of mind are recruited for moral judgment, and demonstrates that such structures are differentially engaged depending on how much another's mental states factor into moral judgment. And there is some evidence that the right TPJ is critical for using information about intentions to make moral judgments, as transcranial magnetic stimulation to this region disrupts the influence of beliefs on moral judgment (Young, Camprodon, et al., 2010).

An important avenue for further research will be the potentially different contributions of cognitive and affective theory of mind to socio-moral functions (Shamay-Tsoory & Aharon-Peretz, 2007). In contrast to cognitive theory of mind, which is typically studied with false-belief tasks and requires understanding that a person has particular thoughts, affective theory of mind has been studied primarily with faux pas tasks, which additionally require understanding that a person is in a particular emotional state. Whereas cognitive theory of mind relies on the dmPFC and TPJ, affective theory of mind has been shown to be impaired in patients with damage to the vmPFC, a critical neural substrate for moral cognition (Shamay-Tsoory & Aharon-Peretz, 2007; Shamay-Tsoory, Tomer, Berger, Goldsher, & Aharon-Peretz, 2005). Also, considering the importance of affective processing in moral functioning, along with the overlap in neural substrates, affective theory of mind might arguably be more relevant to socio-moral functioning than its cognitive counterpart.

Indeed, evidence for the role of affective theory of mind in moral judgment was found in a lesion study conducted by Young, Bechara, et al. (2010). Using the stimuli from a previous examination of theory of mind in moral judgment (Young et al., 2007), the researchers showed that patients with damage to the vmPFC were significantly more likely than comparison participants to judge failed attempts to harm (neutral outcome, negative intent) as permissible. This suggests that affective theory of mind, underpinned by the vmPFC, plays a critical role in moral judgment under certain conditions—specifically,

(a)

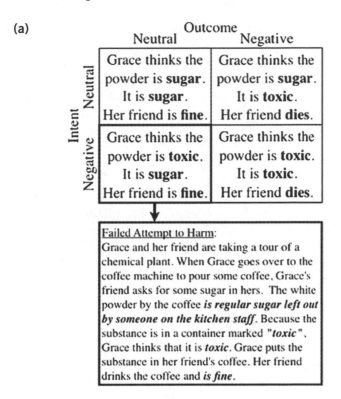

Outcome

	Neutral	Negative
Neutral Intent	Grace thinks the powder is **sugar**. It is **sugar**. Her friend is **fine**.	Grace thinks the powder is **sugar**. It is **toxic**. Her friend **dies**.
Negative Intent	Grace thinks the powder is **toxic**. It is **sugar**. Her friend is **fine**.	Grace thinks the powder is **toxic**. It is **toxic**. Her friend **dies**.

<u>Failed Attempt to Harm</u>:
Grace and her friend are taking a tour of a chemical plant. When Grace goes over to the coffee machine to pour some coffee, Grace's friend asks for some sugar in hers. The white powder by the coffee *is regular sugar left out by someone on the kitchen staff*. Because the substance is in a container marked *"toxic"*, Grace thinks that it is *toxic*. Grace puts the substance in her friend's coffee. Her friend drinks the coffee and *is fine*.

(b)

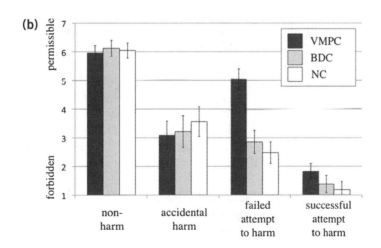

FIGURE 4.2. (a) Example stimuli used in Young, Bechara, et al. (2010). (b) Moral judgments for all four conditions in the study, revealing increased ratings of permissibility on failed attempts to harm for the group with vmPFC damage (VMPC) relative to comparison groups (same as in Figure 4.1). From Young, Bechara, et al. (2010). Copyright 2010 by Cell Press. Reprinted by permission.

when information about an agent's intentions is needed to determine the moral status of his or her actions.

Furthermore, empathy, which plays an integral role in moral functioning (Ambady, Spinrad, & Sadovsky, 2006; Batson, Klein, Highberger, & Shaw, 1995; Hoffman, 2000; Pizarro, 2000), relies on affective theory of mind in combination with emotional contagion to produce an affective response congruent with someone else's situation (Decety & Lamm, 2006; Hoffman, 1987). Psychopathic persons have been described as callous and lacking empathy toward victims of their aggressive behavior, and an underlying lack of empathy may be the cause of some of the symptoms observed in psychopathic disorder (e.g., the lack of inhibition of violent action, reduced empathic behavior, and reduced expression of moral emotion; Blair, 1995). The neural basis of empathy critically relies on the vmPFC (Shamay-Tsoory, Aharon-Peretz, & Perry, 2009) and includes a multicomponent, distributed network that is highly context-dependent, involving neural structures specific to the relevant emotion (Decety & Jackson, 2004; Singer & Lamm, 2009). For example, Singer and colleagues (2004) have shown that empathy for pain involves the anterior insula, which was also activated for the experience of pain.

Facial Information Acquisition and Processing

Facial information is important for basic social processes such as emotion recognition and trait judgment (Hassin & Trope, 2000; Willis & Todorov, 2006), which, like theory of mind and empathy, are key contributors to social functional processes. This makes sense, as faces contain and convey critical information about the emotional and psychological state of a person (Smith, Cottrell, Gosselin, & Schyns, 2005), and the ability to decode information from faces is important for social interaction (Nachson, 1995). The amygdala has been shown to be a critical neural substrate for facial emotion processing (Adolphs, Tranel, Damasio, & Damasio, 1994; Todorov, Baron, & Oosterhof, 2008), perhaps due to a more general function in processing salient emotional information (Adolphs, 2010). Patients with unilateral or bilateral amygdala damage are impaired in recognizing social emotions, such as guilt, from facial stimuli (Adolphs, Baron-Cohen, & Tranel, 2002). Facial emotion processing and amygdala dysfunction have been shown to play a role in psychological disorders associated with deficits in social and emotional functioning, such as autism and anxiety (Baron-Cohen et al., 1999, 2000; Charney, Grillon, & Bremner, 1998; Davidson, 2002; Davis, 1992; Rauch, Shin, & Wright, 2006). People with autism, for example, show deficits in amygdala processing particularly when viewing facial stimuli (Baron-Cohen et al., 1999, 2000).

The amygdala is involved not only in processing facial information, but also in driving the acquisition of such salient information from the environment. Amygdala dysfunction in directing attention to acquire information about salient and potentially threatening stimuli is believed to play a critical role in anxiety (Davis, 1992; Davis & Whalen, 2001). Adolphs et al. (2005) conducted a study of the patient S. M. (who has bilateral amygdala lesions), in which participants' eyes were tracked while they made emotion ratings of facial stimuli. The authors found that, relative to comparison participants, S. M.'s gaze fell on the eyes of the face stimuli significantly less. Considering the importance of the eye region for transmitting information about fear (Smith et al., 2005), Adolphs and colleagues conclude that S. M.'s fear recognition impairment is due to her impairment in spontaneously directing her gaze toward the critical eye region of facial stimuli.

Conclusively demonstrating that SM's fear recognition difficulties are due to impaired acquisition of information from the eye region of facial stimuli (rather than purely a deficit in processing facial information), her fear recognition ability is temporarily recovered when she is instructed to fixate on the eyes of the facial stimuli (Adolphs et al., 2005).

In subsequent studies of S. M.'s face-directed gaze, she displayed impaired eye contact during real conversations with other people (Spezio, Huang, Castelli, & Adolphs, 2007). Moreover, the information acquisition role of the amygdala has been found relevant in clinical contexts where amygdala dysfunction has been hypothesized to contribute to psychological disorders as in autism and anxiety. People with autism show an absence of preferential looking to the eyes of approaching adults, and this predicts social impairment in 2-year-old toddlers with autism spectrum disorders (Jones, Carr, & Klin, 2008). Abnormal gaze fixation and amygdala activity have been found in individuals with autism, and greater amygdala activity has corresponded to greater time spent fixating the eye region of facial stimuli (Dalton et al., 2005). Even the parents of autistic children, who may pass on a genetic predisposition to autism spectrum disorders, show decreased processing of the eyes of facial stimuli (Adolphs, Spezio, Parlier, & Piven, 2008). In another study, atypical gaze fixation and reduced amygdala activity and volume were found in unaffected siblings of autistic individuals (Dalton, Nacewicz, Alexander, & Davidson, 2007). Finally, like the patient S. M., high-functioning individuals with autism show decreased use of eye information and a preference for directing gaze to the mouth when viewing faces (Spezio, Adolphs, Hurley, & Piven, 2006).

Basic Emotional and Motivational Processes

Consistent with the SMH, basic motivational and phylogenetically ancient mechanisms play a fundamental role in human socio-moral functioning (Bechara & Damasio, 2005). Emotions are essential to moral functioning, and seem to "acquire functional meaning from a more basic motivational structure" that includes biological needs (Blasi, 1999, p. 5). Accordingly, the brainstem and limbic system have been described as the "platform" on which moral cognition is built (Casebeer, 2003). As described earlier, the vmPFC is extensively connected to limbic structures—the amygdala, insula, and hippocampus—as well as the hypothalamus and brainstem (Öngür & Price, 2000; Rolls, 2000). Through these connections, it maintains higher-order control over the execution of bodily components of emotions and the encoding of emotional values and the values of the consequences of actions (Öngür & Price, 2000; Rangel et al., 2008; Rolls, 2000). These functions form the basis of somatic marker processing, in which potential behavioral options are marked with affective value (Bechara et al., 1997).

Future Directions

There are limitations to current cognitive neuroscience research on moral reasoning that are worth noting and point to future research directions. Some notable limitations, for example, are the reliance on contrived and unrealistic moral dilemmas for stimuli; the lack of assessment of the neural bases of reasoning and justification processes (i.e., the neural systems involved in understanding *why* something is right or wrong, rather than *that* it is so); and the lack of a clear and consistent operationalization of *morality* (Killen

& Smetana, 2007; Turiel, 2009). To deal with these challenges, future research would do well to be informed by relevant research in philosophy and in developmental and social psychology. Along these lines, one of the most pressing areas for future research is in the developmental cognitive neuroscience of social and moral functioning. So far, very little cognitive neuroscience research has directly tackled the connection between neural and moral development. Studies of patients with childhood-onset brain injury have been instructive (Anderson et al., 1999), as have studies of individuals with autism (Baron-Cohen et al., 2000) or psychopathy (Blair, 1995; Koenigs, Baskin-Sommers, Zeier, & Newman, 2010). And, of course, an expansive literature exists on developmental moral psychology (Killen & Smetana, 2006). However, we agree with Killen and Smetana (2008) that greater collaboration between cognitive neuroscientists and developmental psychologists is needed.

The neural basis of the role of basic emotions (such as anger, fear, and happiness) in moral cognition is another intriguing area for future research. Whereas most moral cognition research has focused on moral emotions, basic emotions also play an important role in social and moral behavior (Eisenberg, Fabes, Guthrie, & Reiser, 2000; Kochanska, Murray, & Coy, 1997), and have relatively well established neural substrates, at least for some of the basic emotions (Murphy, Nimmo-Smith, & Lawrence, 2003).

One important limitation of experimental cognitive neuroscience research on social and moral functioning to date is the use of hypothetical rather than real-world tasks. Case studies, studies of clinical populations, and studies employing neuroeconomic games have provided important data on the neural basis of real-world moral behavior, but there is a need for controlled experiments designed to tests specific questions of the neural basis of socio-moral functioning in the real world. The lesion method will be an important tool as research shifts toward the use of real-world moral tasks, since participants are not constrained to an MRI scanner. The lesion method has proven useful for examining real-world behavior in other domains, such as executive functioning. For example, one study demonstrated that patients with vmPFC injury manifest impairments in the application of advantageous behavioral navigation strategies on unstructured tasks performed in a real-world setting (a shopping mall; Tranel, Hathaway-Nepple, & Anderson, 2007). Neuroscientists should consider a similarly ecological approach in the moral domain of moral cognition.

Other topics on the cutting edge of moral-cognitive neuroscience are individual differences (Prehn et al., 2008), cross-cultural factors (Han & Northoff, 2008), the role of specific neurotransmitter systems (Crockett, Clark, Hauser, & Robbins, 2010), and the implications of moral-cognitive neuroscience research for the society and law (Farah, 2007).

Conclusions

Social and moral functioning rely on a frontolimbic neural system that supports the hierarchical integration of multiple processes. The vmPFC has been shown to be a central higher-order region in this network, critical for integrating affective, somatic information into decision making and the organization of behavior. More basic processes such as theory of mind and face processing are supported by temporal, parietal, and limbic structures, while basic motivational and emotional processes are supported by the limbic system, hypothalamus, and brainstem. It is important to recognize that at this point, this neural processing system does not seem to be specific to the socio-moral domain; the functions of most of the brain regions involved are similar to those observed in nonsocial domains, especially domains that are also complex, ambiguous, risky, and (perhaps most importantly) highly salient (Adolphs, 2010).

Acknowledgments

This work was supported by Grant No. P50 NS19632 from the National Institute of Neurological Disorders and Stroke, and Grant No. DA022549 from the National Institute on Drug Abuse.

References

Adolphs, R. (2003). Cognitive neuroscience of human social behaviour. *Nature Reviews Neuroscience*, 4(3), 165–178.

Adolphs, R. (2010). What does the amygdala contribute to social cognition? *Annals of the New York Academy of Sciences*, 1191(1), 42–61.

Adolphs, R., Baron-Cohen, S., & Tranel, D. (2002). Impaired recognition of social emotions following amygdala damage. *Journal of Cognitive Neuroscience*, 14(8), 1264–1274.

Adolphs, R., Gosselin, F., Buchanan, T. W., Tranel, D., Schyns, P., & Damasio, A. R. (2005). A mechanism for impaired fear recognition after amygdala damage. *Nature, 433*(7021), 68–72.

Adolphs, R., Spezio, M. L., Parlier, M., & Piven, J. (2008). Distinct face-processing strategies in parents of autistic children. *Current Biology*, 18(14), 1090–1093.

Adolphs, R., Tranel, D., Damasio, H., & Damasio, A. R. (1994). Impaired recognition of emotion in facial expressions following bilateral damage to the human amygdala. *Nature, 372*(6507), 669–672.

Allison, T., Puce, A., & McCarthy, G. (2000). Social perception from visual cues: Role of the STS region. *Trends in Cognitive Sciences*, 4(7), 267–278.

Ambady, N., Spinrad, T. L., & Sadovsky, A. (2006). Empathy-related responding in children. In M. Killen & J. G. Smetana (Eds.), *Handbook of moral development* (pp. 517–549). Mahwah, NJ: Erlbaum.

Anderson, S. W., Barrash, J., Bechara, A., & Tranel, D. (2006). Impairments of emotion and real-world complex behavior following childhood- or adult-onset damage to ventromedial prefrontal cortex. *Journal of the International Neuropsychological Society*, 12(2), 224–235.

Anderson, S. W., Bechara, A., Damasio, H., Tranel, D., & Damasio, A. R. (1999). Impairment of social and moral behavior related to early damage in human prefrontal cortex. *Nature Neuroscience*, 2(11), 1032–1037.

Astington, J. W. (2004). Bridging the gap between theory of mind and moral reasoning. *New Directions for Child and Adolescent Development*, 2004(103), 63–72.

Baron-Cohen, S., Ring, H. A., Bullmore, E. T., Wheelwright, S., Ashwin, C., & Williams, S. C. R. (2000). The amygdala theory of autism. *Neuroscience and Biobehavioral Reviews*, 24(3), 355–364.

Baron-Cohen, S., Ring, H. A., Wheelwright, S., Bullmore, E. T., Brammer, M. J., Simmons, A., et al. (1999). Social intelligence in the normal and autistic brain: An fMRI study. *European Journal of Neuroscience*, 11(6), 1891–1898.

Batson, C. D., Engel, C. L., & Fridell, S. R. (1999). Value judgments: Testing the somatic-marker hypothesis using false physiological feedback. *Personality and Social Psychology Bulletin*, 25(8), 1021–1032.

Batson, C. D., Klein, T. R., Highberger, L., & Shaw, L. L. (1995). Immorality from empathy-induced altruism: When compassion and justice conflict. *Journal of Personality and Social Psychology*, *68*(6), 1042–1054.

Beauchamp, M. H., & Anderson, V. (2010). SOCIAL: An integrative framework for the development of social skills. *Psychological Bulletin*, *136*(1), 39–64.

Bechara, A., & Damasio, A. R. (2005). The somatic marker hypothesis: A neural theory of economic decision. *Games and Economic Behavior*, *52*(2), 336–372.

Bechara, A., Damasio, A. R., Damasio, H., & Anderson, S. W. (1994). Insensitivity to future consequences following damage to human prefrontal cortex. *Cognition*, *50*(1–3), 7–15.

Bechara, A., Damasio, H., Tranel, D., & Damasio, A. R. (1997). Deciding advantageously before knowing the advantageous strategy. *Science*, *275*(5304), 1293–1295.

Bechara, A., Tranel, D., Damasio, H., & Damasio, A. R. (1996). Failure to respond autonomically to anticipated future outcomes following damage to prefrontal cortex. *Cerebral Cortex*, *6*(2), 215–225.

Beer, J. S., Heerey, E. A., Keltner, D., Scabini, D., & Knight, R. T. (2003). The regulatory function of self-conscious emotion: Insights from patients with orbitofrontal damage. *Journal of Personality and Social Psychology*, *85*(4), 594–604.

Beer, J. S., John, O. P., Scabini, D., & Knight, R. T. (2006). Orbitofrontal cortex and social behavior: Integrating self-monitoring and emotion–cognition interactions. *Journal of Cognitive Neuroscience*, *18*(6), 871–879.

Berthoz, S., Grèzes, J., Armony, J. L., Passingham, R. E., & Dolan, R. J. (2006). Affective response to one's own moral violations. *NeuroImage*, *31*(2), 945–950.

Blair, R. J. R. (1995). A cognitive developmental approach to morality: Investigating the psychopath. *Cognition*, *57*(1), 1–29.

Blasi, A. (1999). Emotions and moral motivation. *Journal for the Theory of Social Behaviour*, *29*(1), 1–19.

Bolton, G. E., & Zwick, R. (1995). Anonymity versus punishment in ultimatum bargaining. *Games and Economic Behavior*, *10*(1), 95–121.

Buckner, R. L., Andrews?Hanna, J. R., & Schacter, D. L. (2008). The brain's default network. *Annals of the New York Academy of Sciences*, *1124*(1), 1–38.

Bunge, S. A. (2004). How we use rules to select actions: A review of evidence from cognitive neuroscience. *Cognitive, Affective, and Behavioral Neuroscience*, *4*(4), 564–579.

Camille, N., Coricelli, G., Sallet, J., Pradat-Diehl, P., Duhamel, J.-R., & Sirigu, A. (2004). The involvement of the orbitofrontal cortex in the experience of regret. *Science*, *304*(5674), 1167–1170.

Casebeer, W. D. (2003). Moral cognition and its neural constituents. *Nature Reviews Neuroscience*, *4*(10), 840–847.

Charney, D. S., Grillon, C., & Bremner, J. D. (1998). Review: The neurobiological basis of anxiety and fear: Circuits, mechanisms, and neurochemical interactions (Part I). *The Neuroscientist*, *4*(1), 35–44.

Ciaramelli, E., Muccioli, M., Làdavas, E., & di Pellegrino, G. (2007). Selective deficit in personal moral judgment following damage to ventromedial prefrontal cortex. *Social Cognitive and Affective Neuroscience*, *2*(2), 84–92.

Crockett, M. J., Clark, L., Hauser, M., & Robbins, T. W. (2010). Serotonin selectively influences moral judgment and behavior through effects on harm aversion. *Proceedings of the National Academy of Sciences USA*, *107*(40), 17433–17438.

Dalgleish, T. (2004). The emotional brain. *Nature Reviews Neuroscience*, *5*(7), 583–589.

Dalton, K. M., Nacewicz, B. M., Alexander, A. L., & Davidson, R. J. (2007). Gaze-fixation, brain activation, and amygdala volume in unaffected siblings of individuals with autism. *Biological Psychiatry*, *61*(4), 512–520.

Dalton, K. M., Nacewicz, B. M., Johnstone, T., Schaefer, H. S., Gernsbacher, M. A., Goldsmith,

H. H., et al. (2005). Gaze fixation and the neural circuitry of face processing in autism. *Nature Neuroscience, 8*(4), 519–526.

Damasio, A. R. (1994). *Descartes' error: Emotion, reason, and the human brain.* New York: Grosset/Putnam.

Damasio, A. R. (1995). Toward a neurobiology of emotion and feeling: Operational concepts and hypotheses. *The Neuroscientist, 1*(1), 19–25.

Damasio, A. R. (1996). The somatic marker hypothesis and the possible functions of the prefrontal cortex. *Philosophical Transactions of the Royal Society of London: Series B. Biological Sciences, 351*(1346), 1413–1420.

Damasio, A. R., Anderson, S. W., & Tranel, D. (2011). The frontal lobes. In K. M. Heilman & E. Valenstein (Eds.), *Clinical neuropsychology* (4th ed., pp. 417–465). New York: Oxford University Press.

Damasio, A. R., Tranel, D., & Damasio, H. (1990). Individuals with sociopathic behavior caused by frontal damage fail to respond autonomically to social stimuli. *Behavioural Brain Research, 41*(2), 81–94.

Damasio, A. R., Tranel, D., & Damasio, H. (1998). Somatic markers and the guidance of behavior. In J. M. Jenkins, K. Oatley, & N. L. Stein (Eds.), *Human emotions: A reader* (pp. 122–135). Malden, MA: Blackwell.

Damasio, H., Grabowski, T., Frank, R., Galaburda, A. M., & Damasio, A. R. (1994). The return of Phineas Gage: Clues about the brain from the skull of a famous patient. *Science, 264*(5162), 1102–1105.

Davidson, R. J. (2002). Anxiety and affective style: Role of prefrontal cortex and amygdala. *Biological Psychiatry, 51*(1), 68–80.

Davis, M. (1992). The role of the amygdala in fear and anxiety. *Annual Review of Neuroscience, 15*(1), 353–375.

Davis, M., & Whalen, P. J. (2001). The amygdala: Vigilance and emotion. *Molecular Psychiatry, 6*(1), 13–34.

Decety, J., & Jackson, P. L. (2004). The functional architecture of human empathy. *Behavioral and Cognitive Neuroscience Reviews, 3*(2), 71–100.

Decety, J., & Lamm, C. (2006). Human empathy through the lens of social neuroscience. *TheScientificWorldJOURNAL, 6,* 1146–1163.

de Waal, F. (2006). *Primates and philosophers: How morality evolved.* Princeton, NJ: Princeton University Press.

Dimitrov, M., Phipps, M., Zahn, T. P., & Grafman, J. (1999). A thoroughly modern Gage. *Neurocase, 5*(4), 345–354.

Edwards, W. (1954). The theory of decision making. *Psychological Bulletin, 51*(4), 380–417.

Eisenberg, N. (2000). Emotion, regulation, and moral development. *Annual Review of Psychology, 51*(1), 665–697.

Eisenberg, N., Fabes, R. A., Guthrie, I. K., & Reiser, M. (2000). Dispositional emotionality and regulation: Their role in predicting quality of social functioning. *Journal of Personality and Social Psychology, 78*(1), 136–157.

Eslinger, P. J., & Damasio, A. R. (1985). Severe disturbance of higher cognition after bilateral frontal lobe ablation. *Neurology, 35,* 1731–1741.

Farah, M. J. (2007). Social, legal, and ethical implications of cognitive neuroscience: "Neuroethics" for short. *Journal of Cognitive Neuroscience, 19*(3), 363–364.

Fellows, L. K. (2007). Advances in understanding ventromedial prefrontal function. *Neurology, 68*(13), 991–995.

Frith, C. D., & Frith, U. (1999). Interacting minds—a biological basis. *Science, 286*(5445), 1692–1695.

Gottfried, J. A., O'Doherty, J., & Dolan, R. J. (2003). Encoding predictive reward value in human amygdala and orbitofrontal cortex. *Science, 301*(5636), 1104–1107.

Greene, J. D. (2007). Why are VMPFC patients more utilitarian?: A dual-process theory of moral judgment explains. *Trends in Cognitive Sciences, 11*(8), 322–323.

Greene, J. D. (2009). The cognitive neuroscience of moral judgment. In M. Gazzaniga (Ed.), *The cognitive neurosciences* (4th ed., pp. 987–1002). Cambridge, MA: MIT Press.

Greene, J. D., & Paxton, J. M. (2009). Patterns of neural activity associated with honest and dishonest moral decisions. *Proceedings of the National Academy of Sciences USA, 106*(30), 12506–12511.

Greene, J. D., Sommerville, R. B., Nystrom, L. E., Darley, J. M., & Cohen, J. D. (2001). An fMRI investigation of emotional engagement in moral judgment. *Science, 293*(5537), 2105–2108.

Haidt, J. (2001). The emotional dog and its rational tail: A social intuitionist approach to moral judgment. *Psychological Review, 108*(4), 814–834.

Han, S., & Northoff, G. (2008). Culture-sensitive neural substrates of human cognition: A transcultural neuroimaging approach. *Nature Reviews Neuroscience, 9*(8), 646–654.

Harlow, J. M. (1868). Recovery from the passage of an iron rod through the head. *Publications of the Massachusetts Medical Society, 2*, 327–347.

Hassin, R., & Trope, Y. (2000). Facing faces: Studies on the cognitive aspects of physiognomy. *Journal of Personality and Social Psychology, 78*(5), 837–852.

Hauser, M., Cushman, F., Young, L., Kang-Xing Jin, R., & Mikhail, J. (2007). A dissociation between moral judgments and justifications. *Mind and Language, 22*(1), 1–21.

Heekeren, H. R., Wartenburger, I., Schmidt, H., Schwintowski, H.-P., & Villringer, A. (2003). An fMRI study of simple ethical decision-making. *NeuroReport, 14*(9), 1215–1219.

Hoffman, M. L. (1987). The contribution of empathy to justice and moral judgment. In N. Eisenberg & J. Strayer (Eds.), *Empathy and its development* (pp. 47–80). New York: Cambridge University Press.

Hoffman, M. L. (2000). *Empathy and moral development: Implications for caring and justice.* New York: Cambridge University Press.

James, W. (1884). What is an emotion? *Mind, 9*(34), 188–205.

Jones, W., Carr, K., & Klin, A. (2008). Absence of preferential looking to the eyes of approaching adults predicts level of social disability in 2–year-olds with autism spectrum disorder. *Archives of General Psychiatry, 65*(8), 946–954.

Kable, J. W., & Glimcher, P. W. (2009). The neurobiology of decision: Consensus and controversy. *Neuron, 63*(6), 733–745.

Kemper, T. D. (1987). How many emotions are there?: Wedding the social and the autonomic components. *American Journal of Sociology, 93*(2), 263–289.

Kennerley, S. W., Dahmubed, A. F., Lara, A. H., & Wallis, J. D. (2009). Neurons in the frontal lobe encode the value of multiple decision variables. *Journal of Cognitive Neuroscience, 21*(6), 1162–1178.

Killen, M., & Smetana, J. G. (2006). *Handbook of moral development.* Mahwah, NJ: Erlbaum.

Killen, M., & Smetana, J. G. (2007). The biology of morality: Human development and moral neuroscience. *Human Development, 50*(5), 241–243.

Killen, M., & Smetana, J. G. (2008). Moral judgment and moral neuroscience: Intersections, definitions, and issues. *Child Development Perspectives, 2*(1), 1–6.

Kochanska, G., Murray, K., & Coy, K. C. (1997). Inhibitory control as a contributor to conscience in childhood: From toddler to early school age. *Child Development, 68*(2), 263–277.

Koenigs, M., Baskin-Sommers, A., Zeier, J., & Newman, J. P. (2010). Investigating the neural correlates of psychopathy: A critical review. *Molecular Psychiatry, 16*(8), 792–799.

Koenigs, M., & Tranel, D. (2007). Irrational economic decision-making after ventromedial prefrontal damage: Evidence from the Ultimatum Game. *Journal of Neuroscience, 27*(4), 951–956.

Koenigs, M., Young, L., Adolphs, R., Tranel, D., Cushman, F., Hauser, M., et al. (2007). Damage to the prefrontal cortex increases utilitarian moral judgements. *Nature*, *446*(7138), 908–911.

Kohlberg, L. (1969). Stage and sequence: The cognitive-developmental approach to socialization. In D. A. Goslin (Ed.), *Handbook of socialization theory and research* (pp. 347–480). Chicago: Rand McNally.

Krajbich, I., Adolphs, R., Tranel, D., Denburg, N. L., & Camerer, C. F. (2009). Economic games quantify diminished sense of guilt in patients with damage to the prefrontal cortex. *Journal of Neuroscience*, *29*(7), 2188–2192.

Kringelbach, M. L., & Rolls, E. T. (2004). The functional neuroanatomy of the human orbitofrontal cortex: Evidence from neuroimaging and neuropsychology. *Progress in Neurobiology*, *72*(5), 341–372.

Lezak, M. D., Howieson, D. B., Loring, D. W., & Tranel, D. (2012). *Neuropsychological assessment* (5th ed.). New York: Oxford University Press.

Lieberman, M. D. (2003). Reflexive and reflective judgment processes: A social cognitive neuroscience approach. In J. P. Forgas, K. D. Williams, & W. von Hippel (Eds.), *Social judgments: Implicit and explicit processes*. (pp. 44–67). New York: Cambridge University Press.

Mitchell, J. P., Banaji, M. R., & Macrae, C. N. (2005). The link between social cognition and self-referential thought in the medial prefrontal cortex. *Journal of Cognitive Neuroscience*, *17*(8), 1306–1315.

Moll, J., & de Oliveira-Souza, R. (2007). Response to Greene: Moral sentiments and reason: Friends or foes? *Trends in Cognitive Sciences*, *11*(8), 323–324.

Moll, J., de Oliveira-Souza, R., Bramati, I. E., & Grafman, J. (2002). Functional networks in emotional moral and nonmoral social judgments. *NeuroImage*, *16*(3), 696–703.

Moll, J., de Oliveira-Souza, R., Eslinger, P. J., Bramati, I. E., Mourão-Miranda, J., Andreiuolo, P. A., et al. (2002). The neural correlates of moral sensitivity: A functional magnetic resonance imaging investigation of basic and moral emotions. *Journal of Neuroscience*, *22*(7), 2730–2736.

Moll, J., de Oliveira-Souza, R., Moll, F. T., Ignácio, F. A., Bramati, I. E., Caparelli-Dáquer, E. M., et al. (2005). The moral affiliations of disgust: A functional MRI study. *Cognitive and Behavioral Neurology*, *18*(1), 68–78.

Moll, J., de Oliveira-Souza, R., & Zahn, R. (2008). The neural basis of moral cognition. *Annals of the New York Academy of Sciences*, *1124*(1), 161–180.

Moll, J., de Oliveira-Souza, R., Zahn, R., & Grafman, J. (2008). The cognitive neuroscience of moral emotions. In W. Sinnott-Armstrong (Ed.), *Moral psychology: Vol. 3. The neuroscience of morality: Emotion, brain disorders, and development* (pp. 1–17). Cambridge, MA: MIT Press.

Moll, J., Zahn, R., de Oliveira-Souza, R., Krueger, F., & Grafman, J. (2005). The neural basis of human moral cognition. *Nature Reviews Neuroscience*, *6*(10), 799–809.

Murphy, F., Nimmo-Smith, I., & Lawrence, A. (2003). Functional neuroanatomy of emotions: A meta-analysis. *Cognitive, Affective, and Behavioral Neuroscience*, *3*(3), 207–233.

Nachson, I. (1995). On the modularity of face recognition: The riddle of domain specificity. *Journal of Clinical and Experimental Neuropsychology*, *17*(2), 256–275.

Naqvi, N., Shiv, B., & Bechara, A. (2006). The role of emotion in decision making. *Current Directions in Psychological Science*, *15*(5), 260–264.

Nauta, W. J. H. (1971). The problem of the frontal lobe: A reinterpretation. *Journal of Psychiatric Research*, *8*(3–4), 167–187.

Northoff, G., & Bermpohl, F. (2004). Cortical midline structures and the self. *Trends in Cognitive Sciences*, *8*(3), 102–107.

Ochsner, K. N., Ray, R. D., Cooper, J. C., Robertson, E. R., Chopra, S., Gabrieli, J. D. E., et al. (2004). For better or for worse: Neural systems supporting the cognitive down- and up-regulation of negative emotion. *NeuroImage, 23*(2), 483–499.

Öngür, D., & Price, J. L. (2000). The organization of networks within the orbital and medial prefrontal cortex of rats, monkeys and humans. *Cerebral Cortex, 10*(3), 206–219.

Parrott, W. G. (Ed.). (2001). *Emotions in social psychology: Essential readings.* New York: Psychology Press.

Pizarro, D. (2000). Nothing more than feelings? The role of emotions in moral judgment. *Journal for the Theory of Social Behaviour, 30*(4), 355–375.

Prehn, K., Wartenburger, I., Mériau, K., Scheibe, C., Goodenough, O. R., Villringer, A., et al. (2008). Individual differences in moral judgment competence influence neural correlates of socio-normative judgments. *Social Cognitive and Affective Neuroscience, 3*(1), 33–46.

Rangel, A., Camerer, C., & Montague, P. R. (2008). A framework for studying the neurobiology of value-based decision making. *Nature Reviews Neuroscience, 9*(7), 545–556.

Ratiu, P., Talos, I. F., Haker, S., Lieberman, D., & Everett, P. (2004). The tale of Phineas Gage, digitally remastered. *Journal of Neurotrauma, 21*(5), 637–643.

Rauch, S. L., Shin, L. M., & Wright, C. I. (2006). Neuroimaging studies of amygdala function in anxiety disorders. *Annals of the New York Academy of Sciences, 985*(1), 389–410.

Rolls, E. T. (2000). The orbitofrontal cortex and reward. *Cerebral Cortex, 10*(3), 284–294.

Rolls, E. T. (2004). The functions of the orbitofrontal cortex. *Brain and Cognition, 55*(1), 11–29.

Rolls, E. T., Hornak, J., Wade, D., & McGrath, J. (1994). Emotion-related learning in patients with social and emotional changes associated with frontal lobe damage. *Journal of Neurology, Neurosurgery and Psychiatry, 57*(12), 1518–1524.

Rudrauf, D., Lachaux, J.-P., Damasio, A. R., Baillet, S., Hugueville, L., Martinerie, J., et al. (2009). Enter feelings: Somatosensory responses following early stages of visual induction of emotion. *International Journal of Psychophysiology, 72*(1), 13–23.

Sanfey, A. G., Rilling, J. K., Aronson, J. A., Nystrom, L. E., & Cohen, J. D. (2003). The neural basis of economic decision-making in the Ultimatum Game. *Science, 300*(5626), 1755–1758.

Saver, J. L., & Damasio, A. R. (1991). Preserved access and processing of social knowledge in a patient with acquired sociopathy due to ventromedial frontal damage. *Neuropsychologia, 29*(12), 1241–1249.

Saxe, R., & Powell, L. R. (2006). It's the thought that counts: Specific brain regions for one component of theory of mind. *Psychological Science, 17*(8), 692–699.

Schaich Borg, J., Hynes, C., Van Horn, J., Grafton, S., & Sinnott-Armstrong, W. (2006). Consequences, action, and intention as factors in moral judgments: An fMRI investigation. *Journal of Cognitive Neuroscience, 18*(5), 803–817.

Schaich Borg, J., Lieberman, D., & Kiehl, K. A. (2008). Infection, incest, and iniquity: Investigating the neural correlates of disgust and morality. *Journal of Cognitive Neuroscience, 20*(9), 1529–1546.

Schnall, S., Haidt, J., Clore, G. L., & Jordan, A. H. (2008). Disgust as embodied moral judgment. *Personality and Social Psychology Bulletin, 34*(8), 1096–1109.

Shamay-Tsoory, S. G., & Aharon-Peretz, J. (2007). Dissociable prefrontal networks for cognitive and affective theory of mind: A lesion study. *Neuropsychologia, 45*(13), 3054–3067.

Shamay-Tsoory, S. G., Aharon-Peretz, J., & Perry, D. (2009). Two systems for empathy: A double dissociation between emotional and cognitive empathy in inferior frontal gyrus versus ventromedial prefrontal lesions. *Brain, 132*(3), 617–627.

Shamay-Tsoory, S. G., Tomer, R., Berger, B. D., Goldsher, D., & Aharon-Peretz, J. (2005). Impaired "affective theory of mind" is associated with right ventromedial prefrontal damage. *Cognitive and Behavioral Neurology, 18*(1), 55–67.

Silk, J. B., Alberts, S. C., & Altmann, J. (2003). Social bonds of female baboons enhance infant survival. *Science, 302*(5648), 1231–1234.

Singer, T., & Lamm, C. (2009). The social neuroscience of empathy. *Annals of the New York Academy of Sciences, 1156*(1), 81–96.

Singer, T., Seymour, B., O'Doherty, J., Kaube, H., Dolan, R. J., & Frith, C. D. (2004). Empathy for pain involves the affective but not sensory components of pain. *Science, 303*(5661), 1157–1162.

Smith, M. L., Cottrell, G. W., Gosselin, F., & Schyns, P. (2005). Transmitting and decoding facial expressions. *Psychological Science, 16*(3), 184–189.

Sokol, B. W., Chandler, M. J., & Jones, C. (2004). From mechanical to autonomous agency: The relationship between children's moral judgments and their developing theories of mind. *New Directions for Child and Adolescent Development, 2004*(103), 19–36.

Solomon, R. C. (2002). Back to basics: On the very idea of "basic emotions." *Journal for the Theory of Social Behaviour, 32*(2), 115–144.

Spence, S. A., Farrow, T. F. D., Herford, A. E., Wilkinson, I. D., Zheng, Y., & Woodruff, P. W. R. (2001). Behavioural and functional anatomical correlates of deception in humans. *NeuroReport, 12*(13), 2849–2853.

Spence, S. A., Hunter, M. D., Farrow, T. F. D., Green, R. D., Leung, D. H., Hughes, C. J., et al. (2004). A cognitive neurobiological account of deception: Evidence from functional neuroimaging. *Philosophical Transactions of the Royal Society of London: Series B. Biological Sciences, 359*(1451), 1755–1762.

Spezio, M. L., Adolphs, R., Hurley, R. S. E., & Piven, J. (2006). Abnormal use of facial information in high-functioning autism. *Journal of Autism and Developmental Disorders, 37*(5), 929–939.

Spezio, M. L., Huang, P.-Y. S., Castelli, F., & Adolphs, R. (2007). Amygdala damage impairs eye contact during conversations with real people. *Journal of Neuroscience, 27*(15), 3994–3997.

Stuss, D. T., & Benson, D. F. (1984). Neuropsychological studies of the frontal lobes. *Psychological Bulletin, 95*(1), 3–28.

Tangney, J. P., Stuewig, J., & Mashek, D. J. (2007). Moral emotions and moral behavior. *Annual Review of Psychology, 58*(1), 345–372.

Thomas, B. C., Croft, K. E., & Tranel, D. (2011). Harming kin to save strangers: Further evidence for abnormally utilitarian moral judgments after ventromedial prefrontal damage. *Journal of Cognitive Neuroscience 23*(9), 2186–2196.

Thomson, J. J. (1985). The trolley problem. *Yale Law Journal, 94*, 1395–1415.

Todorov, A., Baron, S. G., & Oosterhof, N. N. (2008). Evaluating face trustworthiness: A model based approach. *Social Cognitive and Affective Neuroscience, 3*(2), 119–127.

Tranel, D., Hathaway-Nepple, J., & Anderson, S. W. (2007). Impaired behavior on real-world tasks following damage to the ventromedial prefrontal cortex. *Journal of Clinical and Experimental Neuropsychology, 29*(3), 319–332.

Turiel, E. (2009). The relvance of moral epistemology and psychology for neuroscience. In P. D. Zelazo, M. J. Chandler, & E. Crone (Eds.), *Developmental social cognitive neuroscience* (pp. 313–332). New York: Psychology Press.

Valdesolo, P., & DeSteno, D. (2006). Manipulations of emotional context shape moral judgment. *Psychological Science, 17*(6), 476–477.

Wallis, J. D., & Kennerley, S. W. (2010). Heterogeneous reward signals in prefrontal cortex. *Current Opinion in Neurobiology, 20*(2), 191–198.

Wheatley, T., & Haidt, J. (2005). Hypnotic disgust makes moral judgments more severe. *Psychological Science, 16*(10), 780–784.

Willis, J., & Todorov, A. (2006). First impressions: Making up your mind after a 100-ms exposure to a face. *Psychological Science, 17*(7), 592–598.

Wundt, W. M. (1902). *Outlines of psychology.* London: Engelmann.

Young, L., Bechara, A., Tranel, D., Damasio, H., Hauser, M., & Damasio, A. R. (2010). Damage to ventromedial prefrontal cortex impairs judgment of harmful intent. *Neuron, 65*(6), 845–851.

Young, L., Camprodon, J. A., Hauser, M., Pascual-Leone, A., & Saxe, R. (2010). Disruption of the right temporoparietal junction with transcranial magnetic stimulation reduces the role of beliefs in moral judgments. *Proceedings of the National Academy of Sciences USA, 107*(15), 6753–6758.

Young, L., Cushman, F., Hauser, M., & Saxe, R. (2007). The neural basis of the interaction between theory of mind and moral judgment. *Proceedings of the National Academy of Sciences USA, 104*(20), 8235–8240.

Young, L., & Koenigs, M. (2007). Investigating emotion in moral cognition: A review of evidence from functional neuroimaging and neuropsychology. *British Medical Bulletin, 84*(1), 69–79.

Young, L., & Saxe, R. (2011). An fMRI investigation of spontaneous mental state inference for moral judgment. *Journal of Cognitive Neuroscience, 21*(7), 1396–1405.

Zajonc, R. B. (1980). Feeling and thinking: Preferences need no inferences. *American Psychologist, 35*(2), 151–175.

CHAPTER 5

Environmental Contributions to the Development of Social Competence

Focus on Parents

Amy E. Root, Paul D. Hastings, and Kari L. Maxwell

Researchers of child and adolescent development have concentrated on the role of parents in the process of *socialization*, or the "processes whereby naive individuals are taught the skills, behavior patterns, values, and motivations needed for competent functioning in the culture in which the child is growing up" (Maccoby, 2007, p. 13). Agents of socialization include nonparental figures, including siblings, peers, teachers, and policy makers (Blunt Bugental & Grusec, 2006), but it has been argued that parents are primary in the socialization process, especially in the early years of life (Grusec & Davidov, 2007). Targets of socialization include academic achievement, ethnic identity, cognition, gender, and social competence (see Grusec & Hastings, 2007, for a recent volume on socialization processes). The socialization of social competence is thought to be particularly important for promoting adaptive functioning throughout the life course (Maccoby, 2007).

The *socialization of social competence* refers to the processes whereby children learn and internalize the rules and mores for competent social behavior, as determined by the larger culture. Parents have historically been assigned the primary responsibility for transferring the morals, emotions, and behaviors that constitute social competence to children in the early years of life. The socialization of social competence involves a variety of parenting processes, including the quality of the relationship between parent and child, general parenting styles, and specific parenting behaviors (for reviews, see Denham, Bassett, & Wyatt, 2007; Hastings, Utendale, & Sullivan, 2007). Thus, in conjunction with the lessons learned from other socialization agents (e.g., teachers, peers) and children's own active involvement in interpreting and implementing their social lessons, many components of ongoing parent–child interactions contribute to the development of social competence.

Social competence in childhood encompasses a range of behaviors that support adaptive and successful functioning during interactions with other people. Socially competent children are able to attain their own social goals during interactions with others, while also being aware of the goals held by their social partners and not overriding or interfering with those goals (Howes & James, 2002; Rubin & Rose-Krasnor, 1992). Thus social competence reflects the maintenance of positive and mutually satisfying engagement in reciprocal social exchanges and relationships. Acting with social competence promotes the aims, satisfaction, and well-being of oneself and one's social partners, and builds affection, trust, and stability within social relationships.

Children's social competence is evident in a variety of behaviors (Caldarella & Merrell, 1997; Wentzel & Looney, 2007). These include emotional and behavioral self-regulation; interpersonal perspective taking; being assertive without aggression; helpfulness and compassion (prosocial behavior); and social skills like effective group entry, communication, cooperativeness, social problem solving, and conflict resolution. These and other behaviors allow a child to initiate and maintain satisfying social interactions, and to improve or disengage from unsatisfying ones. Although these might seem like a challenging set of skills that children would rarely master, social competence is actually the norm, and most children can be characterized as socially competent (Masten et al., 1999; Rubin, Bukowski, & Parker, 2006). Some children are quite skilled with their peers and other social partners, whereas other children show mild or marked social deficits. One of the primary goals of developmental scientists is identifying the sources of these individual differences in social competence.

The suggestion that parents influence children's development of social competence represents a belief in the influence of socialization on development. Socialization represents lasting impacts of external agents or forces on a child's beliefs, behaviors, or functioning (Grusec & Hastings, 2007). A major premise of socialization theory is that the environment influences a child's development, yet children are not the passive recipients of these socialization influences. *Internalization* is the process whereby a child takes in the explicit or implicit messages being delivered by socialization, evaluates and possibly reinterprets them, and incorporates them into his or her developing sense of self (Grusec & Goodnow, 1994). Socialization is also a bidirectional process (Hastings & Rubin, 1999): Children act upon the socializing agents in their social environments, and change the agents' beliefs and behaviors through the children's own actions and development. However, our focus in the present chapter is upon the actions and influences of parents, as socializing agents, upon children.

In this chapter, we examine the theoretical and empirical work linking parenting processes to children's social competence. We begin by offering a brief overview of the current theoretical perspectives that guide our own work, as well as the work of others in this field; each of these theories directly underscores the role of parents in the development of social competence. We then move to highlight empirical work that underscores the important role parents play in the development of their children's social competence. Next we highlight what we feel are current trends in the study of parental socialization of social competence. We close with suggestions for integrating socialization research into the study of social neuroscience.

Parental Socialization of Social Competence: Theoretical Foundations

Attachment Theory

The understanding of parental socialization of children's social competence is rooted in the theories that emphasize the role of caregivers in children's development. Attachment theory posits that parents who are warm, sensitive, and responsive to their infants' and young children's needs will have children who develop secure attachment relationships in the first years of life and go on to form competent social relationships with others throughout the life course. This is because children who develop secure attachment relationships with their primary caregivers are thought to form these cognitive schemas: (1) They have a safe base from which they can explore the world; (2) their relationship partners are trustworthy and good people; and (3) they themselves are worthy of receiving love and positive attention from others (Bowlby, 1969). Thus, according to attachment theory, children with warm, responsive parents will view themselves as competent and worthy of positive social relationships, and they will view the world as a safe place; these are necessary "ingredients" for the development of social competence. However, when parents are inconsistent and/or unresponsive to their children's needs, attachment theory states that these children are likely to view themselves in a negative light (e.g., as unworthy and incompetent), and to view the world as an unpredictable, dangerous place; these views place them at risk for the development of social difficulties, including socially incompetent behavior.

Cognitive Social Learning Theory

Although rooted in classic behaviorism and the shaping influences of punishment or reward on behavior, Bandura's cognitive social learning theory extends the classic behaviorist view by proposing that direct reinforcement or punishment is not needed to learn how to behave in the social world (Bandura, 1977, 1997, 2006). Specifically, children model behaviors they witness in the environment, although not in an automatic way. Rather, children selectively attend to potential models. They are most likely to imitate the behaviors of those with whom they are close and have established relationships, and also the behaviors of others who have been rewarded or successful. If the behaviors are retained and rehearsed, and if the children are motivated to imitate the actors, then the behaviors are likely to be reproduced. Indeed, children's first models for behavior are family members, including parents.

Children who are successful in their actions, receiving praise or other positive responses from their social partners, are likely to develop greater confidence in their abilities (or *self-efficacy*) and to continue these learned patterns of behavior. The implications for the influences of parents on their children's development of social competence can be grasped immediately. Children are motivated to attend to their parents, and have many opportunities to see their parents' more or less effective social behaviors in the spousal, parent–child, and parent–peer contexts. Children then selectively emulate those behaviors in their own interactions with others, and, based on the feedback they receive, internalize a view of themselves as more or less socially competent individuals.

Development Systems Theory

More recent understandings of the role of parents in the development of children's social competence often adopt a systems approach. Systems theories of development examine multiple influences on development (including genes, biology, psychology, and context) and stress that the developing individual is active in his or her growth, rather than passive. Furthermore, the systems that surround a developing individual are also active or dynamic, and include parents and the parent–child relationships. From a systems perspective, a parent and child are constantly changing and being changed by one another. Lerner, Rothbaum, Boulos, and Castellino (2002) quoted Bronfenbrenner (1979) in proposing a developmental contextual view of parenting in which the individuals (parent and child), as well as the relationship, are "part of a larger, enmeshed system of fused relations among multiple levels that compose the ecology of human life" (Lerner et al., 2002, p. 318). This "ecology" encompasses more than just parent and child; it also involves siblings, peers, teachers, workplace environments, community, and culture. All of these have an impact on the nature of the relationship between parent and child, as well as on how parents influence the development of their children. These developmental models of behavior, including social competence, assert that the first place where children learn about the world around them is in the home (Bronfenbrenner & Morris, 2006). Furthermore, parents not only have a direct impact on children's development via their interactions with their children, but they also modulate what external influences (e.g., peers, community) the children are exposed to, especially in the early years of life. Thus the parents' role in the development of social competence is primary, via their own interactions with their children, the types of behaviors they model for their children, and the type of environments outside the home to which they expose their children.

These theories and others (e.g., Bronfenbrenner's bioecological model; Bronfenbrenner & Morris, 2006) form a broad basis on which current socialization research rests. However, it would be accurate to say that for the past few decades, the majority of research on family types, parental socialization, and children's social competence has been conducted without explicitly or exclusively drawing from any specific theory. Rather, a more integrative and multifaceted perspective seems to guide most empirical research. A newer omnibus theory, or "macroparadigm," that reflects this eclectic approach to research is *developmental psychopathology* (Luthar, Burack, & Cicchetti, 1997). Rather than adhering to the tenets of a single theoretical perspective, developmental psychopathology acknowledges that all theories are likely to contribute important ideas, and that drawing from each will give researchers the greatest opportunities for understanding development. This perspective also denies that psychology, or any other field of study, has a unique disciplinary advantage. Child development will be best understood by combining the strengths of work in psychology, education, sociology, anthropology, psychiatry, neuroscience, and other fields.

In the next section, we discuss the literature that supports the role of parents in the development of social competence, including studies that have examined general parenting practices and the quality of the parent–child relationship as correlates and predictors of children's social competence in three developmental periods: early childhood, middle childhood, and adolescence. We provide evidence in this review of parents' contribution to the development of socially competent behavior (as defined above), as well as studies empirically linking parenting to socially incompetent behaviors (e.g., aggressive behavior, anxious behavior). We feel it is important, however, to stress that children who display low levels of particular socially incompetent behaviors are not necessarily highly socially competent; for instance, low levels of aggression could be indicative of high levels of other socially incompetent behaviors (e.g., anxiety), rather than of prosocial or altruistic behavior.

In addition, and in keeping with a systems perspective of human development, we ask the reader to bear in mind that children make significant contributions to their own socialization experiences. We note where children's own dispositional traits (temperament, gender) appear to moderate parents' socialization of social competence and where children's abilities or characteristics (e.g., social cognition, self-regulation) appear to mediate parental influences. This notion of child effects is not novel, and an extensive literature demonstrates the impact of children's characteristics on their caregiving experiences (Bell, 1971; Bell & Chapman, 1986), as well as the moderating role of children's characteristics on the relation between parenting and the development of socially competent behaviors (e.g., Kochanska, 1995, 1997). For instance, in our own work, we have found that children's physiological regulation (Hastings & De, 2008; Kennedy, Rubin, Hastings, & Maisel, 2004), temperament (Hastings & Rubin, 1999; Kennedy Root & Stifter, 2010; Rubin, Burgess, & Hastings, 2002; Rubin, Hastings, Stewart, Henderson, & Chen, 1997), and gender (Kennedy Root & Rubin, 2010) contribute to parenting attitudes and beliefs about child rearing, and alter how parental socialization is associated with children's social competence.

Thus it is important to note that research and interest in socialization both implicitly and explicitly incorporate the roles of genetic and biological factors in development. Indeed, humans are biological beings; there are no aspects of functioning and development that can occur without biology. Yet it is equally true that there can be no biological activity without environment. Humans are social creatures, as well as biological, and social environments are the contexts within which development occurs. Experiences in social environments interact with children's genetic predispositions and biological functions to generate individual developmental trajectories. Understanding when and how the social environment makes its contributions to development is the essence of studying socialization.

Parental Socialization of Social Competence: Empirical Evidence

Overview

One of the most influential programs of research on general parenting styles was initiated by Diana Baumrind (1967, 1971) and refined by other socialization researchers (e.g., Darling & Steinberg, 1993; Maccoby & Martin, 1983). Aspects of child rearing are seen along two dimensions of behavior, *control* and *responsiveness*; when crossed, these generate a typology of four parenting styles. The two styles representing high control are most often studied in relation to the development of social competence. *Authoritative* parents are high in both control and responsiveness. They have rules and guidelines for their children, set limits, and expect compliance, but this is balanced by providing explanations for rules, being sensitive to children's needs and wants, a willingness to listen to children's perspectives, and flexibility in the application of control depending on situational factors. *Authoritarian* parents are high in control but low in responsiveness. They apply rules and set limits strictly and without explanations or flexibility, engage in more punitive control and discipline, and do not take their children's perspectives into account when making decisions. When these two typologies of parenting have been examined in relation to the development of children's social competence, it has consistently been reported that children from homes where parents endorse or engage in parenting styles that are indicative of authoritative parenting are more socially competent, and are generally socially and emotionally better adjusted, in the preschool (Baumrind, 1971), elementary school (Domitrovich & Bierman, 2001), and adolescent (Lamborn, Mounts, Steinberg & Dornbusch, 1991) years.

In addition, there is a considerable amount of work linking the quality of the caregiver–child relationship to the development of social competence from early childhood through adolescence. As noted above, attachment theory places primary importance on the sensitivity of parents to the signals and needs of their infants. Parents who respond contingently and appropriately to their infants' needs, and who provide a supporting, warm, and safe context for early development, will facilitate their children's optimal emotional and social functioning. Their children will internalize mental models of "secure" attachment relationships. Conversely, because of their experiences with less sensitive, supportive, and warm parents, children with mental models of "insecure" attachment relationships lack these adaptive belief systems and are at risk of experiencing social difficulties. Indeed, the quality of a child's attachment to his or her primary caregivers has links to social competence in the early childhood (e.g., Fagot, 1997), middle childhood (e.g., Booth-LaForce et al., 2006), and adolescent years (e.g., Laible, 2007).

The Toddler and Preschool Period

Studies have shown that preschool-age children are more socially competent when they have secure attachment relationships with their mothers. Fagot (1997) observed peer play and turn taking between toddlers who had had the quality of their attachment relationships measured 6 months previously. When toddlers with secure attachments made positive and friendly initiations toward their peers, they were more likely to receive positive responses and less likely to receive negative responses, compared to toddlers with insecure attachments. These findings have been supported in longitudinal examinations. Belsky and Fearon (2002) found that 3-year-old children were more socially competent when they had secure attachment relationships in infancy, and their mothers had been sensitive and unobtrusive at 24 months. More recent evidence has extended these findings into the early childhood years. The National Institute of Child Health and Human Development (NICHD) Early Child Care Research Network (2006) reported that attachment security at 15 months was related to mothers' ratings of children's social skills and teachers' ratings of internalizing and externalizing behaviors in the first grade, whereas insecurely attached children were less socially competent and had more internalizing and externalizing difficulties than their securely attached counterparts.

Rather than examining attachment security specifically, some other researchers have demonstrated that the parenting behaviors of sensitivity, responsiveness, and warmth are themselves associated with their children's development of social competence. Parental sensitivity, warmth, and involvement with infants and toddlers have been found to predict children's subsequent empathic and prosocial responses toward adults (Kiang, Moreno, &

Robinson, 2004; Kochanska, 1997). Moreover, repeated observations of mothers' sensitivity, warmth, and involvement with their children from 6 to 36 months predicted both mothers' and teachers' reports of the children's social competence at 4.5 years (NICHD Early Child Care Research Network, 2002).

Parental engagement, warmth, and informal instruction in the form of play also seem to be important components of the socialization of social competence for preschoolers (Lindsey & Mize, 2001a), particularly for fathers. During play, parents can interact with their children much as peers can, with matching of affective states (Robinson, Little, & Biringen, 1993), turn taking (Fiese, 1990), synchronous exchanges (Mize & Pettit, 1997), joint determination of the content and direction of play (Black & Logan, 1995), and mutual compliance (Lindsey, Mize, & Pettit, 1997). Children mimic their parents' styles of play during their own interactions with peers (Lindsey & Mize, 2001b), and parents who are more actively engaged in peer-like play have preschoolers with better social skills and greater social competence (Barth & Parke, 1993; Carson, Burks, & Parke, 1993; Dunn & Brown, 1994; Lindsey et al., 1997; Lindsey & Mize, 2001b; Russell, Mize, & Bissaker, 2002; Russell & Saebel, 1997). Fathers who become more actively involved in play and maintain play engagement for longer periods, have children who are described as more socially skilled by teachers, and who are more popular and less rejected by peers (Carson et al., 1993; Pettit, Glyn Brown, Mize, & Lindsey, 1998). Thus father–child and mother–child play both appear to be important opportunities for children to practice and refine their abilities to get along well with peers.

In addition to directly shaping children's social competence, parenting behaviors support children's development of other skills and abilities that underlie their social competence. For example, Spinrad and colleagues (2007) examined the role of toddlers' development of *effortful control*, or voluntary self-control of their emotions and behaviors, in the association between having a positive, supportive maternal relationship at 18 months and demonstrating social competence at 30 months of age. Toddlers' effortful control was found to mediate this association: More supportive mothers had toddlers who evidenced greater effortful control, and in turn, toddlers with greater effortful control were seen as more socially competent and less aggressive, disruptive, and inattentive by their teachers and parents. In addition, Hastings, Nuselovici, et al. (2008) reported that children's parasympathetic regulation mediated the relation between maternal negative control and externalizing difficulties and self-regulation during the preschool years. Specifically, mothers who used more negative control—including criticism, displays of anger, and punishment—had preschoolers who evidenced less parasympathetic regulation of cardiac arousal during play with peers. In turn, children with lower parasympathetic regulation had more externalizing difficulties and poorer behavioral self-regulation. In sum, these studies indicate that parents play an important role in priming children for the development of social competence via indirect pathways, specifically self-regulatory capabilities, as well as in directly shaping children's development.

Maternal parenting also influences the continuity or discontinuity of temperamental dispositions known to be precursors for socially incompetent behavior. For instance, Rubin, Burgess, Dwyer, and Hastings (2003) found that children who had "difficult" temperaments (low behavioral self-control, high anger proneness) at 2 years only evidenced higher rates of externalizing behaviors at 4 years if they also had mothers who were highly negative and controlling at 2 years. In the same sample of children, Rubin et al. (1997) found that toddlers who were more temperamentally fearful only exhibited shy or reticent behaviors with a same-age peer if they also had mothers who exhibited intrusive, overprotective behaviors. When these children were seen again 2 years later, Rubin et al. (2002) again found that socially inhibited (or shy and reticent) toddlers only continued to display reticent behaviors with unfamiliar peers at 4 years if their mothers had been either overprotective or critical and derisive. Together, these studies show how parenting that deviates from the positive forms of sensitive and supportive behavior described earlier can foster or strengthen behavioral tendencies that would undermine social competence in temperamentally vulnerable children.

Similarly, and in accord with a goodness-of-fit perspective (Thomas & Chess, 1977), the social competence of children with various temperamental vulnerabilities may be bolstered by different parenting approaches. In a series of studies, Kochanska (1995, 1997) examined the development of conscience in children with varying degrees of fearfulness. She found that mothers' gentle discipline (reasoning, low power assertion) was related to prosocial, moral behavior in fearful children during the preschool years. However, for children who were relatively fearless, maternal responsiveness and security of attachment appeared to promote the development of conscience (Kochanska, 1995; 1997; Kochanska, Aksan, & Joy, 2007; see also Dennis, 2006, for similar associations). Thus children with differing dispositional needs may benefit from distinct components of parenting behavior that are typically encompassed within an overall authoritative style.

The Elementary School Years

Socialization researchers have tended to focus their attention on young children and adolescents, and the middle childhood period is relatively understudied. However, there is some evidence linking parenting factors to school-age children's social competence, typically supporting the associations documented in the literature on early childhood. For instance, studies examining the development of prosocial behavior during the elementary school years also provide evidence for parents' contribution to the development of social competence. Janssens and Dekovic (1997) looked at the links between 6- to 10-year-old children's helpful/cooperative behaviors at school and their parents' socialization practices. Both mothers' and fathers' supportive, authoritative parenting were correlated with teachers' and peers' descriptions of children as helpful. Similarly, in a study of 10-year-olds and their parents, parental warmth, responsiveness, and use of reasoning during discussions predicted children's social competence at school, according to teacher and peer reports (McDowell, Kim, O'Neil, & Parke, 2002). Laible, Carlo, Torquati, and Ontai (2004) found that mothers who were warmer had 6-year-old children who generated more prosocial solutions to hypothetical conflicts, and who were described by their mothers as more socially competent, whereas more harshly punitive mothers had children who generated fewer prosocial and more aggressive solutions. Chao and Willms (2002) reported similar findings with older school-age children: Firm, rational, and responsive parents had more prosocial 11-year-old children. Thus there is ample evidence that positive parenting practices are concurrently associated with school-age children's prosocial and socially competent behaviors.

Furthermore, there is longitudinal evidence indicating that earlier parenting practices or quality of parent–child relationships contributes to the continuity of socially competent behavior from early childhood to the elementary school years. For example, mothers who were more authoritative and less authoritarian with preschoolers had children who displayed more sympathetic, compassionate, and helpful behaviors 2 years later, as reflected in the children's observed behaviors, self-descriptions, and characterizations by mothers and teachers (Hastings, Zahn-Waxler, Robinson, Usher, & Bridges, 2000). Similarly, following the development of 600 families, the NICHD Early Child Care Network (2004) reported that mothers and fathers who were more sensitive and supportive of their preschoolers' autonomy had children who were characterized as more cooperative, self-regulated, and appropriately assertive by their early elementary school teachers.

A parallel body of literature shows that less positive and appropriate parental behaviors also contribute to school-age children's development of socially incompetent behaviors. Park and colleagues (2005) reported that mothers who displayed more negativity, involving high levels of criticism, disapproval, instrusiveness, and displeasure, had school-age children who were more physically and relationally aggressive. Similarly, the use of an authoritarian parenting style has been associated with relational and overt aggression (Sandstrom, 2007). In a study examining the style with which parents gave advice to their 9-year-old children about handling difficult situations, parents who were more controlling (telling the children what to do without explaining why) had children who showed more negative social behaviors at school, according to teachers and peers; furthermore, fathers' control and lack of warmth when giving advice predicted more negative and less positive social behavior 1 year later (McDowell, Parke, & Wang, 2003). Finally, Booth-LaForce and Oxford (2008) reported that mothers who were more hostile and unsupportive of their preschoolers had children who displayed more social withdrawal, were less popular, and were excluded from peer activities across the elementary school years, from grades 1 through 6.

As in studies on the socialization of social competence in the preschool period, some researchers have used the attachment perspective to examine how the quality of the parent–child relationship is associated with the development of socially competent behaviors in the elementary school years. Laible and colleagues (2004) reported that school-age children who reported a positive family relationship quality were rated by teachers and parents as socially competent. Demonstrating the converse, Al-Yagon (2008) reported that school-age boys who were insecurely attached to their mothers were rated as highly aggressive by their teachers.

As has been demonstrated within the preschool period, there is also evidence that parenting behaviors and relationship quality can moderate the relation between children's earlier characteristics and their later social competence. In a Swedish sample, Bohlin, Hagekull, and Andersson (2005) showed that children who were temperamentally inhibited during infancy were at greater risk for poor social competence at 8 years only if they also had insecure attachment relationships with their primary caregivers in infancy; however, inhibited infants with secure attachments were more socially competent at school age. Analogously, it was only if preschoolers had mothers who demonstrated negative control and little positive affect (Hane, Fox, Henderson, & Marshall, 2008), or who were intrusive and overprotective (Degnan, Henderson, Fox, & Rubin, 2008), that shy and wary behaviors at 4 years were found to predict more withdrawn and less socially engaged behavior at 7 years.

These studies clearly show that many of the previously identified features of parental socialization, including warmth, involvement, sensitivity, and authoritative parenting, continue to support children's social competence in the elementary school years, whereas overly controlling, hostile, and critical parenting undermine social competence.

In addition, teaching through providing explanations and nondirective advice appears to emerge as salient. This may be due to children's increasing cognitive capacities and ability to understand and apply parents' lessons in their interactions with others. These latter strategies are likely to be reflections of parental support for children's autonomy, which may be of increasing importance because of children's ever-increasing amounts of time spent away from their parents during the elementary school years.

The Adolescent Years

Many studies of socialization in the adolescent years have examined broad parenting styles, and these studies continue to mirror the associations between parenting and social competence reported in samples of younger children. Indeed, these studies consistently show that youth with more authoritative parents show greater social competence, prosocial behavior, self-esteem, and resistance to peer pressure than youth with authoritarian, permissive, or neglectful parents do (Baumrind, 1991; Lamborn et al., 1991). Longitudinal studies show that authoritative parenting fosters adolescents' subsequent social competence, autonomy, and positive orientation toward work, which in turn improves their academic performance (Steinberg, Elmen, & Mounts, 1989; Steinberg, Lamborn, Darling, Mounts, & Dornbusch, 1994). Conversely, other parenting styles, and particularly neglectful parenting, appear to undermine adolescents' well-being.

During adolescence, the parent–child relationship undergoes important changes related to the particular developmental tasks of this age period, most notably the need for youth to achieve autonomy and explore their identity while maintaining a connection with the family (Laursen & Collins, 2009). Supporting this developmental task requires parents to modify their parenting goals and behaviors to be more age-appropriate for their adolescent children. This change in the parent–child relationship can be challenging for parents and adolescents alike. Specifically, parents need to be willing and able to shift the balance of power in the relationship to allow increasing autonomy, while fostering the maintenance of appropriate connectedness with their children (Allen, Hauser, Bell, & O'Connor, 1994). Research has shown that adolescents appear to be better adjusted when they maintain a strong emotional bond with their parents and families, while also gaining new freedoms to explore their individuality (McElhaney & Allen, 2001). Adolescents who do not strike an effective balance between the appropriate amount of autonomy and connectedness experience social and emotional difficulties (Allen, Hauser, Eickholt, Bell, & O'Connor, 1994).

Indeed, the empirical evidence supports the notion that parenting with balanced levels of involvement and connectedness bolsters adolescent social competence. Adolescents who are more socially skilled and have better friendships describe their parents as warm, supportive, and flexible, and as both encouraging their children's autonomy while still monitoring their children's activities, without being enmeshed or intrusive (Engels, Dekovic, & Meeus, 2002). In a study that tracked the development of adolescents' self-confidence from 13 to 15 years, the teens' initial descriptions of their parents as highly critical, emotionally manipulative, and controlling were found to predict lower levels of self-confidence, especially for boys (Conger, Conger, & Scaramella, 1997). Similarly, when young adolescents perceived their parents as more strict and offering them little opportunity to participate in decision making in the home, they were more likely to turn to their peers for advice and to blindly follow peers' directions, even to their own detriment (Fuligni & Eccles, 1993).

Parental monitoring of adolescents' activities and social relationships also appears to be a critically important aspect of effective socialization during this developmental period (Dishion & McMahon, 1998; Patterson & Stouthamer-Loeber, 1984). Allowing autonomy and involvement in decision making does not mean that parents should withdraw from their adolescents' lives. Several studies have shown that adolescents are more likely to get involved with deviant peer groups and engage in socially incompetent behaviors when their parents do not keep track of where and with whom their children are, whereas parents who effectively supervise or monitor have adolescent children who engage in less deviant, risky behaviors (Ary et al., 1999; Cottrell et al., 2003; Stattin & Kerr, 2000). Ary and colleagues (1999) reported that parental monitoring was negatively predictive of adolescent risky behavior (e.g., risky sexual activity) and problem behavior (e.g., academic difficulties, antisocial behavior) over a 6-month period. Indeed, adolescents often need assistance from their parents in order to manage their peer relationships competently. Parents who talk with their adolescent children about their friendships, mediate when there are social difficulties, and grant their children some autonomy have children who report more positive and less conflicted friendships, as well as less engagement in delinquency and drug use (Mounts, 2004).

Examinations of the quality of parent–adolescent relationships also reveal associations with adolescents' social competence. Adolescents who have secure attachments to their mothers (reflecting affection and trust) also have secure attachments to their friends, which are associated with higher-quality friendships (Markiewicz, Doyle, & Brendgen, 2001). Secure attachments to parents also seem to support adolescents' abilities to form satisfying and appropriate romantic relationships (Furman & Wehner, 1994). Adolescents who maintain positively connected relationships with their parents, based in cohesion, supportiveness, and mutual reciprocity, report that they are more capable of making autonomous decisions (McElhaney & Allen, 2001). In addition, Laible (2007) found that youth with greater attachment security manifested more positive emotionality and empathic concern for others. In turn, youth who were more positive and empathic engaged in more prosocial behaviors and less aggression. Attachment to fathers has a similar impact on the development of aggression in adolescence (Gomez & McLaren, 2007). Thus adolescents' autonomy, confidence, and social competence appears to be supported by maintaining positive relationships with their parents, rather than by being either staunchly independent or needy and dependent.

In sum, adolescence is a critical period for the development of autonomy, maturity, and personal identity. Social competence requires balancing the personal goals of emerging adulthood with the increasing social demands of peers while maintaining positive engagement with the family. These studies clearly show that parental socialization has powerful impacts on adolescents' successful accomplishments of these developmental tasks. Granting more autonomy; appropriately monitoring adolescents' activities and friendships; talking with their children; and providing warm, supportive, and secure relationships provide the necessary family context for adolescents' social competence to flourish.

Current Trends: The Contribution of Parental Emotion Socialization to Social Competence

Taken together, the literature reviewed above strongly supports the argument that maternal warmth and sensitivity, and secure attachment relationships, support the early development of social competence. However, current research in the study of socialization has moved beyond examining broad indices of parenting and has adopted specific measurements of discrete child-rearing practices and behaviors (Grusec & Davidov, 2007). One area of specific socialization that has important implications for understanding parents' role in the development of children's social competence is the socialization of emotional competence (Denham et al., 2007). This area seems particularly relevant, given the links between children's emotional functioning and social competence (e.g., Denham et al., 2003; Garner & Estep, 2001).

Children's *emotional competence* encompasses their emotion understanding, emotion expression, and emotion regulation (Denham et al., 2003, 2007; Eisenberg, Cumberland, & Spinrad, 1998). Of these, emotion regulation has received the most attention from researchers studying the development of social competence. Children's emotion regulation has been associated with social competence and likeability in the peer group (Denham et al., 2003; Eisenberg et al., 1996; Shields & Cicchetti, 2001). Conversely, children with poor emotion regulation (or emotion *dys*regulation) are at greater risk for demonstrating social reticence or shyness (Henderson, Marshall, Fox, & Rubin, 2004; Rubin, Cheah, & Fox, 2001; Rubin, Coplan, Fox, & Calkins, 1995), externalizing difficulties (Cole, Zahn-Waxler, Fox, Usher, & Welsh, 1996; Eisenberg et al., 2001; Rubin et al., 1995; Rydell, Berlin, & Bohlin, 2003), internalizing difficulties (Cole et al., 1996; Eisenberg et al., 2001; 1996; Rubin et al., 1995; Rydell et al., 2003), and other indicators of social problems or psychopathology (Cicchetti, Ackerman, & Izard, 1995; Cummings & Davies, 1996; Southam-Gerow & Kendall, 2002). Indeed, understanding how parents contribute to children's development of emotion regulation is critical to understanding parents' role in the development of social competence.

Emotion regulation is thought to be partly dispositional, but also is known to be greatly influenced by interactions within the family during the early years of life (Denham et al., 2007; Eisenberg et al., 1998; Halberstadt, Crisp, & Eaton, 1999). Parents socialize children's understanding, expression, and regulation of emotion in both *direct* and *indirect* ways (Denham, Mitchell-Copeland, Strandberg, Auerbach, & Blair, 1997; Eisenberg et al., 1998). *Indirect socialization* is thought to occur from the emotional climate within the family unit, or family expressiveness of emotion (Halberstadt et al., 1999) and via parents' own expressiveness of emotion during family interaction (Valiente, Fabes, Eisenberg, & Spinrad, 2004). *Direct socialization* is thought to occur via parental reactions to children's emotion displays or parental discussion of emotions with their children. Although we acknowledge the importance of indirect emotion socialization practices, we focus this review on direct emotion socialization practices employed by parents, as this area of work has received more empirical study by ourselves and others.

Emotion Socialization: An Overview

Children learn about emotions via three primary modes of socialization: (1) witnessing others' feelings and emotions; (2) having their emotional displays responded to; and (3) the ways they are taught about their feelings and emotions (Denham et al., 2007). Many research studies focus on parents' reactions or responses to children's emotions. Generally, responses to children's emotions are characterized as either supportive or nonsupportive. Supportive reactions to children's emotion encompass a wide range of responses, including acknowledgment and acceptance of emotional expressions, as well as suggestions for constructively regulating emotions. These types of reactions to children's emotions validate children's feelings and expression of affect, while also providing the scaffolding needed to express emotions in ways that are socially appropriate. Nonsupportive reactions to children's emotions are typically described as punitive, critical, or minimizing responses. These types of reactions not only suggest that emotional experience and expression are unacceptable, but fail to teach children about constructive ways to regulate emotions, potentially exacerbating emotional dysregulation (e.g., Buck, 1984). Importantly, parents' responses to emotions vary according to a number of factors, including children's temperamental characteristics, parents' own emotional competence, the interaction context, and the type of emotion children are expressing (see Denham et al., 2007, and Kennedy Root & Denham, 2010). To date, though, most research has focused on characterizing parents' typical reactions to children's expression of negative emotions, and the links between parents' responses and children's social and emotional adjustment.

Links between Emotion Socialization and Social Competence

In general, supportive emotion socialization practices have been associated with greater social competence (Denham & Grout, 1993; Denham et al., 1997), and nonsupportive reactions have been associated with behavioral dysregulation (Eisenberg & Fabes, 1994) and with socially incompetent behavior (Jones, Eisenberg, Fabes, & MacKinnon, 2002). In a study of 4- to 6-year-old children, Eisenberg and Fabes (1994) found that maternal reports of minimizing and punitive responses to children's negative emotions were associated with children's incompetent social skills with peers. Similarly, Fabes, Leonard, Kupanoff, and Martin (2001) also found that mothers who reacted with minimizing and punitive responses to their preschool-age children's negative emotions had children who were less socially competent, were more highly emotionally reactive, and displayed negative emotions more intensely, according to teachers.

Parallel findings have been documented in samples of older children and adolescents. Elementary school-age children with mothers who minimized and punished their negative emotions were rated as less socially competent by their teachers (Jones et al., 2002). In a longitudinal study across five time points from preschool to late childhood, nonsupportive emotion socialization strategies were negatively predictive of parental reports of emotion regulation and positively predictive of teacher-reported behavioral difficulties (both internalizing and externalizing; Eisenberg et al., 1999).

As was noted with other aspects of parental socialization, the links between emotion socialization and children's social competence are moderated by children's dispositional traits. Hastings and De (2008) reported that preschool-age children with weak parasympathetic regulation, as evidenced by low cardiac vagal tone, were more socially competent if fathers were supportive of displays of fear and sadness, but were the least socially competent if fathers were unsupportive. Fathers' emotion socialization was not associated with social competence for children who had good physiological self-regulatory capacities. In addition, Kennedy Root and Stifter (2010) reported that children who both were highly exuberant and had mothers who reported nonsupportive responses to their negative emotions at 4.5 years were rated by their first-grade teachers as being the most disruptive in class.

In general, the literature on parental socialization of emotion during middle childhood and adolescence is scanty at best. There is some empirical evidence to underscore the significant role parents play through the adolescent years; however, this literature is largely focused on socially incompetent behaviors, rather than social competence. For instance, O'Neal and Magai (2005) reported that adolescents had fewer externalizing problems if they perceived that their parents responded supportively to their expressions of shame, and did not become angry in response to the youth's own expressions of anger. Furthermore, Brand and Klimes-Dougan (2010) reported that adolescents (ages 11–16) who indicated that both their mothers and fathers responded in an unsupportive fashion to their emotions had more internalizing and externalizing difficulties during adolescence. Thus, while empirical support is still growing, it is clear that parents' reactions to their adolescents' emotions continue to contribute to their social and emotional adjustment.

Future Directions and Significance to the Field of Social Neuroscience

In sum, these and other studies point to a number of aspects of parental socialization that support or undermine the development of social competence. Children and adolescents are more likely to be socially competent when they have secure attachment relationships with their parents; when their parents are more sensitive, warm, and engaged without being intrusive or domineering; when parents play more often and more positively; when parents are supportive of their children's emotions and avoid reacting negatively to children's aversive behaviors; and when parents' demands are reasoned, well balanced, and directed toward fostering a child's autonomous behavior. Although the behavioral literature extant provides us with insight into the ways in which parents contribute to the development of social competence, there are several areas that require further empirical investigation; there is also a growing interest in exploring the links between the socialization of social competence and the emerging field of developmental social neuroscience. We close the chapter by briefly outlining these perspectives and offering suggestions for future research.

Integrative Models of Biological and Environmental Contributions to Social Competence

At least since the introductions of Thomas and Chess's (1977) goodness-of-fit model, Beck's (1967, 1983) diathesis–stress model of vulnerability to psychopathology, and Sameroff's (1983, 2010), transactional model of development, developmental scientists have been interested in identifying the ways in which internal and external factors jointly contribute to children's social and emotional development (Bates & Pettit, 2007). Internal factors have typically been assessed as dispositional or biologically based characteristics, such as temperament, and increasingly have included direct assessments of genotypes or neurophysiological functioning, such as markers of adrenocortical activity (e.g., cortisol) or autonomic regulation (e.g., respiratory sinus arrhythmia, electrodermal activity). Typically, children who differed on one of these internal factors were portrayed as likely to differ in their behavioral responses to a given external factor, such as parenting style, so that they would manifest different psychosocial adjustment despite having had similar experiences.

Arguably the most commonly invoked model was the *diathesis–stress* model, which holds that some individuals have a dispositional vulnerability or weakness (diathesis) that is only manifested as maladjustment when they experience disadvantageous events (stress). When environmental conditions are favorable—say, when children have sensitive, warm, and supportive parents—children with and without the vulnerability are similarly likely to reach adaptive developmental outcomes. When conditions are unfavorable, however— for example, when children have harshly punitive and rejecting parents—those children with the vulnerability will be at greatly elevated risk of manifesting problems, whereas children without the vulnerability will be resilient and reach positive developmental outcomes. The diathesis–stress model has typically been used to predict children's social incompetence or behavioral and emotional problems. For example, Rubin and colleagues (2002) found that it was only when they experienced overprotective or critical parenting that socially inhibited toddlers evidenced high levels of reticence with peers 2 years later; socially inhibited toddlers were no more likely to become reticent than were average or uninhibited toddlers if their mothers had been appropriately supportive and structuring. Similarly, Caspi and colleagues (2003) reported that individuals with the short (less functional) allele of the serotonin transporter promoter gene 5HTTLPR were only at elevated risk of developing major depressive disorder by early adulthood if they had experienced maltreatment or abuse during childhood; in the absence of childhood abuse, the rates of depression were identical for individuals with and without the short allele. Predicting a clearer marker of social competence, prosocial behavior in children in kindergarten, Obradovic, Bush, Stamperdahl, Adler, and Boyce (2010) examined the joint contributions of home adversity, which included such things as low income, harsh parenting, parental depression, and children's parasympathetic reactivity to challenge. In the absence of adversity, more and less reactive children had similarly high levels of prosocial behavior; in more adverse homes, however, highly reactive children had strongly reduced levels of prosocial behavior, whereas less reactive children continued to appear highly prosocial.

These and similar studies have clearly shown that children with differing dispositional and biological characteristics react to their environmental experiences in ways that can take them along divergent developmental pathways and make them increasingly distinct. In recent years, however, two teams of developmental scientists (Belsky, Bakermans-Kranenburg, & van IJzendoorn, 2007; Ellis & Boyce, 2008) have proposed convergent hypotheses to account for several observations that the diathesis–stress model cannot explain. These hypotheses have been merged into the *differential susceptibility to the environment* (DSE) model (Ellis et al., 2011). The DSE model contends that rather than marking solely a vulnerability to negative outcomes, the dispositional and biological factors labeled as "diatheses" actually should be seen as "sensitivities" or "flexibilities," making children more subject to influence from external factors, for better or for worse. Thus DSE stipulates that individuals with certain temperamental or physiological sensitivities are going to be those who suffer the most when raised in hostile or adverse environments, but also are going to be those who benefit the most when raised in positive or optimal environments (Ellis et al., 2011). Conversely, less sensitive or reactive children are also less susceptible to influence; in all but the most extreme of conditions, they are likely to do reasonably well, manifesting neither maladaptive problems nor outstanding competencies. Empirically, this framework has largely been examined via correlational design, and has examined interactions between temperament (e.g., Bradley & Corwyn, 2008) or physiological stress reactivity (e.g., Ellis, Essex, & Boyce, 2005) and environmental contexts including parenting environment, family socioeconomic conditions, and other factors.

Although there is a growing body of literature supporting the notion of DSE (see Belsky & Pluess, 2010, for a review), few studies have utilized *physiological or biological functioning* as a measure of susceptibility and *parenting* as a measure of environment, and even fewer of these have focused on the development of socially competent behavior. Our own research in this area has offered mixed support for the DSE model. In a study of emotion socialization, Hastings and De (2008) reported that maternal and paternal emotion socialization practices were associated with young children's social competence and internalizing and externalizing problems most consistently for those children who had weak parasympathetic control, as reflected by low basal respiratory sinus arrhythmia. Most of these effects conformed to diathesis–stress models, such as the observation that children manifested high externalizing problems only if they both had low vagal tone and had fathers who minimized or dismissed children's angry feelings; children with only one, or neither, of these factors showed few externalizing problems.

Hastings, Sullivan, and colleagues (2008) also reported that parenting was more strongly associated with preschool-age children's adjustment if children had weaker parasympathetic regulation. For example, if fathers engaged in high levels of protective overcontrol, children who exhibited low vagal suppression to a cognitive challenge manifested high levels of temperamental inhibition (Figure 5.1a), but if fathers were highly supportive, children with low vagal suppression manifested the fewest internalizing problems (Figure 5.1b). Those children with stronger vagal suppression (i.e., better parasympathetic control) displayed average levels of inhibition and internalizing problems, regardless of paternal parenting. These effects are consistent with the DSE framework. Most recently, Hastings and colleagues (2011) conducted a study using salivary cortisol levels to assess adrenocortical functioning, and found a parallel effect that also illustrated the DSE framework (see Figure 5.2). The positive associations between maternal punishment and preschoolers' externalizing problems were stronger for boys with high adrenocortical stress reactivity than for boys with lower reactivity. These findings are provocative in pinpointing how aspects of children's physiology and socialization experiences combine to shape the development of social behaviors in children. It should be noted, though, that each of these papers also identified interactions between physiology and parenting that did not conform to the DSE model.

Other researchers also have pursued studies of the joint contributions of physiological regulation and parental experiences to children's development of social competence and incompetence (e.g., Calkins & Fox, 1992; El-Sheikh, 2005; Obradovic et al., 2010). As in our own research, some of their results support the DSE model; other findings conform more closely to the diathesis–stress model; and occasionally the joint contributions of physiology and parenting to development do not appear to conform to either framework. In terms of our understanding of the socialization of social competence, however, what these empirical studies and theoretical models indicate is that the effects of parenting on development are dependent upon context, and children themselves constitute a critical component of the socialization context. Whether a given parenting practice should be expected to support the development of social competence depends upon the characteristics of the children who are experiencing that practice; some may benefit from it more than others. This area of investigation is ripe with opportunity for the field of developmental social neuroscience. Indeed, to better understand the development of social competence, examinations testing DSE or other transactional models will need to shed light on both (1) the specific neurobiological markers that confer susceptibility

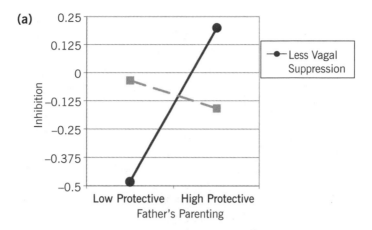

FIGURE 5.1a. Alone, this interaction is an illustration of diathesis–stress, but when accompanied by Figure 5.1b it is an illustration of differential susceptibility. Adapted from "Parental Socialization, Vagal Regulation, and Preschoolers' Anxious Difficulties: Direct Mothers and Moderated Fathers," by P. D. Hastings, C. Sullivan, et al., 2008, *Child Development*, 79, p. 56. Copyright 2008 by the Society for Research in Child Development. Adapted by permission of Wiley-Blackwell.

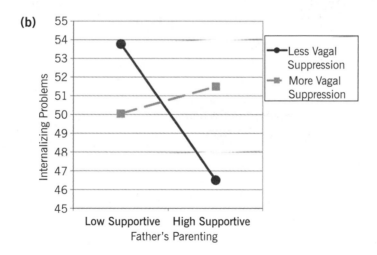

FIGURE 5.1b. Illustration of differential susceptibility when accompanied by Figure 5.1a. Adapted from "Parental Socialization, Vagal Regulation, and Preschoolers' Anxious Difficulties: Direct Mothers and Moderated Fathers," by P. D. Hastings, C. Sullivan, et al., 2008, *Child Development*, 79, p. 56. Copyright 2008 by the Society for Research in Child Development. Adapted by permission of Wiley-Blackwell.

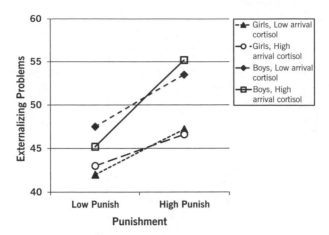

FIGURE 5.2. Illustration of differential susceptibility to environment. Adapted from "Adrenocortical Responses to Strangers in Preschoolers: Relations with Parenting, Temperament, and Psychopathology," by P. D. Hastings et al., 2011, p. 703. *Developmental Psychobiology*. Copyright 2011 by Wiley-Blackwell. Adapted by permission.

and (2) the types of caregiving environments that act upon those susceptibilities to allow these children to thrive. To hark back to the goodness-of-fit model, it may well be the case that distinct neurobiological markers confer susceptibility to different caregiving experiences.

Final Remarks

In sum, this chapter provides evidence indicating that caregiving experiences play a significant role in the development of social competence from early childhood through adolescence. We hope that researchers in the field of social neuroscience can draw from the developmental literature and consider the role of parents in the development of social competence in future studies. It is clear that this area is ripe with opportunity for researchers in social neuroscience. By utilizing the frameworks that examine person by environment interactions, we are now positioned to obtain a better understanding of how social competence develops.

References

Al-Yagon, M. (2008). Maternal personal resources and children's socioemotional and behavioral adjustment. *Child Psychiatry & Human Development, 39,* 283–298.

Allen, J. P., Hauser, S. T., Bell, K. L., & O'Connor, T. G. (1994). Longitudinal assessment of autonomy and relatedness in adolescent–family interactions as predictors of adolescent ego development and self esteem. *Child Development, 65,* 179–194.

Allen, J. P., Hauser, S. T., Eickholt, C., Bell, K. L., & O'Connor, T. G. (1994). Autonomy and relatedness in family interactions as predictors of expressions of negative adolescent affect. *Journal of Research on Adolescence, 4,* 535–552.

Ary, D. V., Duncan, T. E., Biglan, A., Metzler, C. W., Noell, J. W., & Smolkowski, K. (1999).

Development of adolescent problem behavior. *Journal of Abnormal Child Psychology, 27,* 141–150.

Bandura, A. (1977). *Social learning theory.* Englewood Cliffs, NJ: Prentice-Hall.

Bandura, A. (1997). *Self-efficacy: The exercise of control.* New York: Freeman.

Bandura, A. (2006). Social learning theory of aggression. *Journal of Communication, 28,* 12–29.

Barth, J. M., & Parke, R. D. (1993). Parent–child relationship influences on children's transition to school. *Merrill–Palmer Quarterly, 39,* 173–195.

Bates, J. E., & Pettit, G. S. (2007). Temperament, parenting, and socialization. In J. E. Grusec & P. D. Hastings (Eds.), *Handbook of socialization* (pp. 153–177). New York: Guilford Press.

Baumrind, D. (1967). Child-care practices anteceding three patterns of preschool behavior. *Genetic Psychology Monographs, 75,* 43–88.

Baumrind, D. (1971). Current patterns of parental authority. *Developmental Psychology, 4,* 1–103.

Baumrind, D. (1991). The influence of parenting style on adolescent competence and substance use. *Journal of Early Adolescence, 11,* 56–95.

Beck, A. T. (1967). *Depression: Clinical, experimental, and theoretical aspects.* New York: Harper & Row.

Beck, A. T. (1983). Cognitive theory of depression: New perspectives. In P. J. Clayton & J. E. Barrett (Eds.), *Treatment of depression: Old controversies and new approaches.* New York: Raven Press.

Bell, R. Q. (1971). Stimulus control of parent or caretaker behavior by offspring. *Developmental Psychology, 4*(1), 63–72.

Bell, R. Q., & Chapman, M. (1986). Child effects in studies using experimental or brief longitudinal approaches to socialization. *Developmental Psychology, 22,* 595–603.

Belsky, J., Bakermans-Kranenburg, M. J., & van IJzendoorn, M. H. (2007). For better and for worse: Differential susceptibility to environmental influences. *Current Directions in Psychological Science, 16,* 300–304.

Belsky, J., & Fearon, R. M. P. (2002). Early attachment security, subsequent maternal sensitivity, and later child development: Does continuity in development depend on continuity of caregiver? *Attachment and Human Development, 4,* 361–387.

Belsky, J., & Pluess, M. (2009). Beyond diathesis stress: Differential susceptibility to environmental influences. *Psychological Bulletin, 135,* 885–908.

Black, B., & Logan, A. (1995). Links between communication patterns in mother–child, father–child, and child–peer interactions and children's social status. *Child Development, 66,* 255–271.

Blunt Bugental, D., & Grusec, J. E. (2006). Socialization processes. In W. Damon & R. M. Lerner (Series Eds.) & N. Eisenberg (Vol. Ed.), *Handbook of child psychology: Vol. 3. Social, emotional, and personality development* (6th ed., pp. 366–428). Hoboken, NJ: Wiley.

Bohlin, G., Hagekull, B., & Andersson, K. (2005). Maternal personal resources and children's behavioral inhibition as a precursor of peer social competence in early school age: The interplay with attachment and nonparental care. *Merrill–Palmer Quarterly, 51,* 1–19.

Booth-LaForce, C. L., Oh, W., Kim, A., Rubin, K. H., Rose-Krasnor, L., & Burgess, K. B. (2006). Attachment, self-worth, and peer-group functioning in middle childhood. *Attachment and Human Development, 8,* 309–325.

Booth-LaForce, C. L., & Oxford, M. L. (2008). Trajectories of social withdrawal from grades 1 to 6: Prediction from early parenting, attachment, and temperament. *Developmental Psychology, 44,* 1298–1313.

Bowlby, J. (1969). *Attachment and loss: Vol. 1. Attachment.* New York: Basic Books.

Bradley, R. H., & Corwyn, R. F. (2008). Infant temperament, parenting, and externalizing behavior in first grade: A test of the differential susceptibility hypothesis. *Journal of Child Psychology and Psychiatry, 49,* 124–131.

Brand, A. E., & Klimes-Dougan, B. (2010). Emotion socialization in adolescence: The roles of mothers and fathers. *New Directions for Child and Adolescent Development, 2010*(128), 85–100.

Bronfenbrenner, U. (1979). Contexts of child rearing: Problems and prospects. *American Psychologist, 34,* 844–850.

Bronfenbrenner, U., & Morris, P. A. (2006). Bioecological model of human development. In W. Damon & R. M. Lerner (Series Eds.) & R. M. Lerner (Vol. Ed.), *Handbook of child psychology: Vol. 1. Theoretical models of human development* (6th ed., pp. 793–828). Hoboken, NJ: Wiley.

Buck, R. (1984). *The communication of emotion.* New York: Guilford Press.

Caldarella, P., & Merrell, K. W. (1997). Common dimensions of social skills of children and adolescents: A taxonomy of positive behaviors. *School Psychology Review, 20,* 264–278.

Calkins, S. D., & Fox, N. A. (1992). The relations among infant temperament, security of attachment, and behavioral inhibition at twenty-four months. *Child Development, 63,* 1456–1472.

Carson, J., Burks, V., & Parke, R.D. (1993). Parent–child physical play: Determinants and consequences. In K. MacDonald (Ed.), *Parent–child play: Descriptions and implications* (pp. 197–220). Albany: State University of New York Press.

Caspi, A., Sugden, K., Moffitt, T. E., Taylor, A., Craig, I. W., Harrington, H., et al. (2003). Influence of life stress on depression: Moderation by a polymorphism in the 5-HTT gene. *Science, 301,* 386–389.

Chao, R. K., & Wilms, J. D. (2002). The effects of parenting practices on children's outcomes. In J. D. Wilms (Ed.), *Vulnerable children* (pp. 149–165). Edmonton: University of Alberta Press.

Cicchetti, D., Ackerman, B. P., & Izard, C. E. (1995). Emotions and emotion regulation in developmental psychopathology. *Development and Psychopathology, 7,* 1–10.

Cole, P. M., Zahn-Waxler, C., Fox, N. A., Usher, B. A., & Welsh, J. D. (1996). Individual differences in emotion regulation and behavior problems in preschool children. *Journal of Abnormal Psychology, 105,* 518–529.

Conger, K. J., Conger, R. D., & Scaramella, L. V. (1997). Parents, siblings, psychological control, and adolescent adjustment. *Journal of Adolescent Research, 12,* 113–138.

Cottrell, L., Li, X., Harris, C., D'Alessandri, D., Atkins, M., Richardson, B., & Stanton, B. (2003). Parent and adolescent perceptions of parental monitoring and adolescent risk involvement. *Parenting: Science and Practice, 3,* 179–195.

Cummings, E. M., & Davies, P. (1996). Emotional security as a regulatory process in normal development and the development of psychopathology. *Development and Psychopathology, 8,* 123–139.

Darling, N., & Steinberg, L. (1993). Parenting style as context: An integrative model. *Psychological Bulletin, 113,* 487–496.

Degnan, K. A., Henderson, H. A., Fox, N. A., & Rubin, K. H. (2008). Predicting social wariness in middle childhood: The moderating roles of childcare history, maternal personality and maternal behavior. *Social Development, 17,* 471–487.

Denham, S. A., Bassett, H. H., & Wyatt, T. (2007). The socialization of emotional competence. In J. E. Grusec & P. D. Hastings (Eds.), *Handbook of socialization* (pp. 614–637). New York: Guilford Press.

Denham, S. A., Blair, K. A., DeMulder, E., Levitas, J., Sawyer, K., Auerbach-Major, S., et al. (2003). Preschool emotional competence: Pathway to social competence? *Child Development, 74,* 238–256.

Denham, S. A., & Grout, L. (1993). Socialization of emotion: Pathway to preschoolers' emotional and social competence. *Journal of Nonverbal Behavior, 17,* 205–227.

Denham, S. A., Mitchell-Copeland, J., Strandberg, K., Auerbach, S., & Blair, K. (1997). Parental contributions to preschoolers' emotional competence: Direct and indirect effects. *Motivation and Emotion, 21,* 65–86.

Dennis, T. (2006). Emotional self-regulation in preschoolers: The interplay of child approach reactivity, parenting, and control capacities. *Developmental Psychology, 42,* 84–97.

Dishion, T. J., & McMahon, R. J. (1998). Parental monitoring and the prevention of child and adolescent problem behavior: A conceptual and empirical formulation. *Clinical Child and Family Psychology Review, 1,* 61–75.

Domitrovich, C. E., & Bierman, K. L. (2001). Parenting practices and child social adjustment: Multiple pathways of influence. *Merrill–Palmer Quarterly, 47,* 235–263.

Dunn, J., & Brown, J. (1994). Affect expression in the family, children's understanding of emotions, and their interactions. *Merrill–Palmer Quarterly, 40,* 120–137.

Eisenberg, N., Cumberland, A., & Spinrad, T. L. (1998). Parental socialization of emotion. *Psychological Inquiry, 9,* 241–273.

Eisenberg, N., Cumberland, A., Spinrad, T. L., Fabes, R. A., Shepard, S. A., Reiser, M., et al. (2001). The relations of regulation and emotionality to children's externalizing and internalizing problem behavior. *Child Development, 72,* 1112–1134.

Eisenberg, N., & Fabes, R. A. (1994). Mothers' reactions to children's negative emotions: Relations to children's temperament and anger behavior. *Merrill–Palmer Quarterly, 40,* 138–156.

Eisenberg, N., Fabes, R. A., Guthrie, I. K., Murphy, B. C., Maszk, P., Holmgren, R., et al. (1996). The relations of regulation and emotionality to problem behavior in elementary school children. *Development and Psychopathology, 8,* 141–162.

Eisenberg, N. A., Fabes, R. A., Shepard, S. A., Guthrie, I. K., Murphy, B. C., & Reiser, M. (1999). Parental reactions to children's negative emotions: Longitudinal relations to quality of children's social functioning. *Child Development, 70,* 513–534.

Ellis, B. J., & Boyce, W. T. (2008). Biological sensitivity to context *Current Directions in Psychological Science, 17,* 183–187.

Ellis, B. J., Boyce, W. T., Belsky, J., Bakermans-Kranenburg, M. J., & van IJzendoorn, M. H. (2011). Differential susceptibility to the environment: An evolutionary-neurodevelopmental theory. *Development and Psychopathology, 23,* 7–28.

Ellis, B. J., Essex, M. J., & Boyce, W. T. (2005). Biological sensitivity to context: II. Empirical explorations of an evolutionary–developmental theory. *Development and Psychopathology, 17,* 303–328.

El-Sheikh, M. (2005). The role of emotional responses and physiological reactivity in the marital conflict–child functioning link. *Journal of Child Psychology and Psychiatry, 46,* 1191–1199.

Engels, R. M. E., Dekovic, M., & Meeus, W. (2002). Parenting practices, social skills and peer relationships in adolescence. *Social Behavior and Personality, 30,* 3–18.

Fabes, R. A., Leonard, S. A., Kupanoff, K., & Martin, C. L. (2001). Parental coping with children's negative emotions: Relations with children's emotional and social responding. *Child Development, 72,* 907–920.

Fagot, B. I. (1997). Attachment, parenting, and peer interactions of toddler children. *Developmental Psychology, 33,* 489–499.

Fiese, B. H. (1990). Playful relationships: A contextual analysis of mother–toddler interaction and symbolic play. *Child Development, 61,* 1648–1646.

Fuligni, A. J., & Eccles, J. S. (1993). Perceived parent–child relationships and early adolescents' orientation towards peers. *Developmental Psychology, 29,* 622–632.

Furman, W., & Wehner, E. A. (1994). Romantic views: Toward a theory of adolescent romantic relationships. In R. Montmayor & G. R. Adams (Eds.), *Advances in adolescent development: Vol. 6. Personal relationships during adolescence.* Newbury Park, CA: Sage.

Garner, P. W., & Estep, K. M. (2001). Emotion competence, emotion socialization, and young children's peer-related social competence. *Early Education and Development*, *12*, 29–48.

Gomez, R., & McLaren, S. (2007). The inter-relations of mother and father attachment, self-esteem and aggression during late adolescence. *Aggressive Behavior, 33,* 160–169.

Grusec, J. E., & Davidov, M. (2007). Socialization in the family: The role of parents. In J. E. Grusec & P. D. Hastings (Eds.), *Handbook of socialization* (pp. 284–308). New York: Guilford Press.

Grusec, J. E., & Goodnow, J. J. (1994). Impact of parental discipline methods on the child's internalization of values: A reconceptualization of current points of view. *Developmental Psychology, 30,* 4–19.

Grusec, J. E., & Hastings, P. D. (Eds.). (2007). *Handbook of socialization.* New York: Guilford Press.

Halberstadt, A., Crisp, V. W., & Eaton, K. L. (1999). Family expressiveness: A retrospective and new directions for research. In P. Philippot & R. S. Feldman (Eds.), *Social context of nonverbal behavior* (pp. 109–155). New York: Cambridge University Press.

Hane, A. A., Fox, N. A., Henderson, H. A., & Marshall, P. J. (2008). Behavioral reactivity and approach–withdrawal bias in infancy. *Developmental Psychology, 44*(5), 1491–1496.

Hastings, P. D., & De, I. (2008). Parasympathetic regulation and parental socialization of emotion: Biopsychosocial processes of adjustment in preschoolers. *Social Development*, *17*, 211–238.

Hastings, P. D., Nuselovici, J. N., Utendale, W. T., Coutya, J., McShane, K. E., & Sullivan, C. (2008). Applying the polyvagal theory to children's emotion regulation: Social context, socialization, and adjustment. *Biological Psychology, 79,* 299–306.

Hastings, P. D., & Rubin, K. H. (1999). Predicting mothers' beliefs about preschool-aged children's social behavior: Evidence for maternal attitudes moderating child effects. *Child Development, 70,* 722–741.

Hastings, P. D., Ruttle, P. L, Serbin, L. A., Mills, R. S. L., Stack, D. M., & Schwartzman, A. E. (2011). Adrenocortical responses to strangers in preschoolers: Relations with parenting, temperament, and psychopathology. *Developmental Psychobiology, 53,* 694–710.

Hastings, P. D., Sullivan, C., McShane, K. E., Coplan, R. J., Utendale, W. T., & Vyncke, J. D. (2008). Parental socialization, vagal regulation, and preschoolers' anxious difficulties: Direct mothers and moderated fathers. *Child Development, 79,* 45–64.

Hastings, P. D., Utendale, W. T., & Sullivan, C. (2007). The socialization of prosocial development. In J. E. Grusec & P. D. Hastings (Eds.), *Handbook of socialization* (pp. 638–664). New York: Guilford Press.

Hastings, P. D., Zahn-Waxler, C., Robinson, J., Usher, B., & Bridges, D. (2000). The development of concern for others in children with behavior problems. *Developmental Psychology, 36,* 531–546.

Henderson, H. A., Marshall, P. J., Fox, N. A., & Rubin, K. H. (2004). Psychophysiological and behavioral evidence for varying forms and functions of nonsocial behavior in preschoolers. *Child Development, 75,* 236–250.

Howes, C., & James, J. (2002). Children's social development within the socialization context of childcare and early childhood education. In P. K. Smith & C. H. Hart (Eds.), *Blackwell handbook of childhood social development* (pp. 137–155). Oxford, UK: Blackwell.

Janssens, J. M. A. M., & Dekovic, M. (1997). Child rearing, prosocial moral reasoning, and prosocial behaviour. *International Journal of Behavioral Development, 20,* 509–527.

Jones, S., Eisenberg, N., Fabes, R. A., & MacKinnon, D. P. (2002). Parents' reactions to elementary school children's negative emotions: Relations to social and emotional functioning at school. *Merrill–Palmer Quarterly, 48,* 133–159.

Kennedy, A. E., Rubin, K. H., Hastings, P. D., & Maisel, B. A. (2004). The longitudinal relations between child vagal tone and parenting behavior: 2 to 4 years. *Developmental Psychobiology, 45,* 10–21.

Kennedy Root, A. E., & Denham, S. A. (2010). The role of gender in the socialization of emotion: Key concepts and critical issues. *New Directions for Child and Adolescent Development*, *2010*(128), 1–9.

Kennedy Root, A. E., & Rubin, K. H. (2010). Gender and parents' emotion socialization beliefs during the preschool years. *New Directions for Child and Adolescent Development*, *2010*(128), 51–64.

Kennedy Root, A. E., & Stifter, C. A. (2010). Temperament and maternal emotion socialization beliefs as predictors of early childhood social behavior in the laboratory and classroom. *Parenting: Science and Practice*, *10*, 241–257.

Kiang, L., Moreno, A. J., & Robinson, J. L. (2004). Maternal preconceptions about parenting predict child temperament, maternal sensitivity, and children's empathy. *Developmental Psychology*, *40*, 1081–1092.

Kochanska, G. (1995). Children's temperament, mothers' discipline, and security of attachment: Multiple pathways to emerging internalization. *Child Development*, *66*, 597–615.

Kochanska, G. (1997). Mutually responsive orientation between mothers and their young children: Implications for early socialization. *Child Development*, *68*, 94–112.

Kochanska, G., Aksan, N., & Joy, M. E. (2007). Children's fearfulness as a moderator of parenting in early socialization: Two longitudinal studies. *Developmental Psychology*, *43*(1), 222–237.

Laible, D. (2007). Attachment with parents and peers in late adolescence: Links with emotional competence and social behavior. *Personality and Individual Differences*, *43*, 1185–1197.

Laible, D., Carlo, G., Torquati, J., & Ontai, L. (2004). Children's perceptions of family relationships assessed in a doll story completion task: Links to parenting, social competence, and externalization behavior. *Social Development*, *13*, 551–569.

Lamborn, S. D., Mounts, N. S., Steinberg, L., & Dornbusch, S. M. (1991). Patterns of competence and adjustment among adolescents from authoritative, authoritarian, indulgent, and neglectful families. *Child Development*, *62*, 1049–1065.

Laursen, B., & Collins, W. A. (2009). Parent–child relationships during adolescence. In R. M. Lerner & L. Steinberg (Eds.), *Handbook of adolescent psychology: Vol. 2. Contextual influences on adolescent development* (3rd ed., pp. 3–42). Hoboken, NJ: Wiley.

Lerner, R. M., Rothbaum, F., Boulos, S., & Castellino, D. R. (2002). Developmental systems perspective on parenting. In M. Bornstein (Ed.), *Handbook of parenting: Vol. 2. Biology and ecology of parenting* (2nd ed., pp. 315–344). Mahwah, NJ; Erlbaum.

Lindsey, E. W., & Mize, J. (2001a). Interparental agreement, parent–child responsiveness, and children's peer competence. *Family Relations*, *50*, 348–354.

Lindsey, E. W., & Mize, J. (2001b). Contextual differences in parent–child play: Implications for children's gender role development. *Sex Roles*, *44*, 155–176.

Lindsey, E. W., Mize, J., & Pettit, G. S. (1997). Differential play patterns of mothers and fathers of sons and daughters: Implications for children's gender role development. *Sex Roles*, *37*, 643–661.

Luthar, S. S., Burack, J. A., & Cicchetti, D. (1997). *Developmental psychopathology Perspectives on adjustment, risk, and disorder*. Thousand Oaks, CA: Sage.

Maccoby, E. E. (2007). History: Overview of socialization research and theory. In J. E. Grusec & P. D. Hastings (Eds.), *Handbook of socialization* (pp. 13–41). New York: Guilford Press.

Maccoby, E. E., & Martin, J. A. (1983). Socialization in the context of the family: Parent–child interaction. In P. H. Mussen (Series Ed.) & E. M. Hetherington (Vol. Ed.), *Handbook of child psychology: Vol. 4. Socialization, personality, and social development* (4th ed., pp. 1–102). New York: Wiley.

Markiewicz, D., Doyle, A. B., & Brendgen, M. (2001). The quality of adolescents' friendships: Associations with mothers' interpersonal relationships, attachment to parents and friends, and prosocial behaviors. *Journal of Adolescence*, *24*, 429–445.

Masten, A. S., Hubbard, J. J., Gest, S. D., Tellegen, A., Garmezy, N., & Ramirez, M. (1999). Competence in the context of adversity: Pathways to resilience and maladaptation from childhood to late adolescence. *Development and Psychopathology, 11,* 143–169.

McDowell, D. J., Kim, M., O'Neil, R., & Parke, R. D. (2002). Children's emotional regulation and social competence in middle childhood: The role of maternal and paternal interactive style. *Marriage and Family Review, 34,* 345–364.

McDowell, D. J., Parke, R. D., & Wang, S. J. (2003). Differences between mothers' and fathers' advice-giving style and content: Relations with social competence and psychological functioning in middle childhood. *Merrill–Palmer Quarterly, 49,* 55–76.

McElhaney, K., & Allen, J. P. (2001). Autonomy and adolescent social functioning: The moderating effect of risk. *Child Development, 72,* 220–235.

Mize, J., & Pettit, G. S. (1997). Mothers' social coaching, mother–child relationship style, and children's peer competence: Is the medium the message? *Child Development, 68,* 291–311.

Mounts, N. S. (2004). Adolescents' perceptions of parental management of peer relationships in an ethnically diverse sample. *Journal of Adolescent Research, 19,* 446–467.

National Institute of Child Health and Human Development (NICHD) Early Child Care Research Network. (2002). Early child care and children's development prior to school entry: Results from the NICHD study of early child care. *American Educational Research Journal, 39,* 133–164.

National Institute of Child Health and Human Development (NICHD) Early Child Care Research Network. (2004). Fathers' and mothers' parenting behavior and beliefs as predictors of children's social adjustment in the transition to school. *Journal of Family Psychology, 18,* 628–638.

National Institute of Child Health and Human Development (NICHD) Early Child Care Research Network. (2006). Infant–mother attachment classification: Risk and protection in relation to changing maternal caregiving quality. *Developmental Psychology, 42,* 38–58.

Obradovic, J., Bush, N. R., Stamperdahl, J., Adler, N. E., & Boyce, W. T. (2010). Biological sensitivity to context: The interactive effects of stress reactivity and family adversity on socioemotional behavior and school readiness. *Child Development, 81,* 270–289.

O'Neal, C. R., & Magai, C. (2005). Do parents respond in different ways when children feel different emotions?: The emotional context of parenting. *Development and Psychopathology, 17,* 467–487.

Park, J. H., Essex, M. J., Zahn-Waxler, C., Armstrong, J. M., Klein, M. H., & Goldsmith, H. H. (2005). Relational and overt aggression in middle childhood: Early child and family risk factors. *Early Education and Development, 16,* 233–256.

Patterson, G. R., & Stouthamer-Loeber, M. (1984). The correlation of family management practices and delinquency. *Child Development, 55,* 1299–1307.

Pettit, G. S., Glyn Brown, E., Mize, J., & Lindsey, E. (1998). Mothers' and fathers' socializing behaviors in three contexts: Links with children's peer competence. *Merrill–Palmer Quarterly, 44,* 173–193.

Robinson, J., Little, C., & Biringen, Z. (1993). Emotional communication in mother–toddler relationships: Evidence for early gender differentiation. *Merrill–Palmer Quarterly, 39,* 496–517.

Rubin, K. H., Bukowski, W., & Parker, J. G. (2006). Peer interactions, relationships, and groups. In W. Damon & R. M. Lerner (Series Eds.) & N. Eisenberg (Vol. Ed.), *Handbook of child psychology: Vol. 3. Social, emotional, and personality development* (6th ed., pp. 571–645). Hoboken, NJ: Wiley.

Rubin, K. H., Burgess, K. B., Dwyer, K. M, & Hastings, P. D. (2003). Predicting preschoolers' externalizing behaviors from toddler temperament, conflict, and maternal negativity. *Developmental Psychology, 39,* 164–176.

Rubin, K. H., Burgess, K. B., & Hastings, P. D. (2002). Stability and social-behavioral consequences of toddlers' inhibited temperament and parenting behaviors. *Child Development, 73,* 483–495.

Rubin, K. H., Cheah, C. S. L., & Fox, N. A. (2001). Emotion regulation, parenting, and the display of social reticence in preschoolers. *Early Education and Development, 12,* 97–115.

Rubin, K. H., Coplan, R. J., Fox, N. A., & Calkins, S. D. (1995). Emotionality, emotion regulation, & preschoolers' social adaptation. *Development and Psychopathology, 7,* 49–62.

Rubin, K. H., Hastings, P. D., Stewart, S. L., Henderson, H. A., & Chen, X. (1997). The consistency and concomitants of inhibition: Some of the children, all of the time. *Child Development, 68,* 467–483.

Rubin, K. H., & Rose-Krasnor, L. (1992). Interpersonal problem solving. In V. B. Van Hasselt & M. Hersen (Eds.), *Handbook of social development* (pp. 283–323). New York: Plenum Press.

Russell, A., Mize, J., & Bissaker, K. (2002). Parent–child relationships. In P. K. Smith & C. H. Hart (Eds.), *Blackwell handbook of childhood social development* (pp. 205–222). Oxford, UK: Blackwell.

Russell, A., & Saebel, J. (1997). Mother–son, mother–daughter, father–son, and father–daughter: Are they distinct relationships? *Developmental Review, 17,* 111–147.

Rydell, A., Berlin, L., & Bohlin, G. (2003). Emotionality, emotion regulation, and adaptation among 5- to 8-year-old children. *Emotion, 3,* 30–47.

Sameroff, A. J. (1983). Developmental systems: Contexts and evolution. In P. Mussen (Series Ed.) & W. Kessen (Vol. Ed.), *Handbook of child psychology: Vol. 1. History, theory, and methods* (4th ed., pp. 237–294). New York: Wiley.

Sameroff, A. J. (2010). A unified theory of development: A dialectic integration of nature and nurture. *Child Development, 81,* 6–22.

Sandstrom, M. J. (2007). A link between mothers' disciplinary strategies and children's relational aggression. *British Journal of Developmental Psychology, 25,* 399–407.

Shields, A., & Cicchetti, D. (2001). Parental maltreatment and emotion dysregulation as risk factors for bullying and victimization in middle childhood. *Journal of Clinical Child Psychology, 30,* 349–363.

Southam-Gerow, M. A., & Kendall, P. C. (2002). Emotion regulation and understanding: Implications for child psychopathology and therapy. *Clinical Psychology Review, 22,* 189–222.

Spinrad, T. L., Eisenberg, N., Gaertner, B., Popp, T., Smith, C. L., Kupfer, A., et al. (2007). Relations of maternal socialization and toddlers' effortful control to children's adjustment and social competence. *Developmental Psychology, 43*(5), 1170–1186.

Stattin, H., & Kerr, M. (2000). What parents know, how they know it, and several forms of adolescent adjustment: Further support for a reinterpretation of monitoring. *Developmental Psychology, 36,* 366–380.

Steinberg, L., Elmen, J. D., & Mounts, N. S. (1989). Authoritative parenting, psychosocial maturity, and academic success among adolescents. *Child Development, 60,* 1424–1436.

Steinberg, L., Lamborn, S. D., Darling, N., Mounts, N. S., & Dornbusch, S. M. (1994). Over-time changes in adjustment and competence among adolescents from authoritative, authoritarian, indulgent, and neglectful families. *Child Development, 65,* 754–770.

Thomas, A., & Chess, S. (1977). *Temperament and development.* New York: Brunner/Mazel.

Valiente, C., Fabes, R. A., Eisenberg, N., & Spinrad, T. L. (2004). The relations of parental expressivity and support to children's coping with daily stress. *Journal of Family Psychology, 18,* 97–106.

Wentzel, K. R., & Looney, L. (2007). Socialization in school settings. In J. E. Grusec & P. D. Hastings (Eds.), *Handbook of socialization* (pp. 382–403). New York: Guilford Press.

PART III
ASSESSING SOCIAL FUNCTION

CHAPTER 6

Measuring Social Skills with Questionnaires and Rating Scales

Frank Muscara and Louise Crowe

Much evidence suggests that social function is one of the best indicators of future and current behavioral and emotional problems—even better than diagnosis (Angold, Costello, Farmer, Burns, & Erkanli, 1999). Poor social skills are also associated with poor educational achievement (Elliott, Malecki, & Demaray, 2001) and executive dysfunction (Muscara, Catroppa, & Anderson, 2008), and have been rated as having the greatest impact on quality of life in clinical populations (Anderson, Brown, & Newitt, 2010; Oddy, Coughlan, Tyerman, & Jenkins, 1985). Despite the significant and long-term impact that social difficulties have on children and adolescents (Cattelani, Lombardi, Brianti, & Mazzucchi, 1998; Muscara, Catroppa, Eren, & Anderson, 2009), the clinical measurement of this area of function remains underdeveloped when compared to other aspects of child function and development, such as intelligence, motor development, and language (John, 2001). This can be attributed to the fact that *social competence* is a broad construct, encompassing a number of skills and functions. In fact, social competence can be thought of as an organizational construct, reflecting a child's capacity to integrate behavioral, cognitive, and affective demands to adapt effectively to dynamic social contexts (Bierman & Welsh, 2000). Given the broad and multidimensional nature of this construct, it is not surprising that no single measure encompasses all areas of social function (Saltzman-Benaiah & Lalonde, 2007). Rather, the choice of measure depends greatly on the purpose of the assessment and the specific information required by the clinician or researcher.

This chapter outlines and explores existing techniques employed to measure and assess social competence in children and adolescents, as well as the current issues facing researchers and clinicians in the measurement of this construct. Given its limited scope, this chapter specifically focuses on the most widely used methods of assessing social competence: questionnaires and rating scales.

Defining Social Skills

One of the main difficulties in measuring social competence is that the terminology is inconsistent: Different terms are used interchangeably within the literature and in measurement (John, 2001). To ensure accurate and valid measurement, appropriate definitions are required of these terms, and of the specific construct being measured.

Merrell (1997) defines *social competence* as a multidimensional construct consisting of a number of cognitive and behavioral variables, as well as different aspects of emotional adjustment, which are necessary in developing adequate social relations. Others have defined social competence as the ability to achieve personal goals in social interactions, while maintaining positive relationships with others over time and across situations (Rubin & Rose-Krasnor, 1992; Yeates et al., 2007). Canino, Costello, and Angold (1999) have suggested that social competence refers to a 'normative level of adaptive functioning,' while John (2001) suggests that it refers to the ability to function appropriately in interpersonal interactions. *Social adjustment* is broadly defined as the extent to which an individual's performance in his or her life roles conforms to the norms of the referent group (John & Weissman, 1987). *Social interaction* involves actions that can bring individuals together (such as prosocial behaviors), move people against each other (such as aggressive behaviors), isolate individuals from each other (such as withdrawal), or have no impact on other individuals (neutral behaviors) (Rosen, Furman, & Hartup, 2001; Yeates et al., 2007). *Social performance* is defined as a child's actual behavior in social interactions, and whether this is effective in maintaining positive relationships and in helping the child to achieve goals (Yeates et al., 2007). *Social participation* is engagement in socially interactive play, activities, and situations (Guralnick, 1999; Law, Anaby, Dematteo, & Hanna, 2011). *Social dysfunction* consists of two areas of difficulties: (1) the inability to get along with peers, which is reflected in a pattern of social rejection or victimization; and (2) trouble initiating and maintaining positive friendships (Parker & Asher, 1993).

Others divide the social domain into a number of subdomains. According to Gresham (1986), the multidimensional construct of social competence comprises three domains: social skills, peer relations, and adaptive behavior. *Social skills* have been defined as socially acceptable learned behaviors that lead to desirable social outcomes when initiated (Gresham & Elliott, 1990). *Adaptive behavior* is widely accepted as being the "effectiveness or degree with which the individual meets the standards of personal independence and social responsibility" (Grossman, 1983, p. 1). *Peer acceptance* is generally thought to be a measure or result of the level or maturity of social skills.

Although there is confusion regarding the terminology surrounding the construct of social competence, John (2001) has reported that there is consensus regarding what the construct actually is. For children of all ages in the majority of cultures, this construct is encompassed by the expectations placed upon them to learn to interact appropriately with family members and peers; to take part in home life; to establish friendships; to abide by rules; to attend school; to develop physical, cognitive, and social skills; and to pursue various interests in their spare time (John, 2001). For the purposes of this chapter, we have employed John's description as our definition of *social competence*.

Theoretical Underpinnings of Social Assessments

Given the complexity of the multidimensional construct of *social competence*, models of assessment are required to guide comprehensive measurement. Such models are critical, as they inform the conceptualization and development of specific methods of assessing theoretically derived and empirically supported constructs and domains of social competence.

A Model of Assessment

Yeates, Schultz, and Selman (1990) have formulated an approach for child clinical assessment, which may be used to assist in determining appropriate methods of assessing social competence in children. They describe a "Bermuda Triangle" of child clinical assessment. They identify three categories of assessment that have been used in the past and are currently used to varying degrees. The first and second categories involve the traditional administration of a battery of psychological tests. The first broad category involves assessment of cognitive or intellectual function, and the second involves assessment of personality. The third category incorporates measures that are thought to provide greater ecological validity with regard to day-to-day behavior and functioning. This category of assessment consists of such measures as behavior rating checklists or direct observations. Yeates et al. (1990) have suggested that these categories of assessment techniques fall along two separate dimensions. The first dimension can be described as being a continuum of "internal–external" processes, with "internal" reflecting intrapersonal constructs, and "external" reflecting greater interpersonal and behavioral functioning. The second dimension refers to "hot–cold" processes, with "hot" processes reflecting social and emotional processes that are under less conscious control, whereas "cold" processes are more asocial, less emotional, and under more conscious control.

As can be seen in Figure 6.1, intellectual, cognitive, and achievement assessments fall within the upper left quadrant, as they assess internal mental processes, which are largely asocial and unemotional. In contrast, personality assessments fall within the upper right quadrant, as they also involve internal processes, but are within social and emotional contexts. Finally, behavioral measures are located in the middle of the lower quadrants, as they can be either "hot" or "cold," but measure external, directly observable behaviors. Of interest, in recent years there has been a surge in development of various measures to assess observable behaviors, in an effort to develop tests of behavior and everyday function that have greater ecological validity.

Models for Measuring Social Competence

Sheridan, Hunglemann, and Maughan (1999) provide a model more specific to the measurement of social competence in children. They have proposed that social skills need to be assessed in the social environment and context within which a child functions. This position is increasingly recognized as important for obtaining the most accurate evaluation of social competence (Dooley, Beauchamp, & Anderson, 2010). Sheridan et al. (1999) have outlined seven goals for comprehensive assessment of social competence:

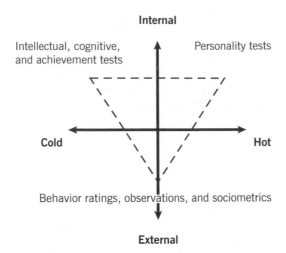

FIGURE 6.1. The "Bermuda Triangle" of child clinical assessment. Adapted from Yeates, Schultz, and Selman (1990). Copyright 1990 by Elsevier B.V. Adapted by permission.

1. Accurately identify behaviors and skills that are important, within a practical and meaningful social environment.
2. Determine the expectations, demands, and norms for behavior in the relevant social environment. This goal is critical to the model: It emphasizes that it is important to measure not only the child's behavior, but also the environment and context in which the behavior occurs, as appropriateness and acceptability of behaviors vary depending on the expectations within different contexts or environments.
3. Analyze conditions within the environment that precipitate, reinforce, or discourage behaviors.
4. Identify the child's skills and deficits with regard to specific social behaviors.
5. Examine the functions served by different behaviors within the different social environments and contexts.
6. Contribute to the development of effective social interventions.
7. Measure outcomes in relation to what those changes mean to the child within specific social environments and contexts.

These goals emphasize the importance of recognizing that social contexts vary and are dynamic, and that stimuli, responses, and the functions of the behaviors/responses are not always consistent. To meet these goals, social competence measures need to incorporate two key elements. First, either they need to mimic real-life situations as closely as possible, or alternatively, they must sample responses in real-life situations (i.e., ecological assessments). Second, they should take into account multiple perspectives—for example, by including multiple informants, such as teachers, parents, peers, and the children themselves. Behavior ratings are typically used to assess this element. Sheridan et al. (1999) suggest that this will lead to more comprehensive and ecologically valid assessment of treatment outcomes, and will also assist with selecting behaviors to target through intervention, leading to meaningful outcomes for children within their specific social environments.

This model is consistent with the views of others, such as Elliott and Gresham (1987) and Sheridan and Walker (1999). Elliott and Gresham (1987) maintain that it is critical to obtain various types of information in the assessment of child social competence, including teacher, parent, and student ratings, as well as interviews, role plays, and sociometrics. Similarly, Sheridan and Walker (1999) suggest that assessment of the child (including informant reports, self-reports, skill-based direct observations, analogue observations, and child interviews), assessment of others (teacher nominations and rankings, sociometric techniques), and assessment of the social context (contextual analysis and ecological observations, functional analysis, performance-based direct observations, interviews) are all important for a comprehensive assessment of social competence.

Measurement of Social Competence in a Clinical Context

Given the models of assessment proposed within the literature, a social competence assessment tool requires a number of critical features to ensure its effectiveness (Bierman & Welsh, 2000; Sheridan et al., 1999). First, the tool needs to be developmentally appropriate. The social behaviors and skills required for success in social interactions vary according to the age or developmental stage of the child (Bierman & Welsh, 2000). Hence these assessments must consider the developmental context of the assessment. In addition, and as previously noted, the assessment must be based on information regarding social function in different social contexts and situations; therefore, information from significant others (such as parents, teachers, and peers) is important. Finally, an assessment of the social situation itself is critical, ensuring that measurement of social behaviors and skills occurs within the context of the dynamic social situation within which they are demonstrated (Sheridan et al., 1999). Since the social environment and context have a significant impact on social competence, measurement using lab-based measures, without input from teachers, peers, or others involved in the social context, lacks external validity (Bierman & Welsh, 2000).

There are two main purposes for assessing social skills in a clinical context: (1) to identify social skill deficits, and (2) to evaluate treatment outcomes (Sheridan & Walker, 1999). Screening and measurement of social and behavioral problems in children and adolescents are essential for effective intervention. Without them, identification of underlying problems would be difficult, and interventions would be disorganized and perhaps even ineffective (Merrell, 2001). Despite suggestions within the literature emphasizing the importance of measuring social competence, the complex nature of this multidimensional construct makes comprehensive and detailed measurement challenging. As a result, various approaches and measurement tools have been developed to assess social competence and are described in the literature, each with different strengths and drawbacks: These include (1) projective/expressive techniques, (2) self-report instruments, (3) interviewing, (4) sociometric techniques, (5) observation, and (6) questionnaires and rating scales (Merrell, 2001). Direct measures of social cognition may also be employed, and these are described in detail by Bruneau-Bhérer, Achim, and Jackson (Chapter 7, this volume). Merrell (2001) argues that although each technique has its own advantages and disadvantages, they cannot be considered equal in their ability to provide information to inform a clinical formulation, and to develop an effective intervention. Various methods are explained in detail below, and are critically evaluated against the framework and criteria described above (see Table 6.1 for a summary). As questionnaires and rating scales are by far the most widely used methods clinically, and the most feasible for use in a clinical context, the remainder of this chapter has a major focus on the benefits and limitations of this method.

Projective/Expressive Techniques

Projective/expressive techniques include drawings, sentence completion tasks, and thematic approaches to assessment. Such techniques are useful for building rapport with the child in a clinical context, and assist in developing hypotheses for further assessment. Merrell (2001) argues, however, that there is little evidence to support the reliability of these measures in classifying social skills and identifying social skill deficits. Furthermore, these techniques fail to satisfy any of the criteria for comprehensive and valid assessment of social competence. Therefore, they should not be considered primary measures for the assessment of social competence in children.

Self-Report Instruments

Self-report instruments consist mainly of objective psychometric measures. That is, children complete questionnaires, and their responses are compared to those of a normative group. Although these have been found to be useful in measuring and identifying personality and internalizing problems such as anxiety (Eckert, Dunn, Guiney, & Codding, 2000; Merrell, 2001), there are few empirical data supporting the use of these instruments to assess social competence (Merrell, 2001). Of the few measures that exist, such as the self-rated Student Form of the Social Skills Rating System (Gresham & Elliott, 1990); which has recently been updated to the Social Skills Improvement System (Gresham & Elliott, 2008); little relationship has been found between self-report instruments and other

TABLE 6.1. Characteristics of the Main Approaches to the Assessment of Social Competence

Approach	Ability to identify skills and behaviors	Developmental relevance	Multiple informants	Ecological validity	Psychometric properties
Projective/ expressive	✓	✓	✓	✓	✓
Self-report	✓✓	✓✓	✓	✓✓	✓
Interviewing	✓✓✓	✓✓✓	✓✓✓	✓✓✓	✓
Sociometric techniques	✓	✓✓	✓✓	✓✓	✓✓
Observation	✓✓✓	✓✓✓	✓✓	✓✓✓	✓✓
Questionnaires/ rating scales	✓✓✓	✓✓✓	✓✓✓	✓✓	✓✓✓

Note. Ratings: ✓ = poor or nonexistent; ✓✓ = fair to good; ✓✓✓ = excellent.

measures, such as behavioral observation and ratings from significant others (Gresham & Elliott, 1990). This method also fails to meet most of the criteria for comprehensive and valid assessment of social competence. Therefore, at this stage, until further data are obtained regarding validity and reliability of self-report measures of social competence, these should also not be considered a primary approach to measuring social competence.

Interviewing

Merrell (2001) argues that interviewing is perhaps the oldest form of socio-emotional assessment. Interviews can be conducted with the child or with parents, teachers, and significant others. This method has many advantages. It allows the collection of information that is functional and relevant to the child, and that can thus be used to inform further assessment and guide intervention. It also satisfies several criteria for assessment of social competence. Namely, it allows information to be collected from multiple informants, which can assist in eliciting skill competencies and deficits, as well as expectations and norms of the various social environments within which the child is engaged; it also allows information to be collected regarding conditions within the environment that reinforce or precipitate behaviors.

Despite its utility in a clinical context, the application of this approach has not been widely researched, and interview techniques have not been examined empirically in formats that are clear, condensed, and able to be replicated and used by clinicians in a consistent and standardized way. Although interviewing is an important approach to assessing social competence, these issues limit its clinical utility. This approach should therefore be considered an important secondary or supportive approach to assessing social competence.

Sociometric Techniques

Sociometric techniques include such methods as peer nomination and peer rankings. In peer nominations, students within a classroom or members of a social group select which other members they like or dislike; in peer rankings, they rank each other member in the group on a list of whom they like most to least. These techniques have been demonstrated to have high levels of reliability and validity, and have been found to be powerful predictors of long-term social outcomes (McElhaney, Antonishak, & Allen, 2008). Merrell (2001), however, suggests that these methods have two limitations. First, they are not considered to be specific to the construct of social competence. Rather, they tend to provide information regarding a child's popularity, which may be considered a consequence of a child's social competence, rather than a measure of a social skill or deficit (Van Hasselt, Hersen, Whitehall, & Bellack, 1978). Rosen et al. (2001) conducted a study investigating the relationship between the amount of social interaction and the results of a sociometric technique where children were asked to rate three children they liked most and three they disliked most. They found that there was little relationship between these two methods; they concluded that sociometric techniques are not adequate measures of social competence when used in isolation.

The second issue associated with these techniques surrounds the process involved in obtaining information. Sociometric assessment is conducted in social groups, such as the classroom or other similar groups in which a child is involved. Practical and ethical issues (e.g., obtaining parental and administrative consent) make implementation challenging. In addition, in order for such an assessment to be valid, participation is required from the entire social group; this is a challenge in itself, especially when individual consent forms are required, and it is not feasible or practical within a clinical context. Even if consent from the entire group is obtained, there is a risk that this approach will further highlight the difficulties of the target child and escalate social rejection (Merrell, 2001). Although these techniques may indirectly address information that is critical for comprehensive assessment of social competence (such as obtaining information from multiple informants), they do not satisfy many of the criteria outlined above. Therefore, although sociometric techniques can provide enormous amounts of information, there are significant limitations that restrict their utility in a clinical context. These limitations suggest that they should be considered only as a secondary approach to assessing social competence in children.

Observation

Unlike the methods previously described, behavioral observation is considered one of the primary methods of assessing social competence (Merrell, 2001). These techniques involve observing children in various social situations, and scoring the nature or maturity of their behavior. There are different forms of behavioral observation, including *naturalistic*, which occurs in the child's natural environment or social setting; and *analogue*, where natural settings are simulated in a laboratory environment. Evidence suggests that the naturalistic form is the preferred method (Elliott & Gresham, 1987; Merrell, 2001). Three key components are important in naturalistic behavioral observation: (1) observing behaviors in their natural setting; (2) using observers that are objective and trained in scoring procedures; and (3) using a behavioral description system that allows the observer to code behaviors with the least amount of subjective interference (Jones, Reid, & Patterson, 1979).

The strength of the observation method is that it is high in ecological validity (Bierman & Welsh, 2000). In a clinical setting, behavioral observation is often used to determine everyday functioning in the child's context, such as familiar home or school environments. Given that there are many advantages to using direct observation, this is considered a primary method of assessing social competence in children. Importantly, this method satisfies several criteria for assessment of social competence. Observation allows the clinician to accurately identify important behavioral competencies and deficits; to assess the environment (such as the expectations of the social environment, and environmental conditions that reinforce and precipitate behaviors); and to examine the function of different behaviors within the environment. Due to the ecological validity of this method, such detailed information will also inform the development of relevant and appropriate interventions, and will allow the measurement of outcomes and changes following the intervention.

Although detailed and ecologically rich information can be obtained from behavioral observation, there are also many drawbacks to using this approach. First, to conduct such an assessment requires a significant time commitment, as does setting up opportunities for the clinician to observe appropriately the child in naturalistic settings. The coding and empirical measurement of the information obtained are also notoriously challenging and time-consuming, as it is difficult to make judgments about observed behaviors without access to normative data. In addition, there is the issue of how many observations are required for reliable assessment (Matson & Wilkins, 2009).

Research suggests that these methods could guide clinicians with regard to interpretation of a child's social competence; however, little research has been done on the clinical usefulness of this information (Bierman & Welsh, 2000; Merrell, 2001). Therefore, given that these techniques (at this stage and in their current forms) are time-consuming, require multiple sessions, necessitate microanalytic scoring systems, and require a group of peers for comparison, they are not practical and feasible within a clinical setting. For these techniques to have greater clinical utility, assessment tools that can explore social difficulties within a single session, with limited involvement of groups or peers, and with rating scales to guide scoring are needed.

Questionnaires and Rating Scales

Questionnaires and rating scales are the most widely used methods for assessment of social competence in children. A recent review of social skills assessments identified 86 measures designed to tap children's social skills, with the majority identified as questionnaires or rating scales (Crowe, Beauchamp, Catroppa, & Anderson, 2011). A rating scale uses a categorical or numerical scale for responses—typically a Likert scale, providing a value on the degree of the behavior being seen. Questionnaires may also employ a set response format or a true–false answer set. Rating scales using a Likert scale are used more frequently than questionnaires in the measurement of social skills (Crowe et al., 2011). Typically, Likert scales common to rating scales use 3 points corresponding to three responses, such as "never," "sometimes" and "always"; some rating scales use a 5-point scale. An odd-numbered scale is usually preferred, as it provides a middle point.

Questionnaires and Rating Scales

A myriad of different rating scales and questionnaires have been developed to measure different aspects and levels of social competence. Some are designed to measure specific social skills, such as emotion awareness (the Emotion Expression Scale for Children; Penza-Clyve & Zeman, 2002), social problem-solving skills (the Social Problem Solving Inventory; D'Zurilla, Nezu, & Maydeu-Olivares, 1998), social participation (the Child and Adolescent Scale of Participation; Bedell, 2009), and peer relationships (the Intimate Friendship Scale; Sharabany, 1974). Many of these questionnaires and rating scales are commonly used in research and clinical environments, and are useful for identifying specific social skill deficits, or for investigating specific areas or domains of social competence. Despite this, they are inadequate to describe and measure the global construct of social competence when used in isolation.

Social competence is also commonly measured by broad assessment tools that are designed to tap behavioral, emotional, or adaptive functions, but also incorporate a social subscale. Examples include the Child Behavior Checklist (Achenbach & Rescorla, 2001); the Adaptive Behavior Assessment System—Second Edition (Harrison & Oakland, 2003); the Vineland Adaptive Behavior Scales, Second Edition (Sparrow, Cicchetti, & Balla, 2005); and the Strengths and Difficulties Questionnaire (Goodman, 1997).

All these measures include subscales on social function, but their focus is not on social competence per se, and hence they do not measure the social domain comprehensively. Although these scales can provide valuable screening information, they can also lead to overinterpretation of a child's social ability through the use of a small pool of items, and they may also not accurately characterize important aspects of social behavior (Drotar, Stein, & Perrin, 1995). Importantly, there is a small group of rating scales that has been designed specifically to measure social competence in children and adolescents. These are described later in the chapter.

There are many advantages to using questionnaires and rating scales to assess social competence: They require relatively little time to administer; are relatively inexpensive; tap into behaviors that are reasonably infrequent; often require minimal training to administer; provide information about people who are not able to provide information about themselves; allow easy comparison of children with others their own age; make use of parent or teacher observations over a long period of time in naturalistic settings (such as at home or at school); and enable the people who know the child best to rate the child's behavior (Merrell, 2008). This method also satisfies many of the requirements for comprehensive assessment of social competence.

Rating scales have been shown to be more objective and reliable than projective techniques or clinical interviews (Edelbrock, 1983), since lab-based measures do not have the external validity to replace teacher and peer ratings of a child (Bierman & Welsh, 2000). And in contrast to data collected via observation, rating scales can provide information on a wider range of behavior, since observation may miss significant behavior that does not appear during the fixed observation period.

The disadvantages of using rating scales and questionnaires for social skills are similar to those related to rating scales and questionnaires designed to measure behavior (Martin, Hooper, & Snow, 1986) and relate to the way that an informant responds to the questionnaire, termed *bias variance*. Bias variance includes the *halo effect*, which refers to inaccurate ratings of a child's function in a particular domain because the child has other redeeming traits. Informants can be either harsh or lenient raters; therefore, the picture of a child's social competence may be inaccurate. *Central tendency* is another form of bias variance, which refers to the tendency for some informants to rate all children at a moderate or average level, shying away from extreme ratings such as "always," "every time," or "never."

Martin and colleagues (1986) highlight difficulties related to the variance of rating scales: source, setting, temporal, and instrument variance. *Source variance* refers to error related to the informant, and includes the issues previously discussed concerning bias. *Setting variance* refers to the fact that a child moves between settings in daily life, such as home and school. It is possible (indeed, highly likely) that a child behaves differently in these environments; therefore, if only a single informant is used, then social behavior can only be confidently assessed within a single setting. *Temporal variance* refers to the potential for behavior to change on a daily basis and suggests that a rating is only valid for a short period of time. *Instrument variance* refers to the finding that children can achieve different scores on two instruments/measurements that are purportedly designed to assess the same behavior.

As previously discussed, it has been argued that in order to generalize social competence to natural settings, assessment of the social environment in which these skills and behaviors occur is necessary. This allows for consideration of the social behaviors within the context of the dynamic social environment (Sheridan et al., 1999). Measurement of social skills and competence with only questionnaires and rating scales limits a clinician's ability to assess the social environment, and hence limits the ecological validity of the assessment. This is a major drawback of this method.

Informants and Rating Scales and Questionnaires

Rating scales and questionnaires can be completed by parents, teachers, or the child being assessed. This flexibility in the number of informants and perspectives that can be obtained regarding a single child is a significant strength of this method, as this is an important factor in the measurement of social competence. Questionnaires and rating scales that use parents as the informants are generally the most popular. Parents are perhaps the greatest "experts" on their child's function, as they have observed their child for a longer time and in a wider range of settings than anyone else, and therefore have the greatest knowledge of their child's typical behavior. Importantly, parents also have a knowledge of important behaviors that are infrequent, and that may not have been seen by others who interact less with the child (Merrell, 2001). Yet the use of the parents to rate their child's social skills assumes that parents have an awareness both of appropriate social skills for the child's age and of how their child compares to peers (Schneider & Byrne, 1989). It is possible that parents' opportunities to view their child in peer interactions may be limited, and/or that their subjectivity may make them reluctant to accept deficits in their child's social ability (Schneider & Byrne, 1989).

Teachers may also be excellent informants, as they are aware of the typical behavior of same-age peers and can compare the target child to other children within the same context and at the same developmental level (Martin et al., 1986). They also tend to be good observers of child behavior and less biased in their ratings than parents.

In contrast, self-report measures are often used as the "gold standard" for assessing social competence. However, reliance on self-report presents difficulties: Self-report assessments have been demonstrated to have low correlations with other assessment types; they can be susceptible to social desirability; and they rely on the young person to comply with instructions (Frankel & Feinberg, 2002). They are also inappropriate for some children and adolescents, such as those who are uncooperative, have limited reading or verbal abilities, exhibit poor attentional skills, or show reduced self-awareness (Eckert et al., 2000). The validity of social behaviors reported by a child or adolescent with significant difficulties in these areas is unclear, and it has been argued that the presence of social deficits may be related to a lack of insight into the presence of a problem (Merrell & Gimpel, 1998). As a consequence, researchers suggest that self-report approaches should be considered an "experimental method" (Merrell, 2001) and need to be supported by information collected from additional sources.

Review of Two Rating Scales

Although numerous rating scales and questionnaires are available (Crowe et al., 2011), few possess robust psychometric properties. In this section, we review two commercially developed rating scales measuring social competence: the Social Skills Rating System (SSRS; Gresham & Elliott, 1990) and the Walker–McConnell Scale of Social Competence and School Adjustment (WMSC; Walker & McConnell, 1989, 1995a, 1995b). Both scales are based on large samples that approximate census data, and are applicable to a wide age range. They also provide both global and specific social outcome indices, tapping into a range of subdomains of social function.

The SSRS (Gresham & Elliott, 1990) is applicable to a wide age range, with an Elementary Level version for children from kindergarten to grade 6 and a Secondary Level version for adolescents from grade 7 to grade 12. The SSRS uses a pencil-and-paper format and a 3-point scale for all versions corresponding to the responses "never," "sometimes," and "very often." An informant is asked to rate statements usually related to behavior. Three questionaires are available: a Child Questionaire, which contains 34–39 items and one scale called Social Skills; a Teacher Questionaire, which contains 40–57 items and three scales called Social Skills, Problem Behaviors, and Academic Competence; and a Parent Questionaire, which contains 49–55 items and two scales called Social Skills and Problem Behaviors. All versions and questionaires were normed on a large sample size that approximated U.S. census data, and all have proven reliability with high internal consistency (Gresham & Elliott, 1990). The validity of the SSRS has been tested against that of other social problem scales, and it has been shown to assess similar concepts. The SSRS is widely used in research (Crowe et al., 2011). A revised version has recently been released, the Social Skills Improvement System Rating Scales (Gresham & Elliott, 2008), which is beginning to be utilized in research.

Criticisms of the SSRS include its limited capability for comparison across versions (child, parent, and teacher), due to a lack of common items (Demaray & Ruffalo, 1995). In addition, its applicability to the social problems described in the context of child brain injury has not yet been established; there is some concern that this tool may be insensitive to the characteristic social difficulties of these children (Muscara et al., 2009).

The WMSC (Walker & McConnell, 1995a, 1995b) is also applicable to a wide age range. It now has an Elementary Version for children from kindergarten to grade 6, which consists of 43 items, and an Adolescent Version for adolescents from grade 7 to grade 12, which consists of 53 items. It uses a pencil-and-paper format and is completed by teachers. Teachers rate a child's behavior when interacting with peers, using a 5-point scale from "never" to "frequently." The scale has strong reliability, with high internal consistency and 1-year test–retest reliability. Validity has been tested by comparing the scale to other well-known behavior scales (including the SSRS), as well as other measures of social skills/problem behaviors and observational measures. A recent review suggests that the WMSC is not as widely used in research as the SSRS (Crowe et al., 2011), most likely because it is only applicable for one informant. However, it has been used with various clinical populations, including children and adolescents with an autism spectrum disorder (ASD) (Cotugno, 2009) or with traumatic brain injury (Greenham, Spencer-Smith, Anderson, Coleman, & Anderson, 2010).

Assessing Social Skills in Special Populations

Some measures have been custom-designed to assess specific clinical populations, in response to the understanding that a number of childhood developmental and acquired conditions are associated with deficits in social competence. For example, the most commonly studied social outcomes are in children with an ASD, for whom deficits in social skills are the defining features of their condition. The Social Responsiveness Scale

(Constantino, 2002) and the Children's Social Behavior Questionnaire (Luteijn, Luteijn, Jackson, Volkmar, & Minderaa, 2000) are both used primarily in the population with ASD, and target areas of social function typically described in these children. The Social Responsiveness Scale includes items pertaining to social awareness, social information processing, capacity for reciprocal social engagement, and social anxiety/avoidance (Constantino, 2002), and it shows good reliability and validity (Constantino & Todd, 2003; Pine, Luby, Abbacchi, & Constantino, 2006). The Children's Social Behavior Questionnaire incorporates items related to social interaction, communication, and understanding social cues. Its psychometric properties with respect to convergent validity, test–retest reliability, interrater reliability, and internal reliability of the scales were reported to be good (Luteijn et al., 2000).

Other rating scales and questionnaires that have been developed to measure social skills in clinical populations include the Adaptive Behavior Assessment System—Second Edition (Harrison & Oakland, 2003) and the Child and Family Follow-Up Survey (Bedell, 2004). The Adaptive Behavior Assessment System includes a subscale tapping social competence and is particularly pertinent to intellectual disability. It has high test–retest reliability, and the Social Composite scale has good reliability and internal consistency (Harrison & Oakland, 2003). The Child and Family Follow-Up Survey looks at participation in children with acquired brain injury, and preliminary evidence has been found regarding its validity and reliability (Bedell, 2004).

Best Practice in Using Rating Scales and Questionnaires

The reliability and validity of the information gathered from rating scales can be improved by (1) testing at multiple time points and within different settings, (2) using more than one rating scale, and (3) utilizing more than one informant. For example, the authors of the SSRS advise users to collect ratings from all possible informants to get the most accurate picture of a child's social competence. Merrell (2001) highlights the value of using both parents and teachers as informants when assessing social competence, since their unique relationships with the child make them ideally placed to collaboratively judge the child's ability. Taken together, parent and teacher reports provide insight into a young person's social competence in a range of environments, and into his or her relationships with peers in these diverse settings (Merrell, 2001). Ideally, an assessment tool that can collect information from more than one source is desirable and increases the validity of findings.

In a clinical context, when an evaluator is considering which questionnaire or rating scale to use, it is critical to consider the factors that contribute to the development of an effective tool. Andresen (2000) outlines a number of factors that are important in determining the quality of a questionnaire or rating scale:

1. The use of a conceptual model on which the tool is based.
2. The development of normative data or standard values for the tool.
3. The specificity and sensitivity of the tool for the population in question.
4. The degree to which bias exists toward specific populations.
5. The administrative and participant burden (e.g., tool length, ease of scoring and interpretation).

6. The reliability and validity of the tool.
7. The responsiveness, which is the tool's ability to identify changes or differences in functioning.
8. The existence of alternate forms for a number of different informants.

This framework allows the clinician and researcher to evaluate the empirical quality of a particular tool, in order to maximize the effectiveness and reduce the limitations inherent in the use of questionnaires and rating scales for measurement of social competence.

Current Best Practice in the Assessment of Social Competence

Despite their popularity, questionnaires and rating scales have a number of drawbacks that limit their capacity, when used in isolation, to provide a comprehensive clinical assessment. Merrell (2001) outlines the "aggregation principle" as a method of best practice in assessing social competence in children and adolescents. The aggregation principle involves obtaining information from a number of different sources and via several methods, such as interviewing, rating scales, and observation. Given the strengths and weaknesses of various methods and tools, each will provide different information and present a slightly different perspective on a child's functioning. In addition, collating information from various perspectives will allow identification of particular social skills or deficits that are consistent across various social contexts or settings. Hence information collected from a number of teachers and from parents will provide the most comprehensive picture of the child's level of social competence. Merrell (2001) argues that differences in ratings should be expected when information is obtained from different perspectives in various social settings, although significant behavioral or social deficits are likely to be consistent across settings and sources.

The aggregation of information using different methods, different sources, and different settings will give the clinician the best chance of identifying social deficits and developing an understanding of the child's social competence. This process also informs the development of targeted intervention for specific problem behaviors or skill deficits. Although this is currently the most comprehensive method of assessing social competence in children and adolescents, such an approach may result in some challenges in interpreting the information, as there will be large amounts of information from various sources and in various forms.

Assessment of Social Competence and the Social Neurosciences

Despite ongoing research and theoretical advances in this area, there remains a lack of appropriate, relevant, and robust measurement tools assessing social competence for particular clinical populations within the pediatric population (Crowe et al., 2011; McCauley et al., 2011). Social neuroscience is a burgeoning field of research, with significant advances occurring within the field. Unfortunately, the measurement tools available to support this research are lagging behind theoretical advances, impeding progress within this area. In fact, a National Institutes of Health working group has examined measurement tools used by pediatric acquired brain injury researchers. This international working group identified a lack of appropriate and psychometrically robust measures of social competence in children with acquired brain injury (McCauley et al., 2011). The lack of comprehensive, theoretically based measures of social competence makes it particularly difficult to assess aberrant social outcomes, such as those seen in children within different clinical populations (e.g., acquired brain injury). The development of psychometrically robust, theoretically based tools is therefore required to better inform clinicians and researchers regarding patterns of social function and dysfunction across particular diagnostic groups, and to support research attempting to describe and characterize factors that may underpin and contribute to these difficulties. Importantly in the context of the social neurosciences, more relevant and reliable measures are important for use within imaging studies, which are continuing to expand our knowledge and to contribute to our understanding of brain correlates of social competence and its many aspects.

Future Directions

We note that the assessment of social competence may have much in common with measurement in the domain of executive function. Specifically, it requires a multifaceted approach to measurement—tapping into a range of cognitive domains (see Anderson & Beauchamp, Chapter 1, this volume), and using a number of assessment methodologies and tools (Saltzman-Benaiah & Lalonde, 2007). Although this may be best practice, it is a time-consuming process. More accurate, psychometrically robust assessment methods and tools are needed that will allow clinicians to measure social difficulties with limited involvement of groups or peers, and with rating scales to guide scoring. Part of this challenge is to develop performance-based instruments with high social and ecological validity, so that they can be used in clinical settings.

This best-practice method of assessment can also be fraught with difficulties in interpreting multiple and aggregated sources of information. The future challenges in this field thus also include finding empirically supported methods of easily aggregating information from various sources to make this information functional. Such methods will permit clinicians to link findings with specific social competence profiles, in order to assist with the development and evaluation of targeted and effective interventions.

Conclusions

Our understanding of the development of social competence in children has increased a great deal over the past few decades. There is increasing evidence that social impairments not only affect children with ASD, but are also associated with a number of other pediatric clinical conditions, such as traumatic brain injury, stroke, learning disabilities, attention-deficit/hyperactivity disorder, chronic illness, and many other neurological problems. Despite rapid development in knowledge, there remain few valid and reliable tools that are relevant for specific clinical groups. Importantly, among the tools that exist, a uniform, empirically based approach to aggregating information from various sources is still lacking. This chapter has outlined the current state of the field, and has discussed the need for further development of valid, reliable, and clinically useful measures of social competence in children and adolescents.

References

Achenbach, T., Rescorla, L. A. (2001). *ASEBA School-Age Forms and Profiles*. Burlington: University of Vermont, Research Center for Children, Youth, and Families.

Anderson, V., Brown, S., & Newitt, H. (2010). What contributes to quality of life in adult survivors of childhood traumatic brain injury? *Journal of Neurotrauma, 27*, 863–870.

Andresen, E. M. (2000). Criteria for assessing the tools of disability outcomes research. *Archives of Physical Medicine and Rehabilitation, 81*, S15–S20.

Angold, A., Costello, E. J., Farmer, E. M., Burns, B. J., & Erkanli, A. (1999). Impaired but undiagnosed. *Journal of the American Academy of Child and Adolescent Psychiatry, 38*, 129–137.

Bedell, G. M. (2004). Developing a follow-up survey focused on participation of children and youth with acquired brain injuries after discharge from inpatient rehabilitation. *NeuroRehabilitation, 19*, 191–205.

Bedell, G. M. (2009). Further validation of the Child and Adolescent Scale of Participation (CASP). *Developmental Neurorehabilitation, 12*, 342–351.

Bierman, K. L., & Welsh, J. A. (2000). Assessing social dysfunction: The contributions of laboratory and performance-based measures. *Journal of Clinical Child Psychology, 29*, 526–539.

Canino, G., Costello, E. J., & Angold, A. (1999). Assessing functional impairment for mental health services research: A review of measures. *Journal of Mental Health Research, 1*, 93–108.

Cattelani, R., Lombardi, F., Brianti, R., & Mazzucchi, A. (1998). Traumatic brain injury in childhood: Intellectual, behavioural and social outcome into adulthood. *Brain Injury, 12*, 283–296.

Constantino, J. N. (2002). *The Social Responsiveness Scale*. Los Angeles: Western Psychological Services.

Constantino, J. N., & Todd, R. D. (2003). Autistic traits in the general population: A twin study. *Archives of General Psychiatry, 60*, 524–530.

Cotugno, A. J. (2009). Social competence and social skills training and intervention for children with autism spectrum disorders. *Journal of Autism and Developmental Disorders, 39*(9), 1268–1277.

Crowe, L. M., Beauchamp, M. H., Catroppa, C., & Anderson, V. (2011). Social function assessment tools for children and adolescents: A systematic review from 1988 to 2010. *Clinical Psychology Review, 31*(5), 767–785.

D'Zurilla, T. J., Nezu, A. M., & Maydeu-Olivares, A. (1998). *Manual for the Social Problem Solving Inventory—Revised (SPSI-R)*. North Tonawanda, NY: Multi-Health Systems.

Demaray, M. K., & Ruffalo, S. L. (1995). Social skills assessment: A comparative evaluation of six published rating scales. *School Psychology Review, 24*, 648–672.

Dooley, J. J., Beauchamp, M. H., & Anderson, V. (2010). The measurement of sociomoral reasoning in adolescents with traumatic brain injury: A pilot investigation. *Brain Impairment, 11*, 152–161.

Drotar, D., Stein, R., & Perrin, E. (1995). Methodological issues in using the Child Behavior Checklist and its related instruments in clinical child psychology research. *Journal of Clinical Child Psychology, 24*, 184–192.

Eckert, T. L., Dunn, E. K., Guiney, K. M., & Codding, R. S. (2000). Self reports: Theory and research in using rating scale measures. In E. S. Shapiro & T. R. Kratochwill (Eds.), *Behavioral assessment in schools: Theory, research, and clinical foundations* (2nd ed.). New York: Guilford Press.

Edelbrock, C. (1983). Problems and issues in using rating scales to assess child personality and psychopathology. *School Psychology Review, 12*, 293–299.

Elliott, S. N., & Gresham, F. (1987). Children's social skills: Assessment and classification practices. *Journal of Counseling and Development, 66*, 96–99.

Elliott, S. N., Malecki, C. K., & Demaray, M. K. (2001). New directions in social skills assessment and intervention for elementary and middle school students. *Exceptionality, 9*, 19–32.

Frankel, F., & Feinberg, D. (2002). Social problems associated with ADHD vs. ODD in children referred for friendship problems. *Child Psychiatry and Human Development, 33*, 125–146.

Goodman, R. (1997). The Strengths and Difficulties Questionnaire: A research note. *Journal of Child Psychology and Psychiatry, 38*, 581–586.

Greenham, M., Spencer-Smith, M. M., Anderson, P. J., Coleman, L., & Anderson, V. (2010). Social functioning in children with brain insult. *Frontiers in Human Neuroscience, 4*, 1–10.

Gresham, F. (1986). Conceptual issues in the assessment of social competence in children. In P. S. Strain, M. J. Guralnick, & H. M. Walker (Eds.), *Children's social behavior: Development, assessment, and modification.* Orlando, FL: Academic Press.

Gresham, F., & Elliott, S. N. (1990). *Social Skills Rating System.* Circle Pines, MN: American Guidance Service.

Gresham, F., & Elliott, S. N. (2008). *Social Skills Improvement System Rating Scales.* Circle Pines, MN: AGS/Pearson Assessment.

Grossman, H. (1983). *Classification in mental retardation.* Washington, DC: American Association on Mental Deficiency.

Guralnick, M. J. (1999). Family and child influences on the peer-related social competence of young children with developmental delays. *Mental Retardation and Developmental Disabilities Research Reviews, 5*, 21–29.

Harrison, P., & Oakland, T. (2003). *Adaptive Behavior Assessment System—Second Edition.* San Antonio, TX: Psychological Corporation.

John, K. (2001). Measuring children's social functioning. *Child Psychology and Psychiatry Review, 6*, 181–188.

John, K., & Weissman, M. M. (1987). The familial and psychosocial measurement of depression. In A. J. Marsella, R. M. A. Hirschfeld, & M. M. Katz (Eds.), *The measurement of depression.* New York: Guilford Press.

Jones, R. R., Reid, J. B., & Patterson, G. R. (1979). Naturalistic observation in clinical assessment. In P. McReynolds (Ed.), *Advances in psychological assessment.* San Francisco: Jossey-Bass.

Law, M., Anaby, D., Dematteo, C., & Hanna, S. (2011). Participation patterns of children with acquired brain injury. *Brain Injury, 25*, 587–595.

Luteijn, E., Luteijn, F., Jackson, S., Volkmar, F. R., & Minderaa, R. B. (2000). The Children's Social Behavior Questionnaire for milder variants of PDD problems: Evaluation of the psychometric characteristics. *Journal of Autism and Developmental Disorders, 30*, 317–330.

Martin, R. P., Hooper, S., & Snow, J. P. (1986). Behavior rating scale approaches to personality assessment in children and adolescents. In H. M. Knoff (Ed.), *The assessment of child and adolescent personality.* New York: Guilford Press.

Matson, J. L., & Wilkins, J. (2009). Psychometric testing methods for children's social skills. *Research in Developmental Disabilities, 30*, 249–274.

McCauley, S. R., Wilde, E. A., Hicks, R., Anderson, V., Bedell, G. M., Beers, S. R., et al. (2011). Recommendations for the use of common outcome measures in pediatric traumatic brain injury research. *Journal of Neurotrauma, 29*(4), 678–705.

McElhaney, K. B., Antonishak, J., & Allen, J. P. (2008). "They like me, they like me not": Popularity and adolescents' perceptions of acceptance predicting social functioning over time. *Child Development, 79*, 720–731.

Merrell, K. W. (1997). Assessing social skills and peer relations. In H. B. Vance (Ed.), *Psychological assessment of children: Best practices for school and clinical settings*. New York: Wiley.

Merrell, K. W. (2001). Assessment of children's social skills: Recent developments, best practices, and new directions. *Exceptionality, 9*, 3–18.

Merrell, K. W. (2008). *Behavioral, social, and emotional assessment of children and adolescents* (3rd ed.). New York: Erlbaum.

Merrell, K. W., & Gimpel, G. A. (1998). *Social skills of children and adolescents: Conceptualisation, assessment, treatment*. Mahwah, NJ: Erlbaum.

Muscara, F., Catroppa, C., & Anderson, V. (2008). Social problem solving skills as a mediator between executive function and long-term social outcome following pediatric traumatic brain injury. *Journal of Neuropsychology, 2*, 445–461.

Muscara, F., Catroppa, C., Eren, S., & Anderson, V. (2009). The impact of injury severity on long-term social outcome following pediatric traumatic brain injury (TBI). *Neuropsychological Rehabilitation, 19*, 541–561.

Oddy, M., Coughlan, T., Tyerman, A., & Jenkins, D. (1985). Social adjustment after closed head injury: A further follow-up seven years after injury. *Journal of Neurology, Neurosurgery and Psychiatry, 48*, 564–568.

Parker, J. G., & Asher, S. R. (1993). Beyond group acceptance: Friendship adjustment and friendship quality as distinct dimensions of children's peer adjustment. In D. Perlman & W. H. Jones (Eds.), *Advances in personal relationships*. London: Kingsley.

Penza-Clyve, S., & Zeman, J. (2002). Initial validation of the Emotion Expression Scale for Children. *Journal of Clinical Child and Adolescent Psychology, 31*, 540–547.

Pine, E., Luby, J., Abbacchi, A., & Constantino, J. N. (2006). Quantitative assessment of autistic symptomatology in preschoolers. *Autism, 10*, 344–352.

Rosen, L. A., Furman, W., & Hartup, W. W. (2001). Positive, negative, and neutral peer interactions as indicators of children's social competency: The issue of concurrent validity. *Journal of Genetic Psychology, 149*, 441–446.

Rubin, K. H., & Rose-Krasnor, L. (1992). Interpersonal problem-solving. In V. B. V. Hassett & M. Hersen (Eds.), *Handbook of social development*. New York: Plenum Press.

Saltzman-Benaiah, J., & Lalonde, C. E. (2007). Developing clinically suitable measures of social cognition for children: Initial findings from a normative sample. *The Clinical Neuropsychologist, 21*, 294–317.

Schneider, B., & Byrne, B. (1989). Parent rating children's social behavior: How focused is the lens? *Journal of Clinical Child Psychology, 18*, 237–241.

Sharabany, R. (1974). *Intimate friendship among kibbutz and city children and its measurement*. Unpublished doctoral dissertation, Cornell University.

Sheridan, S. M., Hungelmann, A., & Maughan, D. P. (1999). A contextualized framework for social skills assessment, intervention, and generalisation. *School Psychology Review, 28*, 84–103.

Sheridan, S. M., & Walker, D. (1999). Social skills in context: Considerations for assessment, intervention, and generalization. In C. R. Reynolds & T. B. Gutkin (Eds.), *The handbook of school psychology* (3rd ed.). New York: Wiley.

Sparrow, S., Cicchetti, D., & Balla, D. (2005). *The Vineland Adaptive Behavior Scales, Second Edition (Vineland-II): Survey Interview Form*. Circle Pines, MN: American Guidance Service.

Van Hasselt, V. B., Hersen, M., Whitehall, M. B., & Bellack, A. S. (1978). Social skill assessment and training for children: An evaluative review. *Behaviour Research and Therapy, 17*, 413–437.

Walker, H., & McConnell, S. (1989). *The Walker–McConnell Scale of Social Competence and School Adjustment: A social skills rating scale for teachers*. Austin, TX: PRO-ED.

Walker, H., & McConnell, S. (1995a). *The Walker–McConnell Scale of Social Competence and School Adjustment (Adolescent Version)* Belmont, CA: Wadsworth/Thomson Learning.

Walker, H., & McConnell, S. (1995b). *The Walker–McConnell Scale of Social Competence and Social Adjustment (Elementary Version)*. Belmont, CA: Wadsworth/Thomson Learning.

Yeates, K. O., Bigler, E., Dennis, M., Gerhardt, C., Rubin, K., Stancin, T., et al. (2007). Social outcomes in childhood brain disorder: A heuristic integration of social neuroscience and developmental psychology. *Psychological Bulletin, 133,* 535–556.

Yeates, K. O., Schultz, L. H., & Selman, R. L. (1990). Bridging the gap in child-clinical assessment: Toward the application of social-cognitive developmental theory. *Clinical Psychology Review, 10,* 567–588.

CHAPTER 7

Measuring the Different Components of Social Cognition in Children and Adolescents

Rosée Bruneau-Bhérer, Amélie M. Achim, and Philip L. Jackson

Humans are faced with complex social interactions right from birth. Social interactions are very intricate, even among children. Accurate perception and interpretation of social cues, as well as the generation of appropriate actions that can lead to social inclusion or exclusion, demand high levels of integration between cognitive and affective processes. Take, for instance, the case of an 8-year-old boy named Brian, being introduced to a new school in the middle of the school year after his family has moved from another city. How is he going to make new friends when peer groups are already formed and children have already known each other for at least several months? This situation represents a challenge for Brian. First, he can try to identify the most welcoming kids and observe them in the schoolyard to collect information about their interests and favorite activities. Then he must initiate interactions based on what he knows about these children, or react promptly if an invitation to join a group is made. Importantly, the complex cognitive processes at work during these first contacts take place in a highly affective context, where emotions are not fully controlled and quite spontaneous.

The development of harmonious social functioning relies on well-tuned social skills that emerge at different stages during childhood and continue to evolve during adulthood. *Social cognition* typically refers to the ensemble of specialized mental processes involved in social functioning (Beauchamp & Anderson, 2010); these are mostly learned, but also depend on some innate processes. Generally speaking, social cognition depends on the integrity of several different yet interacting neural circuits, but their specific functionality remains largely unclear (Beauchamp & Anderson, 2010). It comes as no surprise that such complexity at both the conceptual and cerebral functional levels poses many challenges when it comes to measuring social cognition.

Deficits in social functioning have been linked to poor social cognition, but the specific contributions of different affective and cognitive processes to everyday social functioning is still unknown. The typical developmental trajectories of the distinct processes can be different, but they nevertheless interact. Deficits in one process can directly disrupt social functioning or indirectly prevent the development of one or many other specific skills. Therefore, the origin of a deficit in social functioning is often difficult to pinpoint. Still, the social domain is often disturbed in many childhood disorders. For instance, social functioning is affected in a number of developmental disorders, such as autism spectrum disorders, social phobia, attention-deficit/hyperactivity disorder (ADHD), oppositional and conduct disorders, and schizophrenia (American Psychiatric Association, 2000). It can also become disrupted more suddenly—for instance, after a brain insult, which can affect subsequent social development as well (Catroppa, Anderson, Morse, Haritou, & Rosenfeld, 2008). As social difficulties are more frequently recognized among various clinical populations, it becomes crucial to identify means of assessment focusing on social cognition that can (1) characterize typical development trajectories, and (2) identify atypical functioning at different developmental stages.

There is an increasing interest in social functioning and its determinants in various domains, such as developmental psychology, neuropsychology, cognitive neurosciences, and psychopathology. This extensive literature implies that several different theories of social cognition's development (Baron-Cohen, Leslie, & Frith, 1985; Beauchamp & Anderson, 2010; Burnett, Sebastian, Cohen Kadosh, & Blakemore, 2011; Eisenberg et al., 2007) and several strategies to assess everyday social functioning are emerging. In general, however, assessment approaches focus on one aspect of social cognition and often fall short of explaining social difficulties in many childhood disorders. For example, researchers in the field of developmental psychopathology can explain autism-like social symptoms by a delay in theory-of-mind (ToM) development, but such a delay is not present in children with ADHD, who also present social deficits. Furthermore, a current problem with the various measures of social cognition is their lack of external and internal validity. First, it is difficult to recreate complex social interactions in experimental settings and to make them realistic enough to simulate the emotional aspects that are thought to play a key role in social cognition (Achim, Boutin, Jackson, Guitton, & Monetta, 2011). This is why performance on laboratory tests cannot easily be generalized to real-life events—a difficulty that limits these tests' external validity. Second, the association between performance in experimental settings and everyday social functioning, or between performance in an experimental setting and questionnaires assessing social skills, is modest at best. Last, definitions vary from one domain to another, and measures of social cognition (or its outcome, social functioning) often lack norms and sometimes have been tested only with small samples. These elements limit the possibility of drawing definite conclusions about the source of social deficits, and they raise legitimate questions about the best way to assess social cognition and social skills in order to predict social functioning trajectories.

Although there is as yet no consensus in the scientific literature about how best to assess typical and atypical social cognition developmental trajectories, the aim of this chapter is to review the different tasks that have been used thus far to assess social cognition in children within clinical and experimental settings. We further try to pinpoint the main components that a comprehensive assessment should cover, based on a contemporary theoretical framework (see Figure 7.1).

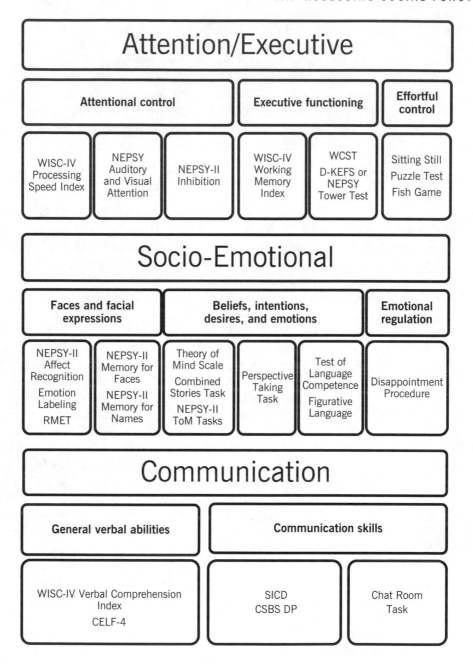

FIGURE 7.1. Suggested assessment tools for the various processes involved in each component of social cognition.

The theoretical framework in relation to which we discuss the different social cognition assessment tools is the biopsychosocial socio-cognitive integration of abilities model (SOCIAL) described by Anderson and Beauchamp (Chapter 1, this volume; see also Beauchamp & Anderson, 2010). This model, based on recent neuroimaging and behavioral studies, suggests that the acquisition of social cognition abilities and social skills is a product of the interaction among the development of three major components: general cognition capacities (the attention/executive component), social and emotional processes (the socio-emotional component), and language and communication abilities (the communication component). As proposed by Beauchamp and Anderson, the assessment of social functioning should include general outcome measures such as the Vineland Adaptive Behavior Scales, Second Edition (Sparrow, Cicchetti, & Balla, 2005) and the Child Behavior Checklist (Achenbach & Rescorla, 2001) to provide indicators of behavioral functioning. In order to interpret general outcome measures and explain everyday social functioning, this chapter identifies a number of tests within each category of processes (attention/executive, socio-emotional, and communication) and regroups them accordingly. In addition to the many questionnaires and scales (reviewed by Muscara & Crowe, Chapter 6, this volume), diverse experimental tasks have been used to measure relations among cognitive processes, socio-emotional processes, communicative abilities, and their contributions to everyday social outcomes. This chapter provides guidelines for building a theory-driven multidimensional assessment battery, based on the combination of information from standardized classic neuropsychological tests, observation paradigms, and experimental tasks.

One advantage of using multiple measures is the possibility of comparing general cognitive processes, such as aspects of *executive function* (EF), with more specific processes like *mentalizing*. EF is a broad concept covering cognitive abilities associated with the conscious control of thoughts and actions (Baddeley & Dell Sala, 1996; Zelazo & Müller, 2002, 2011). In contrast, mentalizing is defined as the capacity to infer the mental states of others from the social cues they produce as well as from contextual information (Achim et al., 2011). Information about general cognition and specific social cognition domains may explain why a child seems to progress satisfactorily at school but has a hard time making and keeping friends. Of course, deficits in general cognitive abilities such as memory or inhibition can also disrupt social cognition, but a specific delay in ToM (a major aspect of mentalizing) can also occur in a child who is showing typical "nonsocial" cognitive performance. Social cognition assessment should tap into crucial skills for social interactions to isolate deficient and preserved functions. Indicating more precisely the levels at which mental processes are compromised is important to guide individualized intervention, based on preserved functions.

Measures of the Attention/Executive Component

Several aspects of general cognition have been identified as having an impact on social skills; as such, classic intelligence and neuropsychological tests can be useful in the assessment of social cognition. The Wechsler Intelligence Scale for Children (WISC-IV; Wechsler, 2003) is an individually administered intelligence test for children from the ages of 6 through 16. The Wechsler Preschool and Primary Scale of Intelligence—Third Edition (WPPSI-III; Wechsler, 2002) is designed for children between 2 years, 6 months and 7 years, 3 months of age. Both tests give a score for general intellectual ability (intelligence quotient) and two indexes reflecting Verbal Comprehension and Perceptual Reasoning. In addition, there are indexes for Processing Speed for children 4 years old and above, and Working Memory for children 6 years and above. The WISC-IV and the WPPSI-III can be used to identify specific cognitive deficits that may lead directly or indirectly to impairments in social functioning. Furthermore, these batteries provide a general measure of cognitive capacities that may serve as a reliable point of comparison for specific performances in social cognition tasks, and as a basis for predicting results according to cognitive capacities.

A few studies have used neuropsychological assessment batteries in combination with specific social understanding measures discussed below in the socio-emotional component, including measures of ToM (Wellman, 2011), to try to isolate the specific processes responsible for ToM task performance (Carlson, Mandell, & Williams, 2004; Carlson & Moses, 2001; Hala, Hug, & Henderson, 2003). Results suggest that tasks targeting both inhibitory control and working memory processes produce the strongest correlations between EF and ToM (Moses & Tahiroglu, 2010). This is consistent with the structure of ToM tasks, which require children to maintain different perspectives on a given situation simultaneously in working memory. Also, inhibitory control is necessary for social functioning, as it allows an individual to avoid impulsive aggressive behaviors (Beauchamp & Anderson, 2010). Inhibition can affect automatic cognitions, behaviors, or emotional responses. Different types of inhibition can be dissociated to pinpoint the potential source of social cognition deficits. An inhibitory problem may occur solely within affective social situations in relation to poor affect regulation, or may be generalized to cognitive and behavioral inhibition as well.

To try to explain the relation between EF and ToM, Moses and Tahiroglu (2010) examined a large set of empirical data to test different hypotheses concerning their intrinsic ties. Their conclusion suggests that executive skills are necessary for the emergence of ToM performance, but that no causal link can be drawn from the literature reviewed. The role of social opportunities as a potential mediator of this relation should also be taken into account, since better EF may lead to more frequent adequate social interactions and consequently facilitate the emergence of ToM. It is also possible that better interactions may facilitate EF development, which in turn would positively influence ToM acquisition (see Hughes, 1998, and Moses & Tahiroglu, 2010, for discussions on this topic). Also, Hughes, Ensor, Wilson, and Graham (2010) specified a model in which performance on measures of planning loaded onto an EF factor separate from working memory and inhibitory control. Therefore, it seems that increasing capacities in inhibitory control, working memory, and planning are especially important for social adjustment (measured in terms of behavioral problems and self-perceived academic competence), and that this relation could be mediated by children's self-esteem (Hughes & Ensor, 2011).

A social cognition battery should include intellectual and neuropsychological tests to assess attentional control (including speed of processing, selective attention, and response inhibition) and EF (including working memory, cognitive flexibility, problem solving, and planning). Also, ecological strategies to assess EF behaviorally, through the concept of effortful control, have been proposed by Eisenberg et al. (2007). Most neuropsychological tests concern broad cognitive abilities and tap into several processes at once, but some tests are better than others at isolating specific processes. The NEPSY-II (Korkman, Kirk, & Kemp, 2007) is used to evaluate children and adolescents between 3 and

16 years old in six functional domains. Some subtests of the NEPSY-II assess attention and EF (described in this section); other subtests assess socio-emotional processes and are reviewed in the section on those processes.

Attentional Control

Attention, defined here as the capacity to focus on important information in the social environment, is crucial for social skills. In a busy schoolyard, children must be able to identify welcoming cues to find new friends, and also to ignore teasing and cues of rejection, in order to make their way through this new social world. Selective attention is important for social cognition, as it allows a child to focus on relevant visual or auditory information and to ignore irrelevant information. Selective attention tasks require participants to respond only to previously defined target stimuli and to ignore other stimuli. Selective attention measures should be included in a social cognition battery, to ensure that this general cognitive ability is not compromised before more specific deficits are interpreted. Also, speed of processing should be controlled for, as it has been associated with good social functioning (Anderson, 2008).

Speed of Processing

Speed of processing, which represents the ability to perceive and process information automatically and rapidly, can be assessed with the Wechsler index that combines the Coding and Symbol Search subtests. This index is available in the WISC-IV (Wechsler, 2003) for children and adolescents from 6 through 16 years of age, and also in the WPPSI-III (Wechsler, 2002) for children between 4 and 7 years old.

Selective Attention

The NEPSY (Korkman, Kirk, & Fellman, 1998) includes two subtests for *selective attention*, the ability to identify target stimuli among distractors—namely, the Auditory Attention and the Visual Attention subtests. These tests can be used to compare performance differences between visual (drawings) and auditory (words) attention, and thus to dissociate specific difficulties in one modality from a generalized attention deficit.

Response Inhibition

Inhibition is the ability to suppress automatic, dominant, or prepotent responses (Baddeley & Della Sala, 1996). This ability is crucial for social cognition (Carlson et al., 2004), as the most automatic response to social stimulation is not always appropriate. Cognitive inhibition assessment is generally included in neuropsychological batteries, but the classic inhibition test is the Stroop test (Stroop, 1935). A Stroop-like test is included in the NEPSY-II (Korkman et al., 2007) as the Inhibition subtest, and it comprises three conditions: rapid naming, inhibition, and flexibility. The third condition demands mental flexibility (see "Cognitive Flexibility and Problem Solving," below) to alternate between two different instructions, and should be used only with children 7 years old and above. Children from the age of 5 can be assessed with the first two conditions. The NEPSY-II gives two versions for each condition. Because the total performance on both versions is used to get the scaled scores, the impact of attention disruption is minimized. It takes longer to administer, but the score thus obtained is more likely to reflect the participant's level of cognitive inhibition, and thus this subtest is preferred over the classic Stroop test.

Executive Functioning

Working Memory

Working memory is the capacity to maintain and manipulate information in short-term memory (Baddeley & Della Sala, 1996). Deficits in working memory can lead to problems with interpreting and using important social information, and thus to behavioral and interactional difficulties (e.g., Kofler et al., 2011). In day-to-day interactions, such deficits may give the impression that a child is not taking into account the information provided by peers or teachers. The WISC-IV (Wechsler, 2003) includes two subtests specifically assessing verbal working memory: Digit Span and Letter–Number Sequencing. For Digit Span, children listen to sequences of numbers and are asked to repeat them in the same or reverse order. Letter–Number Sequencing assesses cognitive manipulation, and sequences must be repeated in a reorganized fashion.

Cognitive Flexibility and Problem Solving

Cognitive flexibility should be measured as part of a social cognition assessment, because social situations can change rapidly and social representations must be adapted to new elements. When Brian (the child in our earlier example) arrives at his new school, he may be interested in befriending a child who seems popular at first, but may then change his mind when he later sees the same kid bullying others. As noted above, the third NEPSY-II Inhibition subtest condition assesses flexibility, or the capacity to alternate easily between two sets of task instructions. Again, only children 7 years old and above can be assessed reliably with the flexibility condition. Another possibility is the Wisconsin Card Sorting Test (WCST; Heaton, 1981), which taps into numerous aspects of EF, such as flexibility and problem solving. This task can be used with children from the age of 6 and requires participants to associate individual cards with one of four stimulus cards. Cards can be classified according to different characteristics (color, shapes, and number of shapes), and participants must figure out the sorting rule from the experimenter's feedback on each trial (correct or incorrect). Successful completion of this task requires that children construct a problem representation, identify relevant dimensions, deduct implicit rules, and carry on a plan long enough. The WCST also evaluates cognitive flexibility, as children are required to change strategies when the rule changes implicitly. This task gives a general measure of EF, but because it taps into many aspects of EF, it should be used in concert with more specific assessment of each aspect contributing to performance.

Planning

The Delis–Kaplan Executive Function System (D-KEFS; Delis, Kaplan, & Kramer, 2001) includes the Tower subtest to evaluate EF and, more specifically, planning abilities in children 8 years old and above. Although the D-KEFS version of the Tower test is preferred for its complexity, children under 8 can be assessed with the easier NEPSY Tower subtest

(Korkman et al., 1998). In the D-KEFS subtest, three wooden pegs, each supporting cylinders of different sizes, are placed in front of the child, who is asked to reproduce the model tower shown on a drawing. The child has to move the cylinders from one stick to another while following two rules: (1) Only one cylinder can be moved at a time, and (2) a big cylinder cannot be placed on top of a smaller one. Participants are told to reproduce the model as fast as they can, using a minimum number of moves, but without breaking the rules. To perform this task successfully, children and adolescents must inhibit impulsive moves and show good planning abilities. Also, this task recruits working memory, because the rules must be maintained online during the task to avoid losing time. Impulsivity (time before the first move), rule breaking, number of moves, and completion time can be scored separately to assess the different aspects of EF necessary to complete the task. Also, a total achievement score gives a general measure of planning capacities.

Effortful Control, or the Ecological Aspects of EF

Effortful control is defined by Rothbart and Bates (2006) as the behavioral expression of classic aspects of EF, such as inhibition and planning. It can be measured by three simple tasks, which can be used in combination with intellectual and neuropsychological tests to increase external validity of cognitive performances. The Sitting Still paradigm was created to assess inhibitory control or the capacity to inhibit behaviors. This very ecological and simple task requires children to sit still for 1 minute while the experimenter is out of the room (for a complete description, see Eisenberg et al., 2007). Children are told not to move so that the material (an electrode is stuck to one of their hands for the purpose of the paradigm) will not be disturbed. They are filmed, and hand and body movements are coded on 4-point scales. This behavioral measure can be used with toddlers, children, and adolescents. In addition to the Sitting Still paradigm, a Puzzle Test is used to assess persistence on tasks. Participants are asked to try to complete a puzzle for 4 or 5 minutes to win a prize, and the proportion of time spent on the task is used as a persistence score. A third behavioral tool used to assess effortful control is the Fish Game, a computerized game in which children need to get as many points as possible by the end of the game. The punishment ratio increases with time (the children lose more points than they win), and children with high EF tend to quit the game before they lose all their points.

Eisenberg and colleagues (2007) showed in a longitudinal study how effortful control can be related to social functioning in children. Indeed, effortful control is thought to be the behavioral expression of EF and indicates the capacity to regulate one's own mental states in order to focus on others' needs and feelings (Eisenberg et al., 1996). Accordingly, effortful control seems to be linked with affective processes. In general, these behavioral measures of effortful control have been modestly related to sympathy as reported by parents, and this is especially true for younger children (Eisenberg et al., 1996). *Sympathy* is defined as feelings of concern for others based on the recognition of their emotional state or situation, and it predisposes children to prosocial behaviors (Eisenberg et al., 1996). Sympathy can emerge from cognitive processes such as perspective taking, and/or from the affective aspect of *empathy*—an automatic emotional response similar to the emotional state observed in another person (Decety & Jackson, 2004). Other studies have also shown how effortful control is predictive of less aggressive behaviors (Rothbart, Ahadi, & Hershey, 1994), higher use of nonhostile verbal methods to deal with anger

(Eisenberg, Fabes, Nyman, Bernzweig, & Pinuelas, 1994), and greater empathy and tendency to experience guilt and shame (Rothbart et al., 1994). These findings confirm the central role of effortful control in children's emotional and social development (Posner & Rothbart, 2000).

In conclusion, it seems important to begin a comprehensive assessment of social cognition with tests of general intellectual abilities, to determine strengths and weaknesses on which potential therapeutic interventions can be based. Also, performance on classic neuropsychological tests—mostly those focusing on selective attention, cognitive inhibition, flexibility, and planning—is useful to understand global difficulties, as these processes are important for both general and social cognition. Finally, behavioral paradigms such as the effortful control tasks are useful to test EF in a more ecological setting than pencil-and-paper tests. Altogether, the information provided by these tests can shed some light on possible disruptions of one or many of the attention/executive processes involved in social cognition. Although other tests tapping into these processes could be used instead of those presented so far, those presented here are suggested because of their validity and their relevance for social cognition. Still, it may not be necessary to administer all of the tests described above to capture an accurate cognitive profile. Tests should be chosen according to specific hypotheses regarding the causes of social disruptions (such as focal brain lesions or specific disorders), but a comprehensive social cognition assessment battery should minimally cover each of the processes listed above. Further assessment of each process can be conducted if difficulties emerge from the global assessment.

Measures of the Socio-Emotional Component

The socio-emotional component of social cognition encompasses processes necessary to perceive and process human faces, emotions, intentions, desires, and beliefs. It is important for Brian, who is attending a new school, to recognize facial expressions—for instance, recognizing sadness in a peer who is picked last in a sports team. Affect recognition can lead to prosocial behaviors; for example, Brian may pick the other kid earlier for the next activity to comfort him. Brian also needs to understand others' intentions and deceiving attempts in his new environment, in order to distinguish nagging or exploitation from sincerely friendly interactions. In this section, we first cover the assessment of the perception and processing of emotional facial expressions, as these expressions often represent the first social contact between humans. Then we extend the discussion to tests that can be used at different ages to measure mentalizing, which includes the ensemble of processes (especially ToM) involved in understanding other people's mental states. In addition, the processes necessary to generate and regulate social skills and social behaviors are covered. For instance, if we assume that Brian has difficulties regulating his negative emotions, mild misunderstandings between classmates may be overwhelming and distressful to him. This may lead to a failure in recognizing others' mental states and ultimately to peer rejection if, for instance, frequent outbursts of anger or crying make Brian's classmates uncomfortable. Thus a deficit in one or more of the processes included in the socio-emotional component of social cognition is bound to create social interaction problems.

The Perception of Faces and Emotional Facial Expression

Basic Emotion Perception

Facial expression recognition improves with age throughout childhood and well into adolescence (Szekely et al., 2011), and its integrity is crucial for social reciprocity (McClure, 2000). The NEPSY-II includes a very interesting Affect Recognition subtest that can be used with children from 3 to 16 years old. The 35-item subtest progressively increases in difficulty, and each age group begins at a level adapted to its capacity. In this task, children are first asked to observe a series of children's faces displaying five basic emotions (happy, sad, fearful, angry, and disgusted), as well as a neutral expression. For each face, they need to identify a picture of another child expressing the same emotion from a series of pictures. The matching picture can express either a different or a similar level of the target emotion (e.g., a very happy face can be matched with a slightly happy face), and the number of distractors (pictures expressing other emotions) varies according to the difficulty level. The task provides a total Affect Recognition score and percentile ranks by age for errors in each specific emotion.

This subtest of social perception is short, easy to administer, and suitable for children from a large age range. However, the conclusions that can be drawn from its scores are limited, as the subtest requires the comparison of different emotional expressions and the identification of similarities, but does not require the identification of the perceived emotions. The ability to recognize which individuals are feeling alike is important, but identifying precisely what these people are feeling and why is also central for social cognition. Adding a simple emotion-naming condition with various emotional intensities can increase the reach of this task. It may also be helpful to assess whether the intensity of the expression influences a child's performance. To compensate for this weakness, the NEPSY-II's contextual ToM task (further discussed in the "Mentalizing" section) offers a good complement to the Affect Recognition task, because the context must be matched with the expressed emotion. Comparing performances in both tasks can clarify the influence of contextual information on affect recognition.

Another method to assess emotion identification is the Emotion Labeling task used with preschoolers (Szekely et al., 2011; Widen & Russell, 2003). In this task, four pictures of different basic facial expressions (happy, sad, angry, and fearful) are shown with a verbal label, and children have to select the picture that matches the label. This task (which has a matching condition similar to the NEPSY-II Affect Recognition paradigm) is also short, but covers only four emotions compared to five emotional expressions plus a neutral expression. In addition, the four pictures (one for each emotion) that a child needs to choose from do not vary in terms of affect intensity and identity of the person expressing the emotion. Because its stimuli cover a greater variety of expressions, the NEPSY-II is preferable to other batteries for the assessment of affect recognition.

Complex Emotion Perception

The five basic emotions included in the NEPSY-II Affect Recognition subtest are typically recognized earlier than more complex emotions are (Widen & Russell, 2008). The Reading the Mind in the Eyes Test (RMET; Baron-Cohen, Jolliffe, Mortimore, & Robertson, 1997; Baron-Cohen, Wheelwright, Hill, Raste, & Plumb, 2001) additionally covers complex emotions such as embarrassment and shame. Participants are presented with a series of pictures showing only the eye region of the face and are asked to select among four labels the one corresponding to the complex emotion expressed in each picture. Norms are available for children 6–10 years old and 12–13 years old. This test can be used to assess the capacity to recognize complex emotions in children from the age of 6 years (Baron-Cohen, Wheelwright, Spong, Scahill, & Lawson, 2001), although younger children may have some trouble understanding verbal labels.

Memory for Faces and Associative Memory for Faces and Names

Face perception and identification offer a great deal of important information for social functioning, and these processes can be dissociated from facial expression recognition (e.g., Vuilleumier & Pourtois, 2007). The NEPSY-II includes a Memory for Faces task, including both immediate and delayed recognition (15–25 minutes later). This task can be used with children from 3 to 16 years old and is fairly quick to administer. Stimuli consist of 16 pictures, each presenting a child's face with all the background and the hair removed, so that only facial features and skin color can be used to distinguish one child from another. Each item is shown for 5 seconds and needs to be memorized. Both the immediate and delayed recognition phases require the identification of the faces that have previously been shown.

This test is interesting for assessing memory for faces without context, but it can be used along with a measure of associative memory like the NEPSY-II Memory for Names, which assesses memory for faces with names. In everyday social interactions, it is very important to recall whether we have seen a person before or not, but it can also be important to remember what his or her name is. In the Memory for Names subtest of the NEPSY-II, eight drawings of children's faces (only six for 5- and 6-year-olds) are first presented to the participants by the experimenter, who also tells them their names. In the three successive recall phases, only the drawings are shown, and participants have to recall each child's name. The correct answer is provided when a mistake is made during the first two recall phases, to ensure that mistakes are not carried over. Improvement in performance between the first and the third recall phases can be used to document associative memory learning capacities.

Mentalizing

Mentalizing is defined as the ability to recognize and to attribute mental states to others in different situations, using contextual information (Achim, Ouellet, Roy, & Jackson, 2012). Children learn gradually to interpret more or less complex social situations in terms of mental states. It has been shown that the understanding of intentions precedes the understanding of desires, which in turn precedes understanding of beliefs, considered the most difficult aspect of mentalizing and the last to emerge in children (Wellman & Liu, 2004). Mentalizing processes include classic ToM, which describes the understanding of agents' mental states and of how behaviors are influenced by these states (Baron-Cohen et al., 1985; Wellman, 2011). Mentalizing is thus conceived as the display of ToM in everyday situations, and requires individuals to infer the specific content of mental states, such as intentions, desires, and thoughts. Classic scenario-based mentalizing tasks allow experimenters to dissociate specific aspects of mental attribution depending on what is inferred (desire, intention, belief, or emotion) and the availability of contextual information. Other strategies can also be used to assess specific facets of mentalizing, such as perspective-taking abilities or understanding of figures of speech.

ToM in Children

A ToM Scale (Wellman, 2011; Wellman & Liu, 2004) has recently been developed to evaluate different steps in the development of ToM, and can be used with children ages 2–6 years. This scale includes seven ToM scenario-based stories and offers an interesting way to assess children's understanding of mental representations by contrasting beliefs and reality (Wellman, Cross, & Watson, 2001). For each of the seven verbally presented short stories (illustrated with figurines and images), a child has to infer or predict the mental states of protagonists who have different points of view (e.g., because they are missing a crucial piece of information). Stories must be presented in a predetermined order; they assess the understanding that other people have desires, beliefs, and knowledge access, which are different from one's own. Beliefs about box content, explicit false beliefs, beliefs about emotions, and the understanding that apparent emotions are not always real are specifically tested. Assessment with these seven stories can indicate whether ToM development follows a typical or atypical course in terms of sequence and age of appearance. In fact, studies using this scale have shown that a child who passes one item tends to succeed on all the previous items (Wellman & Liu, 2004; Wellman, Lopez-Duran, LaBounty, & Hamilton, 2008). As performance on the three stories assessing specific beliefs (beliefs about content, explicit false beliefs, and emotion beliefs) was found to be very similar, the scale can be shortened by keeping only the explicit-false-beliefs story from these three. The five remaining items can be ranked by difficulty level, and the five stories form a highly scalable set (Wellman & Liu, 2004). The five-item scale has been validated with American children (Wellman & Liu, 2004), and Australian children (Peterson, Wellman, & Liu, 2005), confirming the developmental course of different aspects of ToM, which evolves from the attribution of desires to the recognition of complex emotional concealing. Although performance on the three stories assessing specific beliefs seems to be acquired at about the same age, administering all seven items can nonetheless be useful for qualitative observations if time constraints allow this, as they assess different types of beliefs.

The NEPSY-II also offers two ToM tasks that can be used with children and adolescents between 5 and 16 years of age. The Verbal ToM task contains 15 items, some of which are illustrated with pictures or objects, whereas others are presented only verbally. This task assesses a great variety of abilities linked to mentalization, such as the understanding of figures of speech, reality–imagination distinctions, and beliefs about content. Participants are asked to deduce implicit information in order to infer mental states in others—for example, by determining what a child is pretending to do on a picture, or what another child really means by a given figure of speech. In the Contextual ToM task, a drawing of a particular situation (e.g., someone falling from a bicycle or riding a roller coaster), showing a character but not his or her face, is shown next to a set of four pictures of a child expressing different emotions. Children have to determine how the child in the drawing is feeling by choosing the corresponding facial expression. Success at this task implies that a child can perform emotional attribution from the depicted context, in addition to affect recognition by selecting the right picture. Failure can be due to a deficit in either process. This highlights the importance of using a battery of tests, including emotion recognition tasks such as the ones described in previous sections, if the objective of the assessment is to isolate specific deficits.

ToM (Mentalizing) in Adolescents

Although most scenario-based tasks have been designed for children, some also exist for adolescents or young adults (see, e.g., Achim et al., 2012) for a Combined Stories Task with good psychometric properties). These tasks aurally present short stories (read or prerecorded), and for each story one or a few questions are asked that require inferring a character's mental state. The mental states thus assessed can include beliefs, knowledge, intentions, desires, or emotions. First-order mental states (i.e., a character's mental state about the state of the physical world) can be targeted, but second-order mental states (i.e., a character's mental state about another character's mental state) are often assessed for this age group. When an evaluator is selecting a story-based task, it may be interesting to assess several types of mental states or mental state combinations. Although these tasks have been repeatedly used in research, norms are not readily available, which may restrict the choices. One alternative before norms are available may be to use the scores from the published studies' control groups as a reference point.

Perspective Taking

An interesting paradigm has been created to assess another specific facet of mentalizing: *perspective taking*, or the ability to take another person's visual or mental perspective (Dosch, Loenneker, Bucher, Martin, & Klaver, 2010). This task specifically assesses the ability to attribute different mental states to others for different situations, and can be used with children as well as in adults. This task is one example among many of a simple and advantageous task in terms of external validity. It has been used with children 8 years and older (Dosch et al., 2010). In this task, children have to evaluate different leisure activities that do or do not involve social interactions, both from their own point of view and also from an autistic child's perspective. A real picture of the autistic child is shown, and his personal characteristics (such as the fact that he is not very comfortable in large groups) are described in a real short story prior to the experiment, to help the participant infer how much the autistic child would like social or nonsocial activities. This task can be useful to get information on a child's perspective-taking abilities: The discrepancy between the child's ratings in the first-person perspective (his or her own opinion about the activities) and the third-person perspective ratings (what the child thinks the other child would answer) for social activities will be higher if the child being assessed can efficiently use information about the character and inhibit his or her own perspective to adopt someone else's. Second, the evaluation of nonsocial activities should not be different according to the adopted perspective, if the child can efficiently attribute different mental states to others for different situations. Therefore, the difference between discrepancies for social and for nonsocial activities indicates a capacity to attribute desires and preferences to another individual. In typically developing children and adolescents, the response pattern to this task has been found not to differ from that of adults; however, these conclusions should be interpreted with caution, as the sample assessed was very small (Dosch et al., 2010).

Understanding Figures of Speech

The understanding of metaphors and figures of speech provides a wealth of information about the ability to extract the meaning of someone's speech (Landa, 2005) and thus can provide important cues about the ability to infer the meaning behind verbal statements.

This ability could also be classified under the communication component of social cognition, as it is directly linked with language abilities. Nevertheless, because the task we describe here only assesses the comprehension and interpretation of metaphorical expressions and other figures of speech (and not their production), the processes measured by this test are here classified under mentalizing (i.e., understanding of communicative intention behind a person's use of metaphor). The Figurative Language subtest of the Test of Language Competence (Wiig & Secord, 1989) comprises two steps of evaluation, permitting the assessment of children and adolescents from 5 to 18 years old. Step 1 requires children between 5 and 10 years old to point to the picture (out of four possibilities) representing the meaning of a verbally presented expression. In Step 2, children between 10 and 18 years old have to explain the meaning of an expression (aurally presented and written) associated with a specific situation shown in a picture.

Emotional Regulation

How children experience, express, and regulate their own emotions varies across developmental stages (Rothbart & Bates, 1998) and influences social cognition capacities. *Self-regulation* is a broad concept that includes processes of emotional regulation, but also behavioral regulation. Emotional regulation capacities have been argued to be necessary for the development of empathy (Burnett et al., 2011; Decety & Jackson, 2004; Decety & Meyer, 2008), as children have to inhibit the emotion or distress elicited by the observation of someone else's suffering, in order to facilitate the analysis of the situation and to adopt prosocial behaviors (Eisenberg et al., 1998). It has been suggested that emotional regulation helps children conform to social rules and adapt their behaviors to social contexts. Emotional regulation can be distinguished from cognitive regulation, which is implicated in problem solving (Bodrova & Leong, 2006), but can also be seen as one aspect of EF, permitting the conscious control of emotions and behaviors in social contexts (Zelazo & Cunningham, 2007). Self-regulation capacities underlie the perception of others' mental states, as well as production of appropriate social behaviors. Whereas differences in children's reactivity can be observed very early in life, self-regulatory capacities appear at about 3 years of age and continue to develop throughout childhood.

The structured Disappointment Procedure (Saarni, 1984) is a simple behavioral measure designed to assess children's capacity to regulate their expression of negative emotions when such a display is not socially acceptable. This task was developed originally for use with school-age children, but has also been used with preschoolers (Liebermann, Giesbrecht, & Muller, 2007). This task has three parts that should be separated by inserting unrelated tasks (perhaps classic neuropsychological tests) between each. In the first task, participants are asked to rank eight different toys according to their desirability; some items are typically desirable for kids (e.g., a toy car), and others are visibly broken. In the second task, administered after an unrelated task, the experimenter tells the participant that he or she has been awarded a prize and gives a gift bag containing the toy that was the child's first choice. The child is then allowed to explore the gift for 20 seconds while overt behaviors are observed. In the third and last task, the child is again told that the efforts provided on the previous task will be rewarded, and the same gift bag is handed over, but this time containing the last choice. Reactions are also noted. The experimenter keeps a neutral expression and maintains eye contact during each unwrapping period. After 20 seconds, the experimenter tells the child that a mistake has been made, and another attractive toy is offered in exchange. Emotional reactions, verbalizations, and behaviors are scored according to negative and positive dimensions. In typically developing children, the global score for negative affect expression when the unwanted gift is offered tends to be lower in older than in younger children, and this is especially true for girls (Saarni, 1984). This suggests that older kids are better at regulating the expression of their emotions, and also shows a disassociation between internal affect and displayed expressive affect. This may be due to a better awareness of social conventions regarding the expected reaction when receiving a gift, as well as increasing ability and motivation to produce the socially accepted expressive behaviors (Saarni, 1984). In a recent study, emotional regulation assessed with this procedure showed a tendency to correlate with the inhibition component of EF, but failed to show relations with ToM performances (Liebermann et al., 2007).

To summarize this section, once the integrity of perceptual and basic processes related to emotions and facial expressions has been assessed, the evaluator must directly test more complex functions, such as mentalizing, in order to target every facet of social cognition. Likewise, emotion regulation processes should be considered, as they are important to attribution of mental states and to the display of social skills. In short, the socio-emotional component of social cognition includes functions that subserve the perception and understanding of others' mental states from their facial expressions, actions, verbalizations, and contextual cues. The tasks described in this section can be used to document complementary aspects of the socio-emotional component and to characterize specific profiles of weaknesses and strengths. Having a clear view of preserved and deficient processes should help predict whether social cognition development follows a typical or atypical course. Furthermore, distinguishing deficits in basic perceptual processes from difficulties in complex social functions can be of great help in planning rehabilitation, as intervention strategies aimed at improving affect perception (e.g., by indicating which relevant facial features are associated with each emotion) will differ from strategies designed to increase emotion regulation and social skills (which can include interventions based on role playing).

Measures of the Communication Component

Communication abilities are central to social functioning and predispose children to peer approval and good social relationships (Beauchamp & Anderson, 2010). Receptive and expressive verbal skills allow children to understand others, as well as to communicate and interact with them. As previously discussed, it is important for Brian to be able to understand what other children mean when they are trying to establish social relationships, and this is done through mentalizing. In addition, Brian has to be able to exchange factual and emotional information appropriately to be accepted by peers—for instance, to give compliments to children he wants to befriend, and to voice his displeasure with mocking remarks when doing this suits the situation.

Communication includes the understanding and utilization of nonverbal social cues, such as gestures, prosody, and gaze. The combination of verbal content, and nonverbal communication cues, and contextual information is used to infer mental states in others. In a way, the previously described mentalizing tasks could also be classified as communication tasks, but they are mostly limited to the receptive aspect of communication. Thus, in describing assessment strategies for the assessment of social cognition, we focus on tasks that also tap into expressive communication skills. Not surprisingly, general verbal abilities have been related to ToM performance in typically developing children as well as in autistic children (Happé, 1995; Jenkins & Astington, 1996), but no direct causal link has been established (Cutting & Dunn, 1999). Moreover, although most verbal ToM tasks involve a high level of language skills, linguistic demands alone cannot account for ToM performance (e.g., Cutting & Dunn, 1999; Siegal & Varley, 2006).

Social communication skills include joint attention, communicative intent, communicative initiation and responsiveness, and integration of affect and gestures. In preschoolers, these skills are related to verbal abilities, but more specifically to the pragmatic aspects of language (i.e., communicative intentions, presuppositions, and discourse management; Landa, 2005). For children with typically developing language abilities, several strategies are available to examine aspects of language that are specifically related to social cognition, including a comprehensive ecological measure of social communication. It is important to remember, however, that a complete language assessment should be conducted in collaboration with a speech pathologist if language understanding or production is identified in screening assessments as a prominent factor in social difficulties. Below, we cover a few general language assessment tools that can be used to detect cases for which further language assessment is indicated.

General Language Abilities and Social Communication

There is a strong correlation between Verbal IQ (or results on the WISC-IV Verbal Comprehension Index) and language development, but children can have a normal Verbal IQ or Verbal Comprehension Index and still show communicative difficulties (Landa, 2005). Consequently, a battery such as the WISC-IV is insufficient to capture possible contributions of communication skills to social cognition performance, but it can be used to measure language abilities summarily. The Clinical Evaluation of Language Fundamentals— fourth Edition (CELF-4; Semel, Wiig, & Secord, 2003) was designed to measure language skills related to semantics, morphology, syntax, and memory. It takes between 30 and 60 minutes to administer and can be used to identify general language deficits in children and adolescents from ages 5 to 21. The complete CELF-4 proposes a breakdown along seven composite scales: Core Language, Receptive Language, Expressive Language, Language Structure, Language Content, Language Memory, and Working Memory. This battery is useful as a screening tool to identify and characterize language deficits or difficulties that may partially explain social cognition and social skills disruptions.

If specific communication deficits are suspected in a child before the age of 4 years, receptive and expressive communication can be assessed in 4- to 48-month-old children with the Sequenced Inventory of Communication Development (SICD; Hederick,

Prather, & Tobin, 1995), a test recommended by Landa (2005). This tool uses a combination of observed behaviors and parental reports of behaviors at home, which provide a clear advantage over strictly pencil-and-paper tasks, as communicative behaviors are more likely to occur in well-known environments such as home or school. Clinical and experimental settings sometimes fail to elicit key behaviors. This inventory takes between 30 and 75 minutes to administer, and it provides Expressive and Receptive Processing Profiles. Although the SICD can be used with very young children, it does not focus on a broad range of social communication behaviors.

Another possibility for infants and toddlers, the Communication and Symbolic Behavior Scales—Developmental Profile (CSBS DP; Wetherby & Prizant, 2003), is a multi-informant tool designed to assess language and communication predictors in 6- to 24-month-old children. Social cognition skills are not yet developed in infants, but some behaviors (such as joint attention) and symbolic play are strong predictors of social cognition development. In addition to the Infant–Toddler Checklist and the Caregiver Questionnaire, the CSBS DP includes behavior samples observed and rated by a clinician. This last part takes 30 minutes to administer and is used to assess a large range of spontaneous communication behaviors, such as joint attention responding and initiating, symbolic play, sharing, and vocabulary understanding.

Communication Skills

The Chat Room Task has been used for children with ADHD to assess social skills and participants' style of social interactions during online conversations with peers (Mikami, Huang-Pollock, Pfiffner, McBurnett, & Hangai, 2007). This task can also be used for other children with social communication deficits. As virtual communication (e.g., chat rooms, forums, online social networks, and text messaging) has become increasingly important in everyday social interactions, especially among children and adolescents, this task has clear ecological value. It can be used with children between 7 and 12 years old and takes about 20 minutes to administer. This comprehensive task provides an analysis of communication samples, but also assesses memory for social information, tendency for prosocial interaction, and understanding of social cues.

In a validation study for this task (Mikami et al., 2007), children were informed in advance that they would have to interact online with a computer program made to recreate a conversation among four friends who already knew each other. In the task, one of the children is planning a birthday party and participants have to join the conversation and find out what this child would like for his birthday party. Five scores are calculated, reflecting (1) the total number of online responses by the participant; (2) whether those responses are related to the topic or not; (3) whether the participant understands social cues and asks pertinent questions; (4) proportion of prosocial comments; and (5) proportion of hostile comments. After the conversation, the participant's memory of the conversation is assessed with 14 questions such as "What is the birthday kid's favourite color?", and a memory score is calculated (for a complete description, see Mikami et al., 2007). It was shown in the validation study that performance on this task was correlated with parent-reported cooperation, with being positively perceived by peers, and with positive observed social interactions (Mikami et al., 2007). Therefore, this task is thought to be representative of children's everyday social interactions. Of course, this task needs further validation, as it has been used only in one study to date, but the results from the typically developing children in this study may be used as a comparison point. Also, this task may be used to qualify the social profiles of children who have ADHD with or without hyperactivity, as children with these ADHD subtypes differ on specific variables (e.g., proportion of hostile comments).

The Challenge of Measuring Enmeshed Constructs

The development of social cognition can be influenced by numerous biological factors (including neurological development and integrity of brain structures), but also by the interaction opportunities provided by the social environment. Moreover, as this group of functions is complex and relies on distinct components, the use of multiple instruments (intellectual and neuropsychological tests, behavioral observations, and other specific tasks) is highly recommended. Of course, practical constraints—such as limited time for assessment, test availability, and children's endurance—must be considered in both research and clinical settings. It may not be realistic or necessary to use all of the tasks described above. But the proposed battery can serve as a basis from which researchers and professionals can select the appropriate constellation of instruments to cover basic and complex processes.

The performance recorded and the observations made during the proposed assessment battery should be combined with general social outcome indices, as well as specific rating scales or questionnaires completed by the child and other informants such as parents and teachers (see Muscara & Crowe, Chapter 6, this volume). Taken together, the data acquired should provide a more complete profile of preserved and deficient functions, which can then be used to predict trajectories and plan interventions optimizing the developmental course of social cognition. As noted earlier, the different components of social cognition are in constant interrelation in social situations, and their developmental trajectories are intrinsically linked.

A complete review of the interrelations among social cognition processes is beyond the scope of this chapter, but some research conclusions can improve our interpretation of performance patterns. For one thing, moderate to strong correlations have been consistently found between ToM performance, which is crucial for socialization, and EF performance (Carlson & Moses, 2001; Frye, Zelazo, & Palfai, 1995; Hala, et al., 2003; Hughes, 1998; Perner, Lang, & Kloo, 2002). Interestingly, the development of EF and ToM tends to occur during the same period (Hughes, 1998; Hughes & Ensor, 2011). Correlations remain significant even when verbal abilities, age, and sex are controlled for, as well as when other relevant factors such as family size, symbolic understanding, and performance on ToM-like tasks (but without reference to mental states) are taken into account (Carlson & Moses, 2001). Also, longitudinal studies have revealed that EF performance at age 3, or even at age 2, can predict ToM performance at 3–4 years of age (Carlson et al., 2004; Hughes, 1998; Hughes & Ensor, 2007). Even with robust correlations, however, the direction of causality remains unclear. Some studies in adults seem to suggest that EF and ToM are relatively independent constructs and can be impaired in isolation (e.g., Shamay-Tsoory, Shur, Harari, & Levkovitz, 2007), but such evidence, to the best of our knowledge, has not been reported in children. This is also not surprising, considering that both these aspects of social function are not fully developed until late adolescence or early adulthood (Anderson, 2008; Damon & Hart, 1982).

Second, it is important for interpretation to consider that children seem to acquire ToM abilities in the same sequence but not exactly at the same time between 2 and 6 years of age. A meta-analysis confirmed that typically developing children acquire ToM understanding within the first 3 years on average (Wellman et al., 2001; Wellman, 2011). Delays in ToM development have been documented in children with deafness in hearing families (Peterson & Seigal, 1999) and in children with autism (Happé, 1995), but different developmental patterns have been found for each pathology (Peterson et al., 2005). Also, the language level necessary to succeed in a false-belief task differs across developmental psychopathologies. For instance, children with autism typically need at least 11 years of verbal mental age to have an 80% chance of success, compared to 5 years for typically developing children (Happé, 1995).

Considering these empirical conclusions, evaluators can compare children's performance patterns to typical and atypical patterns to suggest or eliminate possible diagnoses. Relations among verbal abilities, socio-emotional processes, and EF performance can be useful to identify trajectories in populations with specific disorders of infancy, but also to predict consequences of brain injury and, most importantly, to plan adapted and individualized interventions. As an example, if the affective aspects of social cognition are disrupted by brain damage, the attention/executive aspects may potentially be preserved and therefore used for rehabilitation. Isolating the primary source of social functioning deficits can rarely be achieved, but this proposed battery can at least point out weaknesses and strengths in cognitive and emotional functioning that need to be addressed in therapy. The bright side is that interventions targeting one process should also have a positive influence on other related processes. For instance, therapy interventions increasing emotion regulation can help children analyze and interpret others' reactions. Because affect intensity is reduced by regulation strategies, children may develop better attentional control to perceive social cues and take the context into account, in order to make sense of others' behaviors.

In conclusion, despite the multiplicity of functions related to social cognition and the complexity of their interrelations, many interesting assessment strategies can be found in the literature. Of course, they vary in terms of internal and external validity, and their association with day-to-day social functioning remains weak. Novel and interesting ideas are emerging, such as strategies based on technology that recreates credible social interactions among participants, confederates, and clinicians (e.g., virtual reality, video game environments). Although great efforts are being made to create ecological measures, the need to amass a large set of normative data for healthy individuals as well as for individuals with neurological disorders remains the main challenge in the assessment of social cognition.

Acknowledgments

This work was made possible with the help of the Canadian Institute of Health Research (salary grant to Philip L. Jackson) and the Centre interdisciplineaire de recherche en réadaptation et intégration sociale (scholarship to Rosée Bruneau-Bhérer). We thank Karine Morasse, PhD, and Fanny Eugène, PhD, for their very useful comments on previous versions of this chapter.

References

Achenbach, T. M., & Rescorla, L. A. (2001). *Manual for ASEBA School-Age Forms and Profiles.* Burlington: University of Vermont, Research Center for Children, Youth, and Families.

Achim, A. M., Boutin, A., Jackson, P. L., Guitton, M. J., & Monetta, L. (2011). *On what ground do we mentalize?: Characteristics of current tasks and sources of information that contribute to mentalizing judgments.* Manuscript submitted for publication.

Achim, A. M., Ouellet O., Roy, M.-A., & Jackson P. L. (2012). Mentalizing in first-episode psychosis. *Psychiatry Research.* [Epub ahead of print]

American Psychiatric Association. (2000). *Diagnostic and statistical manual of mental disorders* (4th ed., text rev.). Washington, DC: Author.

Anderson, P. (2008). Towards a developmental model of executive function. In V. V. Anderson, R. Jacobs, & P. J. Anderson (Eds.), *Executive function and the frontal lobes: A life span perspective* (pp. 3–22). Hove, UK: Psychology Press.

Baddeley, A., & Della Sala, S. (1996). Working memory and executive control. *Philosophical Transactions of the Royal Society of London: Series B. Biological Sciences, 351*(1346), 1397–1404.

Baron-Cohen, S., Jolliffe, T., Mortimore, C., & Robertson, M. (1997). Another advanced test of theory of mind: Evidence from very high functioning adults with autism or Asperger syndrome. *Journal of Child Psychology and Psychiatry, 38*(7), 813–822.

Baron-Cohen, S., Leslie, A. M., & Frith, U. (1985). Does the autistic child have a "theory of mind"? *Cognition, 21*(1), 37–46.

Baron-Cohen, S., Wheelwright, S., Hill, J., Raste, Y., & Plumb, I. (2001). The "Reading the Mind in the Eyes" Test revised version: A study with normal adults, and adults with Asperger syndrome or high-functioning autism. *Journal of Child Psychology and Psychiatry, 42*(2), 241–251.

Baron-Cohen, S., Wheelwright, S., Spong, A., Scahill, V., & Lawson, J. (2001). Are intuitive physics and intuitive psychology independent?: A test with children with Asperger syndrome. *Journal of Developmental and Learning Disorders, 5,* 47–78.

Beauchamp, M. H., & Anderson, V. (2010). SOCIAL: An integrative framework for the development of social skills. *Psychological Bulletin, 136*(1), 39–64.

Bodrova, E., & Leong, D. J. (2006). Self-regulation as a key to school readiness: How early childhood teachers promote this critical competency. In M. Zaslow & I. Martinez-Beck (Eds.), *Critical issues in early childhood professional development* (pp. 203–224). Baltimore: Brookes.

Burnett, S., Sebastian, C., Cohen Kadosh, K., & Blakemore, S. J. (2011). The social brain in adolescence: Evidence from functional magnetic resonance imaging and behavioural studies. *Neuroscience and Biobehavioral Reviews, 35*(8), 1654–1664.

Carlson, S. M., Mandell, D. J., & Williams, L. (2004). Executive function and theory of mind: Stability and prediction from ages 2 to 3. *Developmental Psychology, 40*(6), 1105–1122.

Carlson, S. M., & Moses, L. J. (2001). Individual differences in inhibitory control and children's theory of mind. *Child Development, 72*(4), 1032–1053.

Catroppa, C., Anderson, V. A., Morse, S. A., Haritou, F., & Rosenfeld, J. V. (2008). Outcome and predictors of functional recovery 5 years following pediatric traumatic brain injury (TBI). *Journal of Pediatric Psychology, 33*(7), 707–718.

Cutting, A. L., & Dunn, J. (1999). Theory of mind, emotion understanding, language, and family background: Individual differences and interrelations. *Child Development, 70*(4), 853–865.

Damon, W., & Hart, D. (1982). The development of self-understanding from infancy to adolescence. *Child Development, 53,* 841–864.

Decety, J., & Jackson, P. L. (2004). The functional architecture of human empathy. *Behavioral Cognitive Neuroscience Reviews, 3*(2), 71–100.

Decety, J., & Meyer, M. (2008). From emotion resonance to empathic understanding: A social developmental neuroscience account. *Development and Psychopathology, 20*(4), 1053–1080.

Delis, D., Kaplan, E., & Kramer, J. H. (2001). *Delis–Kaplan Executive Function System.* San Antonio, TX: Psychological Corporation.

Dosch, M., Loenneker, T., Bucher, K., Martin, E., & Klaver, P. (2010). Learning to appreciate others: Neural development of cognitive perspective taking. *NeuroImage, 50*(2), 837–846.

Eisenberg, N., Fabes, R. A., Guthrie, I. K., Murphy, B. C., Maszk, P., Holmgren, R., et al. (1996). The relations of regulation and emotionality to problem behavior in elementary school children. *Development and Psychopathology, 8*(1), 141–162.

Eisenberg, N., Fabes, R. A., Nyman, M., Bernzweig, J., & Pinuelas, A. (1994). The relations of emotionality and regulation to children's anger-related reactions. *Child Development, 65*(1), 109–128.

Eisenberg, N., Fabes, R. A., Shepard, S. A., Murphy, B. C., Jones, S., & Guthrie, I. K. (1998). Contemporaneous and longitudinal prediction of children's sympathy from dispositional regulation and emotionality. *Developmental Psychology, 34*(5), 910–924.

Eisenberg, N., Michalik, N., Spinrad, T. L., Hofer, C., Kupfer, A., Valiente, C., et al. (2007). The relations of effortful control and impulsivity to children's sympathy: A longitudinal study. *Cognitive Development, 22*(4), 544–567.

Frye, D., Zelazo, P. D., & Palfai, T. (1995). Theory of mind and rule-based reasoning. *Cognitive Development, 10*, 483–527.

Hala, S., Hug, S., & Henderson, A. (2003). Executive functioning and false belief understanding in preschool children: Two tasks are harder than one. *Journal of Cognition and Development, 4*, 275–298.

Happé, F. (1995). The role of age and verbal ability in theory of mind task performance for subjects with autism. *Child Development, 66*, 843–855.

Heaton, R. K. (1981). *A manual for the Wisconsin Card Sorting Test.* Odessa, FL: Psychological Assessment Resources.

Hederick, D. L., Prather E.M., & Tobin, A. R. (Eds.). (1995). *Sequenced Inventory of Communication Development.* Seattle: University of Washington Press.

Hughes, C. (1998). Executive function in preschoolers: Links with theory of mind and verbal ability. *British Journal of Developmental Psychology, 16*, 233–253.

Hughes, C., & Ensor, R. (2007). Executive function and theory of mind: Predictive relations from 2 to 4 years. *Developmental Psychology, 43*, 1447–1459.

Hughes, C., & Ensor, R. (2011). Individual differences in growth in executive function across the transition to school predict externalizing and internalizing behaviors and self-perceived academic success at 6 years of age. *Journal of Experimental Child Psychology, 108*(3), 663–676.

Hughes, C., Ensor, R., Wilson, A., & Graham, A. (2010). Tracking executive function across the transition to school: A latent variable approach. *Developmental Neuropsychology, 35*(1), 20–36.

Jenkins, J. M., & Astington, J. W. (1996). Cognitive factors and family structure associated with theory of mind development in young children. *Developmental Psychology, 32*(1), 70–78.

Kofler, M. J., Rapport, M. D., Bolden, J., Sarver, D. E., Raiker, J. S., & Alderson, R. M. (2011). Working memory deficits and social problems in children with ADHD. *Journal of Abnormal Child Psychology, 39*(6), 805–817.

Korkman, M., Kirk, U., & Fellman, V. (1998). *NEPSY: A developmental neuropsychological assessment.* San Antonio, TX: The Psychological Corporation.

Korkman, M., Kirk, U., & Kemp, S. (2007). *NEPSY-II.* San Antonio, TX: Harcourt Assessment.

Landa, R. J. (2005). Assessment of social communication skills in preschoolers. *Mental Retardation and Developmental Disabilities Research Reviews, 11*(3), 247–252.

Liebermann, D., Giesbrecht, G. F., & Muller, U. (2007). Cognitive and emotional aspects of self-regulation in preschoolers. *Cognitive Development, 22,* 511–529.

McClure, E. B. (2000). A meta-analytic review of sex differences in facial expression processing and their development in infants, children, and adolescents. *Psychological Bulletin, 126*(3), 424–453.

Mikami, A. Y., Huang-Pollock, C. L., Pfiffner, L. J., McBurnett, K., & Hangai, D. (2007). Social skills differences among attention-deficit/hyperactivity disorder types in a chat room assessment task. *Journal of Abnormal Child Psychology, 35*(4), 509–521.

Moses, L. J., & Tahiroglu, D. (2010). Clarifying the relation between executive function and children's theory of mind. In S. Bryant, U. Muller, J. Carpendale, A. Young, & G. Iarocci (Eds.), *Self- and social-regulation: Exploring the relations between social interaction, social understanding, and the development of executive functions.* Oxford Scholarship Online. (*www.oxfordscholarship.com/view/10.1093/acprof:oso/9780195327694.001.0001/acprof-9780195327694-chapter-9*).

Perner, J., Lang, B., & Kloo, D. (2002). Theory of mind and self-control: More than a common problem of inhibition. *Child Development, 73*(3), 752–767.

Peterson, C. C., & Seigal, M. (1999). Representing the inner worlds: Theory of mind in autistic, deaf and normal hearing children. *Psychological Science, 10,* 126–129.

Peterson, C. C., Wellman, H. M., & Liu, D. (2005). Steps in theory-of-mind development for children with deafness or autism. *Child Development, 76*(2), 502–517.

Posner, M. I., & Rothbart, M. K. (2000). Developing mechanisms of self-regulation. *Development and Psychopathology, 12*(3), 427–441.

Rothbart, M. K., Ahadi, S. A., & Hershey, K. L. (1994). Temperament and social behavior in childhood. *Merrill–Palmer Quarterly, 40*(1), 21–39.

Rothbart, M. K., & Bates, J. E. (1998). Temperament. In W. Damon (Series Ed.) & N. Eisenberg (Vol. Ed.), *Handbook of child psychology: Vol. 3. Social, emotional, and personality development* (5th ed., pp. 105–176). New York: Wiley.

Rothbart, M. K., & Bates, J. (2006). Temperament. In W. Damon & R. M. Lerner (Series Eds.) & N. Eisenberg (Vol. Ed.), *Handbook of child psychology: Vol. 3. Social, emotional, and personality development* (6th ed. pp. 99–166). Hoboken, NJ: Wiley.

Saarni, C. (1984). An observational study of children's attempts to monitor their expressive behaviors. *Child Development, 55,* 1504–1513.

Semel, L., Wiig, E. H., & Secord, W. (2003). *Clinical Evaluation of Language Fundamentals—Fourth Edition.* San Antonio, TX: Psychological Corporation.

Shamay-Tsoory, S. G., Shur, S., Harari, H., & Levkovitz, Y. (2007). Neurocognitive basis of impaired empathy in schizophrenia. *Neuropsychology, 21*(4), 431–438.

Siegal, M., & Varley, R. (2006). Aphasia, language, and theory of mind. *Social Neuroscience, 1*(3–4), 167–174.

Sparrow, S. S., Cicchetti, D., & Balla, D. (2005). *Vineland Adaptive Behavior Scales, Second Edition (Vineland-II).* Circle Pines, MN: American Guidance Service.

Stroop, J. R. (1935). Studies of interference in serial verbal reactions. *Journal of Experimental Psychology, 18,* 643–662.

Szekely, E., Tiemeier, H., Arends, L. R., Jaddoe, V. W., Hofman, A., Verhulst, F. C., et al. (2011). Recognition of facial expressions of emotions by 3-year-olds. *Emotion, 11*(2), 425–435.

Vuilleumier, P., & Pourtois, G. (2007). Distributed and interactive brain mechanisms during emotion face perception: Evidence from functional neuroimaging. *Neuropsychologia, 45*(1), 174–194.

Wechsler, D. (2002). *Wechsler Preschool and Primary Scale of Intelligence—Third Edition.* San Antonio, TX: Psychological Corporation.

Wechsler, D. (2003). *Wechsler Intelligence Scale for Children—Fourth Edition.* San Antonio, TX: Psychological Corporation.

Wellman, H. M. (2011). Developing a theory of mind. In U. Goswami (Ed.), *The Wiley–Blackwell handbook of childhood cognitive development* (2nd ed., pp. 167–187). Malden, MA: Wiley-Blackwell.

Wellman, H. M., Cross, D., & Watson, J. (2001). Meta-analysis of theory-of-mind development: The truth about false belief. *Child Development, 72*(3), 655–684.

Wellman, H. M., & Liu, D. (2004). Scaling of theory-of-mind tasks. *Child Development, 75*(2), 523–541.

Wellman, H. M., Lopez-Duran, S., LaBounty, J., & Hamilton, B. (2008). Infant attention to intentional action predicts preschool theory of mind. *Developmental Psychology, 44*(2), 618–623.

Wetherby, A., & Prizant, B. M. (2003). *Communication and Symbolic Behavior Scales—Developmental Profile.* Baltimore: Brookes.

Widen, S. C., & Russell, J. A. (2003). A closer look at preschoolers' freely produced labels for facial expressions. *Developmental Psychology, 39*(1), 114–128.

Widen, S. C., & Russell, J. A. (2008). Children acquire emotion categories gradually. *Cognitive Development, 23*, 291–312.

Wiig, E. H., & Secord, W. (1989). *Test of Language Competence* (expanded ed.). New York: Psychological Corporation.

Zelazo, P. D., & Cunningham, W. A. (2007). Executive function: Mechanisms underlying emotion regulation. In J. Gross (Ed.), *Handbook of emotion regulation* (pp. 135–158). New York: Guilford Press.

Zelazo, P. D., & Müller, U. (2002). The balance beam in the balance: Reflections on rules, relational complexity, and developmental processes. *Journal of Experimental Child Psychology, 81*(4), 458–465.

Zelazo, P. D., & Müller, U. (2011). Executive function in typical and atypical development. In U. Goswami (Ed.), *The Wiley–Blackwell handbook of childhood cognitive development* (2nd ed., pp. 574–603). Oxford, UK: Blackwell.

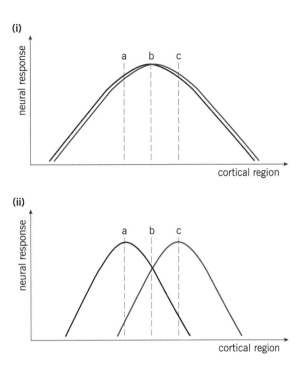

PLATE 3.1. An illustrative schematic that represents the developmental changes associated with the processes of cortical specialization for processing different object categories, as predicted by the interactive-specialization approach to human brain development. In the initial state (i), cortical regions a, b, and c respond equally strongly to faces (red line) and objects (black line). After development (ii), region a becomes specialized for objects (relative to faces), and region c becomes specialized for faces (relative to objects). With regard to localization, all three regions are activated by both classes of stimuli in the initial state. After development, faces no longer activate region a, and therefore the cortical activation generated by this stimulus is more focal than in the initial state. The same applies to objects. Region b represents an intermediate zone of tissue that may respond less selectively.

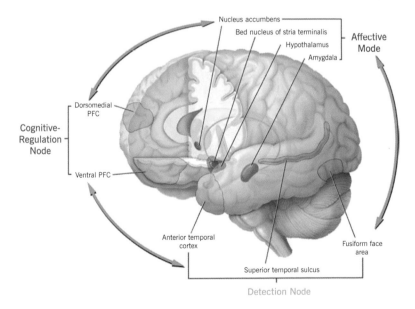

PLATE 3.2. The social-information-processing network (SIPN) model. The "detection node" is involved in carrying out basic perceptual processes on social stimuli; the "affective node" attaches emotional significance to social stimuli; and the cognitive-regulatory node is involved in cognitive control and in mentalizing. The gray arrows indicate that the three nodes interact. From Nelson, Leibenluft, McClure, and Pine (2005). Copyright 2005 by Cambridge University Press. Reprinted by permission.

a) Medial view

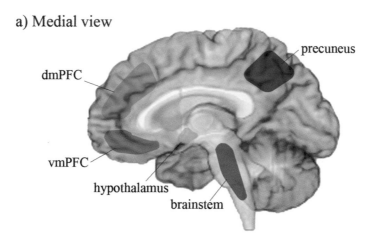

b) Lateral view

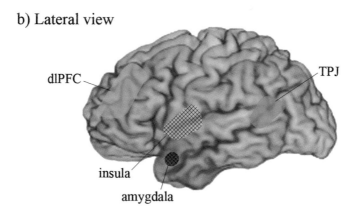

PLATE 4.1. Brain areas involved in social and moral functioning. (a) The medial view highlights the ventromedial prefrontal cortex (vmPFC), dorsomedial prefrontal cortex (dmPFC), precuneus, hypothalamus, and brainstem. (b) The lateral view highlights the dorsolateral prefrontal cortex (dlPFC), insula (a cortical region deep to the sylvian fissure), amygdala (a subcortical anterior temporal lobe structure), and temporoparietal junction (TPJ). Cross-hatching denotes structures deep beneath the surface of the rendering.

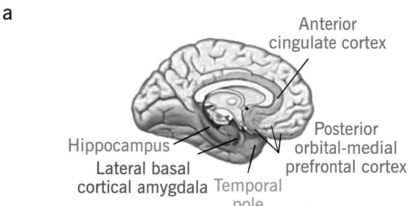

a

Anterior
cingulate cortex

Posterior
orbital-medial
prefrontal cortex

Hippocampus

Lateral basal
cortical amygdala   Temporal
pole

b

Superior temporal
gyrus

Superior temporal
sulcus

Frontal
pole

Fusiform
gyrus

Temporal
pole

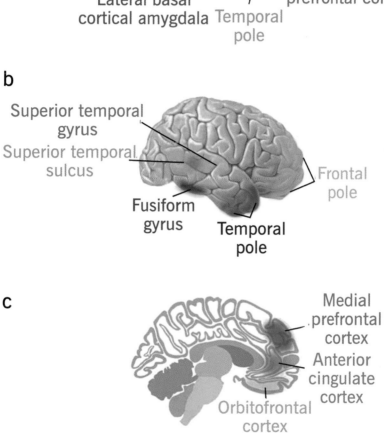

c

Medial
prefrontal
cortex

Anterior
cingulate
cortex

Orbitofrontal
cortex

PLATE 10.1. Brain structures or regions implicated in social cognition and executive function. Top panel (a) shows the medial surface of the right hemisphere, depicting the limbic lobe, laterobasal–cortical amygdala, orbital–medial frontal cortex, and hippocampus. The laterobasal–cortical amygdala and hippocampus are projected on the surface of the parahippocampal gyrus. Middle panel (b) shows the lateral surface of the right hemisphere, depicting the superior temporal sulcus, superior temporal gyrus, fusiform gyrus, and temporal and frontal poles. Bottom panel (c) shows a midsaggital section, depicting orbitofrontal cortex, medial prefrontal cortex, and anterior cingulate cortex.

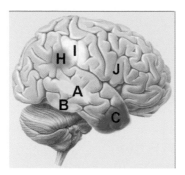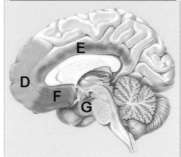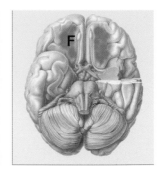

PLATE 11.1. Brain regions contributing to the social brain network: (A) superior temporal sulcus; (B) fusiform gyrus; (C) temporal pole; (D) medial prefrontal cortex and frontal pole; (E) cingulate cortex; (F) orbitofrontal cortex; (G) amygdala; (H) temporoparietal junction; (I) inferior parietal cortex; (J) inferior frontal cortex. Insula not shown. From Beauchamp and Anderson (2010). Copyright 2010 by the American Psychological Association. Reprinted by permission.

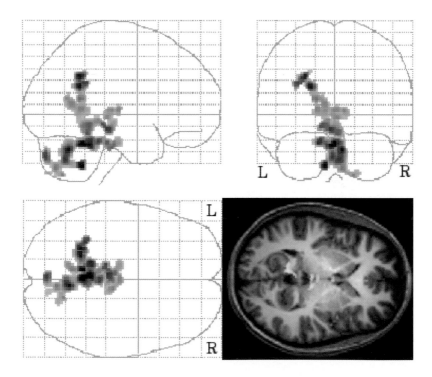

PLATE 12.1. Maximum-intensity projection map and overlay showing regions in which adolescents with TBI demonstrated greater activation than adolescents with TD in the social attributions task for the target × perspective interaction when judgments about the self were made from the perspective of a well-known other. TD > TBI was nonsignificant. Activation was overlaid onto the normalized brain of one adolescent with TD and is shown here in left lingual gyrus (Brodmann's area 18) and bilateral thalami. The right hemisphere of the brain is displayed on the right side of the plate.

CHAPTER 8

Theory-Driven Imaging Paradigms and Social Functions

Implications of Measurement Strategies

Julian J. Dooley, Stefanie Rosema, and Miriam H. Beauchamp

Coinciding with technological developments, the field of socio-cognitive neuroscience has generated important and significant findings to explain social functioning and behaviors. Brain imaging tools have provided a unique opportunity to examine the relationship between neural and social functions. Given that the literature describing the neural underpinnings of social functioning and the impact of brain injuries on social cognition are discussed elsewhere (see Part IV of this volume), this chapter focuses on the measurement of social cognition in brain imaging environments, specifically functional magnetic resonance imaging (fMRI). The issue of measurement of social cognition in fMRI is fundamental to the interpretability (or lack thereof) of research in this area. As Saxe (2006, p. 60) notes, measurement is a "major challenge for social cognitive neuroscientists: how to evoke social cognitive behaviors in an environment so unconducive to social interaction." This is the very crux of the problem: How can we know that we are validly and reliably measuring social cognition in an environment that practically rules out any type of realistic social interaction?

Thus the use of appropriately tested (and validated) stimuli becomes critical to the accurate measurement of neural activation associated with social functions and behaviors. In the following discussion, we focus primarily on methodological topics concerning the measurement of social cognition in fMRI. Ideally, a comprehensive presentation of the psychometric properties (i.e., validity and reliability) of the tasks used in fMRI would be presented. However, our search revealed that in the overwhelming majority of studies, little information on the validity and reliability of tasks is presented.

Given the specificity of fMRI tasks, it is potentially worrisome that little evidence is provided to support the appropriateness and rigor of the tasks being used. This chapter's discussion addresses approaches to measurement in the following areas: emotion processing, theory of mind (ToM), empathy, and moral reasoning. These four areas have been chosen to represent a variety of socio-cognitive functions ranging from relatively basic ones (emotion processing) to more complex and higher-order functions (moral reasoning). Initially, we intended to include fMRI paradigms specific to social information processing (SIP) in this review, given the extent to which this area of research has informed social behaviors (positive and negative; e.g., Crick & Dodge, 1994; Nelson & Crick, 1999). Despite being unable to find any studies that specifically targeted SIP functions (e.g., hostile attribution of intent), we argue that there is much to learn from the evidence describing the various approaches to measuring SIP, and that these approaches can directly inform developments in fMRI. Thus, where appropriate, we also address SIP measurement.

Importance of a Theoretical Approach

The importance of a theoretical approach to measuring social and behavioral impairments after brain injury has previously been highlighted (Dooley, Anderson, Hemphill, & Ohan, 2008). However, few theoretical models of socio-cognitive functioning have been proposed that would provide a heuristic basis for conceptualizing and structuring assessments. Of course, numerous models of social functioning have been developed. Those describing how social information is processed and influences behaviors are among the most widely cited, with the SIP model proposed by Crick and Dodge (1994) being most commonly used. This five-stage cognitive model describes how basic information about sensation and perception is encoded and interpreted, how social goals and responses are generated, and which response that seems most likely to result in goal attainment is chosen. Once this process has occurred, the behavior is enacted.

The Crick and Dodge (1994) SIP model is primarily employed to explain social behaviors in general and aggression in particular. In research using this model, much evidence has been found for patterns of cognitive processing characteristic of particular forms of antisocial behaviors. For example, hostile intent attribution (i.e., interpreting social actions as intentionally hostile) is often observed in young people who engage in volatile and emotionally labile aggression (often called *reactive aggression*). Testing this finding in youth with traumatic brain injury (TBI), Dooley et al. (2008) demonstrated that adolescents with TBI displayed significantly more reactive aggression than their noninjured peers did. Such findings suggest that there is much to be learned about early TBI from adopting a theoretical approach to understanding and measuring socio-cognitive functions.

More recently, Beauchamp and Anderson (2010) have proposed the socio-cognitive integration of abilities model (SOCIAL). Although this model is described elsewhere in the present text in greater detail (see Anderson & Beauchamp, Chapter 1), a few aspects of the model are worth highlighting in the context of the current discussion. The model builds on the work of Crick and Dodge (1994) and Yeates and colleagues (Chapter 10, this volume). It comprises three major components, or mediators, of social function: internal–external factors; brain development and integrity; and cognitive functions (consisting of attention/executive skills, communication skills, and socio-emotional skills).

The benefits of this and similar models are numerous. Most importantly, they provide structure and logic to the assessment of cognitive functions. They also promote purposeful assessment and interpretation of cognitive skills (and consequently of social skills), and can provide a basis for differentiating between cognitive skills that, on the surface, may appear similar. For example, in order to interpret the actions (behavioral and emotional) of another person, it is first necessary to understand that this person has a separate and distinct mental state. Thus intent attribution is dependent on having an intact ToM, and this distinction can have significant implications for measurement, especially in fMRI (Dooley & Beauchamp, 2011). SOCIAL describes an approach differentiation of cognitive skills into lower- and higher-order, which provides an important perspective in determining the precise skills being assessed.

The remainder of this chapter focuses on the various measurement options used in fMRI to assess components of social cognition. Consistent with the structure of the SOCIAL model, we differentiate between the measurement of lower- and higher-order cognitive functions by discussing emotion processing, ToM, empathy, and moral reasoning separately. This is not intended to be an exhaustive review, but rather aims to highlight various approaches to the measurement of socio-cognitive functions in imaging environments. Importantly, less attention is paid to the results of each study (areas of activation are presented for each study in Table 8.1), as the key focus is on the methodology used, and results are described in detail in the relevant articles.

Emotion Processing

Emotion processing involves the capacity to perceive and label emotions and is a "rudimentary stage of processing social cues" (Beauchamp & Anderson, 2010, p. 49). Some of the earliest work in the area of facial and emotion processing is that of Ekman and Friesen (1976), who used the Pictures of Facial Affect (POFA). The POFA consists of 72 pictures of the six basic facial expressions (i.e., anger, disgust, fear, happiness, sadness, and surprise); it has been used in numerous fMRI studies to examine the neural correlates of emotion recognition (e.g., Roelofs, Minelli, Mars, van Peer, & Toni, 2009; Williams, McGlone, Abbott, & Mattingley, 2008).

Using the POFA, Roelofs and colleagues (2009) presented happy, angry, and neutral static pictures to measure approach and avoidant behavior toward these emotions in a sample of men (*N* = 22; *M* age 24 years, *SD* = 3 years, range = 18–32 years). Participants viewed static pictures, which they then categorized by moving a joystick. In the affect-congruent condition, the participants had to pull the joystick in response to a happy face and push the joystick in response to an angry face. The control task required participants to categorize the pictures according to the gender of the person depicted. Participants were given a practice task to begin with, which used similar static pictures that were not included in the experimental design. The psychometric properties of the POFA have been well established, but we were unable to find any evidence of the validity and reliability of this task when it was modified specifically for fMRI. Roelofs and colleagues (2009) reported increased activation in the left lateral orbitofrontal cortex (OFC) in the process of approaching happy emotions and avoiding angry emotions. The authors concluded that the lateral OFC is involved in the voluntary control of social-motivational behavior.

(text resumes on page 173)

TABLE 8.1. Review of Research Employing fMRI to Study Social Cognition

Measure	Stimulus	Standardized measure	Psychometric properties	Age range (years)	Area of interest	Results (areas of activation)
Emotion processing						
Facial expressions (happy and angry) (Roelofs et al., 2009)	Color pictures (Ekman & Friesen, 1976; Lundqvist et al., 1998; Martinez & Benavente, 1998; Matsumoto & Ekman, 1988). Pp used joystick to respond.	Presentation of measure was standardized.	No	18–32	Frontal lobe and whole brain	*Congruent > noncongruent/emotion > gender* • Left lateral OFC
Facial expressions (happy, fearful, and neutral) (Williams et al., 2008)	Pictures of faces (Ekman & Friesen, 1976). Pp used a 2-button keypad to respond.	Presentation of measure was standardized.	No	M = 28; SD = 3.72	• Fusiform gyrus • Amygdala • STS	*Fearful face > no fearful face* • Amygdala • STS
Perceiving emotions (Peelen et al., 2010)	Videos of emotional facial expressions (Banse & Scherer, 1996); videos of emotional body expressions (Atkinson et al., 2004, 2007); audio of emotional expressions (Belin et al., 2008). Pp responded vocally after each stimulus.	Presentation of measure was standardized.	No	20–32	Whole brain	*Disgust vs. other emotions* • Brainstem for face, voice, and face + body + voice *Fear vs. other emotions* • Left middle frontal gyrus for face, body, voice, and face + body + voice *Happiness vs. other emotions* • Right ACC and anterior temporal lobe for body, voice, and face + body + voice • Right occipital for body and face + body + voice
Facial expressions (happy, sad, angry, fearful, and disgusted (Gur et al., 2002)	Color pictures (Gur et al., 2002). Pp used a 2-button keypad to respond.	Presentation of measure was standardized.	Stimulus rated on accuracy and credibility; no properties mentioned.	19–36	From the superior cerebellum up through the frontal lobe	*Emotion > age* • Amygdala bilaterally • Hippocampus • Parahippocampal gyrus • Cingulate gyrus (minimally)

Measure	Source	Standardized	Age	ROI	Results
Emotional expressions: (fear and anger) (Said et al., 2010)	Videos (Said et al., 2010). Pp used a 2-button keypad to respond.	Presentation of measure was standardized.	No	M = 20.6	STS	*Age > emotion* • Cortical, posterior regions *Left STS* • Noncategorical representations of fear and anger *Right STS* • Categorical boundary and stimulus-related gradations of fear and anger
Emotions (sad and neutral) (Lévesque et al., 2003)	Videos (unpublished data).	Presentation of measure was standardized.	No	8–10	—	*Sad > neutral* • Right and left anterior temporal pole • Right and left MPFC • Right and left midbrain • Right VLPFC
Facial expressions (angry, fearful, happy, neutral, sad): imitation and observation (Pfeifer et al., 2008)	Pictures (*www.macbrain. org/faces.index.htm*).	Presentation of measure was standardized.	No	9.6–10.8	—	*Imitation* • Central sulcus • Precentral gyrus • Right medial frontal gyrus • Inferior frontal gyrus • Insula • Postcentral gyrus • Inferior parietal lobule • Fusiform gyrus • Lingual gyrus • Amygdala • Putamen *Observation* • Hippocampus • Inferior frontal gyrus • Insula • Fusiform gyrus • Lingual gyrus • Right amygdala • Left hippocampus

(cont.)

TABLE 8.1. (cont.)

Measure	Stimulus	Standardized measure	Psychometric properties	Age range (years)	Area of interest	Results (areas of activation)
Automatic appraisal of facial expressions (fear and disgust) from adolescence to middle age (Deeley et al., 2008)	Pictures (Young et al., 2002; Surguladze et al., 2003)	Stimulus and presentation of measure were standardized.	No	8–50 (M = 24; SD = 9.6)	—	*Fearful > neutral, young > old* • Medial frontal gyrus • Superior frontal gyrus • Middle frontal gyrus • Precentral gyrus • Cerebellum • Thalamus • Caudate *Old > young* • Lingual gyrus • Fusiform gyrus • Middle temporal gyrus • Superior temporal gyrus • Superior frontal gyrus *Disgust > neutral, young > old* • Middle frontal gyrus • Insula • Superior frontal gyrus • Precentral gyrus • Superior temporal gyrus • Inferior temporal gyrus • Caudate *Old > young* • Uncus • Parahippocampal gyrus • Middle temporal gyrus • Caudate • Inferior parietal lobule • Cingulate gyrus
Facial expressions in children with CD and healthy children (Sterzer et al., 2005)	Pictures (International Affective Picture System; Lang et al., 1995). Pp used a button to respond.	Presentation of measure was standardized.	No	9–15	• Amygdala • OFC • ACC	*Negative vs. neutral emotions* • OFC • Amygdala • Intraparietal sulcus • Ventral occipital cortex

Measure (study)	Task	Standardization	Age	Region	Findings
Facial expression in children with CD and healthy children (Sadler et al., 2007)	Pictures (International Affective Picture System; Lang et al., 1995). Pp used a button to respond.	Presentation of measure was standardized.	No	9–15	Dorsal ACC	*CD vs. healthy* • Right dorsal ACC *Healthy vs. CD* • Left amygdala *Novelty seeking in negative pictures:* *Healthy vs. CD* • Dorsal ACC
Theory of mind					
<u>False belief</u> (Rothmayr et al., 2011)	Cartoon scenarios (used from Baron-Cohen et al., 1985; created by Rothmayr et al., 2011). Pp used a 2-button keypad to respond.	Presentation of measure was standardized.	No	23–24	—	*False belief > true belief* • Left/right middle frontal gyrus • Right supramarginal gyrus, right inferior parietal lobule • Bilateral medial frontal gyrus • Right middle temporal gyrus • Left interior parietal lobule, left supramarginal gyrus • Left middle frontal and temporal gyrus • Left thalamus • Bilateral precuneus • Bilateral medial and superior frontal gyrus • Right middle frontal gyrus • Right middle and superior frontal gyrus
Inference (Mason & Just, 2011)	Written scenarios (Mason & Just, 2011). Pp used a 2-button keypad to respond.	Presentation of measure was standardized.	No	Under-graduate students	The inferior portion of the left interior frontal gyrus, the posterior portion of the left superior temporal gyrus, the superior aspect of the medial frontal gyrus, and the right TPJ	*Inference > control* • Medial and superior frontal areas • Bilateral inferior frontal gyri • Left posterior superior temporal gyrus • Anterior temporal gyri bilaterally

(cont.)

167

TABLE 8.1. *(cont.)*

Measure	Stimulus	Standardized measure	Psychometric properties	Age range (years)	Area of interest	Results (areas of activation)
False belief (Saxe & Kanwisher, 2003)	Written stories and photographs (Fletcher et al., 1995). Pp used 2 response buttons, in the left and right hands.	Presentation of measure was partly standardized; Pp response by button advanced the next screen.	No	—	Coronal slices parallel to the brainstem, covering the occipital lobe and the posterior portion of the temporal and parietal lobes	*Theory of mind > mechanical inference stories* • Left and right TPJ • Left and right anterior STS • Precuneus *False belief > false photograph* • TPJ bilaterally • Precuneus/posterior cingulate • Right anterior STS • Superior frontal gyrus in the frontal pole *Desire > physical people or nonhuman* • Medial TPJ • Right anterior STS
Mental states (Saxe & Wexler, 2005)	Written stories (modeled after Terwoght & Rieffe, 2003; created by Saxe & Wexler, 2005). Pp used 2-button keypad to respond.	Presentation of measure was standardized.	No	—	• Right and left TPJ • MPFC • PC	*Mental state* • Increased activity in right TPJ with mental state of foreign background, but effect was mediated by strong interaction with mental state condition *Background × mental state* • PC *Background* • Left TPJ
Mental states (Saxe & Powell, 2006)	Written stories (Saxe & Powell, 2006). Pp used button to indicate they were finished reading.	Presentation of measure was partly standardized; Pp response by button advanced the next screen.	Stimulus has been piloted.	19–26	• Right TPJ • Left TPJ • PC	*Thoughts > bodily sensations and appearance* • Right TPJ • Left TPJ • PC • Right SMG • Cingulate cortex • Cerebellum

Empathy						
Social perception and empathy (Lawrence et al., 2006)	Video clips (Rosenthal et al., 1979). Pp used 2-button keypad to respond.	Presentation of measure was standardized.	No	M = 32.2	—	*Social perception > nonsocial labeling* • Left middle frontal gyrus • Left middle temporal gyrus • Right middle frontal gyrus • Right superior frontal gyrus • Right fusiform gyrus
Empathy in boys and adult men (Greimel et al., 2010)	Pictures of faces (Greimel et al., 2010). Perspectives of other and self. Pp used 3-button keypad to respond.	Presentation of measure was standardized.	Stimulus has been piloted.	8–27	• MPFC • Precuneus • Fusiform gyrus • IFG	*Other > CO* • Superior occipital gyrus extending into right superior and inferior parietal lobule, increasing with age *Self > CO* • Right intraparietal sulcus extending into the inferior and superior parietal lobule, decreasing with age *Self > other/other > self* • No significant correlations; slightly increasing activity in right posterior cingulate cortex associated with increasing age during other and decreasing age during self
(Carr et al., 2003)	Pictures of faces (Ekman & Friesen, 1976)	Presentation of measure was standardized.	No	21–39	—	*Imitation > observation* • Premotor face area • Dorsal sector of pars opercularis of the inferior frontal gyrus • STS • Insula • Amygdala
Empathy of pain (Decety, 2008)	Pictures (Decety, 2008)	Presentation of measure was standardized.	No	16–18	• Amygdala • PAG • OFC	*CD > CO* • Left anterior insula • SMA • Precentral gyrus *CO > CD* • Left DLPFC • Right superior frontal gyrus

(cont.)

TABLE 8.1. *(cont.)*

Measure	Stimulus	Standardized measure	Psychometric properties	Age range (years)	Area of interest	Results (areas of activation)
Theory of mind in moral judgment (Young & Saxe, 2008)	Written stories (Young & Saxe, 2008). Pp used 2-button keypad to respond.	Presentation of measure was standardized.	No	18–22	• Right TPJ • Left TPJ • PC • Dorsal MPFC • Ventral MPFC	*Moral > nonmoral* • Right TPJ • PC • Dorsal MPFC
Mental states and moral judgment (Young & Saxe, 2009)	Written stories (Young & Saxe, 2009).	Presentation of measure was standardized.	No	18–22	• Right TPJ • Left TPJ • PC • Dorsal MPFC • Ventral MPFC	*Belief and outcome* • Right TPJ • PC • Left TPJ
Moral reasoning						
Emotional engagement in moral judgment (Greene et al., 2001)	Written scenarios (Greene et al., 2001). Pp used 1 button to advance to the next screen; 2-button keypad was used to respond.	Presentation of measure was partly standardized.	Stimulus has been piloted.	—	Anterior commisure–posterior commisure	*Moral–personal > moral–impersonal and nonmoral* • Bilateral medial frontal gyrus • Bilateral posterior cingulate gyrus • Left angular gyrus • Right angular gyrus *Moral–personal < moral–impersonal and nonmoral* • Left parietal lobe
Moral judgment (Greene et al., 2004)	Written scenarios (replication of Greene et al., 2001). Pp used 1 button to advance to the next screen; 2-button	Presentation of measure was partly standardized.	Stimulus has been piloted.	—	• ACC • DLPFC	*Personal > impersonal moral judgments* • MPFC • Posterior cingulate/precuneus • Bilateral STS/inferior parietal lobe

	keypad was used to respond.					• Bilateral anterior DLPFC • Bilateral inferior parietal lobes • ACC
Moral judgment (Moll et al., 2002a)	Written scenarios (Moll et al., 2002a). Pp rated scenarios on right or wrong after the scan.	Presentation of measure was standardized.	M = 30.3; SD = 4.7	Stimulus has been piloted.	• OFC • MPFC • Amygdala • Anterior temporal cortex	*Moral* • Gyrus rectus • Medial orbital gyrus • Left temporal pole • Cortex of the STS *Nonmoral* • Left lateral OFC • Lingual gyri • Inferior occipital gyri • Fusiform gyri *Moral > nonmoral* • Medial OFC • STS *Nonmoral > moral* • Left amygdala • Lateral orbital gyrus • Lingual and fusiform gyrus
Moral judgment (Moll et al., 2002b)	Pictures (International Affective Picture System; Lang et al., 1995). Pp were observing pictures; no response needed.	Presentation of measure was standardized.	M = 27; SD = 3	No	• Anterior and inferior prefrontal cortex • Medial temporal lobe (including amygdala)	*Moral vs. unpleasant* • Right medial fusiform gyrus • Right medial OFC • Right posterior STS • Left posterior middle temporal gyrus • Left lateral occipital gyrus *Unpleasant vs. moral* • Right middle frontal gyrus • Right anterior insula

(cont.)

TABLE 8.1. *(cont.)*

Measure	Stimulus	Standardized measure	Psychometric properties	Age range (years)	Area of interest	Results (areas of activation)
Moral judgment (Moll et al., 2002b) *(cont.)*						*Moral vs. neutral* • Right medial fusiform gyrus • Right medial OFC • Right/left amygdala/midbrain • Right anterior STS • Right/left posterior middle temporal gyri • Right/left fusiform gyri • Right precuneus • Left inferior occipital gyrus
Moral judgment (Borg, 2007)	Written scenarios (Borg, 2007). Pp used 1 button to advance to the next screen; 2-button keypad was used to respond.	Presentation of measure was partly standardized.	No	18–32	Whole brain	*Moral > nonmoral* • Bilateral inferior/superior rostral gyrus • Left frontal pole • Left lingual gyrus • Right and left posterior STS • Left supramarginal gyrus • Left middle frontal, caudal DLPFC *Nonmoral > moral* • Right middle frontal gyrus, caudal DLPFC • Left medial superior frontal gyrus

Note. Pp, participants; CD, conduct disorder; CO, control group; PAG, periaqueductal gray; OFC, orbitofrontal cortex; STS, superior temporal sulcus; TPJ, temporoparietal junction; MPFC, medial prefrontal cortex; PC, posterior cingulate; SMG, supramarginal gyrus; IFG, inferior frontal gyrus; ACC, anterior cingulate cortex; DLPFC, dorsolateral prefrontal cortex; VLPFC, ventrolateral prefrontal cortex; SMA, supplementary motor area.

Other tasks that build on the work of Ekman and Friesen (1976) have been developed. For example, Gur et al. (2002) developed a unique set of stimuli of different facial expressions (happiness, sadness, anger, fear, and disgust) in three levels of intensity and under two conditions: *posed* (correct posing of emotional expressions) and *evoked* (evocation of emotional expression through experiencing the emotion portrayed). Importantly, the authors reported that pilot standardization was conducted with a group of 64 people, who were required to rate the pictures for level of intensity and for how "fake" the expression looked. Although this may be a useful approach, no information was provided about the psychometric properties of the task; as a result, it is not possible to determine the extent to which the pictures were appropriate and effective.

An additional point of interest in the work of Gur and colleagues (2002) is the extent to which actors were trained in specific acting methods designed to enhance the accuracy of the intended emotional display. The *posed* method (or the "English method") focuses on the correct posing of emotional expressions, whereas the *evoked* method (or the "Russian method") focuses on the evocation of emotional expression through experiencing the emotion portrayed. Gur and colleagues (2002) designed their task specifically with accuracy and believability in mind, in an attempt to ensure that the images evoked the intended response. Participants were instructed to discriminate between positive (left button) or negative (right button) emotions in the emotion condition, and to discriminate between whether the face presented looked younger (left button) or older (right button) than 30 years. For the emotion discrimination task (i.e., positive vs. negative), activations were observed in the amygdala, hippocampus, and parahippocampal gyrus, and to a lesser extent in the cingulate gyrus. In the age discrimination task (i.e., younger vs. older), activations were observed in the amygdala as well as cortical and posterior regions. Across both tasks, activations were observed in the fusiform gyrus, thalamus, inferior frontal cortex, and occipital region.

Evidence from SIP research indicates that the more emotionally engaging a task is, the more accurate it will be as an assessment tool (Orobio de Castro, Veerman, Koops, Bosch, & Monschouwer, 2002). For example, the use of video tasks, when constructed properly, can be very beneficial for emotionally engaging participants in social dilemmas (Dooley, Anderson, Ohan, & Hemphill, 2011). Consistently, Peelen, Atkinson, and Vuilleumier (2010) attempted to enhance socio-cognitive measurement accuracy by combining videos of emotional facial expressions (Banse & Sherer, 1996), movies of emotional body expressions (Atkinson, Dittrich, Gemmel, & Young, 2004; Atkinson, Tunstall, & Dittrich, 2007), and emotional voice stimuli (Belin, Fillion-Bilodeau, & Gosselin, 2008). The aim of their study was to investigate the relative activation patterns of the different modalities of emotional states. The stimuli included representations (face, body, and voice) of anger, disgust, fear, happiness, and sadness (and emotional neutrality). Participants rated the intensity of the emotion being displayed (ranging from "a little" to "very much") by using three response buttons (half of participants responded from left to right, while the other half responded from right to left). The participants were not asked to identify the emotion being displayed (emotion recognition), but merely to rate its intensity. They were presented with a fixation cross, the stimulus, a blank screen, and then a response panel that listed the emotion (e.g., "afraid?") with the relevant rating scale. The paradigms consisted of three blocks, each including 12 trials of different emotions presented randomly and counterbalanced across runs. Results showed that the average rating of intensity of perceived emotions was 2.24 (on a 3-point scale) and was generally comparable across modalities (face = 2.13, body = 2.31, voice = 2.27). Although these non-imaging-related data are useful, they do not provide insight into the validity and reliability of the task. The authors reported activation patterns in two neural clusters: the rostral medial prefrontal cortex (PFC) and left posterior superior temporal sulcus (STS). Interestingly, in both areas activation patterns associated with the same emotions were more similar, regardless of stimulus type (body, face, voice), than were patterns associated with different emotions.

The natural progression for facial and emotional recognition tasks is toward engaging and believable stimuli that present realistic and accurate expressions of emotional states. Evidence from SIP research suggests that the more "real-world" stimuli are, the more likely they are to be engaging (Orobio de Castro et al., 2002). Thus moving from static to dynamic images may increase the extent to which participants engage with the task (and, as a result, brain activation patterns may be as close to what they would be outside a laboratory setting). Consistent with this, Said, Moore, Norman, Haxby, and Todorov (2010) developed computer-generated videos of emotional expressions displayed by five different actors. Each video consisted of an image of a neutral expression (presented for 167 milliseconds), which then morphed (over a 767-millisecond period) into an expression of either fear or anger (which remained on the screen for another 767 milliseconds). Participants were then asked to use a button to indicate whether they felt the stimulus was closer to fear or to anger. Response feedback was provided by flashing the form of a red cross after a wrong or missed response was recorded. This novel approach enables the identification of patterns of change (in relation to interpreting emotional expressions) and represents a middle ground between static images and real-time, video-based tasks. Interestingly, Said et al. reported that representations of facial expression in the left STS were graded and noncategorical, whereas in the right STS they were graded but with evidence of a category boundary (i.e., between fear and anger).

Theory of Mind

The capacity of children to understand that someone may have a set of beliefs different from their own is called *theory of mind* (ToM). Various behavioral tasks are commonly used to assess various aspects of ToM. For example, the Sally–Anne test (Baron-Cohen, Leslie, & Frith, 1985) uses two dolls, Sally and Anne, to assess false belief. Sally has a basket into which she places a ball, while Anne has a box. Sally leaves the scene, and while she is gone, Anne places the ball in her box. When Sally returns, participants are asked where they think Sally will look for her ball. Younger children will indicate that Sally will look for the ball where *they* know it to be—namely, in the box (i.e., they have a false belief about what Sally will think). In contrast, older children will predict that Sally will look for the ball where she believes it to be—namely, in the basket.

Saxe has written extensively on the measurement of ToM in fMRI (e.g., Saxe, 2006; Saxe & Powell, 2006), arguing that one of the main fundamental challenges in this domain relates to the nature of stimuli used to investigate ToM in the fMRI environment (Saxe, 2006). She advocates for the use of verbal stimuli to assess ToM, commenting that although the "fMRI scanner environment is inhospitable to natural social interaction" (2006, p. 60), much can be gained from the presentation of verbal (written) information. Saxe notes that much of the work to date in the socio-cognitive domain has focused primarily on nonverbal stimuli (i.e., facial expression, eye gaze, body posture, etc.), and thus overlooks the substantive influence of verbal information in social interactions.

Many tasks used to measure ToM in fMRI incorporate novel and innovative approaches. For example, Saxe and Kanwisher (2003) measured ToM by using stories varying in false belief, mechanical inference, human action, and nonhuman description, and combining these with photographs of whole human bodies in a range of postures (i.e., standing and sitting) and photographs of easily recognizable inanimate objects. When they compared the ToM task to the mechanical inference stories, increasing activation was observed in the left and right temporoparietal junction (TPJ), left and right anterior STS, and precuneus during the ToM task.

Demonstrating the importance of considering additional or related information, Young and Saxe (2008) have also described the use of written stories to investigate the influence of morally relevant facts about an action (Experiment 1) and the relationship between ToM and moral judgment (Experiment 2). They varied the outcome the protagonist produced (negative or neutral outcome) and the belief that the protagonist was causing the outcome (negative and neutral belief). This approach raised an interesting question in relation to how information is processed in social situations—namely, the outcomes associated with particular behaviors and the perceived fault on the part of a protagonist. For example, information may be interpreted in a specific manner if it is believed that a person either contributed to or created a situation through his or her actions. In the second experiment, they investigated whether responses would be influenced by whether explicit beliefs were previously presented (requiring integration) or were not (requiring spontaneous integration).

Using a different technique, Rothmayr et al. (2011) designed cartoon stories consisting of three pictures where participants had to indicate whether the behavior depicted by the child in the image was expected or not. Although it is possible to portray information in hand-drawn cartoons that is not necessarily possible with real-world actors (e.g., showing someone falling off a bridge or a building), evidence from SIP research suggests that the weakest associations between cognitive function and behavior are observed in studies that have used these types of cartoons or other hand-drawn stimuli (Orobio de Castro et al., 2002). Despite the variety of approaches available to measure ToM in fMRI (see Saxe, 2006), no psychometric information could be located to support any of these tasks.

Although many studies have reported that task administration was standardized, validity and reliability were not documented. In terms of activation patterns, Rothmayr et al. (2011) reported that contrasting false-belief with true-belief tasks revealed increased activity in the bilateral TPJ, bilateral middle frontal gyrus, left precentral gyrus, bilateral medial frontal gyrus, bilateral middle temporal gyrus, left thalamus, bilateral precuneus, and bilateral superior frontal gyrus. When go/no-go tasks were contrasted, increased activation was observed in the right inferior and middle frontal gyrus, right middle and superior temporal gyrus, right medial frontal gyrus, right superior frontal gyrus, and bilateral TPJ. Importantly, the areas involved in both belief reasoning and inhibitory control (i.e., activated in both conditions) were the right superior and medial frontal gyrus, right middle temporal gyrus, bilateral middle frontal gyrus, and bilateral TPJ.

Empathy

Empathy has been defined as "the capacity to understand and share the emotional states of other persons with reference to oneself" (Greimel et al., 2010, pp. 781–782). Not surprisingly, there are considerable theoretical, conceptual, and methodological overlaps with ToM, since the former is dependent on the latter's being present. Despite this, a review of the field indicates that ToM functions are generally not taken into account in studies measuring empathy-related patterns of brain activation. Empathy is often measured in the same way as emotion processing—namely, by using pictures or videos of expressed emotions. Despite the similarities, it is anticipated that social situations that involve (higher-order) empathy-related dilemmas should be harder to process and, as a result, be more cognitively taxing. In addition, the accurate portrayal of empathy-related dilemmas may be difficult to achieve, given the complexity of these situations.

Greimel et al. (2010) investigated empathy development in males from childhood to early adulthood by presenting participants with pictures of happy, sad, or neutral faces. Before the presentation of each face, an instruction was displayed on the screen informing the participant to infer the emotional state from the face ("other" task), to judge his or her own emotional response to the face ("self" task), and to decide on the width of the faces (control task). Seventy-two portrait faces were morphed to neutral and emotional faces of varying intensity, and all images were adapted for age (for the boys the stimuli were faces of male children and adolescents, and for the men faces of male adults were used). In the self task versus the control task, activation was observed in the right intraparietal sulcus extending into the inferior and superior parietal lobe. In the self task versus the other task, increased activation was observed in the right posterior cingulate cortex during the other task and (though activity in this area slightly decreased with age) during the self task.

In a related study, Carr, Jacoboni, Dubeau, Mazziotta, and Lenzi (2003) presented participants with three stimulus picture sets, each containing randomly ordered depictions of six emotions (happy, sad, angry, surprised, disgusted, and afraid). One set of pictures contained whole faces, while the faces in the other two sets consisted only of eyes or mouths, which had been cropped from the set of whole-face images. In a novel approach, participants were presented with six pictures of the different emotions and were asked to imitate and internally generate the target emotion being displayed on the computer screen, or simply to observe. This attempt to engage participants and overcome some of the ecological limitations associated with fMRI is, of course, only feasible in the assessment of some socio-cognitive functions. Nonetheless, this approach represents a creative use of the task and the environment to enhance, at a minimum, the ecological validity of the task. Unfortunately, as is consistently reported in this chapter, no psychometric properties were provided for this task. The authors reported greater activation in the premotor face area, the dorsal sector of pars opercularis of the inferior frontal gyrus, the STS, the insula, and the amygdala during imitation of emotions when compared with observation.

Illustrating the extent of challenges to the accurate measurement of social cognition in fMRI, Saxe and Powell (2006) designed written stories (on average 33 words long) to fit three conditions (appearance, bodily sensations, and thoughts). Each story was shown for 10 seconds, followed by a 12-second fixation block. Participants were asked to read the stories carefully and to press a button when they had finished reading. The authors reported activation differences between the false-belief and false-photograph conditions in the posterior cingluate, TPJ (left and right), ventral medial PFC, dorsal medial PFC, right STS, amygdala, and right anterior STS. Although the results of this study are certainly informative, the extent to which reading skills, working memory, and other related cognitive skills may have influenced the results was not considered by the authors. The impact of such critical confounds raises important questions about the appropriateness not only of the task being used, but also of the control task. Furthermore, the use of measures that are highly dependent on particular skills (e.g., reading, attention) can preclude or limit their use with various clinical populations. For example, impairments in attention are often reported after TBI, and these may affect the extent to which information presented in a written scenario is accurately recalled and a judgement is made that accurately reflects nonlaboratory cognitive processing.

Moral Reasoning

Moral development may be described as the process of acquiring appropriate attitudes and behaviors in regard to others based on values, norms, rules, and laws of society and culture. Moral reasoning skills are critical for considering, developing, and engaging in social behaviors (Dooley, Beauchamp, & Anderson, 2010). Traditionally, moral reasoning is measured behaviorally via archetypal dilemmas, such as the "Heinz" or "trolley" dilemmas. In the former case, Heinz discovers that his wife is ill and needs medication; however, the medication is exorbitantly priced, and Heinz must decide whether or not to steal it. In the trolley dilemma, a runaway trolley car is about to hit and kill five people; however, if a switch is flipped, the trolley can be diverted to another track, which will save the five people but kill only one person. Although these dilemmas clearly invoke significant moral conflicts and raise important issues in relation to social justice and the law, the nature and content of these dilemmas is such that clearly, concisely, and accurately communicating them in fMRI is challenging.

Despite the challenges, several innovative measurement approaches have been used to provide important insights into the neural underpinnings of moral reasoning and moral judgment (the instant feeling of approval or disapproval). The work of Greene and colleagues (Greene, Sommerville, Nystrom, Darley & Cohen, 2001) has been particularly influential, especially in the assessment of moral judgments. These authors designed a paradigm to measure emotional engagement in moral judgment in adults. The paradigm involved 60 dilemmas, divided into moral–personal (e.g., stealing one person's organs in order to distribute them to five others, throwing people off a sinking lifeboat), moral–impersonal (e.g., keeping money found in a lost wallet, voting for a policy expected to cause more deaths than its alternatives), and nonmoral conditions. The dilemmas were presented in random order, and each was presented as text in a series of three screens, with the first two describing a scenario and the third posing a question about the appropriateness of an action one might perform in that scenario. Participants could read the scenario at their own pace (to a maximum of 46 seconds per scenario, including response time) and used a button to go to the next screen. They then responded to a question after the third screen by pressing one of two buttons (indicating an "appropriate" or "inappropriate" judgment). Greene and colleagues (Greene et al., 2001; Greene, Nystrom, Engell, Darley & Cohen, 2004) reported that the medial prefrontal gyrus, posterior cingulate

gyrus, precuneus, angular gyrus, bilateral STS, and inferior parietal lobe were significantly more involved in the moral-personal conditions than in the moral-impersonal.

One of the important factors in the dilemmas used by Greene and colleagues (2001)
was the inclusion of several realistic scenarios—for example, keeping money found in a
lost wallet. Many other moral dilemmas described situations that the average person is
unlikely to encounter in the course of daily life. Thus reasoning about these dilemmas,
while complex and cognitively demanding, is unlikely to have as much of an emotional
element as situations that may have been or could be experienced. This emotional element
is crucial, given Greene and Haidt's (2002) suggestion that "moral judgment is more a
matter or emotion and affective intuition than deliberate reasoning" (p. 517).

Some authors have opted to use text-based measures of moral judgment in fMRI.
For example, Young and Saxe (2008) developed eight variations of 48 scenarios (for a
total of 384 stories) with an average of 86 words per story. In the stories, protagonists
produced either a negative outcome (harm to a person) or a neutral outcome (no harm),
and protagonists had the belief that they were causing a negative outcome (negative
belief) or a neutral outcome (neutral belief). Either belief information or information
foreshadowing the outcome was presented first. Stories were presented in four cumulative segments of 6 seconds each, for a total presentation of 24 seconds. Participants (N
= 18; 11 males; ages 18–22 years) saw one variation of each scenario, for a total of 48
stories, presented in a pseudorandom order. The stories were presented and followed by
a question about the moral nature of the action, and the participants responded on a
scale of 1 ("forbidden") to 3 ("permissible") by pressing buttons. The authors reported
that, overall, the activation patterns included rostral TPJ, the precuneus, and the dorsal
medial PFC.

Consistent with the approach to developing dilemmas that enhance emotional
engagement due to the increased likelihood that they might be familiar to participants,
Pujol et al. (2008) consulted with two developmental psychologists and a paediatric neu-
rologist to develop a variety of real-world scenarios that would lend themselves to moral
dilemmas in adolescence. These were displayed in hand-drawn cartoons and included
situations that adolescents might encounter that had a socio-moral component. Examples
included "telling on your friend" versus "allowing the teacher to punish the entire class,"
and "lying to parents about school grades" versus "being honest about them and being
punished." This approach represents an important step toward realistic and (age-)appro-
priate measures that can be used in fMRI.

Due to the lack of psychometric data reported by Pujol and colleagues (2008), it is
not possible to determine the extent to which their task is valid and reliable. Other than
being selected by three professionals, there is no indication that the task was pilot-tested
with an appropriate sample. Furthermore, as noted earlier, the use of hand-drawn cartoons in SIP research has been associated with the weakest relationship between social
behaviors and the measurement of cognitive functions. An additional limitation relates
to the procedure used to assess socio-cognitive functions in fMRI with this task. Participants were familiarized with the task up to 7 days before their fMRI session, with the
authors noting that this was done to ensure that participants comprehended the scenario
and would be able to recall it when presented with it in the fMRI scanner. This was an
unusual approach and may explain the reported variance in results—namely, that 60%
of participants showed deactivation in the posterior cingulate during a cognitive task,
indicating some common functions between posterior cingulate activity during moral

dilemmas and rest. Pujol and colleagues (2008) reported significant activation of the posterior cingulate cortex, medial frontal cortex, and angular gyrus during the moral dilemma condition. Interestingly, these findings were replicated in each subject with the passive viewing task, suggesting that the previous pattern was not specific to moral reasoning or decision making.

In a task that specifically tapped into real-life situations (albeit rarely experienced), Moll et al. (2002b) used emotionally charged pictures with and without moral content (in addition to emotionally neutral pictures). There were six experimental conditions: (1) pictures portraying emotionally charged, unpleasant social scenes depicting moral violations; (2) unpleasant pictures of aversive scenes with no moral connotations; (3) pleasant pictures; (4) interesting pictures, visually arousing but less emotional; (5) neutral pictures; and (6) scrambled pictures. Examples of the moral dilemmas included physical assaults, war scenes, and child neglect. After scanning, participants rated each picture for moral content, emotional valance, and level of arousal on visual analogue scales. These images portrayed realistic dilemmas by using pictures of real people as opposed to cartoons or hand illustrations, reflecting an important enhancement on other tasks, as the use of real people undoubtedly enhanced the emotional valence of the task.

The authors reported that supplementary behavioral investigation of their task revealed that the moral content of the moral dilemmas was rated as significantly higher than that of the nonmoral pictures. Although preliminary, these data support the validity of the task. Moll and colleagues (2002a) reported that moral statements activated the medial OFC, the cortex of the STS, the medial frontal gyrus, and the bilateral amygdala/midbrain.

In a similar approach, Harenski and Hamann (2006) presented first-person perspective pictures depicting unpleasant social scenes involving real people with and without moral content. The authors used two sets of color pictures (32 moral and 32 nonmoral), presented to participants 1 week prior to fMRI scanning. Participants were instructed either to "watch" (view and experience whatever thoughts or feelings were evoked) or to "decrease" (attempt to decrease any emotional response that the picture generated). Overall, activations were observed in the medial PFC, superior frontal gyrus, and amygdala in both the moral and nonmoral conditions. When moral and nonmoral pictures were contrasted, activations were observed in posterior STS and posterior cingulate. Interestingly, the middle frontal gyrus was more active during the nonmoral than during the moral condition.

Recently, Dooley et al. (2010) developed the Socio-Moral Reasoning (So-Moral) task, which employs age-appropriate scenarios to examine socio-moral reasoning skills. The task comprises 16 dilemmas, each consisting of three first-person perspective pictures (introduction–core–dilemma), followed by a moral reasoning question. The So-Moral is intended for use in conjunction with the socio-moral maturity (So-Mature) task, which is based on the theoretical models of Piaget (1932/1965) and Kohlberg (1984), and provides a moral maturity score for responses to the dilemmas presented on So-Moral. So-Moral is currently undergoing extensive standardization and testing before being used in fMRI; however, preliminary evidence indicates the strong psychometric properties of this task (Dooley et al., 2010). Specifically, the task has positive correlations with measures of empathy ($r = .51$, $p = .03$) and prosocial behavior ($r = .50$, $p = .03$). Furthermore, Cronbach's alpha scores range from .83 to .94, indicating good reliability. So-Moral has been used to examine moral reasoning skills after TBI in a sample of adolescents ($N = 25$,

M age = 13.7 years, SD = 2.1 years; 16 males; Full Scale IQ = 98.6, SD = 10.2). Importantly, preliminary evidence was found for a dose–response relationship between moral maturity and TBI severity, with the control and mild-TBI groups displaying the most developmentally mature scores (Dooley et al., 2010). As noted, extensive testing of this task is ongoing, and it will be used in fMRI.

Conclusion

The field of socio-cognitive neuroscience is rapidly emerging, due in part to the significant technological power of fMRI. Nonetheless, it remains a developing field of research and, as is commonly the case, is not without theoretical and methodological issues. Aside from the technical concerns that have been raised, the need for standardized and psychometrically sound instruments is great (Bennett & Miller, 2010; Vul, Harris, Winkielman & Pashler, 2009).

Evidence from SIP research highlights the variance of results when various approaches to measuring social cognition are used. The more "real" and emotionally engaging tasks have the greatest potential for accurately replicating typical (i.e., non-laboratory-based) functions. The same would be expected for fMRI research: The more accurate the task, the more likely it is to generate the sort of neural activation patterns that would occur in the real world. This has significant implications for the ways in which tasks are structured and administered, and for the core constructs they are intended to assess. As discussed, several of the tasks currently being used to assess moral reasoning/judgment in fMRI incorporate many of the elements of engaging and realistic tasks. For example, the use of real people in pictures (in contrast to drawings), the use of first-person perspective, the use of real-world dilemmas, and the use of age-appropriate dilemmas will all contribute to a more engaging and realistic task. Correspondingly, participants' reactions to these tasks are more likely to resemble their reactions if they faced such situations outside the laboratory.

Furthermore, adopting a strong theoretical approach that underlies how socio-cognitive functions operate (and interact) is critical to ensuring that the most conceptually appropriate measures are being included. In order for this to be possible, clear and well-grounded theoretical models such as the SOCIAL model (Beauchamp & Anderson, 2010) are invaluable. Given that we are still in the early days of social cognition research, especially in fMRI, it would be prudent to encourage the use of standardized tasks or, at a minimum, the pilot testing and reporting of the tasks used. Until such time, concerns about the replicability of fMRI research, as noted by Bennett and Miller (2010), will remain.

References

Atkinson, A. P., Dittrich, W. H., Gemmel, A. J., & Young, A. W. (2004). Emotion perception form dynamic and static body expression in point-light and full-light displays. *Perception, 33,* 717–746.

Atkinson, A. P., Tunstall, M. I., & Dittrich, W. H. (2007). Evidence for distinct contributions of form and motion information to the recognition of emotions from body gestures. *Cognition, 104,* 59–72.

Banse, R., & Scherer, K.R. (1996). Acoustic profiles in vocal emotion expression. *Journal of Personality and Social Psychology, 70*, 614–636.

Baron-Cohen, S., Leslie, A. M., & Frith, U. (1985). Does the autistic child have a "theory of mind"? *Cognition, 21*(1), 37–46.

Beauchamp, M. H., & Anderson, V. (2010). SOCIAL: An integrative framework for the development of social skills. *Psychological Bulletin, 136*(1), 39–64.

Belin, P., Fillion-Bilodeau, S., & Gosselin, F. (2008). The Montreal affective voices: A validated set of nonverbal affect bursts for research on auditory affective processing. *Behavior Research Methods, 40*, 531–539.

Bennett, C. M., & Miller, M.B. (2010). How reliable are the results from functional magnetic resonance imaging? *Annals of the New York Academy of Sciences, 1191*, 133–155.

Borg, E. (2007). If mirror neurons are the answer, what was the question? *Journal of Consciousness Studies, 14*(8), 5–19.

Carr, L., Iacoboni, M., Dubeau, M.-C., Mazziotta, J.C., & Lenzi, G.L. (2003). Neural mechanisms of empathy in humans: A relay from neural systems for imitation to limbic areas. *Neuroscience, 100*, 5497–5502.

Crick, N. R., & Dodge, K. A. (1994). A review and reformulation of social-information-processing mechanisms in children's social adjustment. *Psychological Bulletin, 115*, 74–101.

Decety, J., Michalska, K. J., & Akitsuki, Y. (2008). Who caused the pain?: An fMRI investigation of empathy and intentionality in children. *Neuropsychologia, 46*, 2607–2614.

Deeley, Q., Daly, E. M., Azuma, R., Surguladze, S., Giampietro, V., Brammer, M. J., et al. (2008). Changes in male brain responses to emotional faces from adolescence to middle age. *NeuroImage, 40*(1), 389–397.

Dooley, J. J., Anderson, V., Hemphill, S.A., & Ohan, J. (2008). Aggression after paediatric traumatic brain injury: A theoretical approach. *Brain Injury, 22*(11), 836–846.

Dooley, J. J., Anderson, V.A., Ohan, J.L., & Hemphill, S. (2011). *Brain Quest: A novel approach to the measurement of social information processing*. Manuscript submitted for publication.

Dooley, J. J., & Beauchamp, M. H. (2011). *Theory of mind versus intent attribution: Implications for imaging studies*. Manuscript in preparation.

Dooley, J. J., Beauchamp, M., & Anderson, V.A. (2010). The measurement of sociomoral reasoning in adolescents with traumatic brain injury: A pilot investigation. *Brain Impairment, 11*(2), 152–161.

Ekman, P., & Friesen, W. (1976). *Pictures of Facial Affect*. Palo Alto, CA: Consulting Psychologists Press.

Fletcher, P. C., Happé, F., Frith, U., Baker, S. C., Dolan, R. J., Frackowiak, R. S., et al. (1995). Other minds in the brain: A functional imaging study of "theory of mind" in story comprehension. *Cognition, 57*(2), 109–128.

Greene, J., & Haidt, J. (2002). How (and where) does moral judgment work? *Trends in Cognitive Sciences, 6*, 517–523.

Greene, J., Nystrom, L. E., Engell, A. D., Darley, J. M., & Cohen, J. D. (2004). The neural bases of cognitive conflict and control in moral judgment. *Neuron, 44*(2), 389–400.

Greene, J., Sommerville, R. B., Nystrom, L. E., Darley, J. M., & Cohen, J. D. (2001). An fMRI investigation of emotional engagement in moral judgement. *Science, 293*, 2105–2108.

Greimel, E. Schulte-Rüther, M., Fink, G., Piefke, M., Herpertz-Dahlmann, B., et al. (2010). Development of neural correlates of empathy from childhood to early adulthood: An fMRI study in boys and adult men. *Journal of Neural Transmission, 117*, 781–791.

Gur, R. C., Schroeder, L., Turner, T., McGrath, C., Chan, R. M., Turetsky, B. I., et al. (2002). Brain activation during facial emotion processing. *NeuroImage, 16*(3, Pt. 1), 651–662.

Harenski, C.L., & Hamann, S. (2006). Neural correlates of regulating negative emotions related to moral violations. *NeuroImage, 30*, 313–324.

Kohlberg, L. (1984). *Essays on moral development* (Vol. 2). San Francisco: Harper.

Lang, P. J., Bradley, M. M., & Cuthbert, B. N. (1995). *International Affective Picture System (IAPS)*. Bethesda, MD: National Institute of Mental Health Center for the Study of Emotion and Attention.

Lawrence, E. J., Shaw, P., Giampietro, V. P., Surguladze, S., Brammer, M. J., & David, A. S. (2006). The role of "shared representations" in social perception and empathy: An fMRI study. *NeuroImage, 29*(4), 1173–1184.

Lévesque, J., Joanette, Y., Mensour, B., Beaudoin, G., Leroux, J. M., Bourgouin, P., et al. (2003). Neural correlates of sad feelings in healthy girls. *Neuroscience, 121*, 545–551.

Lundqvist, D., Flykt, A., & Öhman, A. (1998). *The Karolinska Directed Emotional Faces—KDEF* (CD ROM). Stockholm: Department of Clinical Neuroscience, Psychology Section, Karolinska Institute.

Martinez, A. M., & Benavente, R. (1998). *The AR face database* (CV Technical Report No. 24). Columbus: Ohio State University.

Mason, R. A., Just, M. A. (2011). Differentiable cortical networks for inferences concerning people's intentions versus physical causality. *Human Brain Mapping, 32*(2), 313–329.

Matsumoto, D., & Ekman, P. (1988). *Japanese and Caucasian facial expressions of emotion (JACFEE)*. San Francisco: University of California, Human Interaction Laboratory.

Moll, J., de Oliveira-Souza, R., Bramati, I. E., & Grafman, J. (2002a). Functional networks in emotional moral and nonmoral social judgments. *NeuroImage, 16*(3), 696–703.

Moll, J., de Oliveira-Souza, R., Eslinger, P. J., Bramati, I. E., Mourao-Miranda, J. I., Andreiuolo, P. A., et al. (2002b). The neural correlates of moral sensitivity: A functional magnetic resonance imaging investigation of basic and moral emotions. *Journal of Neuroscience, 22*(7), 2730–2736.

Nelson, D. A., & Crick, N. R. (1999). Rose-colored glasses: Examining the social information-processing of prosocial young adolescents. *Journal of Early Adolescence, 19*(1), 17–38.

Orobio de Castro, B., Veerman, J. W., Koops, W., Bosch, J. D., & Monshouwer, H. J. (2002). Hostile attribution of intent and aggressive behavior: A meta-analysis. *Child Development, 73*(3), 916–934.

Peelen, M. V., Atkinson, A. P., & Vuilleumier, P. (2010). Supramodal representations of perceived emotions in the human brain. *Journal of Neuroscience, 30*(30), 10127–10134.

Pfeifer, J. H., Iacoboni, M., Mazziotta, J. C., & Dapretto, M. (2008). Mirroring others' emotions relates to empathy and interpersonal competence in children. *NeuroImage, 39*, 2076–2085.

Piaget, J. (1965). *The moral judgement of the child* (M. Gabain, Trans.). Glencoe, IL: Free Press. (Original work published 1932)

Pujol, J., Reixach, J., Harrison, B. J., Timoneda-Gallart, C., Vilanova, J. C., & Pérez-Alvarez, F. (2008). Posterior cingulate activation during moral dilemma in adolescents. *Human Brain Mapping, 29*(8), 910–921.

Roelofs, K., Minelli, A., Mars, R. B., van Peer, J., & Toni, I. (2009). On the neural control of social emotional behavior. *Social Cognitive and Affective Neuroscience, 4*(1), 50–58.

Rosenthal, H., Hall, J. A., DiMatteo, M. R., Rogers, P. L., & Archer, D. (1979). *Sensitivity to nonverbal communication: The PONS test*. Baltimore: Johns Hopkins University Press.

Rothmayr, C., Sodian, B., Hajak, G., Döhnel, K., Meinhardt, J., & Sommer, M. (2011). Common and distinct neural networks for false-belief reasoning and inhibitory control. *NeuroImage, 56*, 1705–1713.

Sadler, C., Sterzer, P., Schmeck, K., Krebs, A., Kleinschmidt, A., & Poustka, F. (2007). Reduced anterior cingulate activation in aggressive children and adolescents during affective stimulation: Association with temperament traits. *Journal of Psychiatric Research, 41*, 410–417.

Said, C. P., Moore, C. D., Norman, K. A., Haxby, J. V., & Todorov, A. (2010). Graded representations of emotional expressions in the left superior temporal sulcus. *Frontiers in Systems Neuroscience, 4*(6), 1–8.

Saxe, R. (2006). Why and how to study Theory of Mind with fMRI. *Brain Research, 1079*(1), 57–65.

Saxe, R., & Kanwisher, N. (2003). People thinking about people: The role of the temporoparietal junction in "theory of mind." *NeuroImage, 19,* 1835–1842.

Saxe, R., & Powell, L. J. (2006). It's the thought that counts. *Psychological Science, 17*(8), 692–699.

Saxe, R., & Wexler, A. (2005). Making sense of another mind: The role of right temporo-parietal junction. *Neuropsychologia, 43*(10), 1391–1399.

Sterzer, P., Sadler, C., Krebs, A., Kleinschmidt, A., & Poustka, F. (2005). Abnormal neural responses to emotional visual stimuli in adolescents with conduct disorder. *Biological Psychiatry, 57*(1), 7–15.

Surguladze, S. A., Brammer, M. J., Young, A. W., Andrew, A., Travis, M. J., Williams, S. C. R., et al. (2003). A preferential increase in the extrastriate response to signals of danger. *NeuroImage, 19,* 1317–1328.

Terwogt, M. M., & Rieffe, C. (2003). Stereotyped beliefs about desirability: Implications for characterizing the child's theory of mind. *New Ideas in Psychology, 21*(1), 69–84.

Vul, E., Harris, C., Winkielman, P., & Pashler, H. (2009). Puzzlingly high correlations in fMRI studies of emotion, personality, and social cognition. *Perspectives on Psychological Science, 4,* 274–290.

Williams, M. A., McGlone, F., Abbott, D. F., & Mattingley, J. B. (2008). Stimulus-driven and strategic neural responses to fearful and happy facial expressions in humans. *European Journal of Neuroscience, 27*(11), 3074–3082.

Young, A. P., Calder, A. J., Sprengelmeyer, R., & Ekman, P. (2002). *Facial Expressions of Emotion: Stimuli and Tests (FEEST)*. Bury St. Edmunds: Thames Valley Test.

Young, L., & Saxe, R. (2008). The neural basis of belief encoding and integration in moral judgment. *NeuroImage, 40*(4), 1912–1920.

Young, L., & Saxe, R. (2009). An fMRI investigation of spontaneous mental state inference for moral judgment. *Journal of Cognitive Neuroscience, 21*(7), 1396–1405.

CHAPTER 9

Measurement of Social Participation

Gary Bedell

S *ocial participation* and the broader concept, *participation,* are relatively new concepts derived from the World Health Organization's (WHO's) *International Classification of Functioning, Disability and Health* (ICF; WHO, 2001). The first part of this chapter provides background information to help readers understand these two concepts and how they might fit within the socio-cognitive integration of abilities model (SOCIAL; Beauchamp & Anderson, 2010; see also Anderson & Beauchamp, Chapter 1, this volume). The second part of the chapter highlights conceptual and methodological issues associated with measuring participation, and then describes selected measures and approaches to assess both participation and social participation. The ultimate aim is to assist readers with making decisions for selection of measures of children's social participation that are feasible for use in their own institutions and consistent with their own information needs and goals (intervention, research, policy).

Participation and Social Participation

Participation is defined very broadly as "involvement in life situations" in both the ICF (WHO, 2001, p. 213) and its newer *Version for Children and Youth* (ICF-CY; WHO, 2007). Participation is considered a multidimensional and universal concept important for all individuals across the lifespan, regardless of disability status, socioeconomic status, gender, age, race, ethnicity, or country of origin (Larson & Verma, 1999; Law, 2002; Michelsen et al., 2009). Greater participation in desired and important activities is an explicit aim of educational, rehabilitation, and community-based interventions and programs across the lifespan for children and adults with disabilities.

It is through participation in activities that children and youth learn to interact, work, and live with others and to function in society (Law, 2002; Larson, 2000; Mahoney, Cairns, & Farmer, 2003). Greater participation with others has been linked to enhanced quality of life, social competence, and educational success of children and youth with disabilities (Bedell & Dumas, 2004; King et al., 2003; Law, 2002; Simeonsson, Carlson, Huntington, McMillen, & Brent, 2001) and children without disabilities (Eccles, Barber, Stone, & Hunt, 2003; Larson, 2000; Mahoney et al., 2003). There is evidence to suggest that participation in activities that promote skill development and provide a sense of accomplishment and enjoyment may protect at-risk children and youth against future social, psychological, and academic problems (Eccles et al., 2003; Fletcher, Nickerson, & Wright, 2003; Mahoney & Cairns, 1997; Mahoney et al., 2003; Rutter, 1987).

Currently, there is little consensus on the appropriate definition of *participation*, given the lack of specificity in the current ICF definition (Dijkers, 2010; Heinemann et al., 2010; Whiteneck & Dijkers, 2009). For example, McConachie, Colver, Forsyth, Jarvis, and Parkinson (2006) have recommended that measures of children's participation should focus on life situations related to the children's well-being. They have classified life situations into four categories: activities that are (1) essential for survival; (2) supportive of child development; (3) discretionary (i.e., chosen by a child according to personal interests and not mandated by the family or society); and (4) educationally enriching at school (e.g., academics, clubs, creative arts). Coster and Khetani (2008) have defined *life situations* as organized sequences of activities that are setting-specific and directed toward a personally or socially meaningful goal. They suggest that goals for children can focus on (1) sustenance and physical health, (2) development of skills and capacities, and (3) enjoyment and emotional well-being. According to Coster and Khetani, participation is reflected by the extent of engagement in the full range of activities that achieve the larger goal. Whiteneck and Dijkers (2009) conceptualize participation as performance or fulfillment of social or societal roles, and emphasize that these roles typically involve interaction with others.

Social participation is not specifically defined in the ICF, and the term is often used interchangeably with the broader term *participation*. In this chapter, social participation is defined as "taking part, involvement, engagement, doing or being *with others*," to differentiate it from the larger concept of participation. Social participation is subsumed under participation, which also reflects involvement in activities that individuals can, should, or may prefer to be done without others, such as personal care, chores, and hobbies. Social participation, in particular, reflects involvement in activities that have a specific social focus or goal in mind, or that individuals participate in for their own sake or sense of enjoyment (e.g., socialization, social play, or leisure activities). As well, social participation can be reflected in involvement in activities that may be done "with others" (e.g., educational, skill-based, domestic, and vocational activities) but focus on different end goals.

Social Participation and SOCIAL

Beauchamp and Anderson (2010) do not explicitly label social participation as such in SOCIAL, but indicate that "social skills are critical for the capacity to participate and function in the community" (p. 39). However, social participation as previously defined is compatible with the definition of *social functioning* they refer to (p. 40), quoted from Yager and Ehmann (2006, p. 48) as "performance across many everyday domains (e.g., independent living, employment, interpersonal relationships, recreation)."

Conceptualizing social participation as *performance* rather than *capacity* is consistent with one suggestion made in the ICF (WHO, 2001): distinguishing the broader concept of *participation* from *activity,* another ICF domain. Activity differs from participation in that it focuses on the individual's ability to execute tasks and actions, rather than on involvement in life situations. Activity is represented in the ICF as functioning at the person level, whereas participation is represented as functioning at the societal level.

This distinction between participation and activity is also useful for differentiating social participation from other social variables in SOCIAL, such as social skills, social competence, social adjustment, and social interaction. Unlike these other variables, social participation typically is not considered a capacity or ability as much as it is considered what a person "does," to the extent to which the person wants and needs to do it. For example, what a child does is highly determined by personal choice, as well as by the *internal* factors (e.g., the child's temperament; motivation; and physical, cognitive, and social abilities) and *external* family, physical, and socio-cultural environmental factors (e.g., parenting, family functioning, values, attitudes, demands, expectations, resources, and opportunities) identified in SOCIAL (Beauchamp & Anderson, 2010) and in other biopsychosocial and person–environment transactional frameworks (Bronfenbrenner, 1979; Gallimore, Weisner, Bernheimer, Guthrie, & Nihira, 1993; Mallinson & Hammel, 2010; Noreau & Boschen, 2010).

The relationships among variables in SOCIAL that interact to influence outcomes, such as social skills and social competence, would appear to have similar effects on social participation as an outcome variable. These variables are likely to have direct effects on social participation outcomes as well. This latter claim seems consistent with that of Yager and Ehmann (2006) (referred to in Beauchamp & Anderson, 2010, p. 39), who have suggested that social functioning can be viewed as an overarching concept mediated by such variables as social skills and social cognition.

Most research on children's participation has been descriptive in nature, has focused on children with physical disabilities, and has been directed at the broader concept of participation as a dependent versus an independent variable (Imms, 2008; King et al., 2006; Michelsen et al., 2009). For example, King and colleagues (2003) used Bronfenbrenner's (1979) *ecology of human development* model to identify child, family, and environmental factors that could have a direct or indirect influence on leisure and recreational participation of school-age children with physical disabilities. In this study, the children's functional abilities and preferences and the family members' involvement in activities were found to be the most significant direct predictors of children's participation. Significant indirect predictors included family cohesion; supportive relationships; and unsupportive physical, social, and attitudinal environmental factors.

Results from other studies with different research designs, populations, and measures have shown similar variables to be associated with participation of children with disabilities. Forsyth, Colver, Alvanides, Woolley, and Lowe (2007) found that the participation of young children with severe disabilities (46% had autism or a behavioral disorder; the remaining children had a learning disability, neurological disorder, or chronic health condition) was influenced to a similar extent by the children's impairments and by physical and social environmental factors. Another study (Bedell & Dumas, 2004) found that functional ability was the strongest predictor of participation of children and youth with acquired brain injuries, followed by the children's extent of impairment and by physical and social environmental problems. In a qualitative study, Heah, Case, McGuire, and Law (2007) also found that the level of functional independence and physical and social features of the environment were perceived by parents and their children with developmental and neurological disabilities as key factors that influenced children's successful participation.

Parents develop strategies and accommodations, and provide the scaffolding needed to support their children's participation and other aspects of functioning in home, school, and community life (Gallimore et al., 1993; Guralnick, 1999; Rogoff, Mistry, Goncu, & Mosier, 1993). In one study, parents of children and youth with acquired brain injuries described how they often based their anticipation of how successful their children's participation would be on how well the features of the specific activity and environment would match their children's preferences and abilities (Bedell, Cohn, & Dumas, 2005). Parents identified strategies that involved creating opportunities, modifying the activity or environment, and helping their children to develop skills needed to participate in a particular activity or setting.

Similar strategies to promote participation were described in another study by parents of children with and without disabilities (Bedell, Khetani, Cousins, Coster, & Law, 2011). When parents were asked to appraise their children's participation, they often did so while simultaneously discussing the features of the environment in which it occurred. In particular, parents of children with disabilities (45% had a child with an autism spectrum disorder; 31% had a child with a mental illness) had to adjust their strategies to the activity, situation, or setting. As well, these parents emphasized the importance of considering both the supportive and hindering features of the environment, and both the children's strengths and challenges, in assessing and promoting participation of their children.

The internal and external factors highlighted in SOCIAL and expanded upon in this section are important to consider for obtaining a more complete view of children's social participation in important life situations across the lifespan. More research is clearly needed to develop interventions that directly or more explicitly address social participation, and to demonstrate the potential direct and indirect effects of other interventions and variables that could have a secondary impact on social participation and participation in general.

Measuring Participation and Social Participation

Conceptual and Methodological Issues

There are several commonly reported conceptual and methodological challenges associated with measuring participation. For instance, the terms *involvement* and *life situations* are not fully defined in the ICF, and thus what is being assessed varies across measures. Examples of how participation is assessed include how frequently a child participates, how involved the child is when participating, and how restricted or limited the child's participation is in general or when compared to same-age peers. It is important to note that a greater extent of participation is not necessarily better unless the activity or life situation is one that is important to the child or family. Some measures address this issue by asking the child or family about subjective aspects of participation, such as degree of satisfaction with the child's participation, level of importance placed on specific activities, and child's level of enjoyment of individual activities.

Another frequently reported measurement challenge is the difficulty distinguishing between the two ICF concepts *participation* and *activity* because the same nine ICF domains (see Table 9.1) used to classify participation are used to classify activity (Bedell & Coster, 2008; Coster & Khetani, 2008; Dijkers, 2010; Whiteneck & Dijkers, 2009). The authors of the ICF (WHO, 2001) have suggested a number of ways to make it easier to distinguish between participation and activity. Two of the suggestions, when combined, seem to offer some guidance for measuring social participation in particular. One suggestion, which has been touched upon in the preceding section, is to use the ICF qualifier *capacity* to assess activity (to reflect the extent of a child's ability) and the ICF qualifier *performance* to assess participation (to reflect the extent to which the child takes part in life situations). Another suggestion is to assess only the first five ICF domains—(1) learning and applying knowledge; (2) general tasks and demands; (3) communication; (4) mobility; and (5) self-care—when measuring activity, and only the last four domains—(6) domestic life; (7) interpersonal interactions and relationships; (8) major life areas; and (9) community, social, and civic life—when measuring participation. These last four domains (particularly 7 and 9) also seem to address social participation more clearly (i.e., taking part, involvement, engagement, doing or being with others).

TABLE 9.1. ICF Participation and Activity Domains

ICF domains	Subdomains
1. Learning and applying knowledge	Purposeful sensory experiences; basic learning; applying knowledge
2. General tasks and demands	Undertaking a single task; undertaking multiple tasks; carrying out daily routines; handling stress and other psychological demands
3. Communication	Communicating—receiving; communicating—producing (spoken and nonverbal); conversation and use of communication devices and techniques
4. Mobility	Changing and maintaining body positions; transferring oneself; carrying, moving, and handling objects; walking and moving; moving around using transportation
5. Self-care	Washing oneself; caring for body parts (grooming); toileting; dressing; eating; drinking; looking after one's health
6. Domestic life	Acquisition of necessities; household tasks; caring for household objects and assisting others
7. Interpersonal interactions and relationships	General (basic and complex) interpersonal interactions; particular interpersonal interactions (relating with strangers, informal and formal social relationships, family relationships, intimate relationships)
8. Major life areas	Education; work and employment; economic life
9. Community, social, and civic life	Community life; recreation and leisure; religion and spirituality; human rights; political life and citizenship

Note. Data from World Health Organization (2007).

There is little consensus on whether participation measures should include comparative or normative standards (Coster, Law, et al., 2011; Dijkers, 2010; Heinemann et al., 2010; Whiteneck & Dijkers, 2009). Some measures of participation use rating scales that compare a child's level of participation to the levels of same-age peers. For some measures, it is not always clear whether comparisons should be made to peers with disabilities or peers without disabilities. Thus it is often recommended that respondents consider the same referent group (the group that is more closely related to the research or clinical question) to increase consistency and interpretability of scores. Measures that are age-referenced may be less responsive to change, because the child may increase his or her level of participation, but never to the levels of same-age peers. Measures that are not age-referenced may be more responsive to change in participation, because the child serves as his or her own control or source of comparison.

None of the standardized measures of participation that have been reviewed in the literature appear to include age-referenced norms. Establishing normative standards of participation is challenging because (as described earlier) it is influenced by multiple factors, including child and family preferences, opportunities and resources, and societal values (Coster, Law, et al., 2011; Dijkers, 2010; Whiteneck & Dijkers, 2009). Coster, Law, et al. (2011) have questioned whether there is too much variation within a culture or even within a given neighborhood to make normative comparisons valid when participation is assessed.

Along the same lines, determining whether there is a minimum threshold of optimal participation and deciding whether this can and should be done with or without normative standards constitute another reported measurement issue (Coster, Law, et al., 2011; Dijkers, 2010; Whiteneck & Dijkers, 2009). It is clear that questions such as what is considered "good enough" participation, and how this might differ across life situations, settings, age groups, and other demographic characteristics, will require further discussion in the field (Coster, Law, et al., 2011; Heinemann et al., 2010).

Selected Participation Measures

Some of the aforementioned measurement issues have also been highlighted in recent reviews of participation measures designed for children and youth (Bedell & Coster, 2008; Bedell, Khetani, Coster, Law, & Cousins, in press; McConachie et al., 2006; Morris, Kurinczuk, & Fitzpatrick, 2005; Sakzewski, Boyd, & Ziviani, 2007; Ziviani, Desha, Feeney, & Boyd, 2010). Four participation measures frequently described in these reviews have been selected for inclusion in this section, given their extent of psychometric study or use in research: (1) the Children's Assessment of Participation and Enjoyment (CAPE; King et al., 2004); (2) the School Function Assessment (SFA) Participation subscale (Coster, Deeney, Haltiwanger, & Haley, 1998); (3) the Assessment of Life Habits (LIFE-H) for Children (Noreau, Fougeyrollas, & Tremblay, 2005); and (4) the Child and Adolescent Scale of Participation (CASP; Bedell, 2009). One newer measure, the Participation and Environment Measure for Children and Youth (PEM-CY; Coster, Bedell, et al., 2011; Coster, Law, et al., in press), has also been selected because it represents a newer approach to assessing participation.

It is important to note that I am the primary author of the CASP and a coauthor of the PEM-CY. My familiarity with these two measures has also influenced my decision to include them in this chapter. The five measures described in this section are illustrative examples that convey a range of ways to measure and understand participation. It will become clear that no single measure of participation can address all stakeholder information needs and goals. Each measure has its strengths and limitations, and readers should consider other measures and approaches that assess participation as they relate to their unique clinical and research questions or purposes.

The measures described assess participation in general, but include items or subsections that target facets of social participation (taking part, involvement, engagement, doing or being with others). Items or subsections might focus on activities that are more specifically social in nature than others. Some of the measures focus on specific settings (SFA) or type of activities (CAPE, SFA); others focus on multiple settings and types of activities (LIFE-H for Children, CASP, PEM-CY). The measures described vary in comprehensiveness; type and length of administration; specified age range of children and youth; and target respondent group (child/youth, parents, professionals). Some of the measures are used for specific purposes or can be used or modified to address multiple purposes, such as individualized intervention planning, program evaluation, and larger-scale multisite or population-based research. Table 9.2 summarizes these five measures.

Children's Assessment of Participation and Enjoyment

The CAPE (King et al., 2004) is a comprehensive tool that measures participation in recreation and leisure activities outside mandated school activities for children and youth with and without disabilities ages 6–21 years. The child or youth either completes a self-administered questionnaire booklet containing 55 items, or is interviewed by an evaluator using activity cards and visual response forms. Assistance from a parent or guardian can be provided if needed for either version. There are three levels of scoring: one overall participation score; two domain scores that reflect participation in two domains (informal activities and formal activities); and five scores related to five types of activities (recreational, physical, social, skill-based, and self-improvement). Social participation is probably best captured in the first three domains. Ordinal-scaled responses to specific items can be examined quantitatively as well (which is true of all measures described in this section).

The CAPE assesses five dimensions of participation: (1) diversity of activities (number of activities/items in which the child participates); (2) intensity, or how often the child participates in each activity (7-point scale); (3) persons with whom the child participates for each activity; (4) places where the child participates for each activity; and (5) extent of enjoyment for each activity (5-point scale). The Preferences for Activities of Children (PAC) scale is a parallel measure used with the CAPE that asks the child to rate (on a 3-point scale) the degree to which he or she would like to be doing the 55 CAPE activities. It takes 30–45 minutes to administer the CAPE, and approximately 15 minutes to administer the PAC.

Evidence of test–retest reliability was demonstrated for the overall participation and two domain scores (intraclass correlation coefficients [ICCs] ranged from .64 to .77). For the five activity type scores, ICCs were generally higher for the diversity and intensity dimensions (.64–.81) than for the enjoyment dimension (.12–.73).

Evidence of internal consistency was highest for the CAPE's informal activities domain score ($\alpha = .77$), as compared to the formal activities domain score ($\alpha = .42$)

TABLE 9.2. Selected Measures of Participation

Children's Assessment of Participation and Enjoyment (CAPE; King et al., 2004)

Measures: Participation in recreational and leisure activities outside mandated school activities.

Age range: 6–21 years.

Items, subsections and scores: There are 55 items that address participation frequency, diversity, and enjoyment (and where and with whom child participates). There is one overall participation score, two domain scores (informal activities and formal activities), and subscores specific to five types of activities: recreational (12 items), physical (13 items), social (10 items), skill-based (10 items), and self-improvement (10 items). The Preferences for Activities of Children (PAC) is a parallel measure used with the CAPE that includes the same 55 activities (items).

Administration: Either self-administered questionnaire or interview (with activity cards and visual response forms, if needed) with children or youth. Both versions use the same rating scale, and family caregivers can provide assistance if needed.

Reliability: Test–retest reliability: For the total and domain scores, ICCs ranged from .64 to .77; for the five activity type subscores, ICCs ranged from .64 to .77 for the diversity and intensity dimensions and from .12 to .73 for the enjoyment dimension. *Internal consistency.* Cronbach's α = .77 (informal activities domain); .42 (formal activities); .30–.62 (five activity types).

Validity: Evidence of construct and criterion-related validity; small to moderate correlations between CAPE scores and variables hypothesized to be associated with participation; limited evidence on the responsiveness of the CAPE.

How to obtain: Available from Pearson Assessments (*psychcorp.pearsonassessments.com*).

School Function Assessment (SFA), Participation subscale (Coster et al., 1998)

Measures: Elementary school-age children's participation in six different school settings.

Age range: 5–12 years.

Items, subsections, and scores: Six items. Each item pertains to one of the six different school settings: (1) classroom (regular or specialized), (2) playground/recess, (3) transportation, (4) bathroom/toileting, (5) transitions, and (6) mealtime/snacktime.

Administration: Rating scale completed by teachers, occupational therapists, or other school personnel familiar with child.

Reliability: Test–retest reliability: ICC = .95. *Internal consistency:* Cronbach's $\alpha \geq$ .95.

Validity: Evidence of construct and discriminant validity. Predicted relationships between participation item ratings and performance of setting relevant activities. Significant differences found in scores between children with cerebral palsy and children without disabilities, and between children with cerebral palsy and with learning disabilities.

How to obtain: Available from Pearson Assessments (*psychcorp.pearsonassessments.com*).

Assessment of Life Habits (LIFE-H) for Children (Noreau et al., 2007)

Measures: Level of accomplishment of and satisfaction with life habits (daily activities and social roles).

Age range: 5 years and older. An adapted version was recently developed for younger children (0–4 years), but no information was found on this version other than what is posted on the website (see below).

(cont.)

TABLE 9.2. *(cont.)*

Items, subsections, and scores: Long form (LF) has 197 items; short form (SF) has 64 items. There are two domain scores (daily activities and social roles) and 11 life habit category scores. Six category scores in the daily activities domain: communication (LF = 21, SF = 8 items); personal care (LF = 35, SF = 8 items); housing (LF = 28, SF = 6 items); mobility (LF = 16, SF = 4 items); nutrition (LF = 21, SF = 4 items); and fitness (LF = 8, SF = 4 items). Five category scores in the social roles domain: recreation (LF = 24, SF = 8 items); responsibility (LF = 16, SF = 7 items); education (LF = 14, SF = 6 items); community life (LF = 2, SF = 2 items); and interpersonal relationships (LF = 10, SF = 5 items).

Administration: Rating scale is designed to be completed by parent or guardian. Children and youth with adequate comprehension can be administered the interview version.

Reliability: Intrarater and interrater reliability: ICCs ≥ .78 for 10 of 11 life habit categories; ICCs = .58 and .63 for interpersonal relationships. *Internal consistency:* Cronbach's α ≥ .82 for item domain (daily activities, social roles) and life habit category scores.

Validity: Evidence of construct validity (convergent and divergent validity). Associations between scores on the LIFE-H for Children and scores from measures of activity performance. Limited evidence on the responsiveness of the LIFE-H for Children.

How to obtain: Available from the International Network on the Disability Creation Process, Québec, Canada (*www.ripph.qc.ca/?rub2=4&rub=16&lang=en*).

Child and Adolescent Scale of Participation (CASP; Bedell, 2004, 2009)

Measures: The extent to which children participate in home, school, and community activities, in comparison to children of the same age.

Age range: Most applicable for school-age children and for youth making the transition to adulthood (5–21 years). Data have been collected on children as young as 2 years and young adults as old as 27 years.

Items, subsections, and scores: There are 20 ordinal scaled items, and four subsections: (1) Home Participation (6 items), (2) Community Participation (4 items), (3) School Participation (5 items), and (4) Home and Community Living Activities (5 items). Open-ended questions about supports and barriers to participation.

Administration: Rating scale completed by parents/guardians via self-administration or interview (in person or home). A somewhat modified youth (young adult) version has been developed, but psychometric findings are not yet available for this version.

Reliability: Test–retest reliability: ICC = .94. *Internal consistency:* Cronbach's α ≥ .96.

Validity: Construct and discriminant validity. Moderate correlations between CASP scores and measures of activity performance, impairment, and environment. Typically developing children had significantly higher scores than children with disabilities. Responsiveness to change has not been examined.

How to obtain: Available free of charge by contacting Dr. Gary Bedell (*gary.bedell@tufts.edu*).

Participation and Environment Measure for Children and Youth (PEM-CY; Coster, Bedell, et al., 2011; Coster, Law, et al., 2011)

Measures: Examines participation (frequency, involvement) of children and youth in three settings (home, school, and community) and parents' desire for change in their children's participation. In addition, environmental factors (extent of supportiveness and availability/adequacy of resources) supporting or hindering participation specific to these three settings are assessed.

Age range: Data have been collected on children ages 5–17 years.

(cont.)

TABLE 9.2. *(cont.)*

Items, subsections, and scores: There are PEM-CY summary scores for each setting: (1) home (participation = 10 items, environment=12 items); (2) school (participation = 5 items, environment = 17 items); and (3) community (participation = 10 items, environment = 16 items). Other composite scores can be created, depending on the specific research or clinical question.

Administration: Rating scale is completed by family caregivers. Only the online survey has been tested. A paper-and-pencil version will be available in the near future.

Reliability: Test–retest reliability (ICC = .58–.95) and *internal consistency* (Cronbach's α = .59–.91) have been examined for the different PEM-CY summary scores across settings.

Validity: Evidence of discriminant validity: Children with disabilities as compared to children without disabilities had significantly lower frequencies of participation in activities, lower general level of involvement when participating in these activities, and less overall environmental supportiveness as perceived by caregivers. Responsiveness has not been examined.

How to obtain: Will be available free of charge in the near future. Please check the following websites that describe the Participation and Environment Project (PEP) for further information:

Boston University (Dr. Wendy Coster, Sargent College, Kids in Context Research Lab: *www.bu.edu/ kidsincontext/pep*)

McMaster University (Dr. Mary Law, CanChild Centre for Childhood Disability Research: *www. canchild.ca/en*)

Note. ICCs, intraclass correlation coefficients.

and five activity type scores (α = .30–.62). Evidence of construct and criterion validity was shown via small to moderate significant correlations among dimension scores within the CAPE and the PAC, and between CAPE intensity scores and scores from measures of child and family functioning and environment factors. Responsiveness to change reportedly requires further investigation; however, one longitudinal study reported on the responsiveness of the frequency scale to patterns of change over a 3-year period for children with physical disabilities (King et al., 2009).

School Function Assessment—Participation Subscale

The SFA (Coster et al., 1998) was designed to assess participation and functional activity performance in elementary school. Its Participation subscale assesses a child's participation in six different school settings: (1) classroom (regular or specialized); (2) playground/ recess; (3) transportation; (4) bathroom/toileting; (5) transitions; and (6) mealtime/snacktime. Social participation is probably best captured in the classroom and playground/ recess settings, and possibly also mealtime/snacktime, depending on the particular school or classroom routine.

The SFA Participation subscale consists of six items. Each item pertains to one of the six settings previously described and includes examples of tasks and activities that are typically part of that setting. Each item is rated on a 6-point scale that reflects the extent to which the child participates in the tasks and activities compared to peers in the same grade (1 = extremely limited; 2 = participation in a few activities; 3 = participation in all aspects with constant supervision; 4 = participation in all aspects with occasional assistance; 5 = modified full participation; 6 = child fully participates in all tasks and activities within each setting). A total participation summary score is computed by summing the ratings of the six items. Responses to individual items can be examined as well. The Participation subscale can be administered in about 5–10 minutes when used separately from the larger SFA.

The SFA Participation subscale has reported evidence of test–retest reliability (ICC = .95) and internal consistency ($\alpha \geq .95$). Evidence of construct and discriminant validity has been reported. Predicted relationships were found between item participation ratings and performance of setting-relevant activities. Rasch analysis confirmed the unidimensionality of the scale. Participation scores, on average, were significantly different between children without disabilities and with cerebral palsy, and between children with learning disabilities and with cerebral palsy. Responsiveness to change reportedly requires further testing.

Assessment of Life Habits for Children

The LIFE-H for Children (Noreau et al., 2005, 2007) is a comprehensive tool initially developed for adults with disabilities and subsequently adapted for use with children ages 5 years and older. Its purpose is to identify disruptions in the accomplishment of *life habits,* "regular activit[ies] or social role[s] valued by the person or his/her socio-cultural context according to his/her characteristics" (Noreau et al., 2005, p. 10). Two domains are assessed: Daily Activities and Social Roles. The LIFE-H for Children was designed to be completed by parent or guardian report. Children and youth with adequate comprehension and communication skills may be administered the interview version.

The LIFE-H for Children has a long form (197 items) and a short form (64 items), both of which include 11 life habit categories. Administration time is about 1–2 hours for the long form and 30–45 minutes for the short form. There are six categories in the daily activities domain (communication, personal care, housing, mobility, nutrition, fitness) and five in the social roles domain (recreation, responsibility, education, community life, interpersonal relationships). More social participation items are located in the social roles domain.

The LIFE-H for Children assesses level of accomplishment on a "difficulty" scale (no difficulty, some difficulty, performed by substitution, not performed, not applicable) and a "type of assistance" scale (no help, technical aid, adaptation, human assistance). The "not applicable" response is used when the life habit is not part of the person's lifestyle by choice. The difficulty and assistance scales are combined to create a 9-point modified life accomplishment scale. The LIFE-H for Children also assesses level of satisfaction for each life habit category on a 5-point scale (very satisfied to very dissatisfied). Level of accomplishment and satisfaction scores can be computed for items, life habit categories, and global scores.

Higher intrarater and interrater reliability was found for 10 of the 11 life habit categories (ICCs of .78 or higher). Somewhat lower intrarater and interrater reliability was found for the interpersonal relationships category (intrarater, ICC = .58; interrater, ICC = .63). Internal-consistency evidence was reported ($\alpha \geq .82$) for the two item domains (daily activities, social roles) and each life habit category score. Convergent and divergent validity evidence was reported via correlations with scores from two measures of activity performance: the Pediatric Evaluation of Disability Inventory (Haley, Coster, Ludlow, Haltiwanger, & Andrellos, 1992) and the Functional Independence

Measure for Children (Msall et al., 1994). Greater associations were found between similar domain scores, and lower associations were found between different domain scores. Little evidence was found on the responsiveness of the LIFE-H for Children. In a recent randomized clinical trial to improve upper-extremity functioning of children with congenital hemipareses, significant improvements were found in scores on the personal care subcategory, but not the other LIFE-H for Children scores examined in this study (Sakzewski et al., 2011).

Child and Adolescent Scale of Participation

The CASP (Bedell, 2009) measures the extent to which children participate in home, school, and community activities, compared to children of the same age. It was developed as part of the Child and Family Follow-Up Survey (CFFS) to monitor outcomes and needs of children with acquired brain injuries; it has since been used separately from the CFFS, as well as to assess children with other diagnoses (Bedell, 2004, 2009; Bedell & Dumas, 2004).

The CASP consists of four subsections: (1) Home Participation, (2) School Participation, (3) Community Participation, and (4) Home and Community Living Activities. Each section includes a subset of social participation items. The first three subsections include subsets of items that are more social in nature, such as socialization and play/leisure activities. The fourth section includes a subset of items that do not have a specific social focus, but often do involve interaction with others in the community, such as shopping and working.

The CASP has 20 items, rated on a 4-point scale (age expected, somewhat restricted, very restricted, and unable). A "not applicable" response is chosen when the item reflects an activity that the child is not expected to do because of age (e.g., work). The majority of items are applicable to children ages 5 and older. The home and community living section is particularly applicable to youth and to youth making the transition to adulthood, because its items focus on activities that are needed for independent living (e.g., work, shopping, and money management).

Each CASP item examines a broad life situation. Most items include examples of activities that fall within the broad life situation. For example, the item addressing participation in "social play or leisure activities with friends in the neighborhood and community" includes the following examples: "casual games, hanging out, going to public places such as a movie theater, park, or restaurant." The CASP also includes open-ended questions that ask about effective strategies and supports, as well as barriers that affect participation. Item, subsection, and total summary scores can be examined. The CASP takes about 10 minutes to administer when it is used separately from the larger CFFS.

The CASP has reported evidence of test–retest reliability (ICC = .94) and internal consistency ($\alpha \geq .96$). Evidence of construct and discriminant validity has been reported. Moderate correlations were found with scores from measures of activity performance (Pediatric Evaluation of Disability Inventory; Haley et al., 1992), extent of child impairment (Child and Adolescent Factors Inventory), and problems in the physical and social environment (Child and Adolescent Scale of Environment). Children without disabilities, as a group, had significantly higher CASP scores than children with disabilities (Bedell, 2009). Results from factor analyses showed that three factors explained 63% of the variance: (1) participation in social, leisure, communication (50%); (2) participation in advanced daily living (7%); and (3) participation in basic daily living and mobility (6%). Results from Rasch analyses demonstrated an expected pattern of life situations that children would find more or less challenging to participate in: Greater limitations were found in school and community activities requiring more complex cognitive and social skills, and lesser limitations were found in more basic and routine home and school activities (e.g., mobility, communication, and personal care). Responsiveness to change has not been examined.

Participation and Environment Measure for Children and Youth

The PEM-CY (Coster, Bedell, et al., 2011; Coster, Law, et al., 2011) is a recently developed caregiver report instrument that examines both participation in home, school, and community activities, and environmental factors that support or hinder participation in home, school, and community settings. The decision to develop participation and environment measures within the same instrument was informed by interviews of parents who described their children's participation while simultaneously considering the social and physical features of the environment in which it occurred (Bedell et al., 2011).

The PEM-CY participation items are broad categories or types of activities that are specific to three settings: home activities (e.g., indoor play/games, watching TV, and household chores); school activities (e.g., classroom activities, school-sponsored teams/clubs/ organizations, and getting together with peers outside of class); and community activities (e.g., neighborhood outings, unstructured physical activities, and classes/lessons). Most of the items are of a social nature, with the exception of items that can be typically done by oneself (personal care, household chores, schoolwork, and some hobbies/games).

Participation items are all rated in three ways: (1) frequency (responses range from 0 = never to 7 = daily); (2) involvement (responses range from 1 = minimally involved to 5 = very involved); and (3) desire for change (responses are yes or no). If the response is yes, a parent is asked to indicate the type of change desired, such as more or less frequency, more or less involvement, and/or more variety.

A separate set of environment items are provided after each set of participation items specific to each setting. Parents are asked questions about whether physical and social features of the environment (e.g., physical layout, social and cognitive activity demands, attitudes of others) help or hinder their children's participation, and whether there are available or adequate resources (e.g., services, equipment, time, and money) to support their children's participation (Coster, Bedell, et al., 2011).

Reliability and validity evidence supporting the use of the PEM-CY for use in population-based research was reported in one study (Coster, Bedell, et al., 2011). This study included data from 576 parents of children with and without disabilities (ages 5–17 years) from the United States and Canada, collected via web-based survey. Test–retest reliability and internal consistency were moderate to very good (ICC = .58–.95; α = .59–.91) for the different PEM-CY summary scores across settings. Evidence of discriminant validity was also reported: Children with disabilities, as compared to children without disabilities, had significantly lower frequencies of participation in activities, lower general level of involvement when participating in these activities, and less overall environmental supportiveness as perceived by caregivers. Responsiveness has not been examined.

The PEM-CY test developers incorporated a clinimetric measurement approach in its design, as well as recommendations for diverse stakeholders (Coster, Law, et al., 2011).

The reader is referred to Whiteneck and Dijkers (2010) and Dijkers (2010), who describe some of the potential advantages of a clinimetric approach (referring to the work of Feinstein, 1987) for measuring participation, in comparison to approaches based on classical test theory (e.g., factor analysis) and item response theory (e.g., Rasch analysis). In a clinimetric approach, the items themselves define the construct and are treated as meaningful indicators in and of themselves, as opposed to representing an underlying or latent trait. Creation of composite or summary scores can be informed by practical considerations. Stakeholders can use subsections or aggregate subsets of items from larger measures to address their specific clinical and research questions. For example, those interested in school participation can use only the school participation and school environment sections of the PEM-CY. Those interested in social leisure participation can create composite scores that include all items addressing this domain within or across settings.

The flexibility of the clinimetric approach makes it a very attractive feature of the PEM-CY. However, stakeholders may also want to consider this approach when using other measures, to ensure that measurement of participation is tailored to their specific information needs and goals.

In summary, five measures have been described that operationalize participation in different ways. The CAPE, LIFE-H for Children, and PEM-CY are the most comprehensive of the measures. The CAPE does not measure school participation, and the SFA Participation subscale (the briefest measure) only measures school participation. Those interested in assessing multiple domains of participation might consider either pairing the CAPE with the SFA or using the LIFE-H for Children, CASP, and PEM-CY, which examine multiple domains including school or educational participation. The CASP is the briefest of the multidomain measures when used separately from the larger CFFS (Bedell, 2004, 2009); because of its brevity, however, it may address participation in less depth than the LIFE-H for Children and PEM-CY.

The LIFE-H for Children focuses on level of accomplishment (degree of difficulty and type of assistance needed), and the rating scales of the CASP and SFA make reference to the degree of assistance needed to participate. Thus the LIFE-H, CASP, and SFA may also be assessing components of the ICF concept of *activity* (WHO, 2001), making it harder to differentiate participation from activity in these measures than in the CAPE and PEM-CY.

Little to no evidence was found on the responsiveness of the selected measures for detecting change over time or due to intervention. The CAPE, LIFE-H for Children, and PEM-CY may be more responsive than the SFA or CASP, because no comparisons are made to age expectations. In addition, the measures that focus on participation in more discrete activities (the CAPE, and the LIFE-H for Children to a lesser degree) may be more responsive than the measures that focus on broader categories of life situations (the SFA, CASP, or PEM-CY). For instance, a child may increase his or her level of participation in one activity within the broader category, but this increase may not be reflected on the scale for the broader category. In general, responsiveness to change is more likely to occur when substantial improvements due to focused intervention, development, or recovery are expected.

Finally, all of the measures described in this section focus primarily on participation of school-age children, with the exception of the LIFE-H for Children, which was adapted from the LIFE-H for adults with disabilities. The Vineland Adaptive Behavior

Scales, Second Edition (Vineland II; Sparrow, Cicchetti, & Balla, 2005), described in the next section, was also designed for use with children and adults across the lifespan. Some items and subsections from the CASP and PEM-CY are applicable for older youth making the transition into adulthood. Also, measures of participation designed for adults should be considered for older youth, particularly as they relate to involvement in independent living, work, and postsecondary education or vocational training activities (Magasi & Post, 2010; Malec, 2004). Measures of participation in younger children that focus on play, preschool, and family activities should be considered as well (Berg & LaVesser, 2006; Dunst, Hamby, Trivette, Raab, & Bruder, 2002; Rosenberg, Jarus, & Bart, 2010).

Other Options for Measuring Social Participation

This section provides an overview of other options for assessing social participation when current measures do not match stakeholders' information needs and goals. One option is to assess whether items and subsections from existing measures of social functioning are measuring facets of social participation of interest; this option is similar to the clinimetric approach described earlier. To illustrate this, two commonly used measures that have undergone extensive psychometric testing are briefly described: the Vineland-II (Sparrow et al., 2005) and the Child Behavior Checklist (CBCL; Achenbach & Rescorla, 2001). Another option is to use tailored approaches, which essentially involve creating items or objectives that are more direct indicators of desired facets of social participation.

The Vineland-II (Sparrow et al., 2005) examines four domains of functioning (Communication, Daily Living Skills, Socialization, and Motor Skills); parent/caregiver and teacher report versions are available. The parent versions can be used to follow children from infancy into older adulthood. A child's performance is rated on a 3-point scale (2 = yes, usually; 1 = sometimes or partially; 0 = no, never; N = no opportunity). A number of scores are available, including a total score as well as scores for each domain. The social participation items are mainly located in the Socialization domain, particularly the two Socialization subdomains: Interpersonal Relationships (friendships, dating, social communication, and responding to others) and Playing and Using Leisure Time (playing and going places with friends). Longitudinal studies of children and youth with traumatic brain injury provide evidence of the responsiveness of the original Vineland's Socialization domain in detecting change over time (Taylor et al., 2002; Yeates et al., 2004).

The Child Behavior Checklist (CBCL; Achenbach & Rescorla, 2001), which is part of the Achenbach System of Empirically Based Assessment (ASEBA), examines multiple domains such as social competence, academic performance, and internalizing–externalizing behaviors. The most recent school-age versions include parent/caregiver report and teacher report forms (for children/youth ages 6–18) and a youth report form (for youth ages 11–18 years). The social competence subscale in the parent and youth report forms includes subsections that measure *social participation* (participation in activities, number of friendships, and frequency of contact with friends). Participation is assessed by asking the parent or youth to list three activities the child or youth most likes to take part in, related to four categories: (1) sports; (2) hobbies, activities, games other than sports; (3) organizations, clubs, teams, or groups; (4) jobs or chores. The respondent can select up to three activities for each category or check "none." Responses for parents and youth are generally the same; however, parents are provided with a "don't know" response, whereas youth are not. All questions ask the respondent to compare the child (or youth) to others of the same age. For sports and hobbies, respondents are asked about how much time the child spends in each activity (less than average, average, more than average) and how well he or she performs the activity (below average, average, above average). For organizations/clubs, they are asked how active the child is during these activities (less active, average, more active). For jobs and chores, they are also asked how well the child performs these activities. Similar to items in other age-referenced measures, these particular CBCL items may be less responsive to change than items from measures that are not age-referenced, except when substantial improvement (due to focused intervention, development, or recovery) is expected.

The two CBCL friendship items seem to have greater potential for detecting change because they are not age-referenced. These items ask about how many friends the child or youth has (none; 1; 2 or 3; 4 or more) and how many times a week the child or youth does things with friends outside regular school hours (less than 1; 1 or 2; 3 or more). Prigatano and Gupta (2006) found that children with more severe brain injury had significantly fewer friends and fewer weekly contacts with friends than children with less severe or no injury, according to the data pertaining to these two questions.

Individualized or tailored measurement approaches can also be used to measure or describe children's participation, in concert with or instead of other participation measures. These approaches often tend to be more responsive to intervention change, because they are individualized to a person's specific needs and goals or are created specifically to detect the effects of particular interventions (Bedell & Coster, 2008).

Examples of tailored approaches include the common practices of developing and monitoring measurable objectives, or of using more systematic goal-setting approaches such as Goal Attainment Scaling (GAS) or the Canadian Occupational Performance Measure (COPM) (Cusik, McIntyre, Novak, Lannin, & Lowe, 2006). The COPM is a client-centered measure, frequently used by occupational therapists; it essentially involves identifying activities that are most important for the person to do and assessing both the person's performance and satisfaction with performance in each activity, using 10-point rating scales. Cusik, Lannin, and Lowe (2007) found the COPM and GAS to be responsive to change specific to intervention in a clinical trial with children with cerebral palsy. COPM scores were also found to be responsive to intervention in a randomized clinical trial to improve upper-extremity functioning of children with congenital hemipareses (Sakzewski et al., 2011).

Social participation can also be assessed by creating observation or rating scales that are tailored to specific life activities, situations, and settings. For example, Glang, Todis, Cooley, Wells, and Voss (1997) used these two methods to examine the effects of the Building Friendship Process intervention for improving social networks of students with traumatic brain injury. The observation scale detected increases in children's number of social contacts, and the rating scale detected improvements in the degree to which children were socially included with others at school.

Finally, asking open-ended questions can be used to fill in the gaps when existing measures are unable to capture more in-depth details about the child's participation. Open-ended questions are typically used in clinical practice as part of the overall evaluation process; in qualitative research; in studies that primarily use quantitative methods, and when a greater understanding of research participants' perspectives is needed. A few questions seem to be particularly useful for collaborative goal setting and intervention planning with children and families, to ensure that the children's participation is supported in the most relevant and meaningful life activities, situations, and settings. My colleagues and I use these basic questions frequently in our clinical practice and research, and they can be directed to or modified for a child, a parent, or others concerned about the child's participation (Bedell et al., 2005, 2011). As will become apparent, elements of these questions are incorporated in some of the measures and approaches already described in this chapter. However, social participation (or participation in general) in these instances would be defined according to specific clinical or research goals:

1. What are the most important activities for the child to participate in?
2. What is the child's current and desired level of participation in these activities?
3. What types of things (internal and external factors) support or hinder the child's participation?
4. What types of strategies, accommodations, or interventions have been used effectively to promote the child's participation? (This question is particularly useful to acknowledge and leverage the expertise of the child, family, and other key people, and to avoid "reinventing the wheel.")

Summary

This chapter has described selected measures and approaches to assess social participation of children and youth, and has discussed related conceptual and methodological considerations. Its ultimate aim is to assist readers with making decisions for selection of measures of children's social participation that are feasible for use in their own institutions and consistent with their own information needs and goals (intervention, research, policy).

More research is clearly needed in the field that directly or more explicitly addresses social participation as well as participation in general. Also, further discussion is warranted to address the ongoing conceptual and methodological issues touched upon in this chapter. For example, what are the most appropriate definitions for participation and social participation? Does it makes sense to use comparative and/or normative standards to assess participation (and, if so, in what specific instances)? What is considered "good enough" participation, and how does this differ across life situations, settings, and sociodemographic factors?

Acknowledgments

I thank the U.S. Department of Education's National Institute on Disability and Rehabilitation for providing funding for two grants that supported the work carried out in this chapter. I also thank the principal investigators on each of these grants with whom I have collaborated: Shari Wade, PhD (Cincinnati Children's Hospital Medical Center, "Rehabilitation, Research and Training Center on Interventions for Children and Youth with Traumatic Brain Injury," Grant No. H133B090010), and Wendy Coster, PhD (Boston University, "Development of Measures of Participation and Environment for Children with Disabilities," Grant No. H133G070140).

References

Achenbach, T. M., & Rescorla, L. A. (2001). *Manual for the ASEBA School-Age Forms and Profiles*. Burlington: University of Vermont, Research Center for Children, Youth, and Families.

Beauchamp, M. H., & Anderson, V. (2010). SOCIAL: An integrative framework for the development of social skills. *Psychological Bulletin, 136*(1), 39–64.

Bedell, G. M. (2004). Developing a follow-up survey focused on participation of children and youth with acquired brain injuries after discharge from inpatient rehabilitation. *NeuroRehabilitation, 19*, 191–205.

Bedell, G. (2009). Further validation of the Child and Adolescent Scale of Participation (CASP). *Developmental Neurorehabilitation, 12*(5), 342–351.

Bedell, G. M., Cohn, E. S., & Dumas, H. M. (2005). Exploring parents' use of strategies to promote social participation of school-age children with acquired brain injuries. *American Journal of Occupational Therapy, 59*(3), 273–284.

Bedell, G., & Coster, W. (2008). Measuring participation of school-aged children with traumatic brain injuries: Considerations and approaches. *Journal of Head Trauma Rehabilitation, 23*(4), 220–229.

Bedell, G. M., & Dumas, H. M. (2004). Social participation of children and youth with acquired brain injuries discharged from inpatient rehabilitation: A follow-up study. *Brain Injury, 18*(1), 65–82.

Bedell, G., Khetani, M. A., Coster, W., Law, M., & Cousins, M. (in press). Measures of community, social and civic life. In A. Majnemer (Ed.), *Measures for children with developmental disabilities: Framed by the ICF-CY*. London: Mac Keith Press.

Bedell, G., Khetani, M. A., Cousins, M. A., Coster, W. J., & Law, M. C. (2011). Parent perspectives to inform development of measures of children's participation and environment. *Archives of Physical Medicine and Rehabilitation, 92*, 765–773.

Berg, C., & LaVesser, P. (2006). The Preschool Activity Card Sort. *Occupational Therapy Journal of Research, 26*(4), 143–151.

Bronfenbrenner, U. (1979). *The ecology of human development*. Cambridge, MA: Harvard University Press.

Coster, W., Bedell, G., Law, M., Khetani, M. A., Teplicky, R., Liljenquist, K., et al. (2011). Psychometric evaluation of the Participation and Environment Measure for Children and Youth (PEM-CY). *Developmental Medicine and Child Neurology, 53*, 1030–1037.

Coster, W. J., Deeney, T., Haltiwanger, J., & Haley, S. M. (1998). *School Function Assessment*. San Antonio, TX: Psychological Corporation.

Coster, W., & Khetani, M. A. (2008). Measuring participation of children with disabilities: Issues and challenges. *Disability and Rehabilitation, 30*(8), 639–648.

Coster, W., Law, M., Bedell, G., Khetani, M. A., Cousins, M., & Teplicky, R. (2011). Development of the Participation and Environment Measure for Children and Youth (PEM-CY): Conceptual basis. *Disability and Rehabilitation* [early online version], 1–9.

Cusik, A., Lannin, N. A., & Lowe, K. (2007). Adapting the Canadian Occupational Performance Measure for use in a paediatric clinical trial. *Disability and Rehabilitation, 29*(10), 761–766.

Cusik, A., McIntyre, S., Novak, I., Lannin, N., & Lowe, K. (2006). A comparison of Goal Attainment Scaling and the Canadian Occupational Performance Measure for paediatric rehabilitation research. *Pediatric Rehabilitation, 9*(2), 149–157.

Dijkers, M. P. (2010). Issues in the conceptualization and measurement of participation: An overview. *Archives of Physical Medicine and Rehabilitation, 91*(Suppl. 1), S5–S16.

Dunst, C. J., Hamby, D. W., Trivette, C. M., Raab, M., & Bruder, M. B. (2002). Young children's participation in everyday family and community activity. *Psychological Reports, 91*, 875–897.

Eccles, J. S., Barber, B.L., Stone, M., & Hunt, J. (2003). Extracurricular activities and adolescent development. *Journal of Social Issues, 59*(4), 865–889.

Feinstein, A. R. (1987). *Clinimetrics*. New Haven, CT: Yale University Press.

Fletcher, A. C., Nickerson, P., & Wright, K. L. (2003). Structured leisure activities in middle childhood: Links to well-being. *Journal of Community Psychology, 31*(6), 641–659.

Forsyth, R., Colver, A., Alvanides, S., Woolley, M., & Lowe, M. (2007). Participation of young severely disabled children is influenced by their intrinsic impairments and environment. *Developmental Medicine and Child Neurology, 49*(5), 345–349.

Gallimore, R., Weisner, T. S., Bernheimer, L. P., Guthrie, D., & Nihira, K. (1993). Family responses to young children with developmental delays: Accommodation activity in ecological and cultural context. *American Journal on Mental Retardation, 98*, 185–206.

Glang, A., Todis, B., Cooley, E., Wells, J., & Voss, J. (1997). Building social networks for children and adolescents with traumatic brain injury: A school-based intervention. *Journal of Head Trauma Rehabilitation, 12*(2), 32–47.

Guralnick, M. J. (1999). Family and child influences on the peer-related social competence of young children with developmental delays. *Mental Retardation and Developmental Disabilities Research Reviews, 5*, 21–29.

Heah, T., Case, T., McGuire, B., & Law, M. (2007). Successful participation: The lived experience among children with disabilities. *Canadian Journal of Occupational Therapy, 74*(1), 38–47.

Haley, S., Coster, W., Ludlow, L., Haltiwanger, J., & Andrellos, J. (1992). *Pediatric Evaluation of Disability Inventory: Development, standardization, and administration manual, version 1.0*. Boston: Trustees of Boston University, Health and Disability Research Institute.

Heinemann, A. W., Tulsky, D., Dijkers, M., Brown, M., Magasi, S., Gordon, W., et al. (2010). Issues in participation measurement in research and clinical applications. *Archives of Physical Medicine and Rehabilitation, 91*(Suppl. 1), S72–S76.

Imms, C. (2008). Children with cerebral palsy participate: A review of the literature. *Disability and Rehabilitation, 30*(24), 1867–1884.

King, G., Law, M., Hanna, S., King, S., Hurley, P., Rosenbaum, P., et al. (2006). Predictors of the leisure and recreation participation of children with physical disabilities: A structural equation modeling analysis. *Children's Health Care, 35*(3), 209–234.

King, G., Law, M., King, S., Hurley, P., Rosenbaum, P., Hanna, S., et al. (2004). *Children's Assessment of Participation and Enjoyment (CAPE) and Preferences for Activities of Children (PAC)*. San Antonio, TX: Psychological Corporation.

King, G., Law, M., King, S., Rosenbaum, P., Kertoy, M. K., & Young, N. (2003). Conceptual model of the factors affecting recreation and leisure participation of children with disabilities. *Physical and Occupational Therapy in Pediatrics, 23*, 63–90.

King, G., McDougall, J., DeWit, D., Petrenchik, T., Hurley, P., & Law, M. (2009). Predictors of change over time in the activity participation of children and youth with physical disabilities. *Children's Health Care, 38*, 321–351.

Larson, R. W. (2000). Toward a psychology of positive youth development. *American Psychologist, 55*(1), 170–183.

Larson, R. W., & Verma, S. (1999). How children and adolescents spend time across the world: Work, play, and developmental opportunities *Psychological Bulletin, 125*(6), 701–736.

Law, M. (2002). Participation in the occupations of everyday life. *American Journal of Occupational Therapy, 56*(6), 640–649.

Magasi, S., & Post, M. W. (2010). A comparative review of contemporary participation measures' psychometric properties and content coverage. *Archives of Physical Medicine and Rehabilitation, 91*(Suppl. 1), S17–S28.

Mahoney, J. L., & Cairns, R. B. (1997). Do extracurricular activities protect against early school dropout? *Developmental Psychology, 33*(2), 241–253.

Mahoney, J. L., Cairns, B. D., & Farmer, T. W. (2003). Promoting interpersonal competence and educational success through extracurricular activity participation. *Journal of Educational Psychology, 95*(2), 409–418.

Malec, J. F. (2004). The Mayo–Portland Participation Index: A brief and psychometrically sound measure of brain injury outcome. *Archives of Physical Medicine and Rehabilitation, 85*(12), 1989–1996.

Mallinson, T., & Hammel, J. (2010). Measurement of participation: Intersecting person, task, and environment. *Archives of Physical Medicine and Rehabilitation, 91*(Suppl. 1), S29–S33.

McConachie, H., Colver, A. F., Forsyth, R. J., Jarvis, S. N., & Parkinson, K. N. (2006). Participation of disabled children: How should it be characterised and measured? *Disability and Rehabilitation, 28,* 1157–1164.

Michelsen, S. I., Flachs, E. M., Uldall, P., Eriksen, E. L., McManus, V., Parkes, J., et al. (2008). Frequency of participation of 8–12-year-old children with cerebral palsy: A multi-centre cross-sectional European study. *European Journal of Paediatric Neurology, 13,* 165–177.

Morris, C., Kurinczuk, J. J., & Fitzpatrick, R. (2005). Child or family assessed measures of activity performance and participation for children with cerebral palsy: A structured review. *Child: Care, Health and Development, 31,* 397–407.

Msall, M. E., DiGaudio, K., Duffy, L. C., LaForest, S., Braun, S., & Granger, C. V. (1994). WeeFIM: Normative sample of an instrument for tracking functional independence in children. *Clinical Pediatrics, 33,* 431–438.

Noreau, L., & Boschen, K. (2010). Intersection of participation and environmental factors: A complex interactive process. *Archives of Physical Medicine and Rehabilitation, 91*(Suppl. 1), S44–S53.

Noreau, L., Fougeyrollas, P., & Tremblay, J. (2005). *Measure of Life Habits (LIFE-H): User's manual.* Québec, Québec, Canada: Reseau International sur le Processus de Production du Handicap (RIPPH).

Noreau, L., Lepage, C., Boissiere, L., Picard, R., Fougeyrollas, P., Mathieu, J., et al. (2007). Measuring participation in children with disabilities using the Assessment of Life Habits. *Developmental Medicine and Child Neurology, 49*(9), 666–671.

Prigatano, G. P., & Gupta, S. (2006). Friends after traumatic brain injury in children. *Journal of Head Trauma Rehabilitation, 21*(6), 505–513.

Rogoff, B., Mistry, J., Goncu, A., & Mosier, C. (1993). Guided participation in cultural activities by toddlers and caregivers. *Monographs of the Society for Research in Child Development, 58*(8, Serial No. 236), 1–174.

Rosenberg, L., Jarus, T., & Bart, O. (2010). Development and initial validation of the Child Participation Questionnaire (CPQ). *Disability and Rehabilitation, 32*(20), 1633–1644.

Rutter, M. (1987). Psychosocial resilience and protective mechanisms. *American Journal of Orthopsychiatry, 57,* 316–331.

Sakzewski, L., Boyd, R., & Ziviani, J. (2007). Clinimetric properties of participation measures for 5- to 13-year old children with cerebral palsy: A systematic review. *Developmental Medicine and Child Neurology, 49,* 232–240.

Sakzewski, L., Ziviani, J., Abbott, D.F., Macdonnell, R.A., Jackson, G.D., & Boyd, R.N. (2011). Participation outcomes in a randomized trial of 2 models of upper-limb rehabilitation for children with congenital hemiplegia. *Archives of Physical Medicine and Rehabilitation, 92,* 531–539.

Simeonsson, R. J., Carlson, D., Huntington, G. S., McMillen, J., & Brent, L. (2001). Students with disabilities: A national survey of participation in school activities. *Disability and Rehabilitation, 23,* 49–63.

Sparrow, S. S., Cicchetti, D. V., & Balla, D. (2005). *Vineland Adaptive Behavior Scales, Second Edition* (Vineland-II). Circle Pines, MN: American Guidance Service.

Taylor, G. H., Yeates, K. O., Wade, S. L., Drotar, D., Stancin, T, & Minich, N. (2002). A prospective study of short- and long-term outcomes after traumatic brain injury in children: Behavior and achievement. *Neuropsychology, 16*, 15–27.

Whiteneck, G., & Dijkers, M. P. (2009). Difficult to measure constructs: Conceptual and methodological issues concerning participation and environmental factors. *Archives of Physical Medicine and Rehabilitation, 90*(11, Suppl. 1), S22–S35.

World Health Organization (WHO). (2001). *International classification of functioning, disability and health.* Geneva: Author.

World Health Organization (WHO). (2007). *International classification of functioning, disability and health: Version for children and youth.* Geneva: Author.

Yager, J. A., & Ehmann, T. S. (2006). Untangling social function and social cognition: A review of concepts and measurement. *Psychiatry, 69*, 47–68.

Yeates, K. O., Swift, E., Taylor, H. G., Wade, S. L., Drotar, D., Stancin, T., et al. (2004). Short- and long-term social outcomes following pediatric traumatic brain injury. *Journal of the International Neuropsychological Society, 2004*(10), 412–426.

Ziviani, J., Desha, L., Feeney, R., & Boyd, R. (2010). Measures of participation outcomes and environmental considerations for children with acquired brain injury: A systematic review. *Brain Impairment, 11*(2), 93–112.

PART IV
DISRUPTED SOCIAL FUNCTION

CHAPTER 10

Theoretical Approaches to Understanding Social Function in Childhood Brain Insults

Toward the Integration of Social Neuroscience and Developmental Psychology

Keith Owen Yeates, Erin D. Bigler, Cynthia A. Gerhardt,
Kenneth H. Rubin, Terry Stancin, H. Gerry Taylor, and Kathryn Vannatta

The nature, basis, and consequences of the social problems associated with neurological dysfunction and brain insults during childhood are not well understood, despite the significant implications of social development for children's functioning at home, in school, and in the community (Parker, Rubin, Erath, Wojslawowicz, & Buskirk, 2006). Until recently, the lack of articulated models of social functioning has limited our ability to study social function in children with brain insults. The development of explicit models of social functioning would help researchers and clinicians to target children with brain insults for further study and intervention.

The emerging field of social neuroscience provides a critical perspective on the social impact of childhood brain disorders. Social neuroscience not only supplies tools needed to better understand the neural substrates and social-cognitive processes associated with social functioning; it also provides a foundation for a multilevel, integrative analysis of the social difficulties arising from neurological insults (Cacioppo, Berntson, Sheridan, & McClintock, 2000; Ochsner & Lieberman, 2001). Social neuroscience to date has focused primarily on adults; however, elegant methods are now available that can inform researchers about brain development and neuropathology in the study of social behavior in children with brain disorders (Gunnar & de Haan, 2009).

We believe that models derived from social neuroscience will be particularly powerful when combined with those associated with the study of social competence in developmental psychology and developmental psychopathology (Parker et al., 2006; Rubin,

Bukowski, & Parker, 2006). The latter approaches reflect a developmental perspective that can enhance the field of social neuroscience. In short, we now have the tools to begin to understand how children's daily functioning in the social world is associated with their abilities to (1) identify, think about, produce, and regulate emotions; (2) consider other people's perspectives, beliefs, and intentions; and (3) solve interpersonal problems in actual interactions and relationships. Furthermore, we can model these associations in terms of developmental processes and brain pathology.

This chapter describes an integrative model of the social outcomes of childhood brain disorders that we have presented in an earlier review (Yeates et al., 2007). The model is grounded in concepts and methods drawn from both the field of social neuroscience and the study of social competence in developmental psychology/psychopathology. It attempts to specify the relations among social adjustment, peer interactions and relationships, social problem solving and communication, social/affective and cognitive/executive processes, and their brain substrates. The model also takes into account the distinct but related developmental trajectories that occur within these domains.

We believe that this model is germane to a wide range of central nervous system abnormalities and insults, both developmental and acquired in origin. These would include neurodevelopmental disorders, such as autism (Baron-Cohen & Belmonte, 2005; see d'Arc & Mottron, Chapter 15, this volume), disorders arising from prenatal exposure to teratogens, such as fetal alcohol syndrome (Schonfeld, Mattson, & Riley, 2005), and acquired brain injuries, such as childhood stroke (Coelho-Mosch, Max, & Tranel, 2005; see Anderson, Rosema, Gomes, & Catroppa, Chapter 11, and Max, Chapter 14, this volume). This does not mean that all childhood brain disorders affect social development to the same degree or in the same way. Rather, the likelihood that any specific disorder will affect social development will be a function primarily of the nature and timing of the brain insult with which it is associated, rather than of the specific etiology involved. For instance, an ischemic stroke and a traumatic brain injury (TBI) arising from closed-head trauma can potentially occur at the same age and affect comparable brain regions and developmental processes, and therefore can give rise to similar social outcomes.

Definitions and Distinctions in the Study of Social Competence

The study of social outcomes in childhood brain disorders depends in part on a definition of *social competence*. Many definitions have been proffered (Rose-Krasnor, 1997). Most suggest that social competence involves the effectiveness of a person's functioning as an individual, in dyadic relationships, and in groups (Rubin, Bukowski, & Parker, 2006). Rubin and Rose-Krasnor (1992) have defined *social competence* as "the ability to achieve personal goals in social interaction while simultaneously maintaining positive relationships with others over time and across situations." This definition places the individual within a social and personal context, and highlights the complex goals that persons seek to attain both as individuals (i.e., satisfying personal goals) and as members of groups (i.e., while maintaining positive relationships).

In this definition, social competence is viewed as a transactional construct. That is, social competence depends on the personal characteristics of the child; the interactions between the child and members of his or her social world; and the interpretations of the self and others that the child's actions are acceptable and successful. Importantly, social competence from this perspective is also viewed as a developmental construct that is both time- and context-dependent (Rubin & Rose-Krasnor, 1992).

The study of social competence can be guided in part by distinguishing among several levels of social complexity: individuals, interactions, and relationships (Rubin, Bukowski, & Parker, 2006). Children bring certain *individual* characteristics to their interactions with others (e.g., the ability to regulate emotion or the capacity to understand the intentions of others). In many ways, these individual characteristics may be thought to constitute children's "social intelligence." In turn, children's *social interactions* also depend on the individual characteristics and behavior of the children and adults with whom they are involved. Social interactions can be characterized as involving actions that bring individuals together (i.e., sociable and prosocial behaviors), actions that move people against each other (i.e., aggression), and actions that isolate individuals from each other (i.e., withdrawal) (Rubin, Bukowski, & Parker, 2006). Finally, interactions are frequently embedded in and give rise to longer-term *relationships*. Relationships are defined in part by the members' individual characteristics and the quality of their interactions, but have distinct properties of their own, such as closeness and commitment. Friendship is a prototypical relationship.

Nassau and Drotar (1997) have made a closely related set of distinctions in their review of social competence in children with chronic health conditions affecting the central nervous system. Drawing on Cavell's (1990) tricomponent model of social competence, they have distinguished among social skills, social performance, and social adjustment. *Social skills* are the individual abilities or characteristics needed to behave competently in social settings. *Social performance* refers to children's actual behavior in social interactions and to whether their responses are effective both in achieving their own goals and in maintaining positive relationships. *Social adjustment*, finally, reflects the extent to which children attain socially desirable and developmentally appropriate goals. Social adjustment encompasses the quality of children's relationships as perceived by others, but it also includes self-perceptions of loneliness, social support, or social self-esteem.

Children whose individual social skills and social interactions engender social success are popular among their peers and are viewed by teachers and parents as well adjusted (Parker et al., 2006). In contrast, children who are less competent are typically rejected by peers and are rated by teachers and parents as maladjusted. For example, children who frequently seek to attain personal goals by means of aggression (i.e., moving against their social partners) are often viewed by teachers and parents as having adjustment problems of an externalizing nature. Children who retreat when others approach them or who attempt to meet their social goals by requesting that adults act on their behalf are often viewed by teachers and parents as having problems of an internalizing nature (Parker et al., 2006).

Previous studies pertaining to social outcomes in childhood brain disorders have focused largely on social adjustment, which in this population has been assessed primarily via parent ratings (Martinez, Carter, & Legato, 2011). Few researchers of childhood brain disorder have examined children's social skills and other individual characteristics that affect social behavior, and investigators have yet to directly examine social interactions and relationships among children with brain disorder. Moreover, in only a handful of studies have researchers investigated the relations among different aspects of social competence. We contend that a comprehensive portrayal of social outcomes in childhood brain disorder must encompass the three levels that characterize contemporary definitions of social competence (i.e., individual characteristics and social skills, social performance and interaction, and social adjustment), as well as the relations between and among the levels. The three levels provide distinct but interrelated windows on social function.

An Integrative, Multilevel Approach to the Study of Social Competence

Research in developmental psychology and developmental psychopathology has provided a more detailed characterization of the individual characteristics and social skills, interactions, and various aspects of social adjustment that constitute social competence. In addition, it has shown how deficits in those areas are linked to poor social adjustment. On the basis of this research, we propose a multilevel, integrative, heuristic model of social competence, as illustrated in Figure 10.1, which details specific components at each level and articulates the relations among levels.

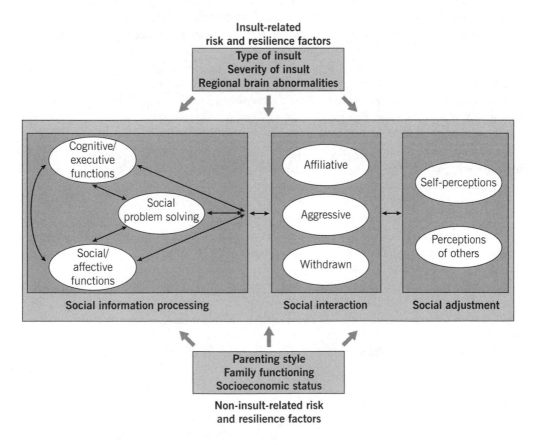

FIGURE 10.1. An integrative, heuristic model of social competence in children with brain disorder.

Components of the Model

At the level of individual characteristics and social skills, *social information processing* is seen as a critical determinant of social competence (Crick & Dodge, 1994; see Rubin, Begle, & McDonald, Chapter 2, this volume). Social information processing is conceived as involving a series of distinct problem-solving steps that are implemented when children are in social situations. Such steps would commonly involve interpreting cues, clarifying goals, generating alternative responses, selecting and implementing a specific response, and evaluating the outcome. Social problem solving is often assessed by asking children to reflect on and answer questions about hypothetical social dilemmas (Dodge, Laird, Lochman, & Zelli, 2002). Children's reasoning about such dilemmas varies systematically as a function of the specific situations presented, such as ones involving peer provocation versus group entry (Burgess, Wojslawowicz, Rubin, Rose-Krasnor, & Booth-LaForce, 2006; Dodge et al., 2002).

Other theorists have recognized that social information processing depends on other cognitive and affective factors, and have incorporated into their models such constructs as language pragmatics, executive function, and emotion regulation (Dodge et al., 2002; Guralnick, 1999). The latter variables are typically treated as stable individual characteristics (i.e., as "latent knowledge" by Dodge et al., 2002; as "foundation processes" by Guralnick, 1999). They are assumed to play a critical role in the implementation of social problem solving, which is seen as a more situation-specific and "online" social skill. The models assume that the effects of these cognitive and affective factors on social interaction and adjustment are mediated in part through their effects on social problem solving.

Research on children's *social interactions* has shown that they vary according to both the type of social situation and the nature of children's relationships with the individuals with whom they interact (Parker et al., 2006; Rubin, Bukowski, & Parker, 2006). For instance, children exhibit different behaviors when attempting to enter a peer group activity than when responding to peer provocation, and they use different strategies when attempting to gain access to objects than when attempting to gain the attention of others. Similarly, children interact differently with friends than with unfamiliar peers (Newcomb & Bagwell, 1995). Notably, the range, flexibility, and adaptability of children's social behaviors across different contexts and relationships are often considered hallmarks of social competence (Rose-Krasnor, 1997).

A detailed understanding of children's social interactions cannot be attained by using conventional rating scales or questionnaires, but instead is best achieved through direct observations in a variety of contexts. Many observational protocols and coding schemes have been developed to study children's social interactions (Rubin, Bukowski, & Parker, 2006). Regardless of the context in which children are observed or with whom they interact, coding schemes frequently focus on the three broad behavioral tendencies noted earlier: (1) moving toward others (i.e., prosocial, affiliative behavior); (2) moving against others (i.e., aggressive or antagonistic behavior); and (3) moving away from others (i.e., socially withdrawn behavior).

Research on *social adjustment* has shown that it too varies along several important dimensions. One critical distinction, consistent with the incorporation of both individual and social goals in contemporary definitions of social competence, is whether social adjustment is evaluated through self-perceptions or the perceptions of others, such as peers, parents, or teachers (Parker et al., 2006; Rubin, Bukowski, & Parker, 2006). This distinction may be especially important for children with brain disorders, who may lack awareness of their own deficits (Prigatano, 1991) and may therefore tend to evaluate their social adjustment more positively than others do.

Social adjustment from the perspective of others can be assessed via classroom peer nominations and ratings of peer acceptance and behavioral reputation. These indices are not independent of one another, but are conceptually and empirically distinct and have different implications for long-term adjustment (Gest, Graham-Bermann, & Hartup, 2001). For instance, in early adolescence, some forms of aggression are linked to perceived popularity among peers; however, they also result in significant constraints on reciprocal friendships (Cillessen & Rose, 2005), which increase in importance as children grow older (Rubin, Bukowski, & Parker, 2006).

Social adjustment also can be measured from the perspective of the self. In early and middle childhood, aggressive children tend to believe that they are well accepted by peers and that they are socially skilled, but their peers think otherwise (Boivin, Vitaro, & Poulin, 2005). Indeed, the friendships of aggressive children are marked by instability and mistrust (Hektner, August, & Realmuto, 2000). In contrast, children who withdraw from social interaction tend to view themselves as lacking in social competence (Rubin, Chen, & Hymel, 1993). They are also inclined to indicate feelings of loneliness and depression (Rubin, Burgess, & Coplan, 2002). However, these socially wary and withdrawn children are like their aggressive counterparts in that they are often unpopular in the peer group and have close relationships with others who have characteristics similar to them (Rubin, Wojslawowicz, Burgess, Rose-Krasnor, & Booth-LaForce, 2006).

Research clearly demonstrates that social information processing, social interactions, and social adjustment are closely interrelated (Parker et al., 2006; Rubin, Bukowski, & Parker, 2006). Children who display deficits in social information processing are more often aggressive or socially anxious and withdrawn in their interactions with other children. Those interactions typically result in peer rejection and in being considered less desirable as friends. In contrast, children whose social information-processing skills are intact tend to be more skilled in initiating and maintaining positive relationships, and they rely on behaviors that are more prosocial. They also are more likely to be socially accepted by peers and to have satisfactory friendships. Thus Figure 10.1 incorporates pathways between the three levels of social competence. The pathways are designated as bidirectional: Social information processing can affect social interactions, which in turn affect social adjustment; conversely, perceptions of self and other can affect social interactions and help to shape social information processing.

Other models of social competence also acknowledge that a variety of risk and resilience factors can hamper or promote social development (Guralnick, 1999; Masten et al., 1999). Some of those factors are intrinsic to the child (e.g., intellectual functioning), while others involve environmental influences (e.g., socioeconomic status, parenting behaviors, and parent–child relationships). For instance, neurological dysfunction or acquired brain injury can be conceptualized as a risk factor that increases the likelihood of deficits in social information processing, atypical social interaction, and poor social adjustment (Janusz, Kirkwood, Yeates, & Taylor, 2002).

On the environmental side, research suggests that parenting beliefs and behaviors and the quality of the parent–child relationship can influence children's social interactions and social adjustment (Carson & Parke, 2008; see Root, Hastings, & Maxwell, Chapter 5, this volume). More general aspects of the family environment, including poverty and parental unemployment, parental conflict, and parents' mental health, also may affect social competence (Du Rocher Schudlich, Shamir, & Cummings, 2004). Even broader sociocultural influences, such as the stigmatization that can result from perceived disability or cultural differences in the value placed on different social behaviors (Rubin, Hemphill, et al., 2006), may have an effect on psychosocial adjustment (Kendall & Terry, 1996; see Anderson & Beauchamp, Chapter 1, this volume).

Risk and resilience factors, whether endogenous or exogenous to the child, can act both as independent predictors of social competence and as moderators of the relations among its various components. For instance, parental warmth and authoritative control tend to predict more appropriate social behavior, which in turn predicts better social adjustment (Ladd & Pettit, 2002; see Root et al., Chapter 5, this volume). In addition, risk and resilience factors do not necessarily operate independently of one another. Indeed, they may interact to predict the social outcomes of childhood brain insults. For example, children of lower socioeconomic status may be more likely to have a TBI (Parslow, Morris, Tasker, Forsyth, & Hawley, 2005). Similarly, the social outcomes of childhood TBI have been found to be moderated by the quality of family functioning, with better outcomes in children from more advantaged backgrounds (Yeates et al., 2004).

Figure 10.1 acknowledges the possibility that both insult-related and non-insult-related variables may act as risk and resilience factors in determining the social outcomes of childhood brain disorder. Because of our emphasis on children with brain insults, the model represented in Figure 10.1 focuses on the influence of the brain and other individual factors on social competence. However, the model also acknowledges the importance of non-insult-related risk and resilience factors (i.e., environmental influences) as potential contributors to or moderators of outcome.

The Emerging Discipline of Social Neuroscience

Until recent years, the study of social competence in children has not been strongly informed by neuroscience. The emerging field of social-cognitive neuroscience, however, now provides a basis for integrating knowledge about brain structure and function into the study of children's social development (de Haan & Gunnar, 2009). Social neuroscience uses such methods as neuroimaging, neurogenetics, neuropsychological assessment, and the study of brain disorders to understand the neural substrates of social functioning. The field promotes integrative, multilevel studies of the links among brain, emotion, cognition, and social behavior (Cacioppo et al., 2000; Gunnar & de Haan, 2009; Ochsner & Lieberman, 2001).

A rapidly growing literature in social neuroscience indicates that a distributed network of interdependent brain regions subserve various social-cognitive and affective processes that gradually become integrated during the course of social development (Adolphs, 2001; Johnson et al., 2005; see Thomas & Tranel, Chapter 4, this volume). Because the network involves multidirectional and recursive connections, the relationship between structure and process is not strictly one-to-one. Any single social-cognitive process typically depends on a variety of brain structures, and a single brain structure can be involved in several processes (Adolphs, 2003). Thus regional specialization is likely to reflect different patterns of activation across structures rather than activity in a single structure.

Nevertheless, many brain regions have been found to play especially important roles in specific processes. For example, the fusiform gyrus and region of the superior temporal sulcus have been implicated in the perception of faces and the movement of living things (Adolphs, 2003; Pascalis et al., 2009); the amygdala plays an especially important role in emotion, particularly fear, and the response to danger or threat (Adolphs, 2002). The anterior cingulate and ventromedial, orbital, and dorsolateral prefrontal cortex are brain structures that appear to play an important role in other aspects of social cognition, such as the understanding of others' mental states (i.e., theory of mind) and emotional regulation (Allman, Hakeem, Erwin, Nimchinsky, & Hof, 2001; Amodio & Frith, 2006; Anderson, Bechara, Damasio, Tranel, & Damasio, 1999; Frith & Frith, 2001; Mah, Arnold, & Grafman, 2004; Siegal & Varley, 2002).

Plate 10.1 on the color insert portrays the various brain regions that have been implicated in social cognition and behavior, and Table 10.1 summarizes some of the links between those regions and specific social-cognitive and affective processes that have been the focus of research to date. As Table 10.1 indicates, most brain regions are involved in multiple functions, and most specific functions draw on multiple brain regions, although some regional specialization is also apparent. Although most of the previous research has been based on adults, the brain regions illustrated in Plate 10.1 and listed in Table 10.1 follow predictable developmental sequences that relate to social development and that can be disrupted or impaired by childhood brain disorders (Johnson et al., 2005; Payne & Bachevalier, 2009).

Notably, the brain regions known to regulate cognitive/executive function overlap substantially with those implicated in social-cognitive and emotional functioning. Indeed, many of these regions play a dual role: They influence not only social cognition, but various aspects of memory and executive function. Thus injury to frontotemporal and limbic regions is likely to affect both the cognitive and emotional aspects of social behavior in children (Levin & Hanten, 2005). On the other hand, early focal lesions to particular regions of the social brain network may have more specific effects. For instance, dorsolateral frontal lesions may lead to cognitive deficits in executive functions without significant emotional or social impairment, whereas damage to the orbital and ventromedial prefrontal cortex often results in profound deficits in self-regulation, emotion, and social behavior (Eslinger, Flaherty-Craig, & Benton, 2004).

The role of anterior brain regions in social behavior may vary as a function of hemispheric specialization. Developmental researchers have interpreted asymmetries in frontal electroencephalographic (EEG) activation in terms of motivational systems of approach and withdrawal (Davidson, 1992; Fox, 1994). The left frontal region appears to facilitate the approach to appetitive stimuli, whereas the right frontal region is thought to evoke withdrawal from aversive stimuli. Fox and colleagues have shown that right frontal EEG asymmetry is associated with high levels of behavioral inhibition and social reticence during infancy and early childhood (Fox et al., 1995). Left frontal asymmetry, on the other hand, has been associated with social approach and positive peer interaction (Fox et al., 1995). Studies of emotional expression in individuals with prefrontal lesions also provide evidence of hemispheric asymmetries in social/affective behavior; right frontal lesions are associated with excessive emotionality and disinhibited behavior, whereas left

TABLE 10.1. Brain Structures and Social Cognition

Brain structure or region	Social/affective and cognitive/executive functions
Temporoparietal junction	Representation of emotional response Viewing others' actions
Fusiform gyrus	Face perception
Superior temporal gyrus	Representation of perceived action Face perception Perception of gaze direction Perception of biological motion
Amygdala	Motivational evaluation Self-regulation Emotional processing Gaze discrimination Linking internal somatic states and external stimuli
Ventral striatum	Motivational evaluation Self-regulation Linking internal somatic states and external stimuli
Hippocampus and temporal poles	Modulation of cognition Memory for personal experiences Emotional memory retrieval
Basal forebrain	Modulation of cognition
Cingulate cortex	Modulation of cognition Error monitoring Emotion processing Theory of mind
Orbitofrontal cortex	Motivational evaluation Self-regulation Theory of mind
Medial frontal cortex	Theory of mind Action monitoring Emotional regulation Emotional responses to socially relevant stimuli Monitoring of outcomes associated with punishment and reward
Dorsolateral frontal cortex	Cognitive executive functions Working memory

frontal lesions are associated with negative emotions, such as depression and fearfulness, as well as withdrawal (Powell & Voeller, 2004).

In summary, social neuroscience provides a detailed picture of the cognitive and affective constructs that also are incorporated in models of social information processing; in addition, it points to potential neural substrates for specific types of social interactions. More broadly, we believe that the cognitive and emotional processes that are the focus of social neuroscience provide a critical bridge between knowledge regarding the brain substrates of social behavior and models of social competence from developmental psychology and developmental psychopathology. Specifically, the cognitive/executive and social/affective functions depicted in Figure 10.1 reflect aspects of social information processing that are linked to a network of specific brain regions (Adolphs, 2001). At the same time, they also represent the stable individual characteristics (i.e., "latent knowledge" or "foundation processes") described in contemporary models of social competence (Dodge et al., 2002; Guralnick, 1999).

Social neuroscience also links research on children's social development to the study of childhood brain disorders. Many childhood brain disorders involve insults to the anterior brain regions implicated in social information processing. Deficits in social information processing, in turn, are known to be associated with atypical social interactions and poor social adjustment across a variety of typical and atypical populations (Parker et al., 2006; Rubin, Wojslawowicz, et al., 2006; Yeates et al., 2004). The insults associated with many childhood brain disorders, therefore, are likely to have negative consequences for children's social competence at multiple levels. By linking a network of specific brain regions to deficits in social-cognitive and emotional processes, social neuroscience provides a foundation for a multilevel analysis of the social problems arising from childhood brain disorders—an analysis that bridges brain, cognition, emotion, and action (Cacioppo et al., 2000).

Developmental Considerations

The brain regions implicated in social behavior are subject to change with age, just as social behavior is itself. The changes are likely to be related, moreover, such that brain maturation correlates with increases in children's capacities for social information processing, which in turn are related to changes in the complexity of their social behavior (Dennis, 2006; Stuss & Anderson, 2004). Understanding the distinct but linked developmental trajectories within these domains, and how they may be altered by childhood brain disorders, will be important for any model of social adaptation and maladaptation.

Brain Development

The anterior regions of the brain that are linked to social behavior undergo gradual development, and the prefrontal cortex is particularly slow to mature. Morphological development of the frontal cortex is not complete until about puberty, with further changes continuing into adulthood (Klingberg, Vaidya, Gabrieli, Moseley, & Hedehus, 1999). Similarly, the prefrontal cortex is not fully myelinated until middle to late adolescence (Giedd et al., 1999; Klingberg et al., 1999; Sowell et al., 1999). Synaptogenesis occurs at the same rate in most cortical regions (Rakic, Bourgeois, Eckenhoff, Zecevic, & Goldman-Rakic, 1986), although the prefrontal cortex may lag behind the rest of the brain (Chugani, Phelps, & Mazziotta, 1987). White matter may also undergo protracted development within anterior brain regions (Klingberg et al., 1999; Sowell et al., 1999).

Magnetic resonance imaging (MRI) studies have shown rapid growth spurts in the frontal lobes relative to the temporal lobes in the first 2 years after birth (Matsuzawa et al., 2001). After age 5, brain volumes remain relatively stable (Reiss, Abrams, Singer, Ross, & Denckla, 1996), but the ratio of gray to white matter lessens with increasing age (Sowell & Jernigan, 1998) as a result of decreases in gray matter volumes between childhood and early adulthood (Gogtay et al., 2004; O'Donnell, Noseworthy, Levine, Brandt,

& Dennis, 2005). Gray matter loss progresses evenly across the brain at an early age; by adolescence, though, the decreases are localized to the frontal and parietal lobes (Sowell et al., 1999). Longitudinal studies of cortical gray matter development have shown that higher-order association cortices mature only after lower-order somatosensory and visual cortices (Gogtay et al., 2004). Within the frontal lobes, maturation proceeds in a back-to-front direction, beginning in the primary motor cortex (precentral gyrus) and spreading anteriorly over the superior and inferior frontal gyri, with the prefrontal cortex developing last. Within the prefrontal cortex, the frontal pole and precentral cortex mature early and the dorsolateral cortex matures last, coinciding with its later myelination.

Development of Social Information Processing

Social information processing also shows developmental changes, in a manner that probably relates to brain development (Anderson, Levin, & Jacobs, 2002; see Burnett, Sebastian, & Cohen Kadosh, Chapter 3, this volume). The executive functions involved in social behavior, particularly inhibitory control and working memory, undergo gradual development. For instance, during the preschool years, children become more able to delay responses, to suppress responses in a go/no-go paradigm, and to respond correctly in the presence of a conflicting response option (Diamond & Taylor, 1996; Kochanska, Murray, Jacques, Koenig, & Vandegeest, 1996). The development of working memory and inhibitory control occurs in tandem (Cowan, 1997), with a close relationship between working memory and inhibitory control beginning to emerge during the preschool years (Dowsett & Livesey, 2000).

Theory of mind is a more specific form of social information processing that also demonstrates ongoing development. Theory of mind involves the ability to think about mental states and to use them to understand and predict what other people know and how they will act. In adults, frontal lesions impair performance on theory-of-mind tasks (Stuss, Gallup, & Alexander, 2001). Theory of mind begins to become apparent early in childhood; infants display expectations about the actions of others, and by 18 months they are able to understand intentions (Meltzoff, Gopnik, & Repacholi, 1999). Children first become able to understand desires and intentions (Bartsch & Wellman, 1989) and later begin to understand false beliefs (Sodian, Taylor, Harris, & Perner, 1991). The emergence of theory of mind appears to be closely related to executive functions, such as working memory and inhibitory control (Moses, 2001). Indeed, the emergence of theory of mind correlates closely with the development of executive skills, although they become less closely coupled at later ages (Carlson & Moses, 2001; Hughes & Ensor, 2005).

The ability to use and understand forms of nonliteral language such as irony and deceptive praise, in which a speaker's affective message does not correspond to the words spoken, also follows a protracted developmental course (Dennis, Purvis, et al., 2001). Early in development, children do not understand the concept of saying one thing while meaning another (Demorest, Meyer, Phelps, Gardner, & Winner, 1984). Later in development, children are able to recognize deliberate falsehoods and take into consideration both the facts of the situation and what they believe the speaker believes (Demorest et al., 1984). By middle childhood, children begin to interpret white lies correctly (Demorest et al., 1984). They also begin to understand ironic criticism and to distinguish it from deceptive intent (Demorest et al., 1984). The ability to understand ironic criticism becomes well established by early adolescence (Winner, 1988).

As they mature, children are increasingly able to think reflectively about more complex social dilemmas, and their growing social problem-solving skills contribute to more successful social function (Crick & Dodge, 1994; Dodge et al., 2002). Young children have knowledge about social problem solving that is not reflected in their spontaneous behavior (Rudolph & Heller, 1997). Children become more skilled at several different aspects of social problem solving, ranging from the retrieval or construction of possible solutions to the evaluation, selection, and enactment of behavioral responses (Yeates, Schultz, & Selman, 1991). These changes may reflect an increasingly sophisticated ability to coordinate social perspectives (Yeates et al., 1991).

Development of Social Behavior

With increasing age and brain maturation, children's social information-processing abilities grow, and their social behavior becomes more diverse, complex, and integrated (Rubin, Bukowski, & Parker, 2006). Changes are apparent both in children's specific interactions and in their relationships (e.g., friendships). For instance, as their motor and language skills grow, toddlers begin to engage in increasingly lengthy interactions with peers, and their play becomes more organized (Eckerman & Stein, 1990). They also display the beginnings of meaningful relationships, preferring to play and engage in complex interactions with familiar as opposed to unfamiliar playmates (Howes, 1988).

Pretend play is a particularly important form of social interaction during the preschool years (Goncu, Patt, & Kouba, 2002). By the third year of life, children are able to share symbolic meanings through social pretense (Howes, 1988). Goncu (1993) has reported quantitative differences in the extent to which the social interchanges of 3- versus 4½-year-olds reflect shared meaning. For example, the social interactions of older preschoolers involve longer sequences or turns. With increasing age, play partners become better able to agree with each other about the roles, rules, and themes of their pretense. They are also better able to maintain their play interactions by adding new dimensions to their expressed ideas. These developments reflect preschoolers' growing capacity to take the perspective of the play partner, and the increasing sophistication of their nascent theory of mind (Watson, Nixon, Wilson, & Capage, 1999).

By middle childhood, children spend significantly more time interacting with peers, and their peer interactions are less supervised. Pretend and rough-and-tumble play becomes less common, and is replaced by games and activities structured by adults (Pellegrini, 2002). Children become increasingly concerned with acceptance by peers during middle childhood (Kuttler, Parker, & La Greca, 2002). Verbal and relational aggression (i.e., insults, derogation, threats, gossip) gradually replace direct physical aggression when conflict occurs. Children's conceptualizations of friendship begin to shift from being more instrumental to more empathic, perhaps contingent on their growing ability to coordinate social perspectives (Selman & Schultz, 1990). Their friendships become more stable and are more likely to be reciprocated (Berndt & Hoyle, 1985).

Many of these trends continue during adolescence. Adolescents spend almost one-third of their waking hours with peers—nearly double what they spend with parents and other adults (Csikszentmihalyi & Larson, 1984). Their interactions are more likely to occur outside adult guidance and control than they were at earlier ages, as well as to involve members of the opposite sex (Brown & Klute, 2003). Friends become increasingly important as sources of support and advice, and friendship begins to involve more intimacy and self-disclosure (Buhrmester & Furman, 1986). Adolescents develop clear conceptions of the properties that distinguish romantic relationships from friendships, and the two kinds of relationships have distinct implications for adolescent adjustment (Collins, 2003).

Developmental Linkages among Brain and Social Behavior

Relatively little is currently known about the association between brain development and social development (see Burnett et al., Chapter 3, this volume). The field of developmental social neuroscience holds substantial promise for linking developmental changes in social information processing and social behavior with those that occur in brain structure and function (de Haan & Gunnar, 2009). Generally speaking, studies of structural and functional brain development suggest that infants and children demonstrate more widely distributed patterns of brain function than adults (Casey, Giedd, & Thomas, 2000), suggesting that regional specialization evolves gradually over the course of development.

Consistent with this general finding, Johnson et al. (2005) have reviewed the development of the social brain network, emphasizing the concept of *interactive specialization*. In contrast to a maturational perspective, which suggests that brain functions emerge once a brain region reaches a certain state of maturity, interactive specialization suggests that functional brain development occurs gradually, as a result of the activation and interaction of multiple brain regions. Over time, organizational changes occur in the neural network, and certain brain regions ascend in their control or primacy over processing and responding to certain stimuli. Thus regional specialization occurs, but "the response properties of a specific region are partly determined by its patterns of connectivity to other regions, and their patterns of activity" (Johnson et al., 2005, p. 600). Johnson et al. (2005) present data from studies of face processing in infants indicating that the entire social brain network is partially active from at least 3 months of age, but it shows less specialized functionality than in adults, so that children display more widespread brain activation to faces than adults do (Passarotti et al., 2003).

One corollary of interactive specialization is that an essential ingredient for typical brain development is connectivity. Connectivity in the brain involves white matter pathways, and hence the development of white matter becomes essential for the emergence of the social brain network. In a postmortem study, Kinney, Brody, Kloman, and Gilles (1988) outlined the staging of myelin development during infancy and used that information to make projections about myelination throughout childhood. Herbert et al. (2004) used the myelination indices developed by Kinney et al. (1988) in the study of autism, a disorder intimately linked to aberrations in the development of the social brain and associated deficits in gaze cueing and joint attention (Johnson et al., 2005). They found white matter volume increases primarily in later or longer-myelinating brain regions.

Another corollary of interactive specialization is that the social brain network may be especially vulnerable to early insults. If brains gradually undergo more regional specialization through interactive processes that depend on connectivity, then early insults may disrupt connectivity in such a way that they have a widespread impact on brain development that may be quite remote from the specific location of the insult itself. Notably, the frontal pole, temporal pole, and corpus callosum are the three brain regions with the most protracted white matter development. The frontal and temporal lobes include key components of the social brain network, and numerous studies have demonstrated the vulnerability of white matter in those regions to certain childhood brain disorders, such as TBI (Tasker et al., 2005; Wilde et al., 2005, 2006). If early insults disrupt the development of the social brain network more than later insults do, then they also are likely to result in more profound consequences for social behavior. Indeed, studies of early focal lesions to the prefrontal cortex suggest that they have more profound effects on social outcomes than similar lesions occurring in adulthood do (Eslinger et al., 2004).

Developmental Dimensions in Childhood Brain Disorders

The outcomes associated with brain insults during childhood are themselves dependent on developmental factors, especially in the case of acquired brain injuries or brain disorders that have their onset after birth. Specifically, outcomes vary along three distinct but interrelated dimensions: the age of the child at the onset of the disorder or time of insult; the amount of time that has passed since the disorder began or insult occurred; and the child's age at the time of outcome assessment (Taylor & Alden, 1997).

Most studies of school-age children and adolescents have not found a strong relationship between age at onset and outcomes. However, recent studies of children with a variety of acquired brain injuries indicate that insults sustained during infancy or early childhood are associated with more persistent deficits than are brain insults occurring during later childhood and adolescence (Anderson et al., 2009; see also Anderson et al., Chapter 11, this volume). The common assumption that early brain insults are associated with greater plasticity has proven incorrect, particularly when insults are diffuse in nature (Anderson, Spencer-Smith, & Wood, 2011).

With regard to time since insult, longitudinal studies indicate that children generally display a gradual recovery over the first few years after acquired brain injuries, with the most rapid improvement occurring during the first year or so after the insult. The initial rate of recovery is often more rapid among children with severe injuries than among those with milder injuries, but severe injuries also are associated with persistent deficits after the rate of recovery slows (Taylor et al., 2002; Yeates et al., 2002). Because relatively few long-term follow-up studies have been completed, we do not know whether children with acquired brain injuries show any progressive deterioration in functioning relative to healthy peers after their initial recovery. However, younger children demonstrate a slower rate of change over time and more significant residual deficits after their recovery plateaus than do older children with injuries of equivalent severity (Anderson, Catroppa, Morse, Haritou, & Rosenfeld, 2005). Similarly, children with other brain insults (such as bacterial meningitis) or with extremely low birth weight acquire some skills more slowly with age, compared with unaffected children (Taylor, Schatschneider, & Minich, 2000; Taylor, Minich, Klein, & Hack, 2004).

The influence of age at testing has been the focus of the least research. The effects of age at testing may be reflected in demonstrations of latent or delayed sequelae resulting from children's failure to meet new developmental demands as a result of a brain disorder. For instance, because adolescence is associated with substantial maturational changes in the frontal lobes, the effects of frontal lesions might not become fully apparent until then, even if they occurred much earlier in life. The phenomena of "growing into a lesion" or time-lagged effects have been reported in case studies showing the delayed onset of social problems in children with early frontal lobe lesions (Eslinger et al., 2004).

Group studies also show evidence for worsening of behavioral adjustment over time in some individual cases following acquired brain injury (Fay et al., 2009). However, true latent effects are difficult to detect, because they require evidence that differences in the consequences of acquired injuries are due specifically to age at testing, as opposed to age at insult or time since insult. Disentangling these dimensions is difficult, even in the context of longitudinal research (Taylor & Alden, 1997).

Summary: An Integrative, Multilevel Model of Social Outcomes

Figure 10.1 represents an integrative, multilevel model of the social outcomes of childhood brain disorder, grounded in concepts and methods drawn from both the field of social neuroscience and the study of social competence in developmental psychology/psychopathology. The model specifies general relationships among social information processing, social interaction, and social adjustment, and reflects the possibility of bidirectional relations among those different levels of social competence (e.g., self-perceptions of adjustment may affect social interactions, and vice versa). The model acknowledges the brain substrates for social cognition and affect regulation, and indicates that factors related directly to the neurological insult, as well as those independent of it (e.g., environmental factors), can influence social competence at all levels and the relations among them. The model as portrayed in Figure 10.1 is largely heuristic in nature, in that it portrays the relationships among the levels of social competence and their association with insult-related and non-insult-related risk and resilience factors in a general fashion that does not necessarily lead to specific predictions. However, when the existing research literature about the individual components of the model is taken into account, the model can give rise to more specific hypotheses.

For instance, when applied to childhood brain insults, the model implies that some disorders—but not all—will be associated with insults predominantly to the temporal cortices, amygdala, anterior cingulate, basal forebrain, and prefrontal cortex (Wilde et al., 2005, 2006). In the presence of such selective damage, children are likely to have difficulties understanding the emotional expressions and mental states of others, as well as regulating their own emotions (Dennis, Purvis, Barnes, Wilkinson, & Winner, 2001). In addition, these children will have difficulty thinking about multiple social perspectives while deciding how to respond to social stimuli (Janusz et al., 2002). Possibly because of associated deficits in executive functions, they will have difficulty thinking flexibly about how to respond (Janusz et al., 2002), and instead may act impulsively because of their poor inhibitory control (Dennis, Guger, Roncadin, Barnes, & Schachar, 2001). The combination of deficits in these cognitive/executive and social/affective functions may influence children's reflective social problem solving (Yeates et al., 2004). They may be predicted to (1) choose instrumental over prosocial goals; (2) misinterpret the intent of others; and (3) generate fewer and less effective responses to social dilemmas.

In their actual interactions, children with brain disorders affecting social information processing may be hypothesized to behave in ways that do not promote social affiliation, but rather involve aggression, social withdrawal, or other inappropriate social behaviors. As a result, they may be expected to be poorly accepted by peers, as are other children who behave this way (Parker et al., 2006; Rubin, Bukowski, & Parker, 2006). They also may be anticipated to have fewer reciprocal friendships, characterized by more avoidance or discord. Peers, teachers, and parents will describe them as less socially competent and as displaying more social problems than other children (Dennis, Guger, et al., 2001; Yeates et al., 2004). In addition, these children may be expected to report lower levels of social self-esteem, higher levels of emotional distress, and more negative social relationships (Bohnert, Parker, & Warschausky, 1997). However, some children with brain disorders may be relatively unaware of their social problems and may actually overestimate their social functioning, as aggressive children have been shown to do (Boivin et al., 2005).

The social outcomes of childhood brain insults are likely to be moderated by the developmental factors outlined earlier, particularly age at insult and time since insult. An earlier age at insult appears to be a risk factor for a number of negative social outcomes, such as persistent deficits in the understanding of social emotions. Social information processing may be particularly vulnerable at early ages, because executive functions and theory of mind are more tightly linked during childhood than during adulthood (Hughes, 1998). Children with early brain insults may tend to show little or no improvement in their social adjustment across time, despite recovery of other cognitive abilities. Indeed, focal frontal lobe lesions in young children appear to result in more persistent social deficits than are sometimes apparent in older children and adults (Eslinger et al., 2004).

The poor social outcomes that occur in association with some childhood brain insults may persist even when they occur later in childhood (Yeates et al., 2004). Deficits in social information processing may potentially limit social experiences and hinder peer interactions, so that social functioning may become more divergent from that of peers with increasing time since insult. Indeed, given the transactional nature of social competence, negative social outcomes may persist even in the face of partial or complete recovery of social information processing. A cascade of negative changes in peer interactions and relationships, and consequently in broader aspects of social adjustment (including the perceptions of peers and adults), could engender a negative developmental spiral leading to chronic social problems that become very difficult to reverse even if children's social information processing improves following a brain insult (Coie, 1990).

Future Directions

Our hope in setting forth the model presented in this chapter is that it will help to spur multilevel, integrative research on social function in children with brain insults and neurological disorders. We believe that the model is potentially applicable to a wide variety of different childhood brain disorders, including TBI (Hanten et al., 2011), brain tumor (Fuemmeler, Elkin, & Mullins, 2002), epilepsy (Hamiwka, Hamiwka, Sherman, & Wirrell, 2011), stroke (Neuner et al., 2011), genetic conditions (see Gray & Cornish, Chapter 13, this volume), and prematurity (Hoy et al., 1992). Our research team is currently working on projects regarding each of these disorders. The projects all have a common goal: a better understanding of how the disorder and its associated neuropathology affects children's social information processing, social interactions, and social adjustment.

One important outcome of this ongoing research will be further refinements in theoretical models of social competence (Beauchamp & Anderson, 2010) and methods for assessing it (Crowe, Beauchamp, Catroppa, & Anderson, 2011; Gunnar & de Haan, 2009). Advances in theories and methods will enable a more sophisticated, interdisciplinary approach to understanding both brain and social function in children with brain disorders. Future research that is prospective and longitudinal in nature will help us to understand the postinsult trajectories of social functioning that children display, as well as the risk and resilience factors that moderate those trajectories (Yeates et al., 2002).

In the longer run, we must also attempt to translate our knowledge into interventions designed to promote social function in children with brain disorders. For instance, recent research suggests that web-based interventions designed to promote executive functions may facilitate behavioral adjustment among adolescents with TBI (Wade et al., 2011). However, the literature on interventions to promote psychosocial outcomes is quite scanty (Ross, Dorris, & McMillan, 2011). Future research that adapts existing treatment approaches (DeRosier & Gilliom, 2007; Frankel et al., 2010) to children with specific brain disorders is likely to prove an effective means of facilitating their friendships and peer relationships and of fostering their overall social competence.

Acknowledgments

The preparation of this chapter was supported in part by Grant Nos. K02 HD44099 and R01 HD48946 from the National Institute of Child Health and Human Development to Keith Owen Yeates. Aside from the first author, the order of authorship is alphabetical.

References

Adolphs, R. (2001). The neurobiology of social cognition. *Current Opinion in Neurobiology, 11,* 231–239.

Adolphs, R. (2002). Neural systems for recognizing emotion. *Current Opinion in Neurobiology, 12,* 169–177.

Adolphs, R. (2003). Cognitive neuroscience of human social behaviour. *Nature Reviews Neuroscience, 4,* 165–178.

Allman, J., Hakeem, A., Erwin, J., Nimchinsky, E., & Hof, P. (2001). The anterior cingulate cortex: The evolution of an interface between emotion and cognition. *Annals of the New York Academy of Sciences, 935,* 107–117.

Amodio, D. M., & Frith, C. D. (2006). Meeting of minds: The medial frontal cortex and social cognition. *Nature Reviews Neuroscience, 7,* 268–277.

Anderson, S., Bechara, A., Damasio, H., Tranel, D., & Damasio, H. (1999). Impairment of social and moral behavior related to early damage in human prefrontal cortex. *Nature Neuroscience, 2,* 1032–1037.

Anderson, V., Catroppa, C., Morse, S., Haritou, F., & Rosenfeld, J. (2005). Functional plasticity or vulnerability after early brain injury? *Pediatrics, 116,* 1374–1382.

Anderson, V., Levin, H. S., & Jacobs, R. (2002). Executive functions after frontal lobe injury: A developmental perspective. In D. T. Stuss & R. T. Knight (Eds.), *Principles of frontal lobe function* (pp. 504–527). New York: Oxford University Press.

Anderson, V., Spencer-Smith, M., Leventer, R., Coleman, L., Anderson, P., Williams, J., et al. (2009). Childhood brain insult: Can age at insult help us predict outcome? *Brain, 132,* 45–56.

Anderson, V., Spencer-Smith, M., & Wood, A. (2011). Do children really recover better?: Neurobehavioural plasticity after early brain insult. *Brain, 134,* 2197–2221.

Baron-Cohen, S., & Belmonte, M.K. (2005). Autism: A window onto the development of the social and the analytic brain. *Annual Review of Neuroscience, 28,* 109–126.

Bartsch, K., & Wellman, H. M. (1989). Young children's attribution of action to beliefs and desires. *Child Development, 60,* 946–964.

Beauchamp, M. H., & Anderson, V. (2010). SOCIAL: An integrative framework for the development of social skills. *Psychological Bulletin, 136*(1), 39–64.

Berndt, T.J., & Hoyle, S.G. (1985). Stability and change in childhood and adolescent friendships. *Developmental Psychology, 21,* 1007–1015.

Bohnert, A. M., Parker, J. G., & Warschausky, S. A. (1997). Friendship and social adjustment of children following a traumatic brain injury: An exploratory investigation. *Developmental Neuropsychology, 13,* 477–486.

Boivin, M., Vitaro, F., & Poulin, F. (2005). Peer relationships and the development of aggressive behavior in early childhood. In R. Tremblay, W. W. Hartup, & J. Archer (Eds.), *Developmental origins of aggression* (pp. 376–397). New York: Guilford Press.

Brown, B.B., & Klute, C. (2003). Friends, cliques, and crowds. In G.R. Adams & M. D. Berzonsky (Eds.), *Blackwell handbook of adolescence* (pp. 330–348). Malden, MA: Blackwell.

Buhrmester, D., & Furman, W. (1986). The changing functions of friends in childhood: A neo-Sullivan perspective. In V. J. Derlega & B. A. Winstead (Eds.), *Friendship and social interaction* (pp. 41–62). New York: Springer-Verlag.

Burgess, K.B., Wojslawowicz, J.C., Rubin, K.H., Rose-Krasnor, L., & Booth-LaForce, C. (2006). Social information processing and coping styles of shy/withdrawn and aggressive children: Does friendship matter? *Child Development. 77,* 371–383.

Cacioppo, J. T., Berntson, G. G., Sheridan, J. F., & McClintock, M. K. (2000). Multilevel integrative analyses of human behavior: Social neuroscience and the complementing nature of social and biological approaches. *Psychological Bulletin, 126,* 829–843.

Carlson, S. M., & Moses, L. J. (2001). Individual differences in inhibitory control and children's theory of mind. *Child Development, 72,* 1032–1053.

Carson, J. L., & Parke, R. D. (2008). Reciprocal negative affect in parent–child interactions and children's peer competency. *Child Development, 67,* 1467–8624.

Casey, B. J., Giedd, J. N., & Thomas, K. N. (2000). Structural and functional brain development and its relation to cognitive development. *Biological Psychology, 54,* 241–257.

Cavell, T. A. (1990). Social adjustment, social performance, and social skills: A tri-component model of social competence. *Journal of Clinical Child Psychology, 19,* 111–122.

Chugani, H. T., Phelps, M. E., & Mazziotta, J. C. (1987). Positron emission tomography study of human brain functional development. *Annals of Neurology, 22,* 487–497.

Cillessen, A. H. N., & Rose, A. J. (2005). Understanding popularity in the peer system. *Current Directions in Psychological Science, 14,* 102–105.

Coelho-Mosch, S., Max, J. E., & Tranel, D. (2005). A matched lesion analysis of childhood versus adult-onset brain injury due to unilateral stroke: Another perspective on neural plasticity and recovery of social functioning. *Cognitive and Behavioral Neurology, 18,* 5–17

Coie, J. D. (1990). Toward a theory of peer rejection. In S.R. Asher & J.D. Coie (Eds.), *Peer rejection in childhood* (pp. 365–401). Cambridge, UK: Cambridge University Press.

Collins, W.A. (2003). More than a myth: The developmental significance of romantic relationships during adolescence. *Journal of Research on Adolescence, 13,* 1–24.

Cowan, N. (1997). The development of working memory. In N. Cowan & C. Hulme (Eds.), *The development of memory in childhood* (pp. 163–200). Hove, UK: Psychology Press.

Crick, N. R., & Dodge, K. A. (1994). A review and reformulation of social information-processing mechanisms in children's social adjustment. *Psychological Bulletin, 115,* 74–101.

Crowe, L. M., Beauchamp, M.H., Catroppa, C., & Anderson, V. (2011). Social function assessment tools for children and adolescents: A systematic review from 1988 to 2010. *Clinical Psychology Review, 31,* 767–785.

Csikszentmihalyi, M., & Larson, R. (1984). *Being adolescent.* New York: Basic Books.

Davidson, R. J. (1992). Anterior cerebral asymmetry and the nature of emotion. *Brain and Cognition, 20,* 125–151.

de Haan, M., & Gunnar, M. R. (2009). The brain in a social environment: Why study development? In M. de Haan & M. R. Gunnar (Eds.), *Handbook of developmental social neuroscience* (pp. 3–12). New York: Guilford Press.

Demorest, A., Meyer, C., Phelps, E., Gardner, H., & Winner, E. (1984). Words speak louder than actions: Understanding deliberately false remarks. *Child Development, 55,* 1527–1534.

Dennis, M. (2006). Prefrontal cortex: Typical and atypical development. In J. Risberg & J. Grafman (Eds.), *The frontal lobes: Development, function and pathology* (pp. 128–162). New York: Cambridge University Press.

Dennis, M., Guger, S., Roncadin, C., Barnes, M., & Schachar, R. (2001). Attentional–inhibitory control and social-behavioral regulation after childhood closed head injury: Do biological, developmental, and recovery variables predict outcome? *Journal of the International Neuropsychological Society, 7,* 683–692.

Dennis, M., Purvis, K., Barnes, M.A., Wilkinson, M., & Winner, E. (2001). Understanding of literal truth, ironic criticism, and deceptive praise after childhood head injury. *Brain and Language, 78,* 1–16.

DeRosier, M. E., & Gilliom, M. (2007). Effectiveness of a parent training program for Improving children's social behavior. *Journal of Child and Family Studies, 16,* 660–670.

Diamond, A., & Taylor, C. (1996). Development of an aspect of executive control: Development of the abilities to remember what I said and to "Do as I say, not as I do." *Developmental Psychobiology, 29,* 315–334.

Dodge, K. A., Laird, R., Lochman, J. E., & Zelli, A. (2002). Multidimensional latent-construct analysis of children's social information processing patterns: Correlations with aggressive behavior problems. *Psychological Assessment, 14,* 60–73.

Dowsett, S. M., & Livesey, D. J. (2000). The development of inhibitory control in preschool children: Effects of "executive skills" training. *Developmental Psychobiology, 36,* 161–174.

Du Rocher Schudlich, T. D., Shamir, H., & Cummings, E. M. (2004). Marital conflict, children's representations of family relationships, and children's dispositions towards peer conflict strategies. *Social Development, 13,* 171–192.

Eckerman, C. O., & Stein, M. R. (1990). How imitation begets imitation and toddlers' generation of games. *Developmental Psychology, 26,* 370–378.

Eslinger, P.J., Flaherty-Craig, C.V., & Benton, A.L. (2004). Developmental outcomes after early prefrontal cortex damage. *Brain and Cognition, 55,* 84–103.

Fay, T. B., Yeates, K. O., Wade, S. L., Drotar, D., Stancin, T., & Taylor, H. G. (2009). Predicting longitudinal patterns of functional deficits in children with traumatic brain injury. *Neuropsychology, 23,* 271–282.

Fox, N. A. (1994). Dynamic cerebral processes underlying emotion regulation. In N. A. Fox (Ed.), The development of emotion regulation: Biological and behavioral considerations. *Monographs of the Society for Research in Child Development, 59*(2–3, Serial No. 240), 152–166.

Fox, N. A., Rubin, K. H., Calkins, S. D., Marshall, T. R., Coplan, R. J., Porges, S. W., et al. (1995). Frontal activation asymmetry and social competence at four years of age. *Child Development, 66,* 1770–1784.

Frankel, F., Myatt, R., Sugar, C., Whitham, C., Gorospe, C. M., & Laugeson, E. (2010). A randomized controlled study of parent-assisted children's friendship training with children having autism spectrum disorders. *Journal of Autism and Developmental Disorders, 40,* 827–842.

Frith, U., & Frith, C. (2001). The biological basis of social interaction. *Current Directions in Psychological Science, 10,* 151–155.

Fuemmeler, B. F., Elkin, T. D., & Mullins, L. L. (2002). Survivors of childhood brain tumors: Behavioral, emotional, and social adjustment. *Clinical Psychology Review, 22*, 547–585.

Gest, S., Graham-Bermann, S., & Hartup, W. (2001). Peer experience: Common and unique features of number of friendships, social network centrality, and sociometric status. *Social Development, 10*, 23–40.

Giedd, J. N., Blumenthal, J., Jeffries, N. O., Castellanos, F. X., Lui, J., Zijdenbos, A., et al. (1999). Brain development during childhood and adolescence: A longitudinal MRI study. *Nature Neuroscience, 2*, 861–863.

Gogtay, N., Giedd, J. N., Lusk, L., Hayashi, K. M., Greenstein, D., Vaituzis, A. C., et al. (2004). Dynamic mapping of human cortical development during childhood through early adulthood. *Proceedings of the National Academy of Sciences USA, 101*, 8174–8179.

Goncu, A. (1993). Development of intersubjectivity in the dyadic play of preschoolers. *Early Childhood Research Quarterly, 8*, 99–116.

Goncu, A., Patt, M. B., & Kouba, E. (2002). Understanding young children's pretend play in context. In P. K. Smith & C. H. Hart (Eds.), *Blackwell handbook of childhood social development* (pp. 418–437). Malden, MA: Blackwell.

Gunnar, M. R., & de Haan, M. (2009). Methods in social neuroscience: Issues in studying development. In M. de Haan & M. R. Gunnar (Eds.), *Handbook of developmental social neuroscience* (pp. 13–37). New York: Guilford Press.

Guralnick, M. J. (1999). Family and child influences on the peer-related social competence of young children with developmental delays. *Mental Retardation and Developmental Disabilities Research Reviews, 5*, 21–29.

Hamiwka, L. D., Hamiwka, L. A., Sherman, E. M., & Wirrell, E. (2011). Social skills in children with epilepsy: How do they compare to healthy and chronic disease controls? *Epilepsy and Behavior, 21*, 238–241.

Hanten, G., Cook, L., Orstena, K., Chapman, S. B., Li, X., Wilde, E. A., et al. (2011). Effects of traumatic brain injury on a virtual reality social problem solving task and relations to cortical thickness in adolescence. *Neuropsychologia, 49*, 486–497.

Hektner, J.M., August, G.J., & Realmuto, G.M. (2000). Patterns and temporal changes in peer affiliation among aggressive and nonaggressive children participating in a summer school program. *Journal of Clinical Child Psychology, 29*, 603–614.

Herbert, M. R., Ziegler, D. A., Makris, N., Filipek, P. A., Kemper, T. L., Normandin, J. J., et al. (2004). Localization of white matter volume increase in autism and developmental language disorder. *Annals of Neurology. 55*, 530–540.

Howes, C. (1988). Peer interaction of young children. *Monographs of the Society for Research in Child Development, 53*(1, Serial No. 217), 1–88.

Hoy, E. A., Sykes, D. H., Bill, J. M., Halliday, H. L., McClure, B. G., & Reid, M. M. (1992). The social competence of very-low-birthweight children: Teacher, peer, and self-perceptions. *Journal of Abnormal Child Psychology, 20*, 123–150.

Hughes, C. (1998). Executive function in preschoolers: Links with theory of mind and verbal ability. *British Journal of Developmental Psychology, 16*, 233–253.

Hughes, C., & Ensor, R. (2005). Executive function and theory of mind in 2 year olds: A family affair? *Developmental Neuropsychology, 28*, 645–668.

Janusz, J. A., Kirkwood, M. W., Yeates, K. O., & Taylor, H. G. (2002). Social problem-solving skills in children with traumatic brain injury: Long-term outcomes and prediction of social competence. *Child Neuropsychology, 8*, 179–194.

Johnson, M. K., Griffin, R., Csibra, G., Halit, H., Farroni, T., de Haan, M., et al. (2005). The emergence of the social brain network: Evidence from typical and atypical development. *Development and Psychopathology, 17*, 599–619.

Kendall, E., & Terry, D. J. (1996). Psychosocial adjustment following closed head injury: A model for understanding individual differences and predicting outcome. *Neuropsychological Reha-bilitation, 6*, 101–132.

Kinney, H. C., Brody, B. A., Kloman, A. S., & Gilles, F. H. (1988). Sequence of central nervous system myelination in human infancy: II. Patterns of myelination in autopsied infants. *Journal of Neuropathology and Experimental Neurology, 47*, 217–234.

Klingberg, T., Vaidya, C. J., Gabrieli, J. D. E., Moseley, M. E., & Hedehus, M. (1999). Myelination and organization of the frontal white matter in children: A diffusion tensor MRI study. *NeuroReport, 10*, 2817–2821.

Kochanska, G., Murray, K., Jacques, T. Y., Koenig, A. L., & Vandegeest, K. A. (1996). Inhibitory control in young children and its role in emerging internalization. *Child Development, 67*, 490–507.

Kuttler, A.F., Parker, J. G., & La Greca, A. M. (2002). Developmental and gender differences in preadolescents' judgments of the veracity of gossip. *Merrill–Palmer Quarterly, 48*, 105–132.

Ladd, G. W., & Pettit, G. S. (2002). Parenting and the development of children's peer relationships. In M.H. Bornstein (Ed.), *Handbook of parenting: Vol. 5. Practical issues in parenting* (2nd ed., pp. 269–309). Mahwah, NJ: Erlbaum.

Levin, H. S., & Hanten, G. (2005). Executive functions after traumatic brain injury in children. *Pediatric Neurology, 33*, 77–93.

Mah, L., Arnold, M. C., & Grafman, J. (2004). Impairment of social perception associated with lesions of the prefrontal cortex. *American Journal of Psychiatry, 161*, 1247–1255.

Martinez, W., Carter, J. S., & Legato, L. J. (2011). Social competence in children with chronic illness: A meta-analytic review. *Journal of Pediatric Psychology, 36*(8), 878–890.

Masten, A. S., Hubbard, J. J., Gest, S. D., Tellegen, A., Garmezy, N., & Ramirez, M. (1999). Competence in the context of adversity: Pathways to resilience and maladaptation from childhood to late adolescence. *Development and Psychopathology, 11*, 143–169.

Matsuzawa, J., Matsui, M., Konishi, T., Noguchi, K., Gur, R. C., Bilker, W., et al. (2001). Age-related volumetric changes of brain gray and white matter in healthy infants and children. *Cerebral Cortex, 11*, 335–342.

Meltzoff, A. N., Gopnik, A., & Repacholi, B. M. (1999). Toddlers' understanding of intentions, desires and emotions: Explorations of the dark ages. In P. D. Zelazo, J. W. Astington, & D. R. Olson (Eds.), *Developing theories of intention: Social understanding and self-control* (pp. 17–41). Mahwah, NJ: Erlbaum.

Moses, L. J. (2001). Executive accounts of theory of mind development. *Child Development, 72*, 688–690.

Nassau, J. H., & Drotar, D. (1997). Social competence among children with central nervous system-related chronic health conditions: A review. *Journal of Pediatric Psychology, 22*, 771–793.

Neuner, B., von Mackensen, S., Krumpel, A., Manner, D., Friefeld, S., Nixdorf, S., et al. (2011). Health-related quality of life in children and adolescents with stroke, self-reports, and parent/proxies reports: Cross-sectional investigation. *Annals of Neurology, 70*, 70–78.

Newcomb, A., & Bagwell, C. (1995). Children's friendship relations: A meta-analytic review. *Psychological Bulletin, 117*, 306–347.

Ochsner, K. N., & Lieberman, M. D. (2001). The emergence of social cognitive neuroscience. *American Psychologist, 56*, 717–734.

O'Donnell, S., Noseworthy, M., Levine, B., Brandt, M., & Dennis, M. (2005). Cortical thickness of the frontopolar area in typically developing children and adolescents. *NeuroImage, 24*, 948–954.

Parker, J., Rubin, K.H., Erath, S., Wojslawowicz, J.C., & Buskirk, A. A. (2006). Peer relationships and developmental psychopathology. In D. Cicchetti & D. Cohen (Eds.), *Developmental psychopathology: Vol. 2. Risk, disorder, and adaptation* (2nd ed., pp. 419–493). Hoboken, NJ: Wiley.

Parslow, R. C., Morris, K. P., Tasker, R. C., Forsyth, R. J., & Hawley, C. A. (2005). Epidemiology of traumatic brain injury in children receiving intensive care in the UK. *Archives of Disease in Childhood, 90,* 1182–1187.

Pascalis, O., Kelly, D. J., & Schwarzer, G. (2009). Neural bases of the development of face processing. In M. de Haan & M. R. Gunnar (Eds.), *Handbook of developmental social neuroscience* (pp. 63–86). New York: Guilford Press.

Passarotti, A. M., Paul, B. M., Bussiere, J. R., Buxton, R. B., Wong, E. C., & Stiles, J. (2003). The development of face and location processing: An fMRI study. *Developmental Science, 6,* 100–117.

Payne, C., & Bachevalier, J. (2009). Neuroanatomy of the developing social brain. In M. de Haan & M. R. Gunnar (Eds.), *Handbook of developmental social neuroscience* (pp. 38–61). New York: Guilford Press.

Pellegrini, A.D. (2002). Rough-and-tumble play from childhood through adolescence: Development and possible functions. In P.K. Smith & C.H. Hart (Eds.), *Blackwell handbook of childhood social development* (pp. 438–453). Malden, MA: Blackwell.

Powell, K. B., & Voeller, K. S. (2004). Prefrontal executive function syndromes in children. *Journal of Child Neurology, 19,* 785–797.

Prigatano, G. P. (1991). Disturbances of self-awareness of deficit after traumatic brain injury. In G. P. Prigatano & D. L. Schacter (Eds.), *Awareness of deficit after brain injury: Clinical and theoretical issues* (pp. 111–126). New York: Springer.

Rakic, P., Bourgeois, J. P., Eckenhoff, M. F., Zecevic, N., & Goldman-Rakic, P. S. (1986). Concurrent overproduction of synapses in diverse regions of the primate cerebral cortex. *Science, 232,* 232–235.

Reiss, A.L., Abrams, M.T., Singer, H.S., Ross, J.L., & Denckla, M.B. (1996). Brain development, gender and IQ in children. *Brain, 119,* 1763–1774.

Rose-Krasnor, L. (1997). The nature of social competence: A theoretical review. *Social Development, 6,* 111–135.

Ross, K. A., Dorris, L., & McMillan, T. (2011). A systematic review of psychological interventions to alleviate cognitive and psychosocial problems in children with acquired brain injury. *Developmental Medicine and Child Neurology, 53*(8), 692–701.

Rubin, K. H., Bukowski, W., & Parker, J. (2006). Peer interactions, relationships, and groups. In W. Damon & R. M. Lerner (Series Eds.) & N. Eisenberg (Vol. Ed.), *Handbook of child psychology: Vol. 3. Social, emotional, and personality development* (6th ed., pp. 571–645). Hoboken, NJ: Wiley.

Rubin, K. H., Burgess, K., & Coplan, R. (2002). Social withdrawal and shyness. In P. K. Smith & C. H. Hart (Eds.), *Blackwell handbook of childhood social development* (pp. 330–352). Malden, MA: Blackwell.

Rubin, K. H., Chen, X., & Hymel, S. (1993). Socioemotional characteristics of withdrawn and aggressive children. *Merrill–Palmer Quarterly, 39,* 518–534.

Rubin, K. H., Hemphill, S. A., Chen, X., Hastings, P., Sanson, A., Lo Coco, A., et al. (2006). A cross-cultural study of behavioral inhibition in toddlers: East–West–North–South. *International Journal of Behavioral Development, 30,* 219–226.

Rubin, K. H., & Rose-Krasnor, L. (1992). Interpersonal problem-solving. In V. B. Van Hasselt & M. Hersen (Eds.), *Handbook of social development* (pp. 283–323). New York: Plenum Press.

Rubin, K. H., Wojslawowicz, J. C., Burgess, K. B., Rose-Krasnor, L., & Booth-LaForce, C. L. (2006). The friendships of socially withdrawn and competent young adolescents. *Journal of Abnormal Child Psychology, 34,* 139–153.

Rudolph, K., & Heller, T. (1997). Interpersonal problem solving, externalizing behavior, and social competence in preschoolers: A knowledge–performance discrepancy? *Journal of Applied Developmental Psychology, 18*, 107–117.

Schonfeld, A. M., Mattson, S. N., & Riley, E. P. (2005). Moral maturity and delinquency after prenatal alcohol exposure. *Journal of Studies on Alcohol, 66*, 545–554.

Selman, R. L., & Schultz, L. H. (1990). *Making a friend in youth: Developmental theory and pair therapy*. Chicago: University of Chicago Press.

Siegal, M., & Varley, R. (2002). Neural systems involved in "theory of mind." *Nature Reviews/Neuroscience, 3*, 463–471.

Sodian, B., Taylor, C., Harris, P. L., & Perner, J. (1991). Early deception and the child's theory of mind: False trails and genuine markers. *Child Development, 62*, 468–483.

Sowell, E. R., & Jernigan, T. L. (1998). Further MRI evidence of late brain maturation: Limbic volume increases and changing asymmetries during childhood and adolescence. *Developmental Neuropsychology, 14,* 599–617.

Sowell, E. R., Thompson, P. M., Holmes, C. J., Batth, R., Jernigan, T. L., & Toga, A. W. (1999). Localizing age-related changes in brain structure between childhood and adolescence using statistical parametric mapping. *NeuroImage, 9*, 587–597.

Stuss, D. T., & Anderson, V. (2004). The frontal lobes and theory of mind: Developmental concepts from adult focal lesion research. *Brain and Cognition, 55*, 69–83.

Stuss, D.T., Gallup, G. G., & Alexander, M. P. (2001). The frontal lobes are necessary for 'theory of mind.' *Brain, 124,* 279–286.

Tasker, R. C., Salmond, C. H., Westland, A. G., Pena, A., Gillard, J. H., Sahakian, B. J., et al. (2005). Head circumference and brain and hippocampal volume after severe traumatic brain injury in childhood. *Pediatric Research, 58*, 302–308.

Taylor, H. G., & Alden, J. (1997). Age-related differences in outcome following childhood brain injury: An introduction and overview. *Journal of the International Neuropsychological Society, 3*, 555–567.

Taylor, H. G., Minich, N.M., Klein, N., & Hack, M. (2004). Longitudinal outcomes of very low birth weight. *Journal of the International Neuropsychological Society, 10*, 1–15.

Taylor, H. G., Schatschneider, C., & Minich, N.M. (2000). Longitudinal outcomes of *Haemophilus influenzae* meningitis in school-age children. *Neuropsychology, 14*, 509–518.

Taylor, H. G., Yeates, K. O., Wade, S. L., Drotar, D., Stancin, T., & Minich, N. (2002). A prospective study of short- and long-term outcomes after traumatic brain injury in children: Behavior and achievement. *Neuropsychology, 16*, 15–27.

Wade, S., Walz., N., Carey, J., McMullen, K., Cass, J., Mark, E., et al. (2011). A randomized trial of Teen Online Problem Solving for adolescent TBI: Reductions in behavior problems. *Pediatrics, 128*, e1–e7.

Watson, A. C., Nixon, C. L., Wilson, A., & Capage, L. (1999). Social interaction skills and theory of mind in young children. *Developmental Psychology, 35*, 386–391.

Wilde, E. A., Chu, Z., Bigler, E. D., Hunter, J. V., Fearing, M. A., Hanten, G., et al. (2006). Diffusion tensor imaging fiber tracking in the corpus callosum in children after moderate to severe traumatic brain injury. *Journal of Neurotrauma, 23*, 1412–1426.

Wilde, E. A., Hunter, J. V., Newsome, M. R., Scheibel, R. S., Bigler, E. D., Johnson, J. L., et al. (2005). Frontal and temporal morphometric findings on MRI in children after moderate to severe traumatic brain injury. *Journal of Neurotrauma, 22*, 333–344.

Winner, E. (1988). *The point of words: Children's understanding of metaphor and irony*. Cambridge, MA: Harvard University Press.

Yeates, K. O., Bigler, E. D., Dennis, M., Gerhardt, C. A., Rubin, K. H., Stancin, T., et al. (2007). Social outcomes in childhood brain disorder: A heuristic integration of social neuroscience and developmental psychology. *Psychological Bulletin, 133*, 535–556.

Yeates, K. O., Schultz, L. H., & Selman, K. O. (1991). The development of interpersonal negotiation strategies in thought and action: A social-cognitive link to behavioral adjustment and
social status. *Merrill–Palmer Quarterly, 37,* 369–406.

Yeates, K. O., Swift, E., Taylor, H. G., Wade, S. L., Drotar, D., Stancin, T., et al. (2004). Short-
and long-term social outcomes following pediatric traumatic brain injury. *Journal of the
International Neuropsychological Society, 10,* 412–426.

Yeates, K. O., Taylor, H. G., Wade, S. L., Drotar, D., Stancin, T., & Minich, N. (2002). A prospective study of short- and long-term neuropsychological outcomes after pediatric traumatic
brain injury. *Neuropsychology, 16,* 514–523.

CHAPTER 11

Impact of Early Brain Insult on the Development of Social Competence

Vicki Anderson, Stefanie Rosema, Alison Gomes, and Cathy Catroppa

Social interaction constitutes the fundamental fabric of human existence. Children are continually reading the actions, gestures, and faces of those around them and actively seeking to recognize their underlying mental states and emotions, in order to determine what they are thinking and feeling, and what they will do next. The manner in which a child operates within a social environment, by relying on social skills and interacting with others, is critical for developing and forming lasting relationships and for participating and functioning within the community (Beauchamp & Anderson, 2010; Blakemore, 2010; Cacioppo, 2002). What appears to happen so naturally is, in actual fact, a highly complex process involving the activation of a distributed neural network and the culmination of the individual's life experiences.

Advances in the social neurosciences demonstrate that social skills are intimately linked with neurological and cognitive functions (Adolphs, 2001). For example, to be socially competent, an individual must attend to others and inhibit inappropriate behaviors (executive functions), communicate effectively (language skills), and interpret others' meanings/interactions (social cognition). These specific skills, which contribute to social function, have been linked to specific brain regions (e.g., theory of mind [ToM] has been linked to prefrontal cortices); however, the end products, the social functions we observe in daily behavior, are most likely to be represented by an integrative, distributed neural network. Brain regions identified as contributing to this social network include (among other regions) aspects of the prefrontal cortex and temporoparietal junction, insula, and amygdala (Adolphs, 2001). As has been demonstrated for cognitive functions, it is likely that this network develops and becomes refined through childhood and adolescence (Beauchamp & Anderson, 2010). An injury to the brain, particularly during the formative childhood years, has the potential to disrupt this network and to result in social dysfunction (Yeates et al., 2007).

The importance of the child's environment has been well established in the developmental psychology literature, with distal factors (e.g., socioeconomic status) and more proximal influences (e.g., family environment) all implicated in the development of intact social functions (Ackerman & Brown, 2006; Bowlby, 1962; Bulotsky, Fantuzzo, & McDermott, 2008; Guralnick, 1999; Masten et al., 1999). These links are also supported by studies of children raised in atypical environments. For example, there is a wealth of research describing Romanian children raised in conditions of severe environmental deprivation. These studies clearly illustrate the potential for such environments to have a negative impact on social development, as well as the cognitive skills that mediate this development (Belsky & de Haan, 2011; Bos, Fox, Zeanah, & Nelson, 2009; Raizada & Kishyama, 2010).

Social skills emerge gradually through infancy and childhood, consolidating during adolescence. This progression reflects a dynamic interplay between the individual and his or her environment. In the first few months of life, the infant begins to smile and engage with others, and to imitate the actions of these others in an interactive manner. By 5–8 months of age, infants display evidence of goal-directed social behavior. At 3–4 years children can describe the mental states or beliefs of others as distinct from their own (Saxe, Carey, & Kanwisher, 2004), and by 7–8 years they can begin to predict the behavior of others from past experiences (Rholes, Newman, & Ruble, 1990). Social decision making and judgment emerge later, in early adolescence (Van Overwalle, 2009). During this protracted developmental process, any disruption of typical maturation processes will have the capacity to impair future progress. The influence of both family and environmental factors on social development is well established (Belsky & de Haan, 2011; Bos et al., 2009). Findings emerging from the social neurosciences also illustrate the close association between these social skills and underlying brain function (Adolphs, 2009; Van Overwalle, 2009).

Disruption to social function at any stage across the lifespan may have negative implications for a range of domains, including mental health, academic progress, career achievement, and quality of life. Such disruption in early life may interfere with a child's capacity to acquire and develop social skills. An example of such disruption in childhood is early brain insult (EBI). It is well established that brain insult occurring during childhood can result in physical dysfunction, cognitive and communication deficits, behavioral problems, and poor academic performance (Anderson, Catroppa, Morse, Haritou, & Rosenfeld, 2009). Less research has examined social function in children with EBI, however, given that many social skills are rapidly emerging through childhood, it is highly likely that victims of such insults will have compromised social development.

This chapter aims to review the current literature examining social outcomes from EBI, using the theoretical frameworks described by Anderson and Beauchamp (Chapter 1, this volume; Beauchamp & Anderson, 2010) and Yeates et al. (2007). These authors suggest that the broad domain of social function may be divided into a series of subdomains—for example, social adjustment, social interaction, and social cognition. We use these categories to structure our review of the empirical evidence. Whenever possible, we also discuss the influence of brain-related and environmental factors for social skills in the context of EBI.

Early Brain Insult

EBI refers to injury or insult to the developing central nervous system (CNS). Such insult may occur during the prenatal, perinatal or postnatal periods. The underlying injury mechanisms vary, but can generally be classified into one of the following categories: traumatic, vascular, developmental, infective, or neuroplastic. Location and extent of resultant damage will also vary: unilateral or bilateral; focal or diffuse; frontal or extrafrontal; cortical or subcortical. Similarly, the primary functional consequences of EBI are wide-ranging and include neurological deficits (e.g., seizures), speech and language difficulties (e.g., dysarthria, dyspraxia), motor impairment (e.g., paresis, incoordination), cognitive disabilities (e.g., problems with attention, memory, executive function, social cognition), and behavioral disturbances (e.g., poor self-regulation). Secondary consequences are also common: academic failure, poor vocational success, communication problems, and social dysfunction. Clearly both the etiology and symptomatology are heterogeneous. The unifying factor for all types of EBI is that they occur in the context of a rapidly developing brain, where (1) cerebral organization is likely to be incomplete; (2) neurobehavioral skills are only beginning to emerge; (3) serious damage has the potential to derail the genetic blueprint for typical CNS development; and (4) environmental factors, such as deprivation or enrichment, may have a significant influence on development.

For these children, elevated risk of social dysfunction may be attributed to a number of factors. First, *brain regions and networks* important for social function may be damaged. Current understandings of the biological bases of social function suggest that many brain regions may be involved, and that these may be linked via a distributed neural network. Of particular relevance to the developing child is that many brain regions identified as contributing to social function undergo protracted development (e.g., prefrontal regions) and are thus vulnerable in the context of EBI.

Second, *medical factors* may restrict opportunities for social interaction. For example, a hemiplegia will reduce a child's mobility and thus his or her capacity to interact independently with peers during informal play and sporting activities. Speech difficulties may lead to reduced expressive language fluency, which will reduce opportunities for communication with peers. Seizures may cause those around the child to feel wary or anxious about the child's well-being and thus influence social interactions. Furthermore, many instances of EBI can be conceptualized as chronic illnesses: They are associated with ongoing medical care and health concerns, as well as frequent absences from school, and thus limit typical exposure to social interactions.

Third, a child's *temperament and adjustment* to his or her condition will contribute to the child's social function. It is not uncommon for children experiencing a serious illness to experience depressed self-esteem—in response to "feeling different" from their peers; for the medical reasons noted above; or because of their psychological responses to often life-threatening illness, which can commonly translate into social anxiety and withdrawal. An additional symptom common to early stages of recovery from brain injury is excessive fatigue, which can severely impair a child's motivation and endurance of social interactions, further limiting social exposure. In response to these medical problems, some parents will be overprotective of their vulnerable children, potentially restricting the children's opportunities for engaging independently with peers.

Finally, a *child's environment*, and parents and other family members in particular, can either support or undermine social development following EBI. Not surprisingly, in the wake of EBI, the family routine can be disrupted. Parents may need to attend hospital and outpatient appointments, and some families may experience financial hardship associated with caring for their children. The associated burden may increase stress and family dysfunction. A secondary impact of early brain insult is the elevated risk of clinically significant stress for parents (McCarthy et al., 2010), with up to one-third of parents presenting with such symptoms even 6 months after diagnosis. Such parental psychopathology has been shown to have a negative impact on the quality of the family environment and on a child's well-being, with recent research identifying a clear link between such factors and children's social and behavioral adjustment (V. Anderson et al., 2006; Yeates, Taylor, Walz, Stancin, & Wade, 2010).

Social Function and EBI: What Are the Challenges?

A review of the literature investigating social outcomes following EBI reveals a relative dearth of information. Various factors may contribute to this current situation. First, until recently, health care professionals working with children with EBI have failed to recognize the importance of the social dimension for recovery and reintegration; they have concentrated primarily on physical and neurobehavioral domains. This is well illustrated in a study conducted by Bohnert, Parker, and Warschausky (1997), who reported that when health care professionals and parents were asked to rank the relative importance of health, education, and friendships for an injured child, they agreed that friendships were of least importance. In contrast, children ranked friendships as their top priority. Furthermore, in the past, social function has tended to lack a clearly defined neural substrate, resulting in a tendency for health care professionals to see this area of function as the domain of the family and school. Today, although the social neuroscience framework is not yet fully defined, it is clear that the CNS plays a critical role in subsuming social functions and must be considered in any formulation of social dysfunction.

More recently, there has been an increasing interest in the behavioral consequences of EBI and, related to these, post-EBI social symptoms. As a result, research interest has also turned to this domain; however, as described below, much of the work to date has lacked a theoretical basis and has failed to take advantage of the rich body of knowledge available from other disciplines. These include the fields of developmental psychology and developmental psychopathology, which focus on typical development of social skills and on developmental conditions where social dysfunction is well established (e.g., autism spectrum disorders), respectively.

The final limitation to accumulating knowledge relates to the current dearth of reliable and psychometrically sound measurement tools. Within the social domain, there are few developmentally driven and appropriately age-normed social assessment measures. Most available standardized options are either rating scales or questionnaires, and commonly canvas only parents' or teachers' perceptions. Many of these are global measures of adaptive ability, behavior, or quality of life, which include a small subset of socially relevant items. A smaller group of measures is specific to social skills such as relationships, social interaction, social participation, and loneliness (Crowe, Beauchamp, Catroppa, & Anderson, 2011). An even smaller selection of more empirical measures is available, mostly tapping aspects of social cognition (including empathy, perspective taking, and intent attribution). Although many of these possess good face validity, most have no normative data, and their psychometric properties are largely undocumented. In summary, social measures with clinical applicability are scarce, and this scarcity impedes progress in integrating social assessment into clinical practice and treatment (see Muscara & Crowe, Chapter 6, this volume).

Social Function and EBI: Theoretical Frameworks

In response to a largely atheoretical approach to the investigation of social consequences of EBI, two complementary neuropsychological models of social function have recently been described (Beauchamp & Anderson, 2010; Yeates et al., 2007). Both highlight the importance of typical brain development for social competence. They each propose that a disruption to development, via an injury or insult to a child's brain, can have significant consequences for the child's acquisition of social skills and knowledge and social function.

Yeates et al. (2007; see also Yeates et al., Chapter 10, this volume) present a heuristic describing social outcomes within a developmental psychology framework and with a focus on outcomes from TBI. Three important components of social function are highlighted: social information processing, social interaction, and social adjustment. Social outcomes are conceptualised as susceptible to insult-related risk factors (e.g., type and severity of insult and brain atypicalities), as well as non-insult-related factors (including parenting style, family function, and socioeconomic status).

Beauchamp and Anderson (2010) offer a similar framework, although with less focus on brain insult specifically. These authors place their emphasis on the mediating role of brain (development and integrity) and environment (family, temperament) on neurobehavioral skills (attention/executive function, communication, and social cognition), and consequently on social competence.

The following review of the research examining the social consequences of EBI draws on the models described by Yeates et al. (2007) and Beauchamp and Anderson (2010) as frameworks for available findings.

Social Function and EBI: What Do We Know?

Over the past few years, in keeping with the emergence of social neuroscience and the recognition of the debilitating and persisting impact of these problems, research has begun to describe social outcomes associated with EBI. The limited research available has demonstrated deficits in the cognitive skills central to social function (e.g., executive function and communication skills) (Anderson, Catroppa, et al., 2009; Catroppa & Anderson, 2005; Didus, Anderson, & Catroppa, 1999; Hanten et al., 2008; Janusz, Kirkwood, Yeates, &Taylor, 2002; Long et al., 2011). While not directly assessing links between these deficits and social function, these studies have provided a platform for conceptualizing the presence of social dysfunction. Only a very small number of studies have attempted to examine possible links between specific cognitive domains implicated in social function and social outcomes (Ganesalingam, Sanson, Anderson, & Yeates, 2007; Ganesalingam, Yeates, Sanson, & Anderson, 2007; Ganesalingam et al., 2011; Greenham, Spencer-Smith, Anderson, Coleman, & Anderson, 2010; Muscara, Catroppa, & Anderson, 2008), and these early results support the presence of such relationships.

Social Outcomes from Traumatic Brain Injury

Traumatic brain injury (TBI) is the most common cause of death and morbidity in childhood. It occurs as a result of a blow to the head, which characteristically leads not only to localized brain damage at the site of impact and the *contrecoup* site (i.e., the area opposite the impact site), but also to more diffuse axonal injury. Due to the shape of the skull, and the effects of injury forces, some brain regions are more vulnerable to damage than others. These include the frontal and temporal regions of the brain and white matter. More recently, several additional subcortical structures have also been found to be affected in the context of child TBI, including the hippocampus, corpus callosum, and amygdala (Beauchamp et al., 2011). Many of these regions, highlighted in Plate 11.1 on the color insert, have also been implicated as contributing to the social brain; this suggests that children with TBI may be particularly vulnerable to social difficulties of an organic basis.

The vast majority of research into social development after early insult to the CNS has focused on TBI. Children with TBI have been reported to show lower levels of self-esteem and adaptive behavior than controls, and higher levels of loneliness and behavioral problems (Andrews, Rose, & Johnson, 1998), as well as more difficulties in peer relationships (Bohnert et al., 1997). For example, Yeates and colleagues (2004) examined social functioning in 109 children with TBI (ages 6–12). They showed that parents of children with moderate to severe TBI reported their children to have poor social and behavioral functioning. No substantial recovery in social function was reported over the 4 years after injury, and in some cases levels of social functioning worsened. These findings are consistent with previous studies with smaller samples, which have also examined social functioning in children with TBI (Andrews et al., 1998). Although long-term outcomes are only just emerging as a research focus, studies of such outcomes are also beginning to identify links between poor social function following child TBI on the one hand and persisting social maladjustment and reduced quality of life on the other (Anderson, Brown, & Newitt, 2010; Cattelani, Lombardi, Brianti, & Mazzucchi, 1998).

Social Adjustment

Social adjustment is best described as "the degree to which children get along with their peers; the degree to which they engage in adaptive, competent social behavior; and the extent to which they inhibit aversive, incompetent behavior" (Crick & Dodge, 1994, p. 82). The majority of studies examining social adjustment have done so by administering broad-band parent questionnaires. Some of these studies have used questionnaires tapping behavioral function—for example, the Child Behavior Checklist (Asarnow, Satz, Light, Lewis, & Neumann, 1991; Poggi et al., 2005) and the Strengths and Difficulties Questionnaire (e.g., Tonks, Williams, Yates, & Slater, 2011). Others have used measures tapping adaptive abilities—for example, the Vineland Adaptive Behavior Scales or the Adaptive Behavior Assessment System (e.g., Anderson, Catroppa, Morse, Haritou, & Rosenfeld, 2001; Fletcher, Ewing-Cobbs, Miner, Levin, & Eisenberg, 1990; Ganesalingam et al., 2011). Overall, these studies have been divided in their findings. Some have reported evidence that children and adolescents with TBI display greater social incompetence than control groups, as demonstrated by poorer parent ratings for socialization and communication skills (Fletcher et al., 1990; Max et al., 1998; Levin, Hanten, & Li, 2009; Poggi et al., 2005); in contrast, others have reported no significant group differences (Anderson et al., 2001; Hanten et al., 2008; Papero, Prigatano, Snyder, & Johnson, 1993; Poggi et al., 2005). Whether greater injury severity leads to poorer social adjustment is as yet unclear. Some studies have suggested that children with severe TBI are more impaired in socialization, communication, and/or social competence than children with milder injuries (Asarnow et al., 1991; Fletcher et al., 1990; Max et al., 1998; Yeates et al., 2004), but other studies have failed to find these dose–response relationships (Papero et al., 1993).

A smaller number of studies have incorporated findings from multiple respondents to investigate social adjustment. Ganesalingam and colleagues (Ganesalingam, Sanson, Anderson, & Yeates, 2006, 2007; Ganesalingam, Yeates, et al., 2007) used both parent ratings and direct child measures, and found that children with moderate and severe TBI were rated by parents as more socially impaired than uninjured children. Similarly, on the direct child measures, survivors of child TBI self-reported poorer emotional and behavioral self-regulation and more frequent aggressive, avoidant, or irrelevant solutions to social problems than uninjured children. These authors found no differences between children with severe and moderate TBI.

Our research team has recently conducted a study tracking 10-year functional outcomes from early TBI (age at injury > 7 years). Using parent report measures, we found differences between injury severity groups (mild, moderate, severe) for social skills and adaptive abilities, but with fewer severity effects for behavioral outcomes, as illustrated in Figure 11.1 (Anderson, Godfrey, Rosenfeld, & Catroppa, in press). These findings highlight the persistence of postinjury social dysfunction over the long term.

Social Interaction

Social interaction refers to the social actions and reactions between individuals or groups modified to their interaction partners (Beauchamp & Anderson, 2010). To study this aspect of social function, Bohnert and colleagues (1997) and Prigatano and Gupta (2006)

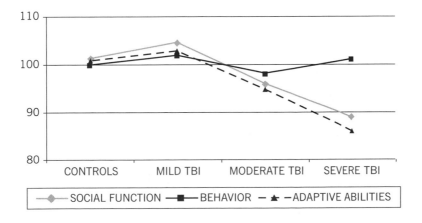

FIGURE 11.1. Functional outcomes 10 years following childhood TBI.

each investigated friendships of children who had sustained a TBI. Bohnert et al. (1997) employed both children and parents as respondents, and found no differences between children with and without TBI in friendship networks or on the Friendship Quality Questionnaire (Parker & Asher, 1993). In contrast, Prigatano and Gupta (2006), using parent ratings, reported results that supported a dose–response relationship. Specifically, children with severe TBI reported less close friendships than children with moderate or mild TBI, and children with moderate TBI had less close friendships than children with mild TBI.

In an early study conducted by Andrews and colleagues (1998), a series of questionnaires tapping various aspects of social interactions was administered to children and parents. Findings showed that children with TBI experienced higher levels of loneliness and had a higher likelihood of aggressive or antisocial behaviors than controls. Similarly, Dooley, Anderson, Hemphill, and Ohan (2008) investigated aggressive responses in adolescents with a history of TBI, compared to a healthy control sample. These authors found that although a frequently used broad-band measure, the Child Behavior Checklist (Achenbach, 1991), detected no group differences in aggressive behavior, the use of an aggression-specific measure showed greater sensitivity, identifying that history of TBI was related to higher rates of both reactive and proactive aggression. Such findings suggest that to accurately identify and characterize the social consequences of EBI, it is important to use tools sensitive to this domain.

These contradictory outcomes are difficult to interpret and are most likely to be explained by methodological differences, including TBI definition, composition of control groups, small sample sizes, and measurement tools. In an attempt to provide some clarity regarding the presence of social interaction difficulties and their persistence into adulthood, we recently conducted a retrospective study and surveyed a sample of 160 survivors of child TBI; we used a quality-of-life scale, a modified version of the Sydney Psychosocial Reintegration Scale (Tate, Hodgkinson, Veerbangsa, & Maggiotto, 1999). As shown in Figure 11.2, adult survivors of severe child TBI reported significant problems in multiple areas of their lives, including relationships and work and leisure activities. Few

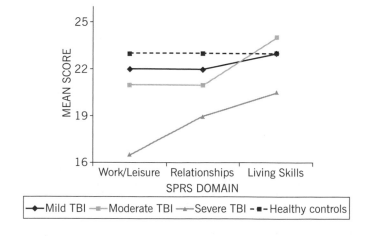

FIGURE 11.2. Quality of life in adult survivors of childhood TBI.

of these individuals reported having a stable group of friends or a life partner, and only a small number were gainfully employed. Furthermore, engagement in leisure activities, such as sports or social groups, was rare. Mild and moderate TBI was related to more typical function in social participation and other domains (Anderson et al., 2010). These findings provide strong support for the lasting effects of TBI (especially severe TBI) on problems in social interaction, and for their secondary repercussions.

Social Cognition

Social cognition refers to the mental processes used to perceive and process social cues, stimuli, and environments (Beauchamp & Anderson, 2010). In contrast to social adjustment and social interaction, measurement within this domain is largely based on direct child assessments, although most of these are currently restricted to experimental tools. This domain of social function appears to have attracted the most recent attention, with a growing number of recent studies investigating outcomes in this area after child TBI. Below, we have grouped this research into studies examining social problem solving, social communication, and social information processing (incorporating ToM and emotion perception).

SOCIAL PROBLEM SOLVING

Several recent studies have investigated social problem solving in children and adolescents after TBI, using direct child measures. Hanten et al. (2008) and Janusz et al. (2002) both used Interpersonal Negotiation Strategies (INS; Yeates, Schulz, & Selman, 1990), a child-based tool, to measure social problem solving. The INS consists of hypothetical interpersonal dilemmas that involve four social-solving problem steps: defining the problem, generating alternative strategies, selecting and implementing a specific strategy, and evaluating outcomes. Hanten et al. (2008) found that children with a TBI scored significantly lower on the INS from baseline through to 1 year after TBI, with no differential improvement in performance 1 year after TBI in both the TBI and control groups. Similarly, Janusz et al. (2002) reported that children with TBI scored significantly lower on social problem solving. Furthermore, children with TBI were able to generate solutions to social problems, but had difficulty choosing the optimal solution. These authors also investigated performance differences between injury severity groups, but detected no severity effects between those with severe and moderate TBI.

Warschausky, Cohen, Parker, Levendosky, and Okun (1997) used a similar paradigm, the Social Problem Solving Measure (Pettit, Dodge, & Brown, 1988), to assess solutions to social problems in children ages 7–13 years. Children with TBI provided significantly fewer peer entry solutions in social engagement situations than control children, but the groups did not differ with regard to the number of solutions to peer provocations. In a study from our team (Muscara et al., 2008), we investigated the relationship between executive function and social function 10 years after child TBI. This study extended the work of Yeates et al. (2004), which had previously proposed that social problem solving is a mediator between neurocognitive function and social skills, rather than a direct link. We found that greater executive dysfunction was associated with less sophisticated social problem-solving skills and poorer social outcomes. Furthermore, the maturity of social problem-solving skills was found to mediate the relationship between executive function and social outcomes in TBI, providing the first empirical evidence for a link between executive and social skills in the context of childhood acquired brain injury.

SOCIAL COMMUNICATION

This domain of social function refers specifically to the child's ability to draw meaning from complex language. Tasks tapping these pragmatic language skills include aspects of cognitive function (e.g., working memory and executive function), as well as abilities more commonly considered as social cognition—such as the identification of irony and sarcasm in conversation, and the ability to draw inferences from linguistic information and to distinguish truth from falsehoods (Turkstra, Dixon, & Baker, 2004; Turkstra, Williams, Tonks, & Frampton, 2008). Using the Video Social Inference Test with a group of adolescents after TBI, Turkstra et al. (2008) demonstrated that child TBI was associated with poorer identification of sarcasm and irony and with greater difficulties interpreting inference in both photographs and stories. Dennis, Guger, Roncadin, Barnes, and Schachar (2001) have reported similar findings with younger children, showing deficits in understanding deceptive emotions, literal truth, irony, and deceptive praise.

SOCIAL INFORMATION PROCESSING

Studies investigating social information processing have focused primarily on ToM and emotion perception in school-age children and adolescents. For example, Turkstra et al. (2004, 2008) measured ToM in adolescents with TBI, using a second-order belief task and a pragmatic judgment test. They found that, in contrast to healthy controls, adolescents with TBI were deficient in judging whether a speaker was talking at the listener's level and in recognizing when an individual was monopolizing a conversation. In contrast, the group with TBI performed similarly to controls on a first-order belief task (identifying a good listener), as well as on a faux pas test and the Strange Stories test.

Walz, Yeates, Taylor, Stancin, and Wade (2010) also examined ToM in a group of children who had sustained TBI between 3 and 5 years of age, and found few differences between the children with TBI and controls. These authors observed that as ToM skills would only be emerging in typically developing children at the time these children sustained their injuries, their results required follow-up. Of note, this group also demonstrated significant problems on ToM tasks, particularly for children with severe TBI. These studies highlight the need for a developmental perspective, as well as the importance of taking into account both age at injury and age at assessment when interpreting study findings.

Children and adolescents with TBI have also been reported to have more difficulty than controls in recognizing emotions. Tonks, Williams, Framton, Yates, and Slater (2007) found that children with TBI were more impaired than control children in recognizing emotions expressed in the eyes, but showed equivalent competence in recognizing facial emotions, suggesting that adding context assisted social information processing.

In summary, the weight of evidence indicates that children sustaining TBI are at elevated risk of experiencing social deficits, including social adjustment, social interaction, and social cognition. These problems persist over the long term after TBI. Further work is needed to describe the potential impact of injury-related factors (e.g., severity, age at insult) and environmental influences on these social consequences.

Social Outcomes from Pediatric Stroke

Pediatric stroke (PS) is a relatively uncommon occurrence, affecting approximately 7 of every 100,000 children. PS is an acute cerebrovascular event that can occur at any stage during childhood, but is perhaps most frequent in the perinatal period. There are two forms of PS: *arterial ischemic stroke* and *hemorrhagic stroke*. Arterial stroke is caused by a blockage or obstruction of an artery due to a clot, resulting in disrupted blood flow and relatively focal damage. Hemorrhagic stroke, in contrast, involves the rupture of an artery, often leading to more diffuse brain damage. Depending on the type of stroke and the artery affected, brain lesions caused by PS will vary in severity, extent, and location (Gomes, Rinehart, & Anderson, 2011). Infarcts in the middle cerebral artery will affect dorsolateral frontal cortex, basal ganglia, and white matter, whereas anterior cerebral artery stroke leads to bilateral lesions in orbitofrontal, temporal, and parietal cortices. Given the distributed nature of the social brain network, it is not surprising that damage from PS may lead to social problems.

Children recovering from stroke often have unique social challenges due to functional and physical impairments. According to the Canadian Paediatric Ischaemic Stroke Registry, there are long-term functional and neurological deficits in 60–85% of cases (Sofranas et al., 2006). Motor impairments are common, with as many as 30–60% of children experiencing effects ranging from mild weakness to severe hemiplegia (Brower, Rollins, & Roach, 1996; Ganesan et al., 2000; Gordon, Ganesan, Towell, & Kirkham, 2002). Hearing and visual impairments may reduce a child's capability to encode and interpret subtle social cues that are based on both verbal and nonverbal behaviors. Specifically, visual difficulties can cause unusual eye contact and lead to subsequent social difficulties, which may mirror the reciprocal communication deficits present in autism spectrum disorders.

In contrast to the range of studies documenting social dysfunction in the context of child TBI, relatively little work has been done investigating social outcomes of PS. There is, however, evidence of disruption to at least some of the cognitive skills underpinning social competence. For example, recent studies have identified significant deficits in executive skills and attention in victims of PS (Long et al, 2011), while the seminal work of Bates and colleagues (2001) has detailed the nature of communication problems in the context of PS.

Three general reviews (deVeber, MacGregor, Curtis, & Mayank, 2000; Ganesan et al., 2000; Goodman & Graham, 1996) have provided some insight into research on social outcomes following PS. As in TBI research, the social measures used in this research have commonly been broad-band in nature, including the Child Behavior Checklist (Achenbach, 1991; Achenbach & Edelbrock, 1983) and the Vineland Adaptive Behavior Scales (Sparrow, Balla, & Cicchetti, 1984). deVeber et al. (2000) focus largely on neurological and physical outcomes, and only mention the social domain briefly. Goodman and Graham (1996) highlight that children with PS require additional assistance for optimal school and home participation, and Ganesan et al. (2000) report that 37% of parents of children with PS were "concerned" about their child's behavior.

Social Interactions

A handful of studies have commented on this area of social function in the context of PS, using specific measures to characterize the quality of social interactions. A recent study by Everts et al. (2008) examined children who suffered stroke from birth to 18 years; qualitative reports indicated low peer acceptance, mood instability, and decreased social support from peers for many participants, though these domains were not explicitly measured. De Schryver, Kappelle, Jennekens-Schinkel, and Boudewyn Peters (2000) also documented changes in social behavior and companionship, as reported by parents of children with PS.

Steinlin, Roellin, and Scroth (2004) studied a small group of 16 survivors of PS and detected significant changes in social interactions, including a qualitative difference in friendships with peers, as reported by parents. Findings emphasised children's difficulties in implementing social skills in real-life situations, linking these problems to the impact of cognitive deficits (e.g., processing speed), though this relationship was not statistically tested. It should be noted that these results were based solely on qualitative questions related to integration with peer groups and family, and so they need to be interpreted with caution.

Social Participation

Social participation is defined by the World Health Organization (2001) as involvement in life situations—for example, sporting and recreational activities. To date, patterns of social participation following PS remain relatively unknown. One study (Hurvitz, Warschausky, Berg, & Tsai, 2004) investigated 29 adults at an average of 12 years following PS. Results demonstrated a high proportion of high school graduations; 90% of participants were employed; and 79% of the adult survivors could drive. Despite these positive outcomes, living skills, communication, and socialization were in the moderately low range. Age at stroke onset was found to have no association with outcomes. The impact of non-insult-related factors was not examined.

Tonks et al. (2011) have also recently studied social participation in children with PS and other acquired brain insults (N = 135). They found that, compared to healthy controls, the children with PS demonstrated a restricted level of diversity and intensity across a range of activities: recreational, social, and self-improvement. This pattern was common across the sample and was not related to severity of brain insult. These authors also highlighted the key role of intact social participation for children's general health and quality of life. In similar studies, parents of children with serious acquired brain injuries have reported that their children show reduced participation in peer-related activities and daily routines (Bedell & Dumas, 2004).

Potential Contributors to Psychosocial Outcomes

BRAIN-RELATED PREDICTORS

Neuropsychological factors and their contribution to social outcomes after stroke have been compared across child and adult participants. Mosch, Max, and Tranel (2005) matched children (n = 29) and adults (n = 29) who had sustained a stroke, with respect to size, location, and hemisphere affected. Impairments in social adjustment (employment/educational status, interpersonal functioning, clinician rating) were measured with the Child Behavior Checklist and the Vineland Adaptive Behavior Scales, as well as a psychiatric diagnostic tool, the Schedule for Affective Disorders and Schizophrenia for School-Age Children—Present and Lifetime Version (Kaufman et al., 1997). Social competence in adults was associated with right-hemisphere lesions, whereas children with stroke displayed mild and moderate social deficits regardless of the side of lesion. Though adults with left-hemisphere lesions had speech and language impairments, their child counterparts obtained average scores. Trauner, Panyard-Davis, and Ballantyne (1996) have also examined social recovery, describing 17 children with a history of stroke prior to 6 months of age (including prenatal, perinatal, and childhood strokes). Using the Personality Inventory for Children (Wirt, Lachar, Klinedinst, Seat, & Broen, 1977). They found that regardless of lesion laterality, children with focal strokes had greater impairment than healthy controls on several subscales, including General Adjustment, Social Skills, and Social Desirability. Extrafrontal lesions were associated with difficulties in emotional communication and processing of affective information. The results from both of these studies (Mosch et al., 2005; Trauner et al., 1996) suggest that regardless of lesion laterality, PS may lead to reduced social competence, even when language skills remain intact. It should be noted that the variable age at stroke onset in these studies may have led to an exaggeration of difficulties in the context of acquired stroke, as deficits may be more severe in children with prenatal/perinatal lesions than in those with childhood stroke (Nass & Trauner, 2004).

Following childhood stroke, severity, often indexed by lesion size, may be a useful predictor of outcome; however, to date relationships among IQ, behavioral and social outcomes, and infarct volume have not been found (Everts et al., 2008; Nass & Trauner, 2004). Similarly, a small body of PS research has demonstrated that localization of infarction is predictive of functional outcome (Long et al., 2011; Roach, 2000). In contrast, the work of Max and his colleagues has consistently documented a relationship between lesion location and psychosocial function following PS. For example, lesions involving the putamen or orbitofrontal cortex have been linked to traits of attention-deficit/hyperactivity disorder in samples with PS (Max et al., 2002, 2005). Overall, however, studies investigating the general impact of lesion size and location have been inconclusive (Nass & Trauner, 2004; Steinlin et al., 2004).

AGE AT STROKE ONSET

Due to the timing of insult and developing brain networks, children suffering from stroke may have greater vulnerabilities to social difficulties and psychopathology than the healthy population. PS presents with unique characteristics, including a defined date of onset, usually a focal location of injury, and often a brain that was premorbidly healthy. As in the cognitive and psychiatric domains (Max, Bruce, Keatley, & Delis, 2010; Westmacott, MacGregor, Askalan, & deVeber, 2009), age at stroke onset may affect social outcomes. Although there have been no studies to date addressing this issue in PS, recent research examining EBI in general suggests that children with injuries sustained in the prenatal period, before 2 years of age (including perinatal stroke), are at greatest risk of social deficits (Greenham et al., 2010). However, the risk of social deficits is elevated following brain insult at any age during childhood (Anderson, Spencer-Smith, et al., 2009).

IMPACT OF CHRONIC ILLNESS

Due to physical limitations, anxiety, absenteeism, parents' concerns, stigmatization, embarrassment, and generally limited opportunities for social interaction, stroke in the early years may disrupt a child's ability to function in his or her typical environment (Middleton, 2001; Nassau & Drotar, 1995). This is consistent with findings in adult stroke research, where withdrawal from professional and social groups, and discontinuation and loss of social identity, are described as leading to decline in quality of life (Haslam et al., 2008). Like their adult counterparts, children are likely to experience a loss of selfhood in the aftermath of PS, with the potential for negative effects on mood, well-being, and overall recovery.

Social Outcomes from Brain Tumor

Childhood cancers, and in particular, brain tumors have been related to higher levels of stress and trauma than most other brain-related conditions (McCarthy et al., 2010). Although brain tumors are relatively uncommon in children, treatment advances in childhood cancers have led to improved survival rates and an increasing focus on quality-of-life outcomes. Psychosocial consequences have received considerable attention, although little research to date has focused on the brain bases of these problems. Rather, the emphasis has been on adjustment to life-threatening disease, extended treatment, and the impact these have on self-concept and quality of life. However, regardless of the assessment approach employed or the social domain under study, findings consistently document long-term social problems in these children (Schulte & Barrera, 2010).

Social Adjustment

Like the studies of children with TBI and PS, research on children with brain tumors has also mainly employed broad-band measurement tools. Using parent- and teacher-based ratings from the Child Behavior Checklist, Aarsen et al. (2006) described poor social adjustment after diagnosis and treatment of brain tumors. Others have replicated this finding, using other measures (Varni, Seid, & Rode, 1999; Bhat et al., 2005; Poretti, Grotzer, Ribi, Schonle, & Bolthauser, 2004; Sands et al., 2005; Upton & Eiser, 2006). Although this finding has not been universal (e.g., Carey, Barakat, Foley, Gyato, & Phillips, 2001), inconsistent results are most likely related to postdiagnosis timing; Mabbott et al. (2005) have noted that social adjustment may appear intact acutely, but problems increase with time since diagnosis.

Social Interaction

Survivors of childhood brain tumors are also reported to struggle with peer interaction and social participation more generally. Studies describe these children as having fewer friends (Barrera, Schulte, & Spiegler, 2008), and as experiencing limited social opportunities, social isolation, peer exclusion, and bullying (Boydell, Stasiulis, Greenberg, Greenberg, & Spiegler, 2008; Upton & Eiser, 2006; Vance, Eiser, & Home, 2004).

Social Cognition

Bonner et al. (2008) have conducted one of the few studies evaluating social cognition in children with brain tumors. These researchers examined facial expression recognition and found that children with brain tumors made more errors than expected when interpreting facial expressions.

Predictors of Social Problems

BRAIN FACTORS

A number of risk and resilience factors have been investigated in the context of childhood brain tumors. Somewhat surprisingly, treatment factors have been shown to have little impact on social outcomes (Schulte, Bouffet, Janzen, Hamilton, & Barrera, 2010). In contrast, there is some evidence that developmental stage is relevant: The greatest social consequences are identified in association with both early diagnosis (social adjustment, social cognition: Foley, Barakat, Herman-Liu, Radcliffe, & Molloy, 2000; Bonner et al., 2008) and diagnosis during adolescence (reduced quality of life: Aarsen et al., 2006). These findings suggest that disturbances in social competence are most likely during "critical periods" of social development. As noted above, and in keeping with observations of increasing brain pathology and neurocognitive impairment with greater time after brain tumor diagnosis, poorer social competence also appears to develop over time; most problems emerge between 7 and 11 years after illness (Kullgren, 2003; Poretti et al., 2004; Aarsen et al., 2006; Mabbott et al., 2005).

INTERNAL AND ENVIRONMENTAL FACTORS

Several studies have addressed child-related contributors to social outcomes after childhood brain tumors. For example, a number of studies have identified poorer social competence in survivors with lower levels of intelligence (Poggi et al., 2005; Carey et al., 2001; Holmquist & Scott, 2003). Interestingly, these studies have reported greater links between nonverbal skills and social skills, with less evidence of a relationship between verbal skills and the social domain. Poorer social adjustment and social participation have also been associated with lower body mass index (Schulte et al., 2010). Perhaps contrary to expectations, links with social and family factors have been less compelling (e.g., Kullgren, 2003).

In a review of this literature, Schulte and Barrera (2010) advise caution in interpreting these findings. They note the heterogeneous nature of samples with brain tumors in terms of age at injury and time since diagnosis, as well as retrospective designs and inadequate assessment tools in this research.

Social Outcomes from Focal EBI

Focal brain insult is relatively rare in infancy and childhood, where the mechanism of insult is more frequently generalized (e.g., TBI, cerebral infection, hypoxia). Types of focal EBI include developmental conditions (e.g., focal dysplasias), as well as those acquired postnatally (e.g., PS, tumor, penetrating head injury). (PS and tumor have been discussed separately above; we focus in this section on focal EBI in general.) Traditionally, it has been argued that in the context of the "plastic" developing brain, the functions subsumed by these focal brain regions may be readily reorganized into other health brain regions, with no observable functional deficits. The earliest data describing the impact of focal EBI on social function comes from several case studies. To date, there have been few group studies of social outcomes following focal brain insults in infancy and early childhood. In a recent study, our team has examined the social consequences of focal brain insult sustained in childhood, with the aim of identifying (1) whether focal EBI is linked to social deficits; (2) whether age at insult is a predictor of social deficits; and (3) what additional factors might predict such deficits (Greenham et al., 2010).

The study compared social outcomes for children sustaining focal EBI at different times from gestation to late childhood, to determine whether the EBI was associated with an increased risk of problems. Children with focal EBI were categorized according to timing of insult: (1) congenital (n = 38), first–second trimester; (2) perinatal (n = 33), third trimester to 1 month after birth; (3) infancy (n = 23), 2 months–2 years after birth; (4) preschool (n = 19), 3–6 years; (5) middle childhood (n = 31), 7–9 years; and (6) late childhood (n = 19), after 10 years. Children's teachers completed questionnaires measuring social function. Results showed that for the total group, children with focal EBI were at a significantly increased risk for social impairment, compared to normative expectations. Somewhat surprisingly, the children with focal EBI did not demonstrate significant deviations from average for self-control, school adjustment, or empathy. Although mean ratings for this group were not severely impaired (i.e., they generally fell within 1 SD of expectations), these children were rated as having fewer prosocial behaviors than typically developing peers, with particular difficulties identified for peer relationships. The children with EBI also experienced significantly more emotional symptoms and hyperactivity than population expectations. Our findings are in line with previous group-based studies examining children with brain injury (Andrews et al., 1998; Bohnert et al., 1997; Ganesalingam et al., 2006; Janusz et al., 2002; Yeates et al., 2007) and with case reports (e.g., S. W. Anderson, Barrash, Bechara, & Tranel, 2006; S. W. Anderson, Bechara, Damasio, Tranel, & Damasio, 1999), both of which have consistently identified persisting social and behavioral problems in children with a variety of diagnoses (e.g., TBI, PS, tumor).

Furthermore, focal EBI before age 2 years was associated with the most significant social impairment, whereas children with focal EBI in the preschool years and in late childhood recorded scores closer to average levels.

The study provided only limited evidence for the role of age of insult in predicting social difficulties. Focal EBI before age 3 years conferred particular risk, but significant effects were confined to the domains of peer relations and emotional symptoms. Examination of impairment ratings yielded additional findings (see Figure 11.3). A third to a half of the children who sustained lesions before age 6 scored in the impaired range for social skills. Pre- and perinatal insults were associated with the greatest social problems, whereas all other groups recorded fewer social difficulties.

Although previous research has indicated that lesion characteristics (location, laterality) contribute to cognitive outcome in children, this did not appear to be the case for social outcomes. Lesion location and laterality were not predictive of social outcome, nor was social risk. In contrast, presence of disability (seizures) and family dysfunction were shown to contribute to poorer social outcomes.

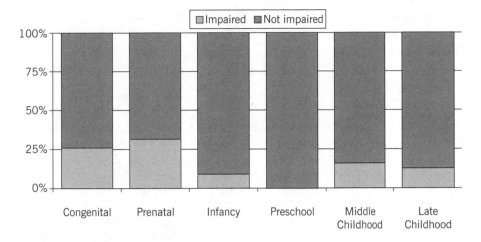

FIGURE 11.3. Rates of social impairment and age at focal EBI. Adapted from Greenham, Spencer-Smith, Anderson, Coleman, and Anderson (2010). Copyright 2010 and adapted by permission of the authors.

Conclusions

There is growing interest in the social consequences of EBI; however, evidence to date is relatively scarce. Not surprisingly, the available literature does indicate that the presence of EBI is associated with an elevated risk of social dysfunction across a range of dimensions—social adjustment, social interaction and participation, and social cognition. The ways in which these domains interact with one another and with a child's other skills remain unclear, and the measures generally employed in such studies are not intended for the assessment of social skills specifically. In addition, it appears that the injury-related risk factors established as predictors of cognitive and physical outcomes from EBI (injury severity, lesion location, age at insult) are unable to predict social outcomes in isolation. Rather, findings suggest that environmental factors play a key role. Social context, family function, and child temperament and adjustment to brain insult are all important in determining a child's social outcome. The recent development of theoretical models of social function and EBI (derived largely from the social neuroscience, developmental, and developmental psychopathology literatures), and the findings emerging from longitudinal studies of EBI, show great promise and will facilitate future research in the field.

References

Aarsen, F., Paquier, P., Reddingius, R., Streng, I., Arts, W., Evera-Preesman, M., et al. (2006). Functional outcome after low grade astrocytoma treatment in childhood. *Cancer, 106,* 396–402.

Achenbach, T. M. (1991). *Manual for the Child Behavior Checklist/4–18 and 1991 profile.* Burlington: University of Vermont, Department of Psychiatry.

Achenbach, T. M., & Edelbrock, C. S. (1983). *Manual for the Child Behavior Checklist and the Revised Behavior Profile.* Burlington: University of Vermont, Department of Psychiatry.

Ackerman, B. P., & Brown, E. D. (2006). Income poverty, poverty co-factors, and the adjustment of children in elementary school. *Advances in Child Development and Behavior, 34,* 91–129.

Adolphs, R. (2001). The neurobiology of social cognition. *Current Opinion in Neurobiology, 11,* 231–239.

Adolphs, R. (2009). The social brain: Neural basis of social knowledge. *Annual Review of Psychology, 60,* 693–716.

Anderson, S. W., Barrash, J., Bechara, A., & Tranel, D. (2006). Impairments of emotion and real-world complex behavior following childhood- or adult-onset damage to ventromedial prefrontal cortex. *Journal of the International Neuropsychological Society, 12,* 224–225.

Anderson, S. W., Bechara, A., Damasio, A. H., Tranel, D., & Damasio, A. (1999). Impairment of social and moral behavior related to early brain damage in human prefrontal cortex. *Nature Neuroscience, 2,* 1032–1037.

Anderson, V., Brown, S., & Newitt, H. (2010). What contributes to quality of life in adult survivors of childhood traumatic brain injury? *Journal of Neurotrauma, 27*(5), 863–870.

Anderson, V., Catroppa, C., Dudgeon, P., Morse, S., Haritou, F., & Rosenfeld, J. (2006). Understanding predictors of functional recovery and outcome thirty-months following early childhood head injury. *Neuropsychology, 20*(1), 42–57.

Anderson, V., Catroppa, C., Morse, S., Haritou, F., & Rosenfeld, J. (2001). Outcome from mild head injury in young children: A prospective study. *Journal of Clinical and Experimental Neuropsychology, 23,* 705–717.

Anderson, V., Catroppa, C., Morse, S., Haritou, F., & Rosenfeld, J. (2009). Intellectual outcome from preschool traumatic brain injury: A 5-year prospective, longitudinal study. *Pediatrics, 124*(6), e1064–e1071.

Anderson, V., Godfrey, G., Rosenfeld, J., & Catroppa, C. (in press). Ten year outcome from childhood traumatic brain injury. *Journal of Developmental Neuroscience.* doi.org/10.10.j.jdevneu. 2011.09.008.

Anderson, V., Spencer-Smith, M., Leventer, R., Coleman, L., Anderson, P., Williams, J., et al. (2009). Childhood brain insult: Can age at insult help us predict outcome? *Brain, 132*(1), 45–56.

Andrews, T. K., Rose, F. D., & Johnson, D. A. (1998). Social and behavioural effects of traumatic brain injury in children. *Brain Injury, 12,* 133–138.

Asarnow, R. F., Satz, P., Light, R., Lewis, R., & Neumann, E. (1991). Behavior problems and adaptive functioning in children with mild and severe closed head injury. *Journal of Pediatric Psychology, 16,* 543–555.

Barrera, M., Schulte, F., & Spiegler, B. (2008). Factors influencing depressive symptoms of children treated for a brain tumor. *Journal of Psychosocial Oncology, 26,* 1–16.

Bates, E., Reilly, J., Wilfeck, B., Donkers, N., Opie, M., Fenson, J., et al. (2001). Differential effects of unilateral lesions on language production in children and adults. *Brain and Language, 79,* 223–265.

Beauchamp, M. H., & Anderson, V. (2010). SOCIAL: An integrative framework for the development of social skills. *Psychological Bulletin, 136,* 39–64.

Beauchamp, M., Ditchfield, M., Maller, J., Cattroppa, C., Godfrey, C., Rosenfeld, J., et al. (2011). Hippocampus, amygdala and global brain change 10 years after childhood traumatic brain injury. *International Journal of Developmental Neuroscience, 29*(2), 137–143.

Bedell, G. M., & Dumas, H. M. (2004). Social participation of children and youth with acquired brain injuries discharged from inpatient rehabilitation: A follow-up study. *Brain Injury, 18*(1), 65–82.

Belsky, J., & de Haan, M. (2011). Parenting and children's brain development: The end of the beginning. *Journal of Child Psychology and Psychiatry, 52,* 409–428.

Bhat, S., Goodwin, T., Burwinkle, T., Lansdale, M., Dahl, G., Huhn, S., et al. (2005). Profile of daily life in children with brain tumors: An assessment of health-related quality of life. *Journal of Clinical Oncology, 23,* 5493–5500.

Blakemore, S. (2010). The developing social brain. *Neuron, 65,* 774–777.

Bohnert, A. M., Parker, J. G., & Warschausky, S. A. (1997). Friendship and social adjustment of children following a traumatic brain injury: An exploratory investigation. *Developmental Neuropsychology, 13,* 477–486.

Bonner, M., Hardy, K., Willard, V., Anthon, K., Hood, M., & Gururangan, S. (2008). Social function and facial expression recognition in survivors of pediatric brain tumors. *Journal of Pediatric Psychology, 33,* 1142–1152.

Bos, K., Fox, N., Zeanah, C., & Nelson, C. (2009). Effects of early psychosocial deprivation on the development of memory and executive function. *Frontiers in Behavioral Neuroscience, 3,* 16.

Bowlby, J. (1962). *Deprivation of maternal care.* Geneva: World Health Organization.

Boydell, K., Stasiulis, E., Greenberg, M., Greenberg, C., & Spiegler, B. (2008). I'll show them: the social construction of (in)competence in survivors of childhood brain tumors. *Journal of Pediatric Oncology Nursing, 25,* 164–174.

Brower, M. C., Rollins, N., & Roach, E. S. (1996). Basal ganglia and thalamic infarction in children. *Archives of Neurology, 53,* 1252–1256.

Bulotsky, R., Fantuzzo, J., & McDermott, P. (2008). An investigation of classroom situational dimensions of emotional and behavioral adjustment and cognitive and social outcomes for Head Start children. *Developmental Psychology, 44,* 139–154.

Cacioppo, J. (2002). Social neuroscience: Understanding the pieces fosters understanding of the whole and vice versa. *American Psychologist, 57,* 819–831.

Carey, M., Barakat, L., Foley, B., Gyato, K., & Phillips, P. (2001). Neuropsychological functioning and social functioning of survivors of pediatric brain tumors: Evidence of nonverbal learning disability. *Child Neuropsychology, 7,* 265–272.

Catroppa, C., & Anderson, V. (2005). A prospective study of the recovery of attention from acute to 2 years post pediatric traumatic brain injury. *Journal of the International Neuropsychological Society, 11,* 84–98.

Cattelani, R., Lombardi R., Brianti, R., & Mazzucchi, A. (1998). Traumatic brain injury in childhood: Intellectual, behavioural and social outcome in adulthood. *Brain Injury, 12,* 283–296.

Crick, N. R., & Dodge, K. A. (1994). A review and reformulation of social information processing mechanisms in children's social adjustment. *Psychological Bulletin, 115,* 74–101.

Crowe, L. M., Beauchamp, M. H., Catroppa, C., & Anderson, V. (2011). Social function assessment tools for children and adolescents: A systematic review from 1988 to 2010. *Clinical Psychology Review, 31,* 767–785.

Dennis, M., Guger, S., Roncadin, C., Barnes, M., & Schachar, R. (2001). Attentional–inhibitory control and social-behavioral regulation after childhood closed head injury: Do biological, developmental, and recovery variables predict outcome? *Journal of the International Neuropsychological Society, 7,* 683–692.

De Schryver, E. L., Kappelle, L. J., Jennekens-Schinkel, A., & Boudewyn Peters, A. C. (2000). Prognosis of ischemic stroke in childhood: A long-term follow-up study. *Developmental Medicine and Child Neurology, 42*(5), 313–318.

deVeber, G. A., MacGregor, D., Curtis, R., & Mayank, S. (2000). Neurologic outcome in survivors of childhood arterial ischemic stroke and sinovenous thrombosis. *Journal of Child Neurology, 15,* 316–324.

Didus, E., Anderson, V., & Catroppa, C. (1999). The development of pragmatic communication skills in head injured children. *Pediatric Rehabilitation, 3,* 177–186.

Dooley, J., Anderson, V., Hemphill, S., & Ohan, J. (2008). Aggression after pediatric traumatic brain injury: A theoretical approach. *Brain Injury, 22*(11), 836–846.

Everts, R., Pavlovic, J., Kaufmann, F., Uhlenberg, B., Seidel, U., Nedeltchev, K., et al. (2008). Cognitive functioning, behavior, and quality of life after stroke in childhood. *Child Neuropsychology, 14*(4), 323–338.

Fletcher, J. M., Ewing-Cobbs, L., Miner, M. E., Levin, H. S., & Eisenberg, H. M. (1990). Behavioral changes after closed head injury in children. *Journal of Consulting and Clinical Psychology, 58*, 93–98.

Foley, B., Barakat, L. P., Herman-Liu, A., Radcliffe, J., & Molloy, P. (2000). The impact of childhood hypothalamic/chiasmatic brain tumors on child adjustment and family function. *Child Health Care, 29*, 209–223.

Ganesalingam, K., Sanson, A., Anderson, V., & Yeates, K. O. (2006). Self-regulation and social and behavioral functioning following childhood traumatic brain injury. *Journal of the International Neuropsychological Society, 12*, 609–621.

Ganesalingam, K., Sanson, A., Anderson, V., & Yeates, K. O. (2007). Self-regulation as a mediator of the effects of childhood traumatic brain injury on social and behavioral functioning. *Journal of the International Neuropsychological Society, 13*, 298–311.

Ganesalingam, K., Yeates, K. O., Sanson, A., & Anderson, V. (2007). Social problem-solving skills following childhood traumatic brain injury and its association with self-regulation and social and behavioral functioning. *Journal of Neuropsychology, 1*, 149–170.

Ganesalingam, K., Yeates, K., Taylor, H., Wade, N., Stancin, T., & Wade, S. (2011). Executive functions and social competence in young children 6 months following traumatic brain injury. *Neuropsychology, 25*, 466–476.

Ganesan, V., Hogan, A., Shack, N., Gordon, A., Isaacs, E., & Kirkham, F. J. (2000). Outcome after ischaemic stroke in childhood. *Developmental Medicine and Child Neurology, 42*, 455–461.

Gomes, A., Rinehart, N., & Anderson, V. (2011). *A review of psychosocial outcomes following stroke in childhood*. Manuscript submitted for publication.

Goodman, R., & Graham, P. (1996). Psychiatric problems in children with hemiplegia: Cross-sectional epidemiological survey. *British Medical Journal, 312*, 1065–1069.

Gordon, A. L., Ganesan, V., Towell, A., & Kirkham, F. J. (2002). Functional outcome following stroke in children. *Journal of Child Neurology, 17*(6), 429–434.

Greenham, M., Spencer-Smith, M. M., Anderson, P. J., Coleman, L., & Anderson, V. A. (2010). Social functioning in children with brain insult. *Frontiers in Human Neuroscience, 4*, 22, doi:10.3389/fnhum.2010.00022.

Guralnick, M. (1999). Family and child influences on the peer-related social competence of young children with developmental delays. *Mental Retardation and Developmental Disabilities Research Reviews, 5*, 21–29.

Hanten, G., Wilde, E. A., Menefee, D. S., Li, X., Lane, S., Vasquez, C., et al. (2008). Correlates of social problem solving during the first year after traumatic brain injury in children. *Neuropsychology, 22*, 357–370.

Haslam, C., Holme, A., Haslam, A., Iyer, A., Jetten, J., & Williams, W. (2008). Maintaining group memberships: Social identity continuity predicts well-being after stroke. *Neuropsychological Rehabilitation, 15*(5–6), 671–691.

Holmquist, L., & Scott, J. (2003). Treatment, age and time-related predictors of behavioral outcome in pediatric brain tumor survivors. *Journal of Clinical Psychology and Medical Settings, 9*, 315–321.

Hurvitz, E., Warschausky, S., Berg, M., & Tsai, S. (2004). Long-term functional outcome of pediatric stroke survivors. *Topics in Stroke Rehabilitation, 11*(2), 51–59.

Janusz, J. A., Kirkwood, M. W., Yeates, K. O., & Taylor, G. (2002). Social problem-solving skills in children with traumatic brain injury: Long-term outcomes and prediction of social competence. *Child Neuropsychology, 8*, 179–194.

Kaufman, J., Birmaher, B., Brent, D., Rao, U., Flynn, C., Moreci, P., et al. (1997). Schedule for

Affective Disorders and Schizophrenia for School-Age Children—Present and Lifetime Version (K-SADS-PL): Initial reliability and validity data. *Journal of the American Academy of Child and Adolescent Psychiatry, 36,* 980–988.

Kullgren, K. (2003). Risk factors associated with long-term social and behavioural problems among children with brain tumors. *Journal of Psychosocial Oncology, 21,* 73–87.

Levin, H. S., Hanten, G., & Li, X. (2009). The relation of cognitive control to social outcome after paediatric TBI: Implications for interventions. *Developmental Neurorehabilitation, 12,* 320–329.

Long, B., Spencer-Smith, M., Jacobs, R., Mackay, M., Leventer, R., Barnes, C., et al. (2011). Executive function following child stroke: The impact of lesion location. *Journal of Child Neurology, 26*(3), 279–287.

Mabbott, D., Spiegler, S., Greenberg, M., Rutka, J., Hyder, D., & Bouffet, E. (2005). Serial evaluation of academic and behavioral outcome after treatment with cranial radiation in childhood. *Journal of Clinical Oncology, 23,* 2256–2263.

Masten, A., Hubbard, J., Gest, S., Tellegen, A., Garmezy, N., & Ramirez, M. (1999). Competence in the context of adversity: Pathways to resilience and maladaptation from childhood to late adolescence. *Development and Psychopathology, 11,* 143–169.

Max, J., Bruce, M., Keaatley, B., & Delis, D. (2010). Pediatric stroke: Plasticity, vulnerability and age at lesion onset. *Journal of Neuropsychiatry and Clinical Neuroscience, 22,* 30–39.

Max, J. E., Fox, P. T., Lancaster, J. L., Kochunov, P., Mathews, K., Manes, F. F., et al. (2002). Psychiatric disorders after childhood stroke. *Journal of the American Academy of Child and Adolescent Psychiatry, 41,* 555–562.

Max, J. E., Koele, S. L., Lindgren, S. D., Robin, D. A., Smith, W. L., Jr., Sato, Y., et al. (1998). Adaptive functioning following traumatic brain injury and orthopedic injury: A controlled study. *Archives of Physical Medicine and Rehabilitation, 79,* 893–899.

Max, J. E., Mathews, K. K., Manes, F. F., Robertson, B. A., Fox, P. T., & Lancaster, J. L. (2005). Prefrontal and executive attention netweork lesions and the development of attention-deficit/hyperactivity symptomatology. *Journal of the American Academy of Child and Adolescent Psychiatry, 44*(5), 443–450.

McCarthy, M. C., Clarke, N. E., Lin Ting, C., Conroy, R., Anderson, V., & Heath, J. A. (2010). Prevalence and predictors of parental grief and depression after the death of a child from cancer. *Journal of Palliative Medicine, 13*(11), 1321–1326.

Middleton, J. A. (2001). Practitioner review: Psychological sequelae of head injury in children and adolescents. *Journal of Child Psychology and Psychiatry, 42*(2), 165–180.

Mosch, C., Max, J., & Tranel, D. (2005). A matched lesion analysis of childhood versus adult-onset brain injury due to unilateral stroke: Another perspective on neural plasticity and recovery of social functioning. *Cognitive and Behavioral Neurology, 18*(1), 5–17.

Muscara, F., Catroppa, C., & Anderson, V. (2008). Social problem-solving skills as a mediator between executive function and long-term social outcome following paediatric traumatic brain injury. *Journal of Neuropsychology, 2,* 445–461.

Nassau, J. H., & Drotar, D. (1995). Social competence in children with IDDM and asthma: Child, teacher, and parent reports of children's social adjustment, social performance, and social skills. *Journal of Pediatric Psychology, 20*(2), 187–204.

Nass, R. D., & Trauner, D. (2004). Social and affective impariments are important recovery after acquired stroke in childhood. *CNS Spectrums, 9*(6), 420–434.

Papero, P. H., Prigatano, G. P., Snyder, H. M., & Johnson, D. L. (1993). Children's adaptive behavioural competence after head injury. *Neuropsychological Rehabilitation, 3,* 321–340.

Parker, J. G., & Asher, S. R. (1993). Friendship and friendship quality in middle childhood: Links with peer group acceptance and feelings of loneliness and social dissatisfaction. *Developmental Psychology, 29,* 611–621.

Pettit, G. S., Dodge, K. A., & Brown, M. M. (1988). Early family experience, social problem solving patterns, and children's social competence. *Child Development, 59,* 107–120.

Poggi, G., Liscio, M., Adduci, A., Galbiati, S., Massimino, M., Sommovigo, M., et al. (2005). Psychological and adjustment problems due to acquired brain lesions in childhood: A comparison between post-traumatic patients and brain tumor survivors. *Brain Injury, 19,* 777–785.

Poretti, A., Grotzer, M., Ribi, K., Schonle, E., & Bolthauser, E. (2004). Outcome of craniopharyngioma in children: Long-term complications and quality of life. *Developmental Medicine and Child Neurology, 46,* 220–229.

Prigatano, G. P., & Gupta, S. (2006). Friends after traumatic brain injury in children. *Journal of Head Trauma Rehabilitation, 21,* 505–513.

Raizada, R., & Kishiyama, M. (2010). Effects of socioeconomic status on brain development, and how cognitive neuroscience may contribute to levelling the playing field. *Frontiers in Human Neuroscience, 4,* 1–11, 3.

Rholes, W., Newman, L., & Ruble, D. (1990). Understanding self and other: Developmental and motivational aspects of perceiving persons in terms of invariant dispositions. In E. Higgins & R. M. Sorentino (Eds.), *Handbook of motivation and cognition: Vol. 2. Foundations of social behavior* (pp. 369–407). New York: Guilford Press.

Roach, E. S. (2000). Stroke in children. *Current Treatment Options in Neurology, 2*(4), 295–304.

Sands, S., Milner, J., Goldberg, J., Mukhi, V., Moliterno, J., Maxfield, C., et al. (2005). Quality of life and behavioral follow-up study of pediatric survivors of craniopharyngioma. *Journal of Neurosurgery, 103,* 302–311.

Saxe, R., Carey, S., & Kanwisher, N. (2004). Understanding other minds: Linking developmental psychology and functional neuroimaging. *Annual Review of Psychology, 55,* 87–124.

Schulte, F., & Barrera, M. (2010). Social competence in childhood brain tumor survivors: A comprehensive review. *Supportive Care in Cancer, 18,* 1499–1513.

Schulte, F., Bouffet, E., Janzen, L., Hamilton, J., & Barrera, M. (2010). Body weight, social competence and cognitive functioning in survivors of childhood brain tumors. *Pediatric Blood and Cancer, 55,* 532–539.

Sofranas, M., Ichord, R. N., Fullerton, H. J., Lynch, J. K., Massicotee, P., Willan, A. R., et al. (2006). Pediatric stroke initiatives and preliminary studies: What is known and what is needed? *Pediatric Neurology, 34*(6), 439–445.

Sparrow, S., Balla, D. A., & Cicchetti, D. V. (1984). *Vineland Adaptive Behavior Scales.* Circle Pines, MN: American Guidance Service.

Steinlin, M., Roellin, K., & Schroth, G. (2004). Long-term follow-up after stroke in childhood. *European Journal of Pediatrics, 163,* 245–250.

Tate, R., Hodgkinson, A., Veerabangsa, A., & Maggiotto, S. (1999). Measuring psychosocial recovery after traumatic brain injury: Psychometric properties of a new scale. *Journal of Head Trauma Rehabilitation, 14,* 543–557.

Tonks, J., Williams, W. H., Framton, I., Yates, P., & Slater, A. (2007). Reading emotions after child brain injury: A comparison between children with brain injury and non-injured controls. *Brain Injury, 21,* 731–739.

Tonks, J., Williams, H., Yates, P., & Slater, A. (2011). Cognitive correlates of psychosocial outcome following traumatic brain injury in early childhood: Comparisons between groups of children aged under and over 10 years of age. *Clinical Child Psychology and Psychiatry, 16,* 185–194.

Trauner, D., Panyard-Davis, J., & Ballantyne, A. (1996). Behavioral differences in school age children after perinatal stroke. *Assessment, 3,* 265–276.

Turkstra, L. S., Dixon, T. M., & Baker, K. K. (2004). Theory of mind and social beliefs in adolescents with traumatic brain injury. *NeuroRehabilitation, 19,* 245–256.

Turkstra, L. S., Williams, W. H., Tonks, J., & Frampton, I. (2008). Measuring social cognition in adolescents: Implications for students with TBI returning to school. *NeuroRehabilitation 23*, 501–509.

Upton, P., & Eiser, C. (2006). School experiences after treatment for a brain tumor. *Child: Care, Health and Development. 32*, 9–17.

Vance, Y., Eiser, C., & Home, B. (2004). Parents' views of the impact of childhood brain tumors and their treatment on young people's social and family functioning. *Clinical Child Psychology and Psychiatry, 9*, 271–288.

Van Overwalle, F. (2009). Social cognition and the brain: A meta-analysis. *Human Brain Mapping, 30*, 829–858.

Varni, J., Seid, M., & Rode, C. (1999). The PedsQL: Measurement model for pediatric quality of life inventory. *Medical Care, 37*, 126–139.

Walz, N., Yeates, K., Taylor, H., Stancin, T., & Wade, S (2010). Theory of mind skills 1 year after traumatic brain injury in 6– to 8–year-old children. *Journal of Neuropsychology, 4*, 181–195.

Warschausky, S., Cohen, E. H., Parker, J. G., Levendosky, A. A., & Okun, A. (1997). Social problem-solving skills of children with traumatic brain injury. *Pediatric Rehabilitation, 1*, 77–81.

Westmacott, R., MacGregor, D., Askalan, R., & deVeber, G. (2009). Late emergence of cognitive deficits after unilateral neonatal stroke. *Stroke, 40*(6), 2012–2019.

Wirt, R. D., Lachar, D., Klinedinst, J. E., Seat, P. D., & Broen, W. E. (1977). *Multidimensional evaluation of child personality: A manual for the Personality Inventory for Children*. Los Angeles: Western Psychological Services.

World Health Organization. (2001). *International classification of functioning, disability and health*. Geneva: Author.

Yeates, K. O., Bigler, E. D., Dennis, M., Gerhardt, C.A., Rubin, K.H., Stancin, T., et al. (2007). Social outcomes in childhood brain disorder: A heuristic integration of social neuroscience and developmental psychology. *Psychological Bulletin, 133*, 535–556.

Yeates, K. O., Schultz, L. H., & Selman, R. L. (1990). Bridging the gaps in child–clinical assessment: Toward the application of social-cognitive development theory. *Clinical Psychology Review, 10*, 657–588.

Yeates, K. O., Swift, E., Taylor, H. G., Wade, S. L., Drotar, D., Stancin, T., et al. (2004). Short- and long-term social outcomes following pediatric traumatic brain injury. *Journal of the International Neuropsychological Society, 10*, 412–426.

Yeates, K. O., Taylor, H., Walz, N., Stancin, T., & Wade, S. (2010). The family environment as a moderator of psychosocial outcome following traumatic brain injury in young children. *Neuropsychology, 24*, 345–356.

CHAPTER 12

Social Development and Traumatic Brain Injury in Children and Adolescents

Gerri Hanten, Harvey S. Levin, Mary R. Newsome, and Randall S. Scheibel

The socio-cognitive integration of abilities model (SOCIAL; Beauchamp & Anderson, 2010; see also Anderson & Beauchamp, Chapter 1, this volume) is aimed at creating a framework within which the development of social skills may be understood, and, by extension, the effect of injury on this development. The ultimate goal is to develop an assessment framework that allows for a complete clinical evaluation. SOCIAL encompasses two classes of mediators that have the capacity to influence development of social function, and that interact bidirectionally. The first class of mediators consists of internal and external factors. Internal factors include genetic determinants, the integrity of brain structure and function, and other interior attributes that can influence the way a person interacts with others. External factors include family environment, socioeconomic status (SES), and culture. Implicit in the role of environment are experiences, such as traumatic experiences, that may alter a child's viewpoint and even neurophysiology (Hayes, Jenkins, & Lyeth, 1992). The second class of mediators, cognitive and affective factors, comprises attention/executive function, communication, and socio-emotional skills.

The model is comprehensive, is broadly supported by a number of studies, and is consistent with the emerging theoretical construct of the *social brain*. This construct attributes complex human social adaptations to development of regions of the brain specialized to process information about people (Adolphs, 2001, 2009; Blakemore, 2008). Such brain regions include (but are not limited to) the ventromedial prefrontal cortex, medial prefrontal cortex, orbitofrontal cortex, frontal pole, inferior frontal gyrus, superior temporal sulcus, temporoparietal junction, amygdala, anterior cingulate cortex, fusiform face area, and the white matter structures that connect them.

In this chapter, we summarize several studies (including a number from our laboratory) that address the influence of traumatic damage, an external factor, on the development of the brain. Our specific focus is on how traumatic brain injury (TBI) affects the capacity for accurately processing intentions and emotions of oneself and others (Frith & Frith, 2003), as exemplified in theory-of-mind (ToM) tasks. The ability to understand the intentions, states, and emotions of others necessarily relies on the mechanism of *metarepresentation*—that is, the ability to form a representation of another person's thoughts or feelings.

After a brief overview of social function after TBI and its relation to the brain, we turn to a discussion of behavioral studies of the cognitive skills that may account for impaired social function after TBI—specifically, skills that rely on the mechanism of metarepresentation. We present studies that address the disruption of metarepresentation after childhood TBI, including ToM, social perspective taking (taking the perspective of another person in a social situation) and mental state attribution (attributing thoughts, feelings, intentions, etc., to another that may be different from one's own). We also present studies on social problem solving, a multilayered area of social competence that rests on the foundations of social perspective taking and social attribution making. Although the precise relation of these skills to one another in terms of development is not clearly understood (and is beyond the scope of this chapter), it is clear that to complete any of these processes successfully, an initial step must be to form a representation of another's mental state distinct from one's own—in other words, to engage in metarepresentation. Finally, we present functional and structural neuroimaging data from studies of the brain regions and structures that support these important socio-cognitive skills.

Metarepresentation after TBI: Overview

After TBI, children and youth may show dysfunction in several domains of functioning, including cognitive and social domains. In fact, poor social outcome, as indicated by measures of social adjustment, is a common consequence of moderate to severe TBI and often persists despite relatively spared cognition (Glang, Todis, Cooley, Wells & Voss, 1997; Taylor et al., 2002). Several studies have attempted to identify the key mechanisms responsible for negative social outcome after childhood TBI and have revealed specific aspects of impaired social processing (Dennis, Guger, Roncadin, Barnes, & Schacher, 2001; Turkstra, Dixon, & Baker, 2004; Turkstra, McDonald, & DePompei, 2001; Yeates, Schultz & Selman, 1991).

Longitudinal studies of children with TBI suggest that impairment in the social domain does not diminish over time (Andrews, Rose, & Johnson, 1998; Bohnert, Parker, & Warschausky, 1997; Dennis, Guger, et al., 2001; Hanten et al., 2008; Janusz, Kirkwood, Yeates, & Taylor, 2002; Yeates et al., 2004), though longitudinal studies of sufficient length to determine whether the resulting social deficits are due to slowed or failure in development or due to direct disruption of function are relatively sparse.

Frontal and temporal regions that subserve social cognition are the most commonly injured areas of the brain following childhood TBI (Graham, Gennerelli, & McIntosh, 2002; Levin et al., 1997). The vulnerability of prefrontal and temporal regions and disruption of related neural circuitry has been implicated in these patients' reduced awareness of their own mental state, impaired processing of the mental states of others, and inability to make social inferences (Cicerone, Levin, Malec, Stuss, & Whyte, 2006).

Behavioral Studies of Metarepresentation after TBI

ToM Studies

ToM, sometime referred to as *mentalizing*, is the ability to attribute mental states (feelings, beliefs, intentions, goals, etc.) to oneself and to others, as well as the ability to understand that others may have mental states different from one's own (Premack & Woodruff, 1978).

Although there is ongoing discussion about what exactly constitutes a ToM task, the essence of ToM is the ability to engage in metarepresentation of others' mental states. Several levels of complexity are assigned to ToM tasks. A first-order level of complexity entails an observer creating a representation of another's mental state (e.g., "Joe is happy"). At a second-order level, awareness of the observer's own representation is applied to the situation (e.g., "I think Joe is happy"). Finally, at the most complex level, the other person's representation of the observer's mental state is added (e.g., "Joe thinks that I think he's happy"). These skills necessarily require the ability to take a perspective other than one's own, and are integral in negotiating favorable solutions to social problems and conflicts.

Adult studies of TBI have reported impairments on tests of ToM when forming representations of others' thoughts or emotional states from stories, pictures, and animations is required (e.g., Henry, Phillips, Crawford, Ietswaart, & Summers, 2006), and when identification of social faux pas is requested (Channon & Crawford, 2010; Geraci, Surian, Ferraro, & Cantagallo, 2010; Muller et al., 2010). Findings generally suggest relative sparing (as compared to uninjured controls) on simple first-order ToM tasks, but impairment that increases with the complexity of the task (please see Martin-Rodriguez & Leon-Carrion, 2010, for a meta-analysis of ToM studies).

In the few studies of children with TBI, a similar pattern has emerged: relative sparing on first-order ToM tasks, but deficits on tasks that require second- or third-order metarepresentation. For example, Snodgrass and Knott (2006) reported that children with moderate to severe TBI showed performance similar to that of control subjects on a simple first-order ToM task that required understanding of a character's false belief, but demonstrated impairment on tests of higher-level ToM. Similarly, Dennis, Purvis, Barnes, Wilkinson, and Winner (2001) examined processing of irony and deceptive praise in children with mild to severe TBI in comparison to a cohort of children with typical development (TD) and without injury, and found no group differences on first order tasks. However, they also found an interaction of group with age on performance, such that older children with severe TBI were significantly impaired on second order intentionality tasks, but group differences for younger children were not significant. Recently, Dennis, Agostino, Roncadin, and Levin (2009) reported that ToM deficits in children with TBI were related to performance on a cognitive inhibition task; this opens the possibility that impairments in social cognition following brain injury may be due, at least in part, to impairments in more general executive functions. Although it is not clear to what extent problems with ToM following TBI are domain-specific for social cognition or whether they reflect more general executive cognitive deficits, it is clear that the adept handling of mental representations is an area of concern for children after TBI.

Studies of Social Perspective Taking

The ability to represent and reflect upon one's own knowledge and states (first-order metarepresentation) is distinct from the ability to represent and reflect upon others' knowledge and states (second- and third-order metarepresentation). Furthermore, the ability to adopt another's perspective is central for effective understanding of others' intentions, beliefs, and feelings, which guide social actions. Although there is some variation within reports in the field, the consensus is that by about the age of 8 years, children with TD realize that their feelings or thoughts may differ from those of others, with self-awareness continuing to develop throughout adolescence (Damon & Hart, 1982). By 9 years of age, children appreciate that traits can be stable over time (Damon & Hart, 1988). In uninjured populations, preadolescents are reported to be less effective in perspective taking than are adolescents, and both preadolescents and adolescents demostrate less effective perspective taking than adults (Choudhury, Blakemore & Charman, 2006).

Studies of Mental State Attributions

The ability to accurately identify or attribute a mental state to another is distinct from the ability to take that person's perspective or to recognize that he or she may have a mental state different from one's own. In one of the few studies of mental state attributions in children after TBI, Levin et al. (2011) compared the inclination of children with TBI and children with orthopedic injury (OI) to make spontaneous mental state attributions in response to a video presentation of moving triangles. The movement of the triangles varied by three conditions that differed in degree of apparent social interaction (Abell, Happé, & Frith, 2000). This approach exploits the finding that observers assign social motivation or response to the movements of interacting shapes presented in animated, nonverbal sequences (Heider & Simmel, 1944). Interpretation of these movements as representing actions of persons has been attributed to the motion of the stimuli rather than to their shape or background (Castelli, Happé, Frith, & Frith, 2000; Heider & Simmel, 1944).

The Levin et al. (2011) study was part of a prospective investigation of children ages 7–17 years who had moderate to severe TBI and a comparison group of children with OI, both of whom were studied 3 months after injury. Subjects viewed animations of a large red triangle and a small blue triangle within a white-framed background on a computer screen. The three movement conditions included (1) random movement (RM), during which the triangles moved around the screen in a random fashion (e.g., floating) but did not interact; (2) goal-directed (GD), during which the triangles moved purposefully and interacted reciprocally (e.g., dancing); and (3) ToM, in which one triangle appeared to react to the other's mental state (e.g., one triangle coaxing the other out of a box). The conditions tested a child's ability to interpret social movement of increasing interaction. In addition to the degree of interaction, level of cueing was varied. In the uncued condition, the child was merely asked to describe what happened on the screen for each of the three conditions (RM, GD, ToM). In the cued condition, the child was again presented with animations in each of the three conditions, but this time was prompted to describe the scenarios as if the triangles were humans. Scoring included accuracy of description, in terms of the movements of the triangles, and mental attributions, which reflected the degree to which mentalizing (i.e., ToM) language was used to describe the triangles' movements.

Levin et al. (2011) found that children with TBI performed similarly to children with OI in the uncued condition. However, in the cued condition, the children with OI were significantly more likely than those with TBI to use mentalizing terms when describing the actions of the triangles only in the ToM condition, as can be seen in Figure 12.1. So, although it appeared that children with TBI were just as likely to engage spontaneously in metarepresentations as children with OI (in the uncued condition), the children with TBI were less likely to be successful when specifically cued to do so (in the cued condition). These findings are consistent with a deficit after TBI arising in more complex, but not simple, levels of social representation. Surprisingly, within the group with TBI, there was no significant relation of Glasgow Coma Scale score to the mental attribution score. With all of the patients having sustained a moderate to severe TBI, it is possible that this finding may be the result of a restricted range of TBI severity.

Studies of Social Problem Solving

To this point, we have discussed studies of the social processing skills of relying on metarepresentation in isolation. Social problem solving is an example of the *in situ* examination of social skills—in this case, the ability to negotiate a favorable outcome to a social conflict with another. It relies on metarepresentation, but only as one component in a complex process. Many studies of social problem solving use narratives to present the

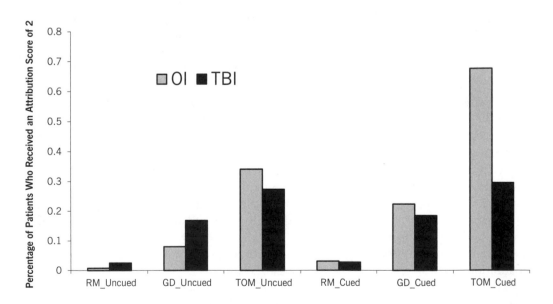

FIGURE 12.1. Performance on the triangles task, by group (TBI or OI), cueing condition (cued or uncued for actions as humans), and movement condition (random, RM; goal directed, GC; theory of mind, ToM). Bars represent the proportion of patients with an attribution score of 2 (highest score for mental state attribution). Based on Levin et al. (2011).

social problem to be solved, and responses are generally in the form of spoken language. Therefore, it may be hypothesized that language comprehension and expression abilities may be involved, as well as memory for spoken language.

Yeates and colleagues studied social problem solving in children with TD and in children with TBI, using a task called Interpersonal Negotiations Strategies (INS; Yeates et al., 1991). The original INS model was based on two theoretically distinct approaches to social development: the structural approach (Kohlberg, 1969), influenced by Piagetian developmental theory, and the functional approach of Dodge (1985), influenced by information-processing theory. Reliability and validity have been shown for key elements of the INS model, with developmental studies demonstrating relationships between children's age or intellectual development and performance on the INS (Beardslee, Schultz, & Selman, 1987; Yeates et al., 1991).

Yeates et al. (2004) used the INS to study social problem solving in 109 children with moderate to severe TBI. Administration of the INS requires the narrative presentation of conflict scenarios that occur between two individuals, after which the participant provides a solution to the conflict based on four problem-solving steps that are included in a semistructured interview: defining the problem (DP), generating solutions (GS), selecting the best solution (SS), and evaluating the likely outcome (EO). Responses are rated on the maturational level of the proposed solutions—that is, how well the proposed solution preserves the relationship between the two people depicted in the conflict. Using the narrative INS, Janusz et al. (2002) reported that children with TBI were impaired relative to children with OI on social problem solving, but only on the SS and EO steps. Using the narrative form of the INS to assess social problem solving longitudinally in children after TBI and in children with OI, Hanten et al. (2008) replicated the findings of Janusz et al. (2002) that children with TBI were significantly impaired on INS performance relative to children with OI. The authors also demonstrated that although the children with TBI showed improvement over time, the degree of improvement was consistent with the presumed developmentally related change over time of the children with OI, rather than with recovery from injury. In addition, Hanten et al. (2008) found that performance on the INS was related to performance on tasks of memory and language, consistent with the findings of Yeates et al. (2004). These studies establish the presence of social problem-solving deficits after TBI in children, but there is room for discussion regarding the ecological validity of the tests used.

Recent advances in virtual reality technology have provided an opportunity to combine the scientific rigor of neuropsychological testing with realism that more closely mimics activities of daily living to achieve a more ecologically valid, and perhaps more sensitive, assessment of cognitive and social skills. In fact, a recent trend in socio-cognitive research has been to make simulating the social environment an important objective, given the need to balance experimental control with ecological validity.

With this in mind, Hanten et al. (2011) created a task called Virtual Reality Social Problem Solving (VR-SPS), which takes advantage of the mundane realism (the extent to which a test of a skill matches the real-world counterpart) of virtual reality technology as well as the proven assessment value of the INS interview (Yeates et al., 1991). In a pilot study, the VR-SPS was used to assess social problem solving in 28 youth ages 12–19 years (15 with moderate to severe TBI, and 13 with TD and no injury). Adolescents viewed animated scenarios depicting social conflict in a virtual microworld environment from an avatar's viewpoint, and were questioned on the four problem-solving steps (DP, GS, SS, and EO). Scenarios always involved four people, of whom two were in conflict, which was depicted via naturalistic dialogue. A screen shot of one scenario is shown in Figure 12.2. To explore the possibility that cognitive processing load would have a disproportionate effect on youth with TBI in the context of social problem solving as compared to youth with TD, processing load was varied in each conflict scenario in terms of number of characters and amount of information presented. In Condition A there were two speakers (of four people visible in the scenario; see Figure 12.2), both of whom contributed only information relevant to the conflict; Condition B had two speakers who contributed both relevant and irrelevant information; and in Condition C, all four people spoke and contributed relevant or irrelevant information. Male and female participants received gender-matched scenarios in which the characters involved in the argument were the same gender as the participant. Scoring was based on a developmental scale on which responses were judged as impulsive, unilateral, reciprocal, or collaborative, in order of increasing score. Each participant received an overall Summary Score for each condition (A, B, or C). In addition, separate scores for each of the four problem-solving steps (DP, GS, SS, and EO) were generated for each condition.

Adolescents with TBI showed significant impairment on the VR-SPS Summary Score in Condition A, in Condition B, and in Condition C, as can be seen in Figure 12.3. Effect sizes (Cohen's d) were large for all three conditions (A = 1.40, B = 0.96, C = 1.23), which did not provide strong support for the idea that processing load would disproportionately affect youth with TBI in this setting. In regard to the problem-solving steps involved, significant group differences were strongest and most consistent in this naturalistic version of social problem solving for DP, which, although the first and simplest step, involves metarepresentation in that to understand the problem fully (and therefore obtain a high score), a participant must be able to take the socio-emotional perspective of each character to understand the conflict in terms of each of the characters' desires. There were also

FIGURE 12.2. Screen shot of a VR-SPS scenario depicting social conflict as dialogue in a microworld environment. From Hanten et al. (2011). Copyright 2011 by Elsevier. Reprinted by permission.

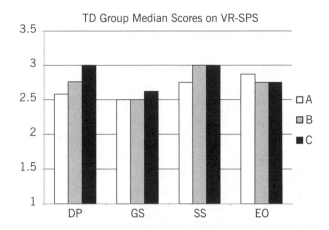

FIGURE 12.3. VR-SPS performance by group, condition, and problem-solving step. Condition A = two speakers, no irrelevant information; Condition B = two speakers, irrelevant information; and Condition C = four speakers, irrelevant information. The problem-solving steps (horizontal axis) are DP, defining problem; GS, generating solutions; SS, selecting best solution; and EO, evaluating outcome. From Hanten et al. (2011). Copyright 2011 by Elsevier. Reprinted by permission.

significant group differences in EO, which requires the ability to foresee the consequences of given actions and to understand their effect on others.

In previously reported studies of social problem solving, using the original narrative form of the INS, injury group differences were more apparent at the more complex stages of social problem solving (GS, SS, EO) than at the initial stage of DP (Hanten et al., 2008; Yeates et al., 2004). With the VR-SPS version, group differences were apparent even at the DP stage, suggesting that the naturalistic approach may be more sensitive to deficits in identifying and defining the social conflict. This is consistent with the growing body of research suggesting that childhood TBI may result in impairments in metarepresentation.

Brain Studies of Metarepresentation after TBI

Functional Neuroimaging

Metarepresentation has been linked to regions of the brain with some specificity in studies using functional brain imaging. For example, using functional magnetic resonance imaging (fMRI), Saxe and Kanwisher (2003) demonstrated that the temporoparietal junction has a selective role in ToM tasks, specific to reasoning about a person's mental state rather than just responding to the presence of another person. In a recent study, Lewis, Rzale, Brown, Roberts, and Dunbar (2011) linked both competence on ToM tasks and social network size (number of social contacts over a given period) to the ventromedial prefrontal cortex and the medial orbitofrontal cortex.

fMRI studies of perspective taking have implicated several brain regions, including the medial prefrontal cortex (mPFC). In an fMRI task, Pfeifer, Lieberman, and Dapretto (2007) reported activation in the mPFC in healthy children and adults when they engaged in first-order self-judgments, such as "I am popular" and "I am a good speller." Activation was greater in children than in adults, suggesting that neural networks differentiating the self from others are in place by about 11 years of age and become more focal with age. In contrast, posterior brain regions, including the precuneus and posterior cingulate, were activated during judgments of another.

D'Argembeau et al. (2007) also reported activation in the mPFC when uninjured adults judged from a first- or third-person perspective whether they or a person well known to them displayed a certain personality trait—a finding consistent with Schmitz and Johnson's (2007) review of dorsal–ventral mPFC regions' involvement in self-appraisal. In this study, trait judgment activated the precuneus, whereas perspective taking activated the temporal pole, the inferior parietal lobe, and the precuneus. MPFC was again implicated when participants had to think about themselves from another person's perspective (i.e., a third-order metarepresentation).

Populations with TBI commonly show deficits in evaluating their own postinjury abilities (e.g., Bach & David, 2006); therefore, it is plausible that self-awareness during perspective taking may be particularly affected in adolescents after TBI. In a recent study, we (Newsome et al., 2010) examined brain activation in nine adolescents with moderate to severe TBI and nine age- and SES-matched typically developing controls while they judged traits from different perspectives in themselves (self) and a well-known other person, such as their best friend or a family member whom they nominated (other). The study was a 2 × 2 design, in which the two independent variables were target (i.e., the person in whom the trait was evaluated—self or other), and perspective (i.e., the person whose viewpoint was queried—self or other). Participants saw statements about themselves or their other and had to endorse or reject them, which allowed for the elucidation of the brain regions associated with lower- and higher-order metarepresentation. Examples of a set of stimuli for a participant making judgments about the trait of being "cheerful" for himself and for a well-known other named Sammy would be as follows:

1. Target = self, perspective = self: "You think you are cheerful."
2. Target = other, perspective = self: "You think Sammy is cheerful."
3. Target = other, perspective = other: "(You think that) Sammy thinks he is cheerful."
4. Target = self, perspective = other: "(You think that) Sammy thinks you are cheerful."

It was hypothesized that disruption of circuitry involving mPFC, medial parietal cortex, and temporal regions would result in more extensive activation of these regions in adolescents with moderate to severe TBI, to compensate for reduced neural resources (Newsome et al., 2010).

Reaction times for judgments about the self were faster than those for judgments about the nominated other, and times for judgments from one's own point of view (self) were faster than those for judgments for which the participant was asked to report from the other's point of view. These patterns were the same for both the group with TBI and the control group.

No regions showed greater activation in the control group than in the group with TBI for the target × perspective interaction, but activation was greater for the group with TBI than for controls in several left-hemispheric regions (posterior cingulate, cuneus, lingual and parahippocampal gyri) and in the thalamus bilaterally as shown in Plate 12.1 on the color insert. This group difference is suggestive of an alteration in brain regions implicated in third-order metarepresentation ("You think that Sammy thinks you are cheerful") after TBI, which is consistent with studies reporting that first-order metarepresentation is relatively spared, but higher-order metarepresentation is impaired after TBI in children (e.g., Dennis, Purvis, et al., 2001). There were no significant between-group differences in the frontal regions for this fMRI analysis.

Functional neuroimaging has been used to directly study mental attributions and its neural substrates, using tasks with animated geometric shapes that move in ways to suggest social interaction and personal agency (Castelli, Frith, Happé, & Frith, 2002; Heider & Simmel, 1944; Schultz et al., 2003). For example, using a triangles task similar to that used by Levin et al. (2011) as described above, Castelli et al. (2000) studied the relation of brain structure to metarepresentation, using positron emission tomography in healthy adults. The authors found greater brain activation for the ToM condition as compared to the random condition in mPFC regions, the temporoparietal junction, the basal region of the temporal pole, and the extrastriate cortex; the degree of activation increased with the degree of social interaction.

In another study, Schultz et al. (2003) acquired fMRI in healthy adults as they viewed movements of animated triangles and were asked to make judgments whether or not the triangles were "friends." This condition was contrasted with a nonsocial condition in which the movements of triangles were judged on physical characteristics, such as heaviness. When the social judgments were made, the task engaged the right and left dorsal mPFC, the right and left inferior frontal gyri, the right and left orbitofrontal cortex, the right temporal pole and amygdala, the right fusiform gyrus, and the area around the right and left superior temporal sulci.

Using a paradigm similar to that of Schultz et al., we (Scheibel et al., 2011) examined structures putatively mediating metarepresentation in nine adolescents with moderate to severe TBI. The study design also included a comparison group of matched adolescents with TD. Using the triangles task to examine mental attributions, we found that the adolescents with TD exhibited an activation pattern similar to that found by Schultz et al. (2003) and in other investigations of mental attribution of intentions (e.g., Castelli et al., 2000), including the mPFC, the anterior cingulate cortex, the fusiform gyrus, and posterior temporal and parietal areas. In contrast, the group with TBI had significant activation within many of these same areas, but their activation was generally more intense and excluded the mPFC. When directly compared to the adolescents with TD, the group with TBI had greater activation within right lateral frontal and parietal areas, as well as bilateral increases within posterior brain regions. These findings are generally consistent with reports of increased activation and altered activation patterns during cognitive paradigms in adults with moderate to severe TBI (Christoudoulou et al., 2001; Scheibel et al., 2009).

Structural Neuroimaging

A question separate from the activity of the brain during task performance is the effect on performance of demonstrable damage to neural structures putatively supporting task performance. To assess the contribution of brain structure to social function, Levin et al. (2011) related performance on the triangles task (described above) to the microstructural integrity of white matter circuitry implicated in social cognition. Specifically, the study examined the relation of white matter microstructure integrity to performance, using diffusion tensor imaging (DTI). The metric of interest, fractional anisotropy (FA), is a measure of the degree of disruption of the white matter fiber bundles as indicated by the relative diffusion of water molecules along and across the fiber bundles; higher values indicate more intact white matter, which is generally expected to coincide with better performance. Consistent with circuitry implicated in networks mediating the processing of others' intentions and emotions (Adolphs, 2009; Amodio & Frith, 2006; Carrington & Bailey, 2009), regions of interest for this study included the corpus callosum (genu, body, and splenium), medial prefrontal white matter (mPFWM), total prefrontal white matter, dorsolateral prefrontal white matter, cingulum bundle, inferior longitudinal fasciculus, uncinate fasciculus, and total temporal white matter. Levin et al. (2011) found that attributions significantly increased in relation to more intact white matter in left dorsolateral frontal region and the left cingulum bundle for both groups and all three conditions, with and without cueing. Importantly, there was also a significant interaction of cueing condition with FA in the mPFWM, indicating that when patients were cued, the number of mental attributions a patient engaged in increased in relation to higher FA in the left prefrontal regions. The relation was not observed in the absence of cueing, suggesting a critical role of left mPFWM in the ability to attribute mental states to inanimate objects after TBI, which may have implications for mental state attributions in humans.

Other structural neuroimaging studies have examined the relation of brain structure to social task performance with measures of brain volume. For example, Hanten et al. (2011) used MRI volumetric data to examine the relation of cortical thickness of the brain to observed deficits in social problem solving (described above). In this study, between-groups whole-brain analyses revealed that adolescents with TBI differed from adolescents with TD in the relation of cortical thickness to the VR-SPS Summary Score. The most obvious group differences were in the medial orbitofrontal region and the cuneus. In the within-group analyses, different patterns emerged for the groups with TD and with TBI. In the group with TBI, on individual steps of problem solving (DP, GS, SS, EO), correlations of cortical thickness with performance revealed that task performance was related to the temporal pole and the cuneus most strikingly in the GS condition, with increases in cortical thickness in these areas related to better performance.

In contrast to the youth with TBI, significant within-group relations for the group with TD were demonstrated in the prefrontal regions, particularly between the EO condition and the frontal pole, with decreases in cortical thickness related to higher scores. This finding is consistent with recent findings of developmental decreases in frontal regions in adolescents with TD (Giedd, 2008).

Multiple Brain Imaging Modalities

Integrating multiple imaging modalities, we (Scheibel et al., 2011) also explored the relationship between fMRI activation during the socio-cognitive condition of the Schultz et al. (2003) animated triangles task (described above) and structural neuropathology after childhood TBI as revealed by MRI, including total lesion volume within areas we considered to be relevant for ToM: the inferior frontal gyrus, middle frontal gyrus, superior frontal gyrus, temporal pole, medial frontal gyrus, gyrus rectus, orbital gyrus, and superior parietal cortex. We found that lesion volume within hypothesized task-relevant areas was related to increased activation, but not within medial frontal areas. In addition to the lesion study, we related fMRI activation patterns on the Schultz et al. task to measures of white matter integrity derived from DTI. The DTI variables included a composite measure of FA, reflecting the degree of white matter disruption across the entire brain, and another composite measure for FA specific to white matter tracts hypothesized to be relevant to social cognition (i.e., genu, bilateral uncinate, and inferior longitudinal fasciculi). Regression analyses for both FA composite scores indicated a relation between decreases in white matter integrity and activation increases within a large posterior portion of the brain, as well as the anterior cingulate cortex and mPFC. In other words, the more intact (or less disrupted) the white matter was, the greater the activation in the specific regions while a participant was performing the task. Thus it appears that both damage to white matter bundles as reflected by reductions in FA, and focal neuropathology (i.e., gray and white matter lesion volumes), contribute to an altered pattern of social-cognition-related activation following TBI.

Summary

Social cognition and social function have been increasingly recognized as important domains of outcome and development to investigate in relation to TBI in children and adolescents (Yeates, Bigler, Dennis, Gerhardt, Rubin, et al., 2007). Developmental data indicate that peer relations assume greater importance during the transition from late childhood to adolescence, imposing demands on social cognition and social skills. From fMRI studies elucidating the social brain and the circuitry connecting relevant regions, it is apparent that focal lesions of the prefrontal cortex and disruption of its connections with parietal and temporal cortex can disrupt development of social cognition and age-appropriate social skills such as social problem solving. Although the social neuroscience of TBI in children and adolescents is a relatively new area of investigation, researchers have shown that the capacity for accurately processing the intentions and emotions of others and responding appropriately is compromised in children and youth following moderate to severe TBI. Moreover, skills in solving social dilemmas are also affected by these injuries, and the gap in performance relative to healthy peers does not diminish over time, indicating that these sequelae of TBI are persistent. Supporting these socio-cognitive findings, children and adolescents tend to become socially isolated following severe TBI (Andrews et al., 1998).

Corresponding brain imaging data show that the functional brain representation of perspective taking and processing the thoughts and actions of others is remodeled in adolescents and older children who have sustained severe TBI. Emerging evidence from structural brain imaging also indicates that reduced integrity of white matter tracts connecting the brain regions implicated in social cognition may contribute to the observed deficits in social cognition. Directions for future research include utilizing ecologically valid virtual technology to develop interventions for socio-cognitive deficits, and corresponding investigations of changes in social function and associated changes in the cortical representation of brain regions engaged in social cognition.

References

Abell, F., Happé, F., & Frith, U. (2000). Do triangles play tricks?: Attribution of mental states to animated shapes in normal and abnormal development. *Cognitive Development, 15*, 1–16.

Adolphs, R. (2001). The neurobiology of social cognition. *Current Opinion in Neurobiology, 11*, 231–239.

Adolphs, R. (2009). The social brain: Neural basis of social knowledge. *Annual Review of Psychology, 60*, 693–716.

Amodio, D. M., & Frith, C. D. (2006). Meeting of minds: The medial frontal cortex and social cognition. *Nature Reviews Neuroscience, 7*, 268–277.

Andrews, T. K., Rose, F. D., & Johnson, D. A. (1998). Social and behavioural effects of traumatic brain injury in children. *Brain Injury, 12*, 133–138.

Bach, L. J., & David, A. S. (2006). Self-awareness after acquired and traumatic brain injury. *Neuropsychological Rehabilitation, 16*, 397–414.

Beardslee, W. R., Schultz, L. H., & Selman, R. L. (1987). Interpersonal negotiation strategies, adaptive functioning, and DSM-III diagnoses in adolescent offspring of parents with affective disorders: Implications for the development of mutuality in relationships. *Developmental Psychology, 23*, 807–815.

Beauchamp, M. H., & Anderson, V. (2010). SOCIAL: An integrative framework for the development of social skills. *Psychological Bulletin, 136*, 39–64.

Blakemore, S. (2008). The social brain in adolescence. *Nature Neuroscience, 9*, 267–277.

Bohnert, A. M., Parker, J. G., & Warschausky, S. A. (1997). Friendship and social adjustment of children following a traumatic brain injury: An exploratory investigation. *Developmental Neuropsychology, 13*, 477–486.

Carrington, S., & Bailey, A. J. (2009). Are there theory mind regions in the brain?: A review of the neuroimaging literature. *Human Brain Mapping, 30*, 2313–2335.

Castelli, F., Frith, C., Happé, F., & Frith, U. (2002). Autism, Asperger syndrome and brain mechanisms for the attribution of mental states to animated shapes. *Brain, 125*, 1839–1849.

Castelli, F., Happé, F., Frith, U., & Frith, C. (2000). Movement and minds: A functional imaging study of perception and interpretation of complex intentional movement patterns. *NeuroImage, 12*, 314–325.

Channon, S., & Crawford, S. (2010). Mentalising and social problem-solving after brain injury. *Neuropsychological Rehabilitation, 20*, 739–759.

Choudhury, S., Blakemore, S.-J., & Charman, T. (2006). Social-cognitive development during adolescence. *Social Cognitive and Affective Neuroscience, 1*, 165–174.

Christodoulou, C., DeLuca, J., Ricker, J. H., Madigan, N. K., Bly, B. M., Lange, G., et al. (2001). Functional magnetic resonance imaging of working memory impairment after traumatic brain injury. *Journal of Neurology, Neurosurgery and Psychiatry, 71*, 161–168.

Cicerone, K., Levin, H., Malec, J., Stuss, D., & Whyte, J. (2006). Cognitive rehabilitation interventions for executive function: Moving from bench to bedside in patients with traumatic brain injury. *Journal of Cognitive Neuroscience, 18*, 1212–1222.

Damon, W., & Hart, D. (1982). The development of self-understanding from infancy through adolescence. *Child Development, 53*, 841–864.

Damon, W., & Hart, D. (1988). *Self-understanding in childhood and adolescence*. New York: Cambridge University Press.

D'Argembeau, A., Ruby, P., Collette, F., Degueldre, C., Balteau, E., & Luxen, A. (2007). Distinct regions of the medial prefrontal cortex are associated with self-referential processing and perspective taking. *Journal of Cognitive Neuroscience, 19*, 935–944.

Dennis, M., Agostino, A., Roncadin, C., & Levin, H. S. (2009). Theory of mind depends on domain-general executive functions of working memory and cognitive inhibition in children with traumatic brain injury. *Journal of Clinical and Experimental Neuropsychology, 31*, 835–847.

Dennis, M., Guger, S., Roncadin, C., Barnes, M., & Schachar, R. (2001). Attentional–inhibitory control and social-behavioral regulation after childhood closed head injury: Do biological, developmental, and recovery variables predict outcome? *Journal of the International Neuropsychological Society, 7*, 683–692.

Dennis, M., Purvis, K., Barnes, M. A., Wilkinson, M., & Winner, E. (2001). Understanding of literal truth, ironic criticism, and deceptive praise following childhood head injury. *Brain and Language, 78*, 1–16.

Dodge, K. A. (1985). Facets of social interaction and the assessment of social competence in children. In B. H. Schneider, H. H. Rubin, & J. E. Ledingham (Eds.), *Children's peer relations: Issues in assessment and intervention* (pp. 3–22). New York: Springer-Verlag.

Frith, U., & Frith, C. D. (2003). Development and neurophysiology of mentalizing. *Philosophical Transactions of the Royal Society of London: Series B. Biological Sciences, 358*, 459–473.

Geraci, A., Surian, L., Ferraro, M., & Cantagallo, A. (2010). Theory of mind in patients with vetromedial or dorsolateral prefrontal lesions following traumatic brain injury. *Brain Injury, 24*, 978–987.

Giedd, J. N. (2008). The teen brain: Insights from neuroimaging. *Journal of Adolescent Health, 42*, 335–343.

Glang, A., Todis, B., Cooley, E., Wells, J., & Voss, J. (1997). Building social networks for children and adolescents with traumatic brain injury: A school-based intervention. *Journal of Head Trauma Rehabilitation, 12*, 32–47.

Graham, D. I., Gennarelli, T. A., & McIntosh, T. R. (2002). Trauma. In D. I. Graham & P. L. Lantos (Eds.), *Greenfield's neuropathology* (7th ed., pp. 823–897). New York: Oxford University Press..

Hanten, G., Cook, L., Orsten, K., Chapman, S. B., Li, X., Wilde, E. A., et al. (2011). Effects of traumatic brain injury on a virtual reality social problem solving task and relations to cortical thickness in adolescents. *Neuropsychologia, 49*, 486–497.

Hanten, G., Wilde, E. A., Menefee, D. S., Li, X., Lane, S., Vasquez, C., et al. (2008). Correlates of social problem solving during the first year after traumatic brain injury in children. *Neuropsychology, 22*, 357–370.

Hayes, R. L., Jenkins, L. W., & Lyeth, B. G. (1992). Neurotransmitter-mediated mechanisms of traumatic brain injury: Acetylcholine and excitatory amino acids. *Journal of Neurotrauma, 9*(Suppl. 1), S173–S187.

Heider, F., & Simmel, M. (1944). An experimental study of apparent behavior. *American Journal of Psychology, 57*, 243–259.

Henry, J. D., Phillips, L. H., Crawford, J. R., Ietswaart, M., & Summers, F. (2006). Theory of mind following traumatic brain injury: The role of emotion recognition and executive dysfunction. *Neuropsychologia, 44*, 1623–1628.

Janusz, J. A., Kirkwood, M. W., Yeates, K. O., & Taylor, H. G. (2002). Social problem-solving skills in children with traumatic brain injury: Long-term outcomes and prediction of social competence. *Child Neuropsychology, 8*, 179–194.

Kohlberg, L. (1969). Stage and sequence: The cognitive-developmental approach to socialization. In D. Goslin (Ed.), *Handbook of socialization theory and research* (pp. 347–480). Chicago: Rand McNally.

Levin, H. S., Mendelsohn, D., Lilly, M. A., Yeakley, J., Song, J., & Scheibel, R. S. (1997). Magnetic resonance imaging in relation to functional outcome of pediatric closed head injury: A test of the Ommaya–Gennarelli model. *Neurosurgery, 40*, 32–40.

Levin, H. S., Wilde, E. A., Hanten, G., Li, X., Chu, Z., Vasquez, A. C., et al. (2011). Mental state attributions and diffusion tensor imaging after traumatic brain injury in children. *Developmental Neuropsychology, 36*, 273–287.

Lewis, P. A., Rzale, R., Brown, R., Roberts, N., & Dunbar, R. I. (2011). Ventromedial prefrontal volume predict understanding of others and social network size. *NeuroImage, 57,* 1624–1629.

Martin-Rodriguez, J. F., & Leon-Carrion, J. (2010). Theory of mind deficits in patients with acquired brain injury: A quantitative review. *Neuropsychologia, 48,* 1181–1191.

Muller, F., Simion, A., Reviriego, E., Glaera, C., Mazaux, J., Barat, M., et al. (2010). Exploring theory of mind after severe traumatic brain injury. *Cortex, 46,* 1088–1099.

Newsome, M. R., Scheibel, R. S., Hanten, G., Steinberg, J. L., Lu, H., Cook, L., et al. (2010). Activation while thinking about the self from another person's perspective after traumatic brain injury in adolescents. *Neuropsychology, 24,* 139–147.

Pfeifer, J. H., Lieberman, M. D., & Dapretto, M. (2007). "I know you are but what am I?!": Neural bases of self- and social knowledge retrieval in children and adults. *Journal of Cognitive Neuroscience, 19,* 1323–1337.

Premack, D. G., & Woodruff, G. (1978). Does the chimpanzee have a theory of mind? *Behavioral and Brain Sciences, 1,* 515–526.

Saxe, R., & Kanwisher, N. (2003). People thinking about people: The role of the temporo-parietal junction in theory of mind. *NeuroImage, 19,* 1835–1842.

Scheibel, R. S., Newsome, M. R., Troyanskaya, M., Steinberg, J. L., Goldstein, F. C., Mao, H., et al. (2009). Effects of severity of traumatic brain injury and brain reserve on cognitive-control related brain activation. *Journal of Neurotrauma, 26,* 1447–1461.

Scheibel, R. S., Newsome, M. R., Wilde, E., McClelland, M., Hanten, G., Krawczyk, D. C., et al. (2011). Brain activation during a social attribution task in adolescents with moderate to severe traumatic brain injury. *Social Neuroscience, 6,* 582–598.

Schmitz, T. W., & Johnson, S. C. (2007). Relevance to self: A brief review and framework of neural systems underlying appraisal. *Neuroscience and Biobehavioral Reviews, 31,* 585–596.

Schultz, R. T., Grelotti, D. J., Klin, A., Kleinman, J., Van der Gaag, C., Marois, R., et al. (2003). The role of the fusiform face area in social cognition: Implications for the pathobiology of autism. *Philosophical Transactions of the Royal Society of London: Series B. Biological Sciences, 358,* 415–427.

Snodgrass, C., & Knott, F. (2006). Theory of mind in children with traumatic brain injury. *Brain Injury, 20,* 825–833.

Taylor, H. G., Yeates, K. O., Wade, S. L., Drotar, D., Stancin, T., & Minich, M. (2002). A prospective study of short-and long-term outcomes after traumatic brain injury in children: Behavior and achievement. *Neuropsychology, 16,* 15–27.

Turkstra, L. S., Dixon, T. M., & Baker, K. K. (2004). Theory of mind and social beliefs in adolescents with traumatic brain injury. *NeuroRehabilitation, 19,* 245–256.

Turkstra, L. S., McDonald, S., & DePompei, R. (2001). Social information processing in adolescents: Data from normally developing adolescents and preliminary data from their peers with traumatic brain injury. *Journal of Head Trauma Rehabilitation 16,* 469–483.

Yeates, K. O., Bigler, E. D., Dennis, M., Gerhardt, C. A., Rubin, K. H., Stancin, T., et al. (2007). Social outcomes in childhood brain disorder: A heuristic integration of social neuroscience and developmental psychology. *Psychological Bulletin, 133,* 535–556.

Yeates, K. O., Schultz, L. H., & Selman, R. L. (1991). The development of interpersonal negotiation strategies in thought and action: A social cognitive link to behavioral adjustment and social status. *Merrill–Palmer Quarterly, 37,* 369–406.

Yeates, K. O., Swift, E., Taylor, H. G., Wade, S. L., Drotar, D., & Stancin, T. (2004). Short- and long-term social outcomes following pediatric traumatic brain injury. *Journal of the International Neuropsychological Society, 10,* 412–426.

CHAPTER 13

Genetic Disorders and Social Problems

Kylie M. Gray and Kim Cornish

Neurodevelopmental disorders with a clear genetic etiology provide important glimpses into the complex world of gene–behavior associations. This knowledge has been facilitated by a decade of groundbreaking research discoveries that have provided critical new insights into how genes interact with other genes to influence early brain maturation and cognitive development, and how genes interact with a rapidly changing and dynamic environment to shape an emerging behavioral phenotype in the first few years of life. New collaborations among the disciplines of molecular genetics, developmental psychology, cognitive and social neurosciences, and brain imaging have driven this new knowledge. Although extremely innovative in design, this research is not without its challenges. Indeed, a key lesson from a decade of research is how elusive the associations between gene(s) and behavior can be, no matter how sophisticated the technological advances— even in single-gene disorders, for which there is increased expectation of gene–behavior correlates.

In the social domain, neurodevelopmental genetic disorders are often associated with a spectrum of social functioning that can range from social aloofness, limited interaction skills, and enhanced anxiety levels in social situations all the way through to "spared" interaction skills, overfriendliness with strangers, and even hypersociability. Given the range of these social problems across different genetic disorders, it becomes imperative to identify potential disorder-specific "signatures" that can facilitate the development of targeted treatment approaches across the lifespan, beginning in early childhood—when, because of the plasticity of the developing neocortex, interventions are most likely to have their most significant impact. Alongside this research, a growing number of studies now highlight the importance of recognizing often striking *individual differences* within a given genetic disorder. That is, multiple "social phenotypes" can be identified in affected individuals with the same genetic dysfunction and from within the same family.

In this chapter, we focus on three neurodevelopmental genetic disorders that have well-documented phenotypes: fragile X syndrome (FXS), Williams syndrome (WS), and Turner syndrome (TS). Specifically, we address the following questions:

1. To what extent do disorders of disparate genetic etiologies exhibit commonalities as well as differences in social functioning?
2. Are individual differences within a genetic disorder simply noise or clues to important gene–gene or gene–environment interactions? We examine the case of FXS and its link to autism.
3. Does development itself play a critical role in guiding targeted interventions, so that the social phenotypes of genetic disorders can be viewed as dynamic profiles that change across developmental time and across environments?
4. Can this new wealth of knowledge bridge the gap between research discovery and the uptake and utilization of these discoveries by clinicians and educators, to improve outcomes for all who live with a genetic disorder, their families, and their communities?

We begin with a brief profile of each of these three genetic disorders before highlighting commonalities and differences in their social phenotypes.

Profile of Three Genetic Disorders

Fragile X Syndrome

FXS was first described in 1943 by Martin and Bell (in fact, it was originally labeled the "Martin–Bell" syndrome) and is the world's most common hereditary cause of developmental delay. Recent estimates indicate a frequency of approximately 1 in 2500 (Hagerman, 2008). The disorder is caused by the silencing of a single gene on the long arm of the X chromosome at q27.3. The gene, named the fragile X mental retardation gene 1 (FMR1), was identified in 1991 and is "turned off" in affected individuals. When this occurs, there is a substantial expansion of a trinucleotide repeat sequence (cytosine–guanine–guanine, or CGG) to above a threshold number. In nonaffected individuals, there are typically between 7 and 54 CGG repeats, with 30 repeats the most common number. However, in some individuals there is a small to medium expansion of CGG repeats to between 55 and 199; this is known as the *premutation status*, and these individuals are referred to as *carriers* of FXS. When the CGG region dramatically expands to over 200 repeats, the gene is switched off, and the protein product (fragile X mental retardation protein, or FMRP) is either deficient or absent. It is the loss of FMRP that results in FXS and its associated intellectual, cognitive, and social phenotype (known as the *full-mutation status*) (see Cornish, Turk, & Hagerman, 2008, for a review). It is also noteworthy that the FMR1 gene is one of the few known single-gene causes of autism, with most recent figures suggesting that 10–25% of individuals diagnosed with autism spectrum disorders will have a genetic mutation on the FMR1 gene (Ey, Leblond, & Bourgeron, 2011).

Because of the X linkage, almost all affected boys will display developmental delay, compared to approximately a third of affected girls. Girls, due to their second unaffected X chromosome, often display less marked intellectual and cognitive impairment. In both boys and girls, however, there is a behavioral profile that includes chronic and pervasive attention problems, marked language impairments, reduced eye contact, social awkwardness, and aloofness; this profile results in severe social anxiety that increases exponentially with age (for reviews, see Cornish & Wilding, 2010; Cornish, Gray, & Rinehart, 2010; Hagerman, 2006).

Williams Syndrome

WS is also known as Williams–Beuren syndrome. Until quite recently, it was assumed to be a rare disorder (1 in 15,000–20,000), but one recent estimate indicates the frequency at approximately 1 in 7500 (Stromme, Bjornstad, & Ramstad, 2002). WS is not usually hereditary and results from the deletion of approximately 25–30 genes on the long arm of chromosome 7 at q11.23. This region encompasses the elastin (ELN) gene, which is responsible for the cardiovascular problems inherent in WS, and the LIM–kinase 1 (LIMK1) gene (Korenberg et al., 2000), which is partially responsible for the known cognitive deficits, especially spatial weaknesses in WS (see Karmiloff-Smith, 2007, for a review).

Intellectual ability in WS can range from average ability to moderate–severe disability, although the behavioral profile demonstrates a distinctive pattern of poor response inhibition (Porter, Coltheart, & Langdon, 2007), spatial cognition impairments (Bellugi & St. George, 2001), distractibility, and inattention (Cornish, Scerif, & Karmiloff-Smith, 2007). Language functions in WS have been the subject of much research and media debate, namely because of the "relative strengths" in specific domains of language (e.g., vocabulary and verbal short-term memory) that exist alongside significant weaknesses in other domains of language (e.g., the social use of language known as *pragmatics* (for reviews, see Karmiloff-Smith, 2007; Mervis & John, 2010). Children with WS can be overly friendly, outgoing, and empathetic, but at the same time can have difficulty initiating and sustaining friendships, as well as the pragmatic difficulty just noted. Both children and adolescents suffer from high rates of social anxiety, have difficulties with changes in routines or schedules, and tend to perseverate on favorite conversational topics (see, e.g., Einfeld, Tonge, & Rees, 2001). This constellation of behaviors has led to speculations of a possible overlap between WS and autism spectrum disorders (Klein-Tasman, Phillips, Lord, Mervis, & Gallo, 2009).

Turner Syndrome

TS is a relatively common genetic disorder associated only with females. It is caused by the lack of all or part of one X chromosome (45,X), resulting in numerous missing genes. The frequency of TS is estimated at 1 in 2500–3000 (Sybert & McCauley, 2004), and the disorder is not hereditary. Studies to identify candidate genes, whose loss on the X chromosome may result in the TS cognitive and social phenotype, have attracted widespread research attention in recent years. Noteworthy are the results of investigations by Skuse and colleagues that postulate a genetic locus on the short arm of the X chromosome at Xp.11.3 associated with increased amygdala size, which in turn affects emotional and social functioning in TS (e.g., Skuse, 2005).

Intellectual ability tends to be unimpaired in TS, but many girls will have learning disabilities centering around their poor nonverbal perceptual–motor and visual–spatial skills (see Hong, Scaletta Kent, & Kesler, 2009, for a review). Behavioral concerns include poor attention and distractibility, poor emotion recognition and gaze perception, difficulties in sustaining peer relationships, poor interpretation of social cues, and social anxiety alongside an inflexibility regarding changes in routine (Sybert & McCauley, 2004). Collectively, these characteristics have led to a speculation of an overlap between TS and autism spectrum disorders (Donnelly et al., 2000; Hou, Wang, & Zhong, 2006).

At first blush, all three of these genetic disorders—FXS, WS, and TS—display weaknesses in social functions that overlap with each other. These include difficulties with social communication, problems in initiating and sustaining peer relationships, social anxiety, and immaturity, alongside social-cognitive deficits in face processing, emotion recognition, and theory of mind. It is this cluster of behaviors that hints of autism or autism spectrum disorders, and one could speculate that there is some common underlying "autism phenotype" that drives these social functions across disparate genetic disorders. An alternative explanation is that so-called "commonalities" in social functions may actually reflect distinct genetic pathways that, although they eventually lead to shared behavioral end states, have quite distinct origins driven by a complex interaction between gene expression in the developing brain and the modifying effects of the environment. The developmental journey to the behavioral end state will therefore be different for each disorder.

In the next two sections, we describe impairments in social function known to be impaired in FXS, WS, and TS, but with a specific emphasis on core disorder-specific weaknesses.

Social Behavior Phenotypes across Genetic Disorders

Social cognition encompasses a broad range of skills and behaviors that commence development from an early age. These skills enable children to think about themselves and others and about how people interact with each other, and to develop social relationships and friendships.

Fragile X Syndrome

In FXS, delays in social communication skills are observed early in development and progress across childhood (e.g., Bailey, Hatton, Skinner, and Mesibov, 2001; Roberts et al., 2009). These can include poor eye contact, social withdrawal and avoidance, and social anxiety. This constellation of atypical features is linked with autism: as stated earlier, a significant minority of children with an autism diagnosis will have FXS, and almost all children with FXS (both boys and girls) will display characteristics of autism. However, a number of important distinctions serve to differentiate these two disorders. First, although children with FXS display reduced social skills, they do tend to enjoy reciprocal relationships, especially with parents; by contrast, children with autism struggle to maintain dyadic relationships. For example, Cohen, Vietze, Sudhalter, Jenkins, and Brown (1989) showed that males with FXS, when compared to males with autism alone, were sensitive to social gaze initiation by their parents but found eye contact aversive. In contrast, the males with autism were insensitive to parent-initiated social gaze and did not find eye contact aversive. However, both groups avoided eye contact with strangers.

In contrast to their specific social strengths, one of the most pervasive and distinctive characteristics in FXS is delayed pragmatic skills, especially in terms of perseverative language. Children with FXS tend to produce exceptionally high rates of repetition of words, conversational phrases, and topics (e.g., Belser & Sudhalter, 2001), far exceeding those of children with autism; these findings suggest that this is a unique and defining feature of FXS. A possible causal mechanism for this striking perseveration in FXS is an underlying weakness in the autonomic nervous system that results in hyperarousal, preventing many individuals with FXS from quickly adapting to a new situation or a change in a conversational topic. This is further accompanied by a severe limitation in response inhibition that prevents a smooth transition along a planned sequence of events, such as a conversation; this exacerbates the perseveration and so reduces the opportunity for successful social communication (for reviews, see Abbeduto, Brady, & Kover, 2007; Cornish, Sudhalter, & Turk, 2004).

Williams Syndrome

In WS, a core defining feature in stark contrast to that observed in both FXS and TS is *hypersociability*, demonstrated by a propensity to be gregarious, engaging, overly friendly, and affectionate. Of particular concern is the tendency for hypersociability to manifest itself as an absence of fear of strangers, with a tendency to approach people indiscriminately. A number of studies have therefore examined approach in individuals with WS. Dodd, Porter, Peters, and Rapee (2010) found that preschool-age children with WS were more likely to approach strangers than either chronological- or mental-age-matched controls. Similar findings were reported by Doyle, Bellugi, Korenberg, and Graham (2004): Parents of children with WS reported that their children were more likely to approach and initiate social interactions with other people, including strangers, compared to typically developing children or children with other neurodevelopmental genetic disorders (e.g., Down syndrome).

Atypical amygdala functioning is seen as a core causal link in these behaviors. For example, a recent study by Martens, Wilson, Dudgeon, and Reutens (2009) directly investigated the proposed link between hypersociability and approachability in WS and amygdala volume. Consistent with previous findings, they found that children, adolescents, and adults with WS rated faces as more approachable than typically developing chronological-age-matched participants rated them. Importantly, these ratings of approachability were not associated with cognitive measures (IQ, visual–spatial tasks). In addition to the higher approachability ratings, increased amygdala volumes were found in the group with WS. Specifically, a significant positive relationship was found between right amygdala volume and approachability ratings of negative faces. As in FXS, a disorder-specific underlying deficit is coupled with poor response inhibition that appears to exacerbate the impairment. It has been proposed that an inability to inhibit social information could result in the atypical approach behaviors of individuals with WS (Jawaid, Schmolck, & Schulz, 2008). Specifically, individuals with WS know that they should not approach strangers; however, their poor response inhibition triggers an approach to strangers anyway. Indeed, converging research now points to impulsive and inattentive behaviors as pervasive in children and adolescents with WS (Cornish, Scerif, et al., 2007; Einfeld, Tonge, & Florio, 1997; Einfeld et al., 2001).

Turner Syndrome

In TS, despite no evidence of impaired use of language in social interactions (Mazzocco et al., 2006), adults with the syndrome rate themselves as more socially withdrawn and less socially competent than their same-age peers (Lagrou et al., 2006; Wide-Boman, Hanson, Hjelmquist, & Möller, 2006). Compared to other short-statured women and to their own typical sisters, adults with TS describe themselves as having social impairment, fewer friends, and fewer social contacts outside the home (Downey, Ehrhardt, Gruen, Bell, & Morishima, 1989). A similar picture is painted for children and adolescents with TS. Compared to other short-statured children and adolescents and to their own sisters, girls with TS are described by parents and teachers as having social difficulties, poor peer relationships, fewer positive social interactions, fewer friends, difficulty in reading social cues, and a need for more structure to socialize (Lagrou et al., 1998; Mazzocco, Baum-gardner, Freund, & Reiss, 1998; McCauley, Ito, & Kay, 1986; Siegel, Clopper, & Stabler, 1998). Interestingly, derivation of the X chromosome has been linked to impairments in social functioning in children with TS. Specifically, girls with a maternally derived X chromosome have been described as having more social problems, difficulties in getting along with other children, and social immaturity, compared to girls with a paternally derived X chromosome (Skuse, Elgar, & Morris, 1999).

In addition to social impairments, low self-esteem, poor self-concept, and a negative body image have been described for both adolescents and adults with TS (McCauley, Ito, et al., 1986; McCauley, Sybert, & Ehrhardt, 1986; Pavlidis, McCauley, & Sybert, 1995; Ross et al., 1996; van Pareren et al., 2005). It has been suggested that poor self-concept may contribute to social impairments and may be linked to short stature, delayed puberty and sexual development, and dysmorphic features (McCauley, Sybert, et al., 1986; Siegel et al., 1998). Hearing abnormalities and being overweight have also been linked to ratings of self-esteem (Carel et al., 2006). Although a correlation between growth rate (after 2 years of growth hormone therapy) and children's perception of themselves as being more attractive, more intelligent, more popular, and less often teased has been demonstrated (Rovet & Holland, 1993), other studies have failed to find an association between height and psychosocial functioning (Carel et al., 2006; Lagrou et al., 2006; van Pareren et al., 2005). In children and adolescents with TS, dysmorphic features were not associated with parent or teacher ratings of social impairment, but were associated with self-ratings of low self-esteem (McCauley, Ito, et al., 1986). In adults with TS, however, no association has been reported between dysmorphic features and either self-esteem or social competence (Carel et al., 2006). In sum, it has been concluded that the impairments in social functioning observed in TS cannot be explained solely by differences in appearance, and are likely to be influenced by deficits in social processing (see Hong et al., 2009).

Social-Cognitive Phenotypes across Genetic Disorders

Facial recognition and emotion processing are recognized as core skills in social interaction. They are key means of communication and have a significant impact on our ability to understand and interact with others in a social environment, as well as to evaluate which behaviors and objects to approach or avoid (Blair, 2003). Children acquire knowledge about emotions from an early age, and are able to identify basic emotions in facial expressions by 3 years of age (Widen & Russell, 2008). Although further longitudinal research is needed, results to date indicate that the development of emotion recognition skills continues through childhood and adolescence (Herba & Phillips, 2004). A further skill important to social interaction is that of *theory of mind*, which is the ability to attribute mental states (e.g., beliefs, desires, intentions) to other people, and thus be able to explain and predict the behavior of others (Baron-Cohen, 1989). Children typically develop theory-of-mind skills from 3 to 5 years of age (Callaghan, 2005).

Fragile X Syndrome

In FXS, social-cognitive skills are relatively understudied. To date, the published research has focused on theory of mind and differences in performance between children with FXS and children with autism. Consistent with intellectual delay, boys with FXS display impaired theory-of-mind skills on the classic first-order belief tasks—notably the Sally–Anne paradigm (e.g., Baron-Cohen, 1989), which requires the ability to understand that others can hold false beliefs about the world and that their behavior can be predicted on the basis of these false beliefs rather than on the basis of what is actually true. However, the severity of impairment on such tasks is much less than commonly reported in autism (<30% success rate) and is comparable that in to other neurodevelopmental genetic disorders not specifically linked with autism, such as Down syndrome (Cornish et al., 2005; Garner, Callias, & Turk, 1999). Of particular interest is a recent study by Grant, Aperley, and Oliver (2007), who compared 15 children with FXS alone, 15 children with FXS and autism, and 15 comparison children with intellectual disability of unknown etiology on two standard first-order belief tasks and two video-based nonverbal false-belief reasoning tasks. At first glance, all three groups demonstrated significant theory-of-mind deficits commensurate with their intellectual disabilities; however, the underlying reasons for the deficits appeared to be syndrome specific. In the case of FXS, the associated deficit suggested that severe difficulties in general information processing were affecting working memory performance (most probably a deficiency in skills requiring attentional, executive capacity), rather than a core deficit in theory of mind per se as in autism.

An earlier study by Cornish et al. (2005) examined the ability to understand one's own mental states—in which one must distinguish between the perception of an object (its appearance) and actual knowledge of it (its real identity)—in children with FXS, children with Down syndrome, and typically developing peers. Findings indicated a similar profile of deficits in the group with FXS and Down syndrome. However, closer inspection of the impairments in these two groups revealed qualitative differences in error types, indicative of distinct disorder-specific "signatures" that possibly reflect different social-cognitive mechanisms underlying these disorders.

Williams Syndrome

In WS, recent pioneering research has begun to explore social cognition in infants. Specifically, infants with WS have been found to spend more time looking at the face of a stranger than typically developing and developmentally delayed infants do (Mervis et al., 2003). Similar results have been reported for children and adults. Using eye-tracking technology, Riby and Hancock (2008) examined attention to pictures of social scenes in children and young people with autism and WS, with both chronological and nonverbal age matches for both groups. Although there was no difference between the group with WS and controls in terms of time spent viewing the social scenes, the children and adolescents with autism spent less time than the typically developing controls did viewing the faces in social scenes. The group with WS spent a greater proportion of time looking at the eyes of the faces, but not the mouths. In addition, Riby, Doherty-Sneddon, and Bruce (2009) demonstrated that children and adolescents with WS performed with greater accuracy on tasks involving matching of unfamiliar faces when only the eye region of the face was visible, compared to the mouth region. In contrast, the group of children and adolescents with autism performed relatively poorly across the face tasks, with significant deficits demonstrated when only the face region was visible.

In a further study, Riby and Hancock (2009) investigated the gaze behavior of children and adolescents with autism, children and young people with WS, and typically developing children matched on nonverbal ability, using photos of scenes with an embedded face and scrambled images containing faces. Whereas the group with autism displayed reduced face gaze, the individuals with WS displayed prolonged face gaze. Porter, Shaw, and Marsh (2010) reported that although the eyes of faces in images did not capture the attention of individuals with WS faster than that of mental-age-matched controls, they spent significantly more time attending to this region once they were attending to the eyes. This pattern of gaze did not differ across different facial expressions. This body of work demonstrates that faces do capture the attention of individuals with WS; however, they tend to fix their gaze on the eyes, focusing on this region for extended periods of time. This preference for attention to the eye region is in direct contrast to individuals with autism and FXS (Baron-Cohen, Campbell, Karmiloff-Smith, Grant, & Walker, 1999; Cornish et al., 2008; Leekam, Lopez, & Moore, 2000).

Despite their atypical face gaze behaviors, individuals with WS have been found to have emotion processing skills commensurate with their developmental level, especially on emotion recognition tasks that require identification of emotions in eyes. For example, the eyes and type of emotion appear to play a critical role in this process. When the eyes of facial images are frozen to a neutral expression, individuals with WS demonstrate impaired emotion recognition—a response not observed in matched controls (verbal, nonverbal, and chronological age matches). The type of emotion may also play a role, with individuals with WS demonstrating an attention bias for happy faces (Dodd & Porter, 2010). Out of keeping with developmentally appropriate skills in emotion recognition, impairment in recognition of anger has been identified. Compared to chronological- and mental-age-matched controls, impairments in the identification of angry faces have been reported in individuals with WS (Porter et al., 2010; Santos, Silva, Rosset, & Deruelle, 2010). In research measuring skin conductance response and heart rate, a decreased responsiveness to emotions expressed in faces has been identified in adolescents and adults with WS, compared to age- and IQ-matched controls (Plesa Skwerer et al., 2009). Although recognition of emotions in the face, with the exception of anger, has been found to be comparable to that of mental-age-matched controls, individuals with WS have been shown to have impairments in identifying vocal expressions of emotion (Plesa-Skwerer, Faja, Schofield, Verbalis, & Tager-Flusberg, 2006). In addition, an impairment in identifying emotions in nonhuman faces has been identified (Santos, Rosset, & Deruelle, 2009). These specific impairments in emotion processing may indicate a failure to recognise threat stimuli in social environments, and has been suggested as contributing to the tendency to approach strangers.

Turner Syndrome

In TS, as noted earlier, girls tend to have fewer friends, engage in fewer social activities, and have lower self-esteem and poorer self-concept than their typically developing peers. Accumulating research also points to significant weaknesses in aspects of social cognition that require emotion and face processing. The seminal research of Skuse and others (cited above) suggests that amygdala dysfunction is likely to be involved in these specific social impairments. One of the most consistent findings is that girls and women with TS appear to have considerable difficulty in interpreting fear from the eye region of the face. Lawrence et al. (2003) compared women with TS to control women on a task that required the ability to infer the mental state and intention of others solely from the information provided in the eye region. On similar tasks, individuals with high-functioning autism perform exceptionally poorly across a range of emotions (e.g., Baron-Cohen, Wheelwright, Raste, & Plumb, 2001). Lawrence et al. found that a specific deficit in the categorization of fear differentiated the two groups. In contrast, recognition of other core emotions (sadness, anger, disgust, surprise, and happiness) was not especially impaired. Subsequent research has replicated this finding. Notably, Mazzola et al. (2006), using eye-tracking technology, found women with TS to be profoundly impaired in recognizing fear in faces, whether viewed either as whole faces or only the eye region. Furthermore, these authors speculate that since women with TS have an enlarged amygdala, one consequence could be an overresponsiveness to direct gaze or, in this case, fear in faces.

Extending these findings, Lawrence et al. (2007) recently reported that a group of women with TS—specifically those who inherited the maternal X—were impaired relative to a group of age- and verbal-IQ-matched control women in attributing mental states to animated shapes (silent cartoons that featured a large red triangle and a smaller blue triangle). They speculate that a fundamental weakness in face processing extends to a more global theory-of-mind deficit in TS (see Burnett, Reutens, & Wood, 2010, for a review).

A summary of social characteristics across these three syndromes is provided in Table 13.1.

TABLE 13.1. Social Characteristics of Fragile X Syndrome (FXS), Williams Syndrome (WS), and Turner Syndrome (TS)

FXS	WS	TS
• Language impairment • Perseverative language • Social withdrawal, avoidance • Aloofness • Social awkwardness • Poor eye contact • Limited response inhibition • Attention problems • Hyperarousal • Social anxiety	• Excessive talk • Impaired conversation, pragmatic skills • Hypersociability • Excessive friendliness • Social disinhibition • Excessive attention to faces (eye region) • Difficulty identifying anger in faces • Inattentiveness • Impulsivity • General anxiety, phobias	• No language impairment • Poor peer relationships • Social withdrawal • Fewer friends • Low self-esteem, poor self-concept • Negative body image • Poor emotion recognition skills • Deficit in recognizing fear in faces • Inattentiveness, distractibility • Social anxiety

Capturing Multiple Social-Cognitive Phenotypes within Genetic Disorders: FXS as an Example

To date, research has only just begun to capture the complexity of within-disorder differences, even in a single-gene disorder such as FXS. Intuitively, it makes sense that not every individual with a given genetic disorder will exhibit identical characteristics with the same degree of severity. However, all too often, researchers can make the false assumption that their findings based on a relatively small, skewed sample are applicable to the wider community of all affected individuals. Moderator variables such as gender, age, IQ, or developmental level can play critical roles by interacting together or acting individually to influence social-phenotypic trajectories and outcomes; therefore, these variables need to be empirically controlled and evaluated.

In the case of FXS, there are currently very few single-gene disorders with a strong probability of the involvement of autism. FXS is one. However, not all individuals with FXS will be diagnosed with autism: Findings vary widely, but approximately 33–67% fulfill the diagnostic criteria (Clifford et al., 2007; Rogers, Wehner, & Hagerman, 2001). Commonalities across core social and language domains define the link between FXS and autism, and it therefore seems highly plausible that similar neurobiological mechanisms are affected in both disorders. Loesch and colleagues (2007) found that a common impairment in verbal skills best described the comorbidity of FXS and autism at the cognitive level. In an earlier study, Philofsky, Hepburn, Hayes, Hagerman, and Rogers (2004) reported a similar link between exceptionally low verbal ability (in this case, receptive language) in children with FXS and a dual diagnosis of autism, compared to children with FXS alone, for whom verbal skills appeared to be a relative strength. Overall, children with a dual diagnosis tend to display more impaired cognitive performance than children with either autism alone or FXS alone. See Cornish, Levitas, and Sudhalter (2007) for more detailed descriptions of the commonalities and differences between FXS and autism across cognitive domains.

Furthermore, given that FXS is an X-linked disorder, it becomes critical for research to control for the differential impact of gender on social functioning. So far, research has tended to focus on cognitive performance, and findings are striking. For example, although intellectual disability and behavioral difficulties characterize many children with FXS, careful examination of the gender profiles reveals overlapping but nonetheless distinct profiles. In almost all boys with FXS, IQ is in the mild to moderate range of impairment, with profiles emerging as young as 3 years of age (Skinner et al., 2005). In contrast, females present with more phenotypic variation. Some girls only show subclinical learning disabilities (Bennetto & Pennington, 2002), while approximately 25% display more significant cognitive impairment (most with borderline to mild mental retardation). Genetic variation in the form of X inactivation (i.e., one of the two X chromosomes remains inactive and the other active) is seen as the major contributor to the heterogeneity of intellectual disability and the broad range of cognitive deficits seen in girls with FXS. For obvious reasons, this is not an issue in boys with FXS, whose impairment, without the protection of X inactivation, shows greater severity.

In view of these findings, it seems highly likely that gender differences will exist in the social domain—possibly producing more enhanced trajectories for girls than for boys, but with both genders displaying a similar range of impairments. This knowledge, if accrued, will facilitate a new generation of gender-specific interventions that target emerging social skills in boys and girls with FXS separately.

Avenues for Future Research

It is clear that research needs to extend its focus to longitudinal investigations instead of relying almost exclusively on cross-sectional data, which, although providing invaluable information, can never really inform us about subtle dynamic changes in development. Due to the rarity of these conditions, samples across studies generally encompass children, adolescents, and adults; such heterogeneous samples do not permit comprehensive exploration of age-specific impairments. To achieve a better understanding of developmental trajectories and age-related changes, we need research on samples with more narrowly defined age bands. For example, following children from early childhood to early adolescence would provide invaluable data. Undoubtedly, subtle age-related changes will emerge at some time points, and yet at others atypical children may perform comparably to typically developing children. Such longitudinal research will give us a fuller understanding of the specific patterns of strengths and weaknesses in these syndromes, providing information critical to the development of targeted interventions, including early interventions, school-based interventions, and supports for adults.

Conclusions

Taken together, the findings discussed in this chapter point to atypical patterns of social functioning in different neurodevelopmental genetic disorders—patterns that extend from the behavioral to the cognitive domains. At first glance, all three of the genetic disorders reviewed here show commonalities in social impairments, and, interestingly, all three have some overlapping behaviors in common with autism. This latter finding has caused a flurry of speculation that the autism phenotype may have a genetic basis across a range of neurodevelopmental genetic disorders, including those specifically linked to the X chromosome (e.g., FXS and TS). However, although the behavioral end states may show some common behaviors (e.g., eye gaze difficulties), the underlying social-cognitive mechanisms that produce those outcomes may be very different, reflecting each disorder's distinct genetic mechanisms and pathways. Collectively, these findings indicate that although many genetic neurodevelopmental disorders appear to show similar social outcomes by late childhood, the cognitive mechanisms by which these similarities emerge will be quite different and therefore need close investigation in studies using cross-disorder comparisons within a developmental framework.

References

Abbeduto, L., Brady, N., & Kover, S. T. (2007). Language development and fragile X syndrome: Profiles, syndrome-specificity, and within-syndrome differences. *Mental Retardation and Developmental Disabilities Research Reviews, 13*(1), 36–46.

Bailey, D. B., Hatton, D. D., Skinner, M., & Mesibov, G. (2001). Autistic behavior, FMR1 protein, and developmental trajectories in young males with fragile X syndrome. *Journal of Autism and Developmental Disorders, 31*(2), 165–174.

Baron-Cohen, S. (1989). The autistic child's theory of mind: A case of specific developmental delay. *Journal of Child Psychology and Psychiatry, 30*(2), 2811–2297.

Baron-Cohen, S., Campbell, R., Karmiloff-Smith, A., Grant, J., & Walker, J. (1999). Are children with autism blind to the mentalistic significance of the eyes? *British Journal of Developmental Psychology, 13*, 379–398.

Baron-Cohen, S., Wheelwright, S., Hill, J., Raste, Y., & Plumb, I. (2001). The "Reading the Mind in the Eyes" Test revised version: A study with normal adults, and adults with Asperger syndrome or high-functioning autism. *Journal of Child Psychiatry and Psychiatry, 42*, 241–252.

Bellugi, U., & St. George, M. (Eds.). (2001). *Journey from cognition to brain to gene: Perspectives from Williams syndrome.* Cambridge, MA: Bradford Books.

Belser, R. C., & Sudhalter, V. (2001). Conversational characteristics of children with fragile X syndrome: Repetitive speech. *American Journal on Mental Retardation, 106*(1), 28–38.

Bennetto, L., & Pennington, B. F. (2002). Neuropsychology. In R. J. Hagerman & P. J. Hagerman (Eds.), *Fragile X syndrome: Diagnosis, treatment, and research* (3rd ed., pp. 206–248). Baltimore: Johns Hopkins University Press.

Blair, R. J. R. (2003). Facial expressions, their communicatory functions and neuro-cognitive substrates. *Philosophical Transactions of the Royal Society of London: Series B. Biological Sciences, 358*, 561–572.

Burnett, A. C., Reutens, D. C., & Wood, A. G. (2010). Social cognition in Turner's syndrome. *Journal of Clinical Neuroscience, 17*(3), 283–286.

Callaghan, T. C. (2005). Cognitive development beyond infancy. In B. Hopkins (Ed.), *The Cambridge encyclopedia of child development* (pp. 204–209). New York: Cambridge University Press.

Carel, J.-C., Elie, C., Ecosse, E., Tauber, M., Leger, J., Cabrol, S., et al. (2006). Self-esteem and social adjustment in young women with Turner syndrome—influence of pubertal management and sexuality: Population-based cohort study. *Journal of Clinical Endocrinology and Metabolism, 91*(8), 2972–2979.

Clifford, S., Dissanayake, C., Bui, Q. M., Huggins, R., Taylor, A. K., & Loesch, D. Z. (2007). Autism spectrum phenotype in males and females with fragile X full mutation and premutation. *Journal of Autism and Developmental Disorders, 37*(4), 738–747.

Cohen, I. L., Vietze, P. M., Sudhalter, V., Jenkins, E. C., & Brown, W. T. (1989). Parent–child dyadic gaze patterns in fragile X males and in non-fragile X males with autistic disorder. *Journal of Child Psychology and Psychiatry, 30*(6), 845–856.

Cornish, K., Burack, J. A., Rahman, A., Munir, F., Russo, N., & Grant, C. (2005). Theory of mind deficits in children with fragile X syndrome. *Journal of Intellectual Disability Research, 49*(5), 372–378.

Cornish, K. M., Gray, K. M., & Rinehart, N. J. (2010). Fragile X syndrome and associated disorders. In J. Holmes (Ed.), *Advances in child development and behavior* (Vol. 39, pp. 211–235). Burlington, MA: Academic Press.

Cornish, K., Levitas, A., & Sudhalter, V. (2007). Fragile X syndrome: The journey from genes to behavior. In M. M. Mazzocco & J. L. Ross (Eds.), *Neurogenetic developmental disorders: Variation of manifestation in childhood* (pp. 73–103). Cambridge, MA: MIT Press.

Cornish, K., Scerif, G., & Karmiloff-Smith, A. (2007). Tracing syndrome-specific trajectories of attention across the lifespan. *Cortex, 43*, 672–685.

Cornish, K., Sudhalter, V., & Turk, J. (2004). Attention and language in fragile X. *Mental Retardation and Developmental Disabilities Research Reviews, 10*(1), 11–16.

Cornish, K., Turk, J., & Hagerman, R. (2008). The fragile X continuum: New advances and perspectives. *Journal of Intellectual Disability Research, 52*, 469–482.

Cornish, K. M., & Wilding, J. (2010). *Attention, genes and developmental disorders.* Oxford, UK: Oxford University Press.

Dodd, H. F., & Porter, M. A. (2010). I see happy people: Attention bias towards happy but not angry facial expressions in Williams syndrome. *Cognitive Neuropsychiatry, 15*(6), 549–567.

Dodd, H. F., Porter, M. A., Peters, G. L., & Rapee, R. M. (2010). Social approach in preschool children with Williams syndrome: The role of the face. *Journal of Intellectual Disability Research, 54*, 194–203.

Donnelly, S. L., Wolpert, C. M., Menold, M. M., Bass, M. P., Gilbert, J. R., Cuccaro, M. L., et al. (2000). Female with autistic disorder and monosomy X (Turner syndrome): Parent-of-origin effect of the X chromosome. *American Journal of Medical Genetics, 96*(3), 312–316.

Downey, J., Ehrhardt, A., Gruen, R., Bell, J., & Morishima, A. (1989). Psychopathology and social functioning in women with Turner syndrome. *Journal of Nervous and Mental Disease, 177*(4), 191–201.

Doyle, T. F., Bellugi, U., Korenberg, J. R., & Graham, J. (2004). "Everybody in the world is my friend" hypersociability in young children with Williams syndrome. *American Journal of Medical Genetics: Part A, 124A*(3), 263–273.

Einfeld, S. L., Tonge, B. J., & Florio, T. (1997). Behavioural and emotional disturbance in individuals with Williams syndrome. *American Journal on Mental Retardation, 102*(1), 45–53.

Einfeld, S. L., Tonge, B. J., & Rees, V. W. (2001). Longitudinal course of behavioral and emotional problems in Williams syndrome. *American Journal on Mental Retardation, 106*(1), 73–81.

Ey, E., Leblond, C. S., & Bourgeron, T. (2011). Behavioral profiles of mouse models for autism spectrum disorders. *Autism Research, 4*(1), 1939–3806.

Garner, C., Callias, M., & Turk, J. (1999). Executive function and theory of mind performance of boys with fragile-X syndrome. *Journal of Intellectual Disability Research, 43*(6), 466–474.

Grant, C. M., Apperly, I., & Oliver, C. (2007). Is theory of mind understanding impaired in males with fragile X syndrome? *Journal of Abnormal Child Psychology, 35*(1), 17–28.

Hagerman, P. J. (2008). The fragile X prevalence paradox. *Journal of Medical Genetics, 45*, 498–499.

Hagerman, R. J. (2006). Lessons from fragile X regarding neurobiology, autism, and neurodegeneration. *Journal of Developmental and Behavioral Pediatrics, 27*(1), 63–74.

Herba, C., & Phillips, M. (2004). Development of facial expression recognition from childhood to adolescence: Behavioural and neurological perspectives. *Journal of Child Psychology and Psychiatry, 45*(7), 1185–1198.

Hong, D., Scaletta Kent, J., & Kesler, S. (2009). Cognitive profile of Turner syndrome. *Developmental Disabilities Research Reviews, 15*(4), 270–278.

Hou, M., Wang, M. J., & Zhong, N. (2006). Principal genetic syndromes and autism: From phenotypes, proteins to genes. *Beijing Da Xue Xue Bao, 38*(1), 110–115.

Jawaid, A., Schmolck, H., & Schulz, P. E. (2008). Hypersociability in Williams syndrome: A role for the amygdala? *Cognitive Neuropsychiatry, 13*(4), 338–342.

Karmiloff-Smith, A. (2007). Williams syndrome. *Current Biology, 17*(24), R1035–R1036.

Klein-Tasman, B. P., Phillips, K. D., Lord, C., Mervis, C. B., & Gallo, F. J. (2009). Overlap with the autism spectrum in young children with Williams syndrome. *Journal of Developmental and Behavioral Pediatrics, 30*(4), 289–299.

Korenberg, J. R., Chen, X.-N., Hirota, H., Lai, Z., Bellugi, U., Burian, D., et al. (2000). Williams syndrome: The search for genetic origins of cognition. *Journal of Cognitive Neuroscience, 12*(Suppl. 1), 89–107.

Lagrou, K., Froidecoeur, C., Verlinde, F., Craen, M., De Schepper, J., François, I., et al. (2006). Psychosocial functioning, self-perception and body image and their auxologic correlates in growth hormone and oestrogen-treated young adult women with Turner syndrome. *Hormone Research in Paediatrics, 66*(6), 277–284.

Lagrou, K., Xhrouet-Heinrichs, D., Heinrichs, C., Craen, M., Chanoine, J.-P., Malvaux, P., et al. (1998). Age-related perception of stature, acceptance of therapy, and psychosocial functioning in human growth hormone-treated girls with Turner's syndrome. *Journal of Clinical Endocrinology and Metabolism, 83*(5), 1494–1501.

Lawrence, K., Campbell, R., Swettenham, J., Terstegge, J., Akers, R., Coleman, M., et al. (2003). Interpreting gaze in Turner syndrome: Impaired sensitivity to intention and emotion, but preservation of social cueing. *Neuropsychologia, 41*(8), 894–905.

Lawrence, K., Jones, A., Oreland, L., Spektor, D., Mandy, W., Campbell, R., et al. (2007). The development of mental state attributions in women with X-monosomy, and the role of mono-amine oxidase B in the sociocognitive phenotype. *Cognition, 102*(1), 84–100.

Leekam, S. R., Lopez, B., & Moore, C. (2000). Attention and joint attention in preschool children with autism. *Developmental Psychology, 36*(2), 261–273.

Loesch, D. Z., Bui, Q. M., Dissanayake, C., Clifford, S., Gould, E., Bulhak-Paterson, D., et al. (2007). Molecular and cognitive predictors of the continuum of autistic behaviours in fragile X. *Neuroscience and Biobehavioral Reviews, 31*(3), 315–326.

Martens, M. A., Wilson, S. J., Dudgeon, P., & Reutens, D. C. (2009). Approachability and the amygdala: Insights from Williams syndrome. *Neuropsychologia, 47*(12), 2446–2453.

Martin, J. P., & Bell, J. (1943). A pedigree of mental defect showing sex-linkage. *Journal of Neurology, Neurosurgery, and Psychiatry, 6*(3–4), 154–175.

Mazzocco, M. M., Baumgardner, T., Freund, L. S., & Reiss, A. L. (1998). Social functioning among girls with fragile X or Turner syndrome and their sisters. *Journal of Autism and Developmental Disorders, 28*(6), 509–517.

Mazzocco, M. M., Thompson, L., Sudhalter, V., Belser, R. C., Lesniak-Karpiak, K., & Ross, J. L. (2006). Language use in females with fragile X or Turner syndrome during brief initial social interactions. *Journal of Developmental and Behavioral Pediatrics, 27*(4), 319–328.

Mazzola, F., Seigal, A., MacAskill, A., Corden, B., Lawrence, K., & Skuse, D. H. (2006). Eye tracking and fear recognition deficits in Turner syndrome. *Social Neuroscience, 1*(3–4), 259–269.

McCauley, E., Ito, J., & Kay, T. (1986). Psychosocial functioning in girls with Turner's syndrome and short stature: Social skills, behavior problems, and self-concept. *Journal of the American Academy of Child Psychiatry, 25*(1), 105–112.

McCauley, E., Sybert, V. P., & Ehrhardt, A. A. (1986). Psychosocial adjustment of adult women with Turner syndrome. *Clinical Genetics, 29*(4), 284–290.

Mervis, C. B., & John, A. E. (2010). Cognitive and behavioral characteristics of children with Williams syndrome: Implications for intervention approaches. *American Journal of Medical Genetics: Part C. Seminars in Medical Genetics, 154C*(2), 229–248.

Mervis, C. B., Morris, C. A., Klein-Tasman, B. P., Bertrand, J., Kwitny, S., Appelbaum, L. G., et al. (2003). Attentional characteristics of infants and toddlers with Williams syndrome during triadic interactions. *Developmental Neuropsychology, 23*(1–2), 243–268.

Pavlidis, K., McCauley, E., & Sybert, V. P. (1995). Psychosocial and sexual functioning in women with Turner syndrome. *Clinical Genetics, 47*(2), 85–89.

Philofsky, A., Hepburn, S. L., Hayes, A., Hagerman, R., & Rogers, S. J. (2004). Linguistic and cognitive functioning and autism symptoms in young children with fragile X syndrome. *American Journal on Mental Retardation, 109*(3), 208–218.

Plesa Skwerer, D., Borum, L., Verbalis, A., Schofield, C., Crawford, N., Ciciolla, L., et al. (2009). Autonomic responses to dynamic displays of facial expressions in adolescents and adults with Williams syndrome. *Social Cognitive and Affective Neuroscience, 4*(1), 93–100.

Plesa-Skwerer, D., Faja, S., Schofield, C., Verbalis, A., & Tager-Flusberg, H. (2006). Perceiving facial and vocal expressions of emotion in individuals with Williams syndrome. *American Journal on Mental Retardation, 111*(1), 15–26.

Porter, M. A., Coltheart, M., & Langdon, R. (2007). The neuropsychological basis of hypersociability in Williams and Down syndrome. *Neuropsychologia, 45*(12), 2839–2849.

Porter, M. A., Shaw, T. A., & Marsh, P. J. (2010). An unusual attraction to the eyes in Williams–Beuren syndrome: A manipulation of facial affect while measuring face scanpaths. *Cognitive Neuropsychiatry, 15*(6), 505–530.

Riby, D. M., Doherty-Sneddon, G., & Bruce, V. (2009). The eyes or the mouth?: Feature salience and unfamiliar face processing in Williams syndrome and autism. *Quarterly Journal of Experimental Psychology, 62*(1), 189–203.

Riby, D. M., & Hancock, P. J. (2008). Viewing it differently: Social scene perception in Williams syndrome and autism. *Neuropsychologia, 46*(11), 2855–2860.

Riby, D. M., & Hancock, P. J. (2009). Do faces capture the attention of individuals with Williams syndrome or autism?: Evidence from tracking eye movements. *Journal of Autism and Developmental Disorders, 39*(3), 421–431.

Roberts, J. E., Mankowski, J. B., Sideris, J., Goldman, B. D., Hatton, D. D., Mirrett, P. L., et al. (2009). Trajectories and predictors of the development of very young boys with fragile X syndrome. *Journal of Pediatric Psychology 34*(8), 827–836.

Rogers, S. J., Wehner, E., & Hagerman, R. J. (2001). The behavioral phenotype in fragile X: Symptoms of autism in very young children with fragile X syndrome, idiopathic autism, and other developmental disorders. *Journal of Developmental and Behavioral Paediatrics, 22*, 409–417.

Ross, J. L., McCauley, E., Roeltgen, D., Long, L., Kushner, H., Feuillan, P., et al. (1996). Self-concept and behavior in adolescent girls with Turner syndrome: Potential estrogen effects. *Journal of Clinical Endocrinology and Metabolism, 81*(3), 926–931.

Rovet, J., & Holland, J. (1993). Psychological aspects of the Canadian randomized controlled trial of human growth hormone and low-dose ethinyl oestradiol in children with Turner syndrome. *Hormone Research, 39*(Suppl. 2), 60–64.

Santos, A., Rosset, D., & Deruelle, C. (2009). Human versus non-human face processing: Evidence from Williams syndrome. *Journal of Autism and Developmental Disorders, 39*(11), 1552–1559.

Santos, A., Silva, C., Rosset, D., & Deruelle, C. (2010). Just another face in the crowd: Evidence for decreased detection of angry faces in children with Williams syndrome. *Neuropsychologia, 48*(4), 1071–1078.

Siegel, P. T., Clopper, R., & Stabler, B. (1998). The psychological consequences of Turner syndrome and review of the National Cooperative Growth Study psychological substudy. *Pediatrics, 102*(2, Pt. 3), 488–491.

Skinner, M., Hooper, S., Hatton, D. D., Roberts, J., Mirrett, P., Schaaf, J., et al. (2005). Mapping nonverbal IQ in young boys with fragile X syndrome. *American Journal of Medical Genetics: Part A, 132*(1), 25–32.

Skuse, D. H. (2005). X-linked genes and mental functioning. *Human Molecular Genetics, 14*(Suppl. 1), R27–R32.

Skuse, D., Elgar, K., & Morris, E. (1999). Quality of life in Turner syndrome is related to chromosomal constitution: Implications for genetic counselling and management. *Acta Paediatrica, 88*(Suppl. 428), 110–113.

Stromme, P., Bjornstad, P., & Ramstad, K. (2002). Prevalence estimation of Williams syndrome. *Journal of Child Neurology, 17*, 269–271.

Sybert, V. P., & McCauley, E. (2004). Turner's syndrome. *New England Journal of Medicine, 351*, 1227–1238.

van Pareren, Y. K., Duivenvoorden, H. J., Slijper, F. M. E., Koot, H. M., Drop, S. L. S., & de Muinck Keizer-Schrama, S. M. P. F. (2005). Psychosocial functioning after discontinuation of long-term growth hormone treatment in girls with Turner syndrome. *Hormone Research in Paediatrics, 63*(5), 238–244.

Wide-Boman, U., Hanson, C., Hjelmquist, E., & Möller, A. (2006). Personality traits in women with Turner syndrome. *Scandinavian Journal of Psychology, 47*(3), 219–223.

Widen, S. C., & Russell, J. A. (2008). Children acquire emotion categories gradually. *Cognitive Development, 23*, 291–312.

CHAPTER 14

Pediatric Brain-Injury-Related Psychiatric Disorders and Social Function

Jeffrey E. Max

The highest-quality child psychiatric assessment involves a comprehensive biological, psychological, social, and spiritual understanding of the patient. Much has been written about the assessment and formulation of child psychiatric cases (Engel, 1978; Nurcombe & Gallagher, 1986). However, the literature is sparse with regard to psychiatric assessment and formulation in cases of pediatric brain injury (Max, 2011). Furthermore, the relationship of pediatric brain-injury-related psychiatric disorders to social function has never been systematically studied. Pediatric brain injury is a heterogeneous group of conditions that includes such conditions as traumatic brain injury (TBI), stroke, very low birth weight, epilepsy, fetal alcohol syndrome, meningitis, and encephalitis. This chapter focuses on pediatric TBI and social function, but includes references to other brain injury conditions. Several psychiatric disorders and their underlying psychosocial and lesion correlates are addressed.

Interview Measures for Assessing Psychiatric Symptoms and Social Dysfunction

Standardized child psychiatric measures that involve a structured or semistructured interview of a parent or caretaker and a child, in addition to considering outside data (including teachers' reports and academic records), constitute the "gold standard" for psychiatric assessment in research settings (Brown, Chadwick, Shaffer, Rutter, & Traub, 1981; Max, Robin, et al., 1997). The diagnosis reached by pursuing this standardized multi-informant approach has been referred to as the "best-estimate" diagnosis (Leckman, Sholomskas, Thompson, Belanger, & Weissman, 1982). The Neuropsychiatric

Rating Schedule (Max, Castillo, Lindgren, & Arndt, 1998) is an important standardized interview that was developed to measure a specific brain-injury-related psychiatric disorder: personality change (PC) due to brain injury (American Psychiatric Association, 2000). More general psychiatric interviews that are standardized and structured or semistructured include the Schedule for Affective Disorders and Schizophrenia for School-Age Children—Present and Lifetime Version (Kaufman et al., 1997) and the Diagnostic Interview for Children and Adolescents (Reich, 2000). These instruments can provide a rich narrative of the antecedents, behavior, and consequences of symptoms and their concomitant social dysfunction.

A semistructured interview conducted by a psychiatric expert can be invaluable in describing and understanding psychiatric symptoms and their social implications. Poor social functioning may be related to varied symptoms, such as irritability, garrulousness, inattention, and social withdrawal. The cause and phenomenology of these symptoms can be understood with careful questioning. For example, irritability may be a symptom of major depressive disorder, oppositional defiant disorder (ODD), psychosis, a manic or hypomanic episode of a bipolar disorder, or PC due to brain injury (labile subtype). Garrulousness may be related to a manic or hypomanic episode or to attention-deficit/ hyperactivity disorder (ADHD). Inattention may be evident in children with ADHD, generalized anxiety disorder, major depressive disorder, PC due to brain injury (apathetic type), posttraumatic stress disorder (PTSD), or petit mal seizures. Social withdrawal may be associated with generalized anxiety disorder, major depressive disorder, psychosis, or PC due to brain injury (apathetic type).

Models of Psychiatric Disorders and Social Competence

Social impairment is listed as a sufficient threshold-of-severity criterion in the definition of many psychiatric disorders (American Psychiatric Association, 2000). However, at this juncture, there are few measures that deal with all the components of social function outcome (Crowe, Beauchamp, Catroppa, & Anderson, 2011). Furthermore, there are few standardized measures of social function that capture specific characteristics of the heterogeneous *mechanisms* leading to varied dysfunctional social outcomes. In this regard, the socio-cognitive integration of abilities model (SOCIAL) provides a systematic framework for considering social function in brain-injured children (Beauchamp & Anderson, 2010). This model, which is described in greater detail elsewhere in this volume (see Anderson & Beauchamp, Chapter 1), attempts to model social skills and social function. SOCIAL includes a consideration of "mediators" such as internal factors (e.g., temperament, personality, physical attributes), external factors (e.g., family environment, socioeconomic status, culture), and brain development and integrity. These mediators purportedly interact reciprocally with cognitive functions, including attention/executive functions, communication abilities, and socio-emotional or affective processes, in determining the level and quality of social competence. This model has obvious overlap with the biopsychosocial model used for psychiatric formulations for many years (Engel, 1978). However, SOCIAL places greater emphasis on the cognitive (psychological) aspects of the biopsychosocial model, and thereby stimulates research into mechanisms of social competence and possibly even psychiatric disorders.

Relationship of Social Functioning to Behavioral and Emotional Symptoms in Brain-Injured Children

There has been no study of detailed social function measures and their relation to psychiatric disorders in brain-injured children. However, detailed measures of peer relationships have been studied in relation to behavioral and emotional symptoms in a sample of children with hemiplegia. These children were more likely than their nondisabled classmates to experience peer rejection, lack of friends, and victimization (Yude, Goodman, & McConachie, 1998). Lower IQ and teacher-reported conduct and/or hyperactivity problems measured soon after school entry predicted peer problems measured 3 years later (Yude & Goodman, 1999). The authors offered potential explanations for this association, including (1) that low IQ and disruptive behavior could lead to peer problems, or (2) that the same underlying brain abnormality could result in low IQ, behavioral disturbance, and constitutional difficulties in social skills and understanding. Interestingly, peer problems were not related to extent of neurological involvement, visibility of the physical disability, or family adversity. Furthermore, peer problems were not related to parent-reported behavioral and emotional ratings, suggesting that peer relationships at school were more closely related to a child's behavior at school than to the child's behavior at home. Another example of the relationship of social functioning and behavior comes from a study of children with TBI and uninjured controls: Assertive solutions in a social problem-solving paradigm were positively related to parent- and teacher-rated social and behavioral outcomes, whereas aggressive solutions were negatively related to the outcomes (Ganesalingam, Yeates, Sanson, & Anderson, 2007).

Social Functioning in Major Categories of Brain Injury-Related Psychiatric Disorders

The important categories of psychiatric disorders that have been studied in relation to pediatric brain injury are (1) PC due to TBI and due to stroke; (2) secondary ADHD (SADHD); (3) ODD and conduct disorder (CD); (4) anxiety disorders (including social phobia, PTSD, generalized anxiety disorder, and panic disorder with agoraphobia); (5) depressive disorders; (6) manic or hypomanic episodes of bipolar disorders; and (7) pervasive developmental disorders. First, however, the description of psychiatric disorders as *novel* or *current* should be clarified.

Novel and Current Psychiatric Disorders

Novel psychiatric disorder is the term given to a psychiatric disorder that emerges after brain injury. Such a disorder can be identified in circumstances where a child has no psychiatric disorder before the injury, but then develops some new disorder. The description can also be applied in a situation where a child with one or more preinjury psychiatric disorders develops a different disorder after the injury (e.g., a child with ADHD before the injury develops a major depressive disorder). Predictors of novel psychiatric disorders include severity of injury, preinjury psychiatric disorders, preinjury family function, socioeconomic status/preinjury intellectual function, family psychiatric history, and preinjury adaptive function (Brown et al., 1981; Max, Robin, et al., 1997).

In a controlled study involving children with severe TBI, children with mild TBI, and controls with orthopedic injury, family functioning, *current psychiatric disorders* (including persisting preinjury disorders and novel psychiatric disorders) in the children, and IQ were significant variables explaining between 22% and 47% of the variance in adaptive functioning outcomes (Max, Koele, Lindgren, et al., 1998). These outcomes included the Vineland Adaptive Behavior Scales (Sparrow, Balla, & Cicchetti, 1984) Adaptive Behavior Composite, Communication, Socialization, and Daily Living Skills ratings, in addition to the Child Behavior Checklist (Achenbach, 1991a) Social Competence rating and the Teacher's Report Form (Achenbach, 1991b) Adaptive Function rating.

PC Due to TBI

PC occurs after various types of brain injury, including TBI (Max et al., 2000, 2006; Max, Robertson, & Lansing, 2001; Max, Levin, et al., 2005) and stroke (5/29 in one cohort) (Max, Mathews, et al., 2002). In fact, PC is the most important psychiatric disorder that occurs after TBI. The name of the disorder is misleading, in that personality per se is not measured. Rather, PC is characterized by clinically significant affective lability, aggression, disinhibition/markedly impaired social judgment, apathy, or paranoia. The first three subtypes are the most common and frequently co-occur. Each of the subtypes is relevant to a discussion of social competence, but the disinhibited subtype is perhaps most directly related. The other subtypes lead to significant social problems as a complication of the poorly regulated expressions of anger and outright aggression (leading to avoidance by others), apathy (leading to social isolation because of indifference to participation), and paranoia (leading to social isolation due to active avoidance of others).

In one study of 94 children with TBI (severe TBI, $n = 37$; mild/moderate TBI, $n = 57$), PC was present in 60% of participants with severe TBI at some postinjury point and was persistent in 38% of these participants an average of 2 years after injury (Max et al., 2000). PC also occurred in 5% of participants with mild/moderate TBI but was transient. Within the group with severe TBI, PC was significantly associated with severity of TBI, impaired consciousness lasting longer than 100 hours, intellectual functioning and adaptive functioning (particularly socialization and daily living skills) decrements, and concurrent diagnosis of SADHD. PC was not associated with any psychosocial variables, such as family function, family life events, socioeconomic status, family psychiatric history, or preinjury lifetime psychiatric disorder in a child (Max et al., 2000). The reader is referred to a detailed description of the phenomenology of PC (Max et al., 2001). The subtypes of PC diagnosed in the cohort with severe TBI were as follows: labile, 49%; aggressive 38%; disinhibited, 38%; apathetic, 14%; and paranoid, 5%. Within the disinhibited category of PC, rates of the following symptoms occurred: uninhibited/disinhibited acts, 12/37 (32%); disinhibited vocalization/verbalization, 15/37 (41%); lack of tact or concern for others/insensitivity to others' feelings/reactions, 8/37% (22%); inability to plan ahead (lack of foresight, inability to judge consequences of actions), 10/37 (27%); and sexually inappropriate behavior, 6/37 (16%) (Max et al., 2001).

A large follow-up prospective study found that 31 of 140 children with mild to severe TBI had PC at some point within the first 6 months after injury (Max, Levin, et al., 2005). The frequency distribution of PC subtypes was similar to that in the earlier study. As in the previous study, severity of injury was associated with PC, whereas none of the psychosocial variables were associated. Lesions of the dorsal prefrontal cortex, particularly the superior frontal gyrus, were associated with PC after severity of injury and the presence of other lesions were controlled for. PC was associated with SADHD and postinjury ODD/CD/disruptive behavior disorder not otherwise specified. The morbidity of PC was reflected in the finding that concurrent adaptive functioning (including socialization) was significantly lower in participants with PC than in those without PC (Max, Levin, et al., 2005).

Further postinjury follow-up through 12 months and 24 months revealed that severity of injury continued to be associated with PC (Max et al., 2006). The association with superior frontal gyrus lesions was significant through 12 months, and there was a significant association with frontal white matter lesions in the second postinjury year. Lower preinjury adaptive function was associated with PC in the second year, supporting the concept of preinjury brain "reserve" (Kesler, Adams, Blasey, & Bigler, 2003) as a protective factor with regard to outcome.

With data from our prospective TBI work, my colleagues and I plan to analyze attention, executive function, and social problem-solving data to determine the relationship of PC or its subtypes to these cognitive functions. In this way, the nature and the mechanisms of PC-related social competence problems may be elucidated.

Secondary ADHD

SADHD refers to a condition in which ADHD develops after TBI. SADHD is associated with increasing severity of injury, and with intellectual and adaptive function deficits as well as family dysfunction, in samples of children with mild to severe TBI (Max, Arndt, et al., 1998; Max et al., 2004). In samples with at least moderate severity of injury, adaptive deficits are still evident, but findings regarding intellectual function outcome are mixed (Gerring et al., 1998; Max et al., 2004). Furthermore, in samples of at least moderate injury severity, SADHD is not associated with family function at the time of assessment, socioeconomic status, family stressors, family psychiatric history, gender, or lesion area. An overlapping study of attention-deficit/hyperactivity symptoms reported a similar relationship with severity and found that, overall, these symptoms were associated with lower preinjury family functioning (Max, Arndt, et al., 1998). A rehabilitation center sample of children with mostly severe TBI had similar findings, and the children with SADHD had greater "premorbid" psychosocial adversity (Gerring et al., 1998). Lesion correlates of SADHD include right putamen or thalamic lesions (Herskovits et al., 1999; Gerring et al., 2000) and orbitofrontal gyrus lesions (Max, Schachar, et al., 2005). These anatomical findings suggest that the clinical syndrome of SADHD may be generated by lesions in varied locations along specific corticostriatal–pallidal–thalamic loops.

One study (Schachar, Levin, Max, Purvis, & Chen, 2004) focused on the relationship of SADHD and inhibition deficit (measured with a stop signal reaction time task) in injured children with mild to severe TBI and uninjured control children. A deficit in inhibition, not unlike that usually seen in developmental ADHD, was found only in children with severe TBI who also had SADHD. An unrelated study found worse memory skills in children with SADHD than in children with TBI and preinjury ADHD (Slomine et al., 2005).

ADHD and/or traits of ADHD are the most common psychopathologies after childhood stroke (Max, Mathews, et al., 2002). ADHD symptomatology was associated with lesions of the combined mesial prefrontal and orbitofrontal regions, as well as with lesions within Posner's executive attention network (Posner & Petersen, 1990)—that is, frontal mesial structures bilaterally, including anterior and posterior cingulate, supplementary motor area, prefrontal region, Rolandic region, and bilateral basal ganglia) (Max, Manes, et al., 2005). ADHD symptomatology after small strokes (< 10 cc) was associated with ventral striatal lesions involving predominantly putamen lesions (Max, Fox, et al., 2002). A neurocognitive study of ADHD symptomatology after stroke found a significant relationship with an "impaired neurocognition" factor (generalized intellectual function, specific reading ability, overall adaptive function, perseveration) and an "inattention/apathy" factor (nonperseverative errors, low motivation/apathy) (Max et al., 2003).

ODD and CD

ODD is characterized by a recurring pattern of hostile, negativistic, and defiant behavior, with specific symptoms including losing one's temper, arguing with adults, disregarding adults' rules, deliberately provoking others, blaming others for one's own mistakes, easily becoming irritated, often being angry, and often being spiteful (American Psychiatric Association, 2000). The symptomatology must cause significant impairment in social, academic, or occupational functioning. From the panoply of symptoms, it is easy to appreciate how social impairment may be a complication of the syndrome. CD is a related and more severe form of disruptive behavior characterized by symptoms within four broad categories: including aggression toward people and animals, destruction of property, theft or deceitfulness, and serious violation of rules (American Psychiatric Association, 2000). As for ODD, a diagnosis of CD is made only if there is a clinically significant impairment in social, academic, or occupational functioning. ODD and CD can occur as complications of TBI.

In one study, ODD symptoms in the first 12 months after TBI was related to preinjury family function, socioeconomic status, and preinjury ODD symptomatology (Max, Castillo, Bokura, et al., 1998). Increased severity of TBI predicted ODD symptoms 24 months after injury. Socioeconomic status influenced change (from before TBI) in ODD symptoms at 6, 12, and 24 months after TBI. Only at 24 months after injury was severity of injury a predictor of change in ODD symptoms. Psychosocial factors appeared to exert a greater influence than severity of injury in accounting for ODD symptoms and change in such symptomatology in the first but not the second year after pediatric TBI. This seems to be related to persistence of new ODD symptoms after more serious TBI.

Study design differences yield vastly different findings, however. For example, relevant findings from a referred brain injury clinic sample were that children who developed ODD/Conduct Disorder after TBI, when compared to children without a lifetime history of the disorder, had significantly greater family dysfunction, showed a trend toward more family history of alcohol dependence/abuse and incurred a milder TBI (Max, Lindgren, et al., 1998). Another study found that new-onset ODD and CD at 1 year after injury in a referred sample from a rehabilitation center occurred in 9% and 8% of participants, respectively. Risk factors for the development of disruptive disorders/symptoms included preinjury psychosocial adversity, delinquency ratings, and affective lability (Gerring et al., 2009).

Anxiety Disorders

Anxiety disorders often occur as complications of brain injury (Gerring et al., 2002; Vasa et al., 2002; Max, Keatley, et al., 2011). Studies in samples of children with TBI have examined biological, psychological, and social correlates and predictors of the development of anxiety disorders and symptoms. Anxiety symptoms tend to distract people from optimal engagement in their endeavors. For example, a child who is worried about school grades and test performance may "go blank" and score lower than he or she might in the absence of worry. Tension headaches and stomachaches may result in being absent from school and missing social events, which can decrease social engagement and social competence. Social phobia is a more discrete anxiety disorder characterized by a strong and persisting fear of at least one social or performance-related situation in which the person is exposed to unknown people or to potential scrutiny by others. The individual is afraid of acting in a way (or showing anxiety symptoms) that will be humiliating or embarrassing (American Psychiatric Association, 2000). Social phobia typically adversely affects social function because the person avoids social engagements (resulting in less lifetime experience) and demonstrates excessive anxiety (leading to lower levels of social success). PTSD related to TBI leads to social function deficits because the child is preoccupied by intrusive thoughts and memories, as well as distractibility and feelings of alienation. Panic disorder, especially if accompanied by agoraphobia, also leads to decreased social function. Children with panic disorder occupy ever-constricted space to feel safer in the event of the next panic attack.

Research has shown that PTSD and subsyndromal posttraumatic stress disturbances occur despite neurogenic amnesia associated with TBI. In one study, few (2/46; 4%) children with TBI developed PTSD (Max, Castillo, Robin, et al., 1998). However, the frequency with which children exhibited at least one PTSD symptom ranged from 68% in the first 3 months afterr injury to 12% at 24 months. The most consistent predictors of PTSD symptomatology were (1) the presence of mood or anxiety disorder at the time of injury and (2) greater injury severity. A second study (Levi, Drotar, Yeates, & Taylor, 1999) found a significant relationship between parent- and child-reported PTSD symptomatology and severe TBI (vs. moderate TBI and orthopedic injury), even after the researchers controlled for ethnicity, social disadvantage, and age at injury. Family socioeconomic disadvantage was associated with greater PTSD symptomatology in all injury groups. A third study found that PTSD 1 year after injury was present in 13% of children with severe TBI from a rehabilitation center (Gerring et al., 2002). PTSD was associated with early postinjury anxiety symptoms and female gender. Posttraumatic symptoms were related to preinjury psychosocial adversity, preinjury anxiety symptoms, and injury severity, as well as early postinjury depression symptoms and non-anxiety-related psychiatric diagnoses. The PTSD criteria for reexperiencing were associated with a lower lesion fraction (ratio of lesion volume within the region of interest relative to the total volume of the region of interest) in the right limbic area—specifically, the cingulum, an area often activated in imaging studies of trauma reexperiencing (Herskovits, Gerring, Davatzikos, & Bryan, 2002). PTSD hyperarousal symptoms were associated with left temporal lesions and absence of left orbitofrontal lesions (Vasa et al., 2004).

Obsessive–compulsive disorder can occur following TBI in adolescents (Max et al., 1995; Vasa et al., 2002). Frontal and temporal lobe lesions may be sufficient to precipitate the syndrome in the absence of striatal injury (Max et al., 1995). New-onset obsessions are associated with female gender, psychosocial adversity, and mesial frontal as well as temporal lesions (Grados et al., 2008).

A wide variety of other anxiety disorders have been reported after pediatric TBI. These include specific phobia, separation anxiety disorder, and overanxious disorder of childhood (which is now subsumed under generalized anxiety disorder) (Max, Robin, et al., 1997; Max, Smith, Sato, Mattheis, Castillo, et al., 1997; Max, Koele, Smith, et al., 1998). One study found no statistically significant increase in any single anxiety disorder compared with preinjury rates, but there was a trend in this regard for overanxious disorder (Vasa et al., 2002). However, anxiety symptoms increased significantly after injury. Younger age at injury and preinjury anxiety symptoms correlated positively with postinjury anxiety symptoms (Vasa et al., 2002). Other investigators reported that postinjury level of stress and severity of injury were associated with mood/anxiety disorders (Luis & Mittenberg, 2002). A lesion study of general anxiety symptoms found a significant inverse relationship with orbitofrontal cortex lesion volume and orbitofrontal cortex total lesions (Vasa et al., 2004). Another study found that lesions of the superior frontal gyrus were significantly associated with postinjury anxiety symptoms, whereas frontal white matter lesions were associated at a trend level (Max, Keatley, et al., 2011).

Depression

Depressive disorders after pediatric TBI may negatively affect social functioning for several reasons: poor motivation and low energy for engaging in activities; being less present in social interactions because of poor concentration and depressive ruminations; and being less socially attractive due to pessimistic attitude, blunted or depressed affect, and irritability.

A substantial proportion of children who manifest depressed mood after TBI have a preinjury personal history of depressive disorders, and most of the remaining children have identifiable risk factors for a new-onset depressive disorder, such as a personal history of preinjury anxiety disorder or a first-degree relative with major depressive disorder (Max, 2011). One study found that 25–33% of children with severe TBI had an ongoing depressive disorder or a postinjury history of a depressive disorder (Max, Koele, Smith, et al., 1998). TBI increased the risk of depressive *symptoms*, especially among more socially disadvantaged children, and depressive symptoms were not strongly associated with postinjury neurocognitive performance (Kirkwood et al., 2000). New-onset depression—including disorders with no comorbid new anxiety disorder (nonanxious depression), as well as disorders with a comorbid new anxiety disorder (anxious depression)—was associated with specific left-sided and right-sided lesions and with older age at injury, compared with children who did not develop depression. Nonanxious depression was associated with left-hemisphere lesions, specifically in the left inferior frontal gyrus and temporal tip; anxious depression was associated with right-hemisphere lesions, specifically in the right frontal lobe white matter, and left parietal lobe lesions (Max et al., 2012). The association of nonanxious depression with left-sided lesions and anxious depression predominantly with right-sided lesions is similar to adult TBI findings (Jorge, Robinson, Starkstein, & Arndt, 1993). Anxious depression was also associated with PC and a family history of anxiety disorder (Max et al., 2011).

Manic or Hypomanic Episodes

A manic or hypomanic episode of a childhood bipolar disorder can cause major social disruption related to inappropriate expansive affect, which frequently progresses to intense irritability. Despite the fact that such episodes may be relatively brief, behavioral indiscretions can frequently stigmatize a patient long after remission is achieved. Several case reports have been published on the emergence of mania or hypomania after pediatric TBI (Sayal, Ford, & Pipe, 2000), but there is only one report of mania or hypomania from a pediatric TBI cohort. Four of 50 children (8%) from a prospective study of TBI developed mania or hypomania (Max, Smith, Sato, Mattheis, Robin, et al., 1997). A psychiatric interview is the optimal method to discern the phenomenology of the overlapping diagnoses of a bipolar disorder, ADHD, and PC and to generate a valid differential diagnosis (Max et al., 2000). Frontal and temporal lobe lesion location, increased severity of injury, and family history of major mood disorder appear to be risk factors for mania/hypomania secondary to TBI. Long-lasting episodes and similar frequency of elation and irritability may be characteristic.

Pervasive Developmental Disorders

Pervasive development disorders or autism spectrum disorders are developmental disorders characterized by symptoms related to qualitative impairments in social interaction and communication, as well as by restricted, repetitive, stereotyped patterns of behavior, interests, and activities (American Psychiatric Association, 2000). These disorders have been studied extensively because at their core is social impairment. The prospect of understanding the biopsychosocial mechanisms of these disorders may have far-reaching benefits for many other conditions characterized by social function problems. There have been no reports of the development of pervasive developmental disorders after TBI. However, cases have been described in relation to brain tumors (Hoon & Reiss, 1992) and childhood hemiplegia (4/149; 3%) (Goodman & Graham, 1996).

Connecting Behavioral Symptoms, Psychiatric Disorders, and Specific Social Outcomes

The challenge facing researchers at the nexus of the fields of social function, behavior, psychiatric disorders, neuropsychology, and brain development is to define the dynamic interrelationships of the constructs these fields have identified. SOCIAL may be helpful in this effort. Beauchamp and Anderson have discussed the state of the science of assessing components of SOCIAL (Beauchamp & Anderson, 2010; see also Anderson & Beauchamp, Chapter 1, this volume).

Essentially, there is a fairly well-developed methodology for the measurement of "mediators" in this model, including the internal factors (e.g., temperament, personality, physical attributes), external factors (e.g., family environment, socioeconomic status, culture), and brain development and integrity. There is also a reasonably well-developed methodology for the measurement of attention/executive functions and communication abilities. However, the assessment of socio-emotional or affective processes needs refinement. Integration of specific behavioral symptomatology and specific psychiatric syndromes into the model is essential to an understanding of social function (see Figure 14.1). This integration will require superimposition of a permeable layer, incorporating behavioral symptoms and psychiatric syndromes, over the other components (mediators and cognitive functions) of SOCIAL. In the model, patterns and bidirectional pathways to and from psychiatric syndromes within the contexts of mediators and cognitive functions can be identified and tracked through to their influence on social skills and ultimately to overall social competence.

The development of a valid and reliable lexicon or compendium of constructs of clinically informed plausible mechanisms linking brain-injury-related behavioral symptoms and/or psychiatric disorders with a validated and reliable set of social outcome measures is essential. Important behavioral symptoms include irritability, aggression, emotional lability, hyperactivity, impulsivity, anxiety, vigilance, avoidance, apathy, depression, callousness, oppositional behavior, and paranoia. Important psychiatric diagnostic categories include PC due to brain injury, SADHD, ODD, CD, anxiety disorders (including social phobia, generalized anxiety disorder, PTSD, and panic disorder with agoraphobia), depressive disorders, manic or hypomanic episodes in bipolar disorders, and pervasive developmental disorders. Empirically testable *mechanisms* affecting social competence will emanate from competing and complementary models of cognitive and psychological functioning. An important mechanism includes approach–avoidance dysregulation, which may occur in individuals with emerging personality disorders (Sharp et al., 2011), certain attachment classifications (Zeanah, Keyes, & Settles, 2003), affective problem-solving deficits (Epstein & Bishop, 1981), and related social information-processing/problem-solving difficulties (Crick & Dodge, 1996). Other mechanisms include aversive distancing of others (via aggression or rudeness), limited approach toward others (via shyness, depression, or low empathy), and active avoidance of others (via paranoia or

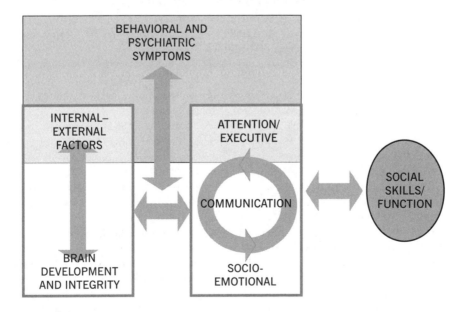

FIGURE 14.1. A version of Anderson and Beauchamp's SOCIAL in which a permeable layer representing behavioral and psychiatric symptoms is superimposed on the other components.

misperception of intent). Each of these measurable mechanisms may be affected in different ways and degrees by the above-noted mediators (internal–external factors and brain development and integrity), by cognitive functions (attention/executive, communication, and socio-emotional), and by behavioral symptoms and psychiatric syndromes. Finally, reliable and valid models measuring social outcome will need to be applied (Crowe et al., 2011). These could include such constructs as peer rejection, lack of friends, victimization, and bullying others (Yude & Goodman, 1999).

Conclusion

As humans, we are social beings. The acquisition of social competence is an ongoing developmental process that is a critical aspect of perceived health and success. Impairment in social functioning is considered a sufficient criterion in the differentiation of subsyndromal behavioral symptoms and a variety of psychiatric disorders. Brain injury in children is an important cause of psychiatric disorders, neurocognitive deficits, and problems with social functioning. The study of social competence in association with new-onset psychiatric disorders after brain injury and neurocognitive function has the potential to enhance knowledge about its biopsychosocial underpinnings. Identification of the mechanisms linking specific psychiatric disorders with social outcomes may provide clues for targeted approaches to the amelioration of social dysfunction, and, by extension, to a reduction of psychiatric morbidity after brain injury.

References

Achenbach, T. M. (1991a). *Manual for the Child Behavior Checklist/4–18 and 1991 Profile*. Burlington: University of Vermont, Department of Psychiatry.

Achenbach, T. M. (1991b). *Manual for the Teacher's Report Form and 1991 Profile*. Burlington: University of Vermont, Department of Psychiatry.

American Psychiatric Association. (2000). *Diagnostic and statistical manual of mental disorders* (4th ed., text rev.). Washington, DC: Author.

Beauchamp, M. H., & Anderson, V. (2010). SOCIAL: An integrative framework for the development of social skills. *Psychological Bulletin, 136*(1), 39–64.

Brown, G., Chadwick, O., Shaffer, D., Rutter, M., & Traub, M. (1981). A prospective study of children with head injuries: III. Psychiatric sequelae. *Psychological Medicine, 11*(1), 63–78.

Crick, N. R., & Dodge, K. A. (1996). Social information-processing mechanisms in reactive and proactive aggression. *Child Development, 67*(3), 993–1002.

Crowe, L. M., Beauchamp, M. H., Catroppa, C., & Anderson, V. (2011). Social function assessment tools for children and adolescents: A systematic review from 1988 to 2010. *Clinical Psychology Review, 31*(5), 767–785.

Engel, G. L. (1978). The biopsychosocial model and the education of health professionals. *Annals of the New York Academy of Sciences, 310*, 169–187.

Epstein, N. B., & Bishop, D. S. (1981). Problem centered systems therapy of the family. In A. S. Gurman & D. P. Kniskern (Eds.), *Handbook of family therapy* (pp. 444–482). New York: Brunner/Mazel.

Ganesalingam, K., Yeates, K. O., Sanson, A., & Anderson, V. (2007). Social problem-solving skills following childhood traumatic brain injury and its association with self-regulation and social and behavioural functioning. *Journal of Neuropsychology, 1*(Pt. 2), 149–170.

Gerring, J., Brady, K., Chen, A., Quinn, C., Herskovits, E., Bandeen-Roche, K., et al. (2000). Neuroimaging variables related to development of secondary attention deficit hyperactivity disorder after closed head injury in children and adolescents. *Brain Injury, 14*(3), 205–218.

Gerring, J. P., Brady, K. D., Chen, A., Vasa, R., Grados, M., Bandeen-Roche, K. J., et al. (1998). Premorbid prevalence of ADHD and development of secondary ADHD after closed head injury. *Journal of the American Academy of Child and Adolescent Psychiatry, 37*(6), 647–654.

Gerring, J. P., Grados, M. A., Slomine, B., Christensen, J. R., Salorio, C. F., & Cole, W. R. (2009). Disruptive behaviour disorders and disruptive symptoms after severe paediatric traumatic brain injury. *Brain Injury, 23*(12), 944–955.

Gerring, J. P., Slomine, B., Vasa, R. A., Grados, M., Chen, A., Rising, W., et al. (2002). Clinical predictors of posttraumatic stress disorder after closed head injury in children. *Journal of the American Academy of Child and Adolescent Psychiatry, 41*(2), 157–165.

Goodman, R., & Graham, P. (1996). Psychiatric problems in children with hemiplegia: Cross sectional epidemiological survey. *British Medical Journal, 312*(7038), 1065–1069.

Grados, M. A., Vasa, R. A., Riddle, M. A., Slomine, B. S., Salorio, C., Christensen, J., et al. (2008). New onset obsessive–compulsive symptoms in children and adolescents with severe traumatic brain injury. *Depression and Anxiety, 25*(5), 398–407.

Herskovits, E. H., Gerring, J. P., Davatzikos, C., & Bryan, R. N. (2002). Is the spatial distribution of brain lesions associated with closed-head injury in children predictive of subsequent development of posttraumatic stress disorder? *Radiology, 224*(2), 345–351.

Herskovits, E. H., Megalooikonomou, V., Davatzikos, C., Chen, A., Bryan, R. N., & Gerring, J. P. (1999). Is the spatial distribution of brain lesions associated with closed-head injury predictive of subsequent development of attention-deficit/hyperactivity disorder?: Analysis with brain-image database. *Radiology, 213*(2), 389–394.

Hoon, A. H., & Reiss, A. L. (1992). The mesial-temporal lobe and autism: Case report and review. *Developmental Medicine and Child Neurology, 34*(3), 252–259.

Jorge, R. E., Robinson, R. G., Starkstein, S. E., & Arndt, S. V. (1993). Depression and anxiety following traumatic brain injury. *Journal of Neuropsychiatry and Clinical Neurosciences, 5*(4), 369–374.

Kaufman, J., Birmaher, B., Brent, D., Rao, U., Flynn, C., Moreci, P., et al. (1997). Schedule for Affective Disorders and Schizophrenia for School-Age Children—Present and Lifetime Version (K-SADS-PL): Initial reliability and validity data. *Journal of the American Academy of Child and Adolescent Psychiatry, 36*(7), 980–988.

Kesler, S. R., Adams, H. F., Blasey, C. M., & Bigler, E. D. (2003). Premorbid intellectual functioning, education, and brain size in traumatic brain injury: An investigation of the cognitive reserve hypothesis. *Applied Neuropsychology, 10*(3), 153–162.

Kirkwood, M., Janusz, J., Yeates, K. O., Taylor, H. G., Wade, S. L., Stancin, T., et al. (2000). Prevalence and correlates of depressive symptoms following traumatic brain injuries in children. *Child Neuropsychology, 6*(3), 195–208.

Leckman, J. F., Sholomskas, D., Thompson, W. D., Belanger, A., & Weissman, M. M. (1982). Best estimate of lifetime psychiatric diagnosis: A methodological study. *Archives of General Psychiatry, 39*(8), 879–883.

Levi, R. B., Drotar, D., Yeates, K. O., & Taylor, H. G. (1999). Posttraumatic stress symptoms in children following orthopedic or traumatic brain injury. *Journal of Clinical Child Psychology, 28*(2), 232–243.

Luis, C. A., & Mittenberg, W. (2002). Mood and anxiety disorders following pediatric traumatic brain injury: A prospective study. *Journal of Clinical and Experimental Neuropsychology, 24*(3), 270–279.

Max, J. E. (2011). Children and adolescents: Traumatic brain injury. In J. M. Silver, T. W.

McAllister, & S. C. Yudofsky (Eds.), *Textbook of traumatic brain injury* (2nd ed., pp. 438–450). Washington, DC: American Psychiatric Publishing.

Max, J. E., Arndt, S., Castillo, C. S., Bokura, H., Robin, D. A., Lindgren, S. A., et al. (1998). Attention-deficit hyperactivity symptomatology after traumatic brain injury: A prospective study. *Journal of the American Academy of Child and Adolescent Psychiatry, 37*(8), 841–847.

Max, J. E., Castillo, C. S., Bokura, H., Robin, D. A., Lindgren, S. D., Smith, W. L., Jr., et al. (1998). Oppositional defiant disorder symptomatology after traumatic brain injury: A prospective study. *Journal of Nervous and Mental Disease, 186*(6), 325–332.

Max, J. E., Castillo, C. S., Lindgren, S. D., & Arndt, S. (1998). The Neuropsychiatric Rating Schedule: Reliability and validity. *Journal of the American Academy of Child and Adolescent Psychiatry, 37*(3), 297–304.

Max, J. E., Castillo, C. S., Robin, D. A., Lindgren, S. D., Smith, W. L., Sato, Y., et al. (1998). Post-traumatic stress symptomology after childhood traumatic brain injury. *Journal of Nervous and Mental Disease, 186,* 589–596.

Max, J. E., Fox, P. T., Lancaster, J. L., Kochunov, P., Mathews, K., Manes, F. F., et al. (2002). Putamen lesions and the development of attention-deficit/hyperactivity symptomatology. *Journal of the American Academy of Child and Adolescent Psychiatry, 41*(5), 563–571.

Max, J. E., Keatley, E., Wilde, E. A., Bigler, E. D., Levin, H. S., Schachar, R. J., et al. (2011). Anxiety disorders in children and adolescents in the first six months after traumatic brain injury. *Journal of Neuropsychiatry and Clinical Neurosciences, 23*(1), 29–39.

Max, J. E., Keatley, E., Wilde, E. A., Bigler, E. D., Schachar, R. J., Saunders, A. E., et al. (2012). Depression in children and adolescents in the first six months after traumatic brain injury. *International Journal of Developmental Neuroscience.* [Epub ahead of print]

Max, J. E., Koele, S. L., Castillo, C. C., Lindgren, S. D., Arndt, S., Bokura, H., et al. (2000). Personality change disorder in children and adolescents following traumatic brain injury. *Journal of the International Neuropsychological Society, 6*(3), 279–289.

Max, J. E., Koele, S. L., Lindgren, S. D., Robin, D. A., Smith W. L., Jr., Sato, Y., et al. (1998). Adaptive functioning following brain injury and orthopedic injury: A controlled study. *Archives of Physical Medicine and Rehabilitation, 79,* 893–899.

Max, J. E., Koele, S. L., Smith, W. L., Jr., Sato, Y., Lindgren, S. D., Robin, D. A., et al. (1998). Psychiatric disorders in children and adolescents after severe traumatic brain injury: A controlled study. *Journal of the American Academy of Child and Adolescent Psychiatry, 37*(8), 832–840.

Max, J. E., Lansing, A. E., Koele, S. L., Castillo, C. C., Bokura, H., Schachar, R., et al. (2004). Attention deficit hyperactivity disorder in children and adolescents following traumatic brain injury. *Developmental Neuropsychology, 25*(1–2), 159–177.

Max, J. E., Levin, H. S., Landis, J., Schachar, R., Saunders, A., Ewing-Cobbs, L., et al. (2005). Predictors of personality change due to traumatic brain injury in children and adolescents in the first six months after injury. *Journal of the American Academy of Child and Adolescent Psychiatry, 44*(5), 434–442.

Max, J. E., Levin, H. S., Schachar, R. J., Landis, J., Saunders, A. E., Ewing-Cobbs, L., et al. (2006). Predictors of personality change due to traumatic brain injury in children and adolescents six to twenty-four months after injury. *Journal of Neuropsychiatry and Clinical Neuroscience, 18*(1), 21–32.

Max, J. E., Lindgren, S. D., Knutson, C., Pearson, C. S., Ihrig, D., & Welborn, A. (1998). Child and adolescent traumatic brain injury: Correlates of disruptive behavior disorders. *Brain Injury, 12*(1), 41–52.

Max, J. E., Manes, F. F., Robertson, B. A., Mathews, K., Fox, P. T., & Lancaster, J. (2005). Prefrontal and executive attention network lesions and the development of attention-deficit/

hyperactivity symptomatology. *Journal of the American Academy of Child and Adolescent Psychiatry, 44*(5), 443–450.

Max, J. E., Mathews, K., Lansing, A., Robertson, B., Fox, P., Lancaster, J., et al. (2002). Psychiatric disorders after childhood stroke. *Journal of the American Academy of Child and Adolescent Psychiatry, 41*(5), 555–562.

Max, J. E., Mathews, K., Manes, F. F., Robertson, B. A., Fox, P. T., Lancaster, J. L., et al. (2003). Attention deficit hyperactivity disorder and neurocognitive correlates after childhood stroke. *Journal of the International Neuropsychological Society, 9*(6), 815–829.

Max, J. E., Robertson, B. A. M., & Lansing, A. E. (2001). The phenomenology of personality change due to traumatic brain injury in children and adolescents. *Journal of Neuropsychiatry and Clinical Neurosciences, 13*(2), 161–170.

Max, J. E., Robin, D. A., Lindgren, S. D., Smith, W. L., Jr., Sato, Y., Mattheis, P. J., et al. (1997). Traumatic brain injury in children and adolescents: Psychiatric disorders at two years. *Journal of the American Academy of Child and Adolescent Psychiatry, 36*(9), 1278–1285.

Max, J. E., Schachar, R. J., Levin, H. S., Ewing-Cobbs, L., Chapman, S. B., Dennis, M., et al. (2005). Predictors of attention-deficit/hyperactivity disorder within 6 months after pediatric traumatic brain injury. *Journal of the American Academy of Child and Adolescent Psychiatry, 44*(10), 1032–1040.

Max, J. E., Smith, W. L., Lindgren, S. D., Robin, D. A., Mattheis, P., Stierwalt, J., et al. (1995). Case study: Obsessive–compulsive disorder after severe traumatic brain injury in an adolescent. *Journal of the American Academy of Child and Adolescent Psychiatry, 34*(1), 45–49.

Max, J. E., Smith, W. L., Jr., Sato, Y., Mattheis, P. J., Castillo, C. S., Lindgren, S. D., et al. (1997). Traumatic brain injury in children and adolescents: Psychiatric disorders in the first three months. *Journal of the American Academy of Child and Adolescent Psychiatry, 36*(1), 94–102.

Max, J. E., Smith, W. L., Sato, Y., Mattheis, P. J., Robin, D. A., Stierwalt, J. A. G., et al. (1997). Mania and hypomania following traumatic brain injury in children and adolescents. *Neurocase, 3*, 119–126.

Nurcombe, B., & Gallagher, R. M. (1986). *The clinical process in psychiatry.* Cambridge, UK: Cambridge University Press.

Posner, M. I., & Petersen, S. E. (1990). The attention system of the human brain. *Annual Review of Neuroscience, 13*, 25–42.

Reich, W. (2000). Diagnostic Interview for Children and Adolescents (DICA). *Journal of the American Academy of Child and Adolescent Psychiatry, 39*(1), 59–66.

Sayal, K., Ford, T., & Pipe, R. (2000). Bipolar disorder after head injury. *Journal of the American Academy of Child and Adolescent Psychiatry, 39*(4), 525–528.

Schachar, R., Levin, H. S., Max, J. E., Purvis, K., & Chen, S. (2004). Attention deficit hyperactivity disorder symptoms and response inhibition after closed head injury in children: Do pre-injury behavior and injury severity predict outcome? *Developmental Neuropsychology, 25*(1–2), 179–198.

Sharp, C., Pane, H., Ha, C., Venta, A., Patel, A. B., Sturek, J., et al. (2011). Theory of mind and emotion regulation difficulties in adolescents with borderline traits. *Journal of the American Academy of Child and Adolescent Psychiatry, 50*(6), 563–573.

Slomine, B. S., Salorio, C. F., Grados, M. A., Vasa, R. A., Christensen, J. R., & Gerring, J. P. (2005). Differences in attention, executive functioning, and memory in children with and without ADHD after severe traumatic brain injury. *Journal of the International Neuropsychological Society, 11*(5), 645–653.

Sparrow, S., Balla, D., & Cicchetti, D. (1984). *Vineland Adaptive Behavior Scales.* Circle Pines, MN: American Guidance Service.

Vasa, R. A., Gerring, J. P., Grados, M., Slomine, B., Christensen, J. R., Rising, W., et al. (2002).

Anxiety after severe pediatric closed head injury. *Journal of the American Academy of Child and Adolescent Psychiatry, 41*(2), 148–156.

Vasa, R. A., Grados, M., Slomine, B., Herskovits, E. H., Thompson, R. E., Salorio, C., et al. (2004). Neuroimaging correlates of anxiety after pediatric traumatic brain injury. *Biological Psychiatry, 55*(3), 208–216.

Yude, C., & Goodman, R. (1999). Peer problems of 9- to 11-year-old children with hemiplegia in mainstream schools: Can these be predicted? *Developmental Medicine and Child Neurology, 41*(1), 4–8.

Yude, C., Goodman, R., & McConachie, H. (1998). Peer problems of children with hemiplegia in mainstream primary schools. *Journal of Child Psychology and Psychiatry, 39*(4), 533–541.

Zeanah, C. H., Keyes, A., & Settles, L. (2003). Attachment relationship experiences and childhood psychopathology. *Annals of the New York Academy of Sciences, 1008*, 22–30.

CHAPTER 15

Social Cognition in Autism

Baudouin Forgeot d'Arc and Laurent Mottron

In this chapter, we cover various cognitive processes plausibly linked to autism, in relation to their neural substrates. The word *autism* refers to an apparent lack of interest in conspecifics, and since the first description of this syndrome (Kanner, 1943), it has been defined as a primary disorder of social interaction. The physical integrity of a large proportion of the autistic population, as well as the high level of intelligence and the astonishing performances of some autistics, have resulted in autism's being considered as a circumscribed and possibly specific deficit in social cognition. The primacy of a social view of autism has found support in the cognitive domain with the explosion of research on mentalizing in the 1980s. The coexistence of a major alteration of the typical social phenotype with preserved processing in many other domains is still taken as a core argument for a modularist approach to social cognition (e.g., Leslie, Friedman, & German, 2004). Furthermore, the purported social impairment in autism is often considered a model in the study of social cognition in other disorders, such as schizophrenia (Frith, 1992), anorexia nervosa (Zucker et al., 2007), Turner syndrome (El Abd, Patton, Turk, Hoey, & Howlin, 1999), and focal brain lesions.

A chapter on social cognition in autism provides an opportunity to show how the scholarly view of autism as resulting from a primary and specific deficit of social cognition does not correspond to the current knowledge on this developmental condition. First, the social phenotype of autistics has not yet been completely characterized. Second, *social symptoms* of autism cannot be equated with *impaired social cognition* in a straightforward way, as behavior indexing social performance should be systematically distinguished from social competence and actual processing of social information. Third, an atypical social phenotype represents only a subset among autistic differences, which also encompass reasoning, perception, and attention, sometimes at quite elementary levels. In this chapter, after a brief general description of autism, we present the current state of knowledge on social cognition in autism and discuss how this aspect of information processing relates to other characteristic autistic differences.

A Clinical Definition of Autism

In the first clinical description of "autistic disturbances of affective contact," Kanner (1943) identified as pathognomonic "the children's inability to relate themselves in the ordinary way to people and situations from the beginning of their life" (p. 242), together with "an anxiously obsessive desire for the maintenance of sameness" (p. 245). Kanner's inaugural description therefore emphasized both social and nonsocial features. This duality is still represented in the co-presence of socio-communicative signs and of restricted interests and repetitive behaviors (RIRB) in the American Psychiatric Association's definitions of autism. Socio-communicative signs constitute the two first areas of the autism definition in the *Diagnostic and Statistical Manual of Mental Disorders*, fourth edition, text revision (DSM-IV-TR; American Psychiatric Association, 2000), and the first area in the DSM-5 draft definition (American Psychiatric Association, 2011). RIRB constitute the third area in the DSM-IV-TR definition, and the second area in the DSM-5 draft definition.

The DSM-IV-TR definition of autistic disorder gives a descriptive primacy to signs related to the social phenotype. Six of 12 possible criteria are explicitly related to the social aspect of behavior: four signs in the social domain (failure to use nonverbal behaviors to regulate social interaction, failure to develop appropriate peer relationships, lack of shared enjoyment, and absence of socio-emotional reciprocity); one sign in the social aspect of communication (relative failure to initiate or sustain conversational interchange); and one sign in the social aspect of imagination (absence of spontaneous varied make-believe or social imitative play). Moreover, the diagnostic algorithm in DSM-IV-TR requires at least two signs in the social domain, against only one in the communication domain and one in the RIRB domain. When these criteria operationalized in the standardized instruments for diagnosis—for example, the Autism Diagnostic Interview—Revised (ADI-R; Lord et al., 1997) and the Autism Diagnostic Observation Schedule—Generic (ADOS-G, Lord et al., 2000)—this primacy becomes still more evident. The ADOS-G, for example, did not until recently require a sign in the RIRB area to provide a diagnosis. The ADI-R provides a DSM-IV diagnosis through, among other signs, the scoring of atypical direct gaze; social smiling; range of facial expressions used to communicate; imaginative play with peers; interest in children; response to approaches of other children; group play with peers; friendships; showing and directing attention; offering to share; seeking to share enjoyment with others; offering comfort; quality of social overtures; appropriateness of social responses; lack of pointing to express interest; imitative social play; social verbalization/chat; and reciprocal conversation.

As currently defined, autism spectrum disorders (ASD) affect about 1 in 150 individuals, with a majority of males. Within ASD, clinical presentation can be subject to important variations, depending on age, language abilities, and the presence of associated disorders—among which intellectual disability, learning disorders, epilepsy, anxiety disorders, mood disorders, and attention-deficit/hyperactivity disorder (ADHD) are the most frequent. Family aggregation and twin studies provide epidemiological evidence for a major role of genetic factors in ASD. However, no major gene has been observed to be relevant for the majority of ASD cases, in which rare mutations of larger effect and a few common variants of small effect in several different genes seem to be causal (Zhao et al., 2007). Genes currently associated with ASD are involved in (1) cell–cell interaction and synaptic function and development; (2) neuronal migration and growth; or (3) excitatory and inhibitory neurotransmission. However, the specificity of several genetic findings in ASD has been questioned and must be further assessed through multilevel phenotypic assessment.

At a microscopic, structural level, modifications of the structure of brain tissue have been reported—mainly in frontotemporal areas, where cell bodies are smaller and their density inside microcolumns is increased. At a macroscopic level, the most frequently replicated findings regarding structure are (1) diminished volume of the corpus callosum (mostly in its anterior part); and (2) increased head and brain volume at the end of the first year that approach average levels by adolescence, but remain significant in transmission pathways (Radua, Via, Catani, & Mataix-Cols, 2011) and in cortical areas related to autistic symptoms (Hyde, Samson, Evans, & Mottron, 2010; Yu, Cheung, Chua, & McAlonan, 2011). At a functional level, constant hyperactivity in visual expertise areas has been shown to occur along with diminished frontal activity, but the relationship between these regions is as yet unknown. This task-independent feature is robust to differences in level of functioning and assessment procedures (Samson, Mottron, Soulieres, & Zeffiro, 2011). Multiple studies have been dedicated to synchrony between cortical areas in functional magnetic resonance imaging (fMRI), and diminished anterior–posterior functional connectivity has been repeatedly reported, although sometimes with very limited amplitude (Just, Cherkassky, Keller, & Minshew, 2004; Wicker et al., 2008). White matter presents clear atypicalities, which currently lack a systematized classification (Boddaert et al., 2009). Moreover, other measures, such as electroencephalography (EEG), suggest stronger anterior–posterior coherence (Leveille et al., 2010).

Cognitive Characterization

The DSM-IV-TR duality between social and nonsocial signs in defining the phenotype of autism has a historical counterpart. Despite initial hypotheses involving perception (Ornitz & Ritvo, 1968) and language (Rutter, 1968), the first efforts to systematize the cognitive differences underlying autistic signs were focused primarily on social cognition with the *theory-of-mind* (ToM) movement in the 1980s. This was followed by models of nonsocial, high-level processing in the 1990s (e.g., executive dysfunction, weak central coherence), and by models of perceptual, nonsocial processing in the 2000s (enhanced perceptual functioning).

Approaches to Explaining Social Cognition in Autism

Whatever the place of social features in the global autistic phenotype, the characterization of social cognition in autism is part of the puzzle. Different approaches have attempted to account for this. They focus on comprehension, perception, and motivation, respectively, as limitations in social functioning.

Social Comprehension: The ToM Account

Attributing thoughts and goals to others—the ability we call ToM (Premack & Woodruff, 1978)—is central to our human social life. Teaching, convincing, manipulating, and perhaps even developing a language or pretend play would be impossible if we were totally unable to infer what our conspecifics think and desire. The questions of whether ToM really is a theory or involves an implicit simulation of others' thoughts (Apperly, 2008), whether it involves explicit reasoning or intuition (Forgeot d'Arc & Ramus, 2011), and what its developmental course is (Onishi & Baillargeon, 2005; Wellman & Cross, 2001) have received much attention and given rise to a great deal of controversy and empirical work.

The idea that social difficulties in ASD may result from a disruption in ToM processes originated from a seminal paper by Baron-Cohen, Leslie, and Frith (1985), which had a major influence on research programs during the 1990s. In this view, the lack of pretend play or difficulties in understanding nonliteral language (e.g., irony or lies) are not consequences of withdrawal or indifference, but point toward an inability to represent others' thoughts. The main empirical argument in support of the ToM model relies on the false-belief task (FBT). Several versions of this task have been designed, each based on a narrative in which a character has a belief (e.g., about the location of an object) incongruous with reality (e.g., the real location of the object). To predict the character's behavior correctly ("Where will Sally search for her marble?"), the participant must correctly attribute a false belief ("She thinks that . . . "). Numerous studies have shown that children with ASD perform more poorly on the FBT than IQ-matched controls do. However, failure on the FBT is far from a universal hallmark of autism, since many able/older children with autism pass the task (Bowler, 2007; see Happé, 1995, for a review). Ecological validity of the FBT has been corroborated by correlations between individual differences in FBT performance and individual differences in social behavior in the real world (for a review, see Astington, 2003). However, questions have been raised about the relevance of this test as a measure of ToM (Bloom & German, 2000). First, the FBT generally imposes high verbal and executive demands, which make failure almost impossible to interpret as a specific deficit. Furthermore, the FBT tests only a subset of ToM abilities; thus succeeding on the FBT is not proof of intact ToM. Interpretation of FBT performance is debated not only in the ASD literature, but also in relation to typically developing children (Onishi & Baillargeon, 2005; Senju, Southgate, White, & Frith, 2009).

During the 1990s, reports of ceiling effects in FBT performances led to the search for more advanced tests of social comprehension. Their increased difficulty relies on the need to take into account more complex mental states (such as guilt and shame), as well as on the use of much more elaborate narratives ("faux pas"; Baron-Cohen, O'Riordan, Stone, Jones, & Plaisted, 1999) or impoverished perceptive cues ("reading the mind in the eyes"; Baron-Cohen, Wheelwright, Hill, Raste, & Plumb, 2001). These mentalizing tasks have repeatedly shown group differences between participants with and without ASD, illustrating atypical understanding of complex social situations in high-functioning adults with ASD. However, these sophisticated tasks may stretch the link between autism and basic concepts relating to others' minds to the point of rupture (Baron-Cohen, 2009).

Various studies have further explored different aspects of the understanding of other people's minds by persons with ASD. Atypicalities have been reported in distinguishing physical from mental entities and in knowing that seeing leads to knowing (Baron-Cohen, 2001). Recent work has also pointed out that ASD may be associated with a diminished tendency to attribute agency to conspecifics (Gray, Jenkins, Heberlein, & Wegner, 2011), to integrate mental state information for moral judgment (Moran et al., 2011), and to use recursive mental representations during cooperation (Yoshida et al., 2010).

The classical fMRI finding that brain regions typically involved in belief attribution, such as perigenual anterior cingulate cortex, show less activation in adults with ASD when these adults are performing social tasks (Castelli, Frith, Happé, & Frith, 2002) has received support from a recent meta-analysis (Di Martino et al., 2009). Moreover, hypoactivation in other regions (e.g., the temporoparietal junction or temporal pole) has repeatedly been reported in studies involving mental state attribution (Castelli et al., 2002; Kana, Keller, Cherkassky, Minshew, & Just, 2009). Some authors (Decety & Lamm, 2007) suggest that areas involved in ToM processing are also involved in lower-level mechanisms (e.g., sense of agency and reorienting attention to salient stimuli). However, a recent meta-analysis argues for the high-level specialization of socio-cognitive processes (Van Overwalle, 2011).

Social Perception

A large body of research using behavioral, EEG, or fMRI measures has been devoted to the exploration of the processing of stimuli involved in social interactions (i.e., faces, emotions, biological motion).

FACE PERCEPTION

One of the most persistent clichés regarding autistic social cognition is that autistics "do not recognize faces." Early studies of autistics with an associated phenotype of intellectual disability reported impaired recognition of familiar (Boucher, Lewis, & Collis, 1998) or unfamiliar (Boucher & Lewis, 1992; Gepner, de Gelder, & de Schonen, 1996; Hauck, Fein, Maltby, Waterhouse, & Feinstein, 1998) faces. In contrast, studies of autistics with average IQ strongly indicate that face identification and manipulation (on recognition and matching tasks) may reach typical levels (Barton et al., 2004; Behrmann, Thomas, & Humphreys, 2006; Lahaie et al., 2006; Lopez, Donnelly, & Hadwin, 2004; Rouse, Donnelly, Hadwin, & Brown, 2004).

Following Langdell's (1978) first studies reporting superior face recognition from atypical face parts, a whole branch of research on autism has tracked a deficit in configural aspects of face perception. Multiple studies have thus investigated global processing of faces, typically characterized by low spatial frequencies, second-order relations (grouping effects), and holistic properties (metric relations between face parts). Empirical strategies include comparison of low-pass versus high-pass filtering on face processing, as well as various face manipulations aimed at testing perception of global and holistic properties (inversion effect, Thatcher illusion, composite face effect, natural vs. non natural segmentation). The relative role of configuration and details in autistic face perception is atypical (Deruelle, Rondan, Gepner, & Tardif, 2004), up to the point that local face parts have an identification effect that has no equivalent in typical individuals (Lahaie et al., 2006; Langdell, 1978). In children with ASD as young as 3 to 4 years, neural response to high spatial frequencies of faces is enhanced, and neural response to facial expression seems to be determined by details rather than by configuration (Vlamings, Jonkman, van Daalen, van der Gaag, & Kemner, 2010). In autistic adults, midrange spatial frequencies, characterizing most features in face processing, appear to evoke the same activity as high spatial frequencies, which is not the case in non-autistics (Jemel, Mimeault, Saint-Amour, Hosein, & Mottron, 2010). Two reviews of face processing in autism conclude that autistics use different strategies in term of orientation to faces, scanning, whole–part relationships, and hierarchical use of face parts. However, Jemel, Mottron, and Dawson (2006, p. 101) have concluded that "the versatility and abilities of face processing in persons with autism have been underestimated." Simmons et al. (2009, p. 2723) have added that "The literature on face processing in ASD presents quite a confusing picture, with very few clear results to hang potential theories on."

Investigations of the brain correlates of face perception in autism have followed the same trend. The available studies indicate that autistics display a typical event-related potential response while perceiving faces, with possible differences in the specialization of brain structures associated with face perception. In fMRI, whereas an absence of activity in the fusiform gyrus (FG), a temporoparietal region involved in face processing, was initially reported (Schultz et al., 2000), this result was secondarily attributed to task conditions, particularly variation in attention, as other studies found no difference between participants with and without autism once attention was controlled for (Hadjikhani et al., 2004). However, three using activation likelihood estimation meta-analyses found an atypical recruitment of activity in lower-order processing systems while autistics were processing faces. Di Martino et al. (2009) reported consistent patterns of activation in the posterior lateral portion of the FG, typically associated with physical aspects of face processing. Samson et al. (2011) demonstrated that this pattern is actually found for any visually presented material in autism. This meta-analysis, as well as the anatomical one by Yu et al. (2011), indicate that the FG is disproportionally more activated in autistics, in comparison to the activity associated with perception of objects and words. Similarly, very preliminary results in processing of voices (the equivalent of faces in the auditory modality) were initially interpreted as impairments in voice processing (Gervais et al., 2004), but warrant reexamination in the light of face studies.

EMOTION PERCEPTION

Following the same trend as research on face perception in autism, an impairment in the processing of facial emotional expressions (Celani, Battacchi, & Arcidiacono, 1999; Gross, 2004; Hobson, 1986; Tantam, Monaghan, Nicholson, & Stirling, 1989) has been reported. However, these results are challenged by multiple findings that show typical intact facial expression processing in individuals with ASD (Adolphs, Sears, & Piven, 2001; Baron-Cohen, Wheelwright, & Jolliffe, 1997; Braverman, Fein, Lucci, & Waterhouse, 1989; Castelli, 2005; Gepner, Deruelle, & Grynfeltt, 2001; Grossman, Klin, Carter, & Volkmar, 2000; Ozonoff, Pennington, & Rogers, 1990; Pelphrey et al., 2002; Volkmar, Sparrow, Rende, & Cohen, 1989). The more recent and larger studies on this topic reliably affirm that there is no gross deficit in emotion perception in autism (Jones et al., 2011), despite an atypical balance between the effect of positive and negative emotions (Kleinhans et al., 2010).

fMRI findings regarding emotion processing are as contradictory as those involving face perception. Amygdala hypoactivation (Baron-Cohen, Ring et al., 1999; Critchley et al., 2000; Pierce, Muller, Ambrose, Allen, & Courchesne, 2001) was initially reported, followed by a series of studies documenting typical activation (Pierce, Haist, Sedaghat, & Courchesne, 2004; Piggot et al., 2004). Recent findings support amygdala hyperactivation (Dalton et al., 2005; Weng et al., 2011), but marginally significant atypical modulation between tasks, in autistics (Wang, Dapretto, Hariri, Sigman, & Bookheimer, 2004).

However, the finding that regions typically involved in emotion processing, such as the amygdala, showed a lower probability of activation for social stimuli in autistics received support from a recent meta-analysis (Di Martino et al., 2009). This meta-analysis also showed less activation in anterior insula, a region previously associated with empathy (Singer, 2006). Furthermore, increased white matter volume in autistics in tracts involved in language and social cognition (e.g., right arcuate fasciculus and left inferior fronto-occipital and uncinate fasciculi) has also been reported in a recent meta-analysis (Radua et al., 2011).

BIOLOGICAL MOTION

Research on biological motion perception in autism is also characterized by conflicting results. Early reports of impairment have been followed by failure to replicate these findings in the majority of recent studies (Congiu, Schlottmann, & Ray, 2009; Murphy, Brady, Fitzgerald, & Troje, 2009; Saygin, Cook, & Blakemore, 2010, but see Atkinson, 2009). Biological motion tasks activate different networks (Freitag et al., 2008); in particular, the temporal sulcus, central to biological motion perception in non-autistics, may not be animated by motion-specific activity in autistics, although this result needs to be replicated.

In sum, a definitive conclusion on perception of social stimuli in ASD is not yet possible, due to the inconsistent findings in research conducted with apparently similar paradigms. Besides atypicalities in performance and processing, there are now numerous indications of typical gross performance in face perception, and even in perception of faces' configural aspects. This discrepancy may reflect heterogeneity in paradigms or in the autistic population regarding measured IQ or intensity of autistic phenotype; alternatively, it may reflect an atypical access to explicit and implicit processes, compensatory mechanisms, or differences in maturation of attention processes. For example, verbal explicit paradigms may fail to show any differences despite atypical underlying processes, at least in a population with ASD and a sufficiently high level of speech to be given tests requiring oral responses.

Social Motivation and Orienting: The Enactive Mind

Beside its perceptual aspects and conceptual basis, social cognition can be studied from the perspective of motivation, or drive toward the social environment (e.g., gaze, faces, and actions).

SOCIAL STIMULI AS THE FOCUS OF ATTENTION

The first descriptions of autism emphasized that precocious signs of interest in the social environment are lacking or grossly atypical in autism, to the point of providing a label for this condition. Reduced orienting to faces has been reported both in eye-tracking studies with adults of average IQ (Klin, Jones, Schultz, Volkmar, & Cohen, 2002b) and in retrospective studies examining the home movies of infants prior to ASD diagnosis (Osterling & Dawson, 1994; Osterling, Dawson, & Munson, 2002; Werner, Dawson, Osterling, & Dinno, 2000). In regard to within-face scanning, where non-autistics tend to focus their fixations on the eye region, atypical direct gaze is part of the diagnostic criteria for autism. Preferential fixation on the mouth region, associated with decreased fixation in the eye region, has been reported in adults with autism and average IQ while viewing movie sequences (Klin et al., 2002b); preferential fixation on outer-face versus inner-face features has been reported in young children with ASD, associated with diminished identity recognition (Chawarska & Shic, 2009). However, Back, Ropar, and Mitchell (2007) selectively manipulated the information provided by different parts of the face in a facial expression recognition paradigm. In contradiction of Klin, Jones, Schultz, Volkmar, and Cohen's (2002a) interpretation and of previous findings (Baron-Cohen et al., 1997), adolescents with ASD showed decreased performance both when information from eye and from mouth regions was withheld, providing evidence that both mouth and eye regions provide information to autistics when they are attributing mental states to facial expressions.

In contrast to mere deficit models, other studies suggest a different degree of competition between social and nonsocial information in autism. For example, diminished action monitoring and enhanced fixation of scene background are evident in 20-month-old children with ASD (Shic, Bradshaw, Klin, Scassellati, & Chawarska, 2010). In another study (Klin, Lin, Gorrindo, Ramsay, & Jones, 2009), infants were presented with point-light displays (made by fixing lights to body parts of actors moving in the dark). Whereas typically developing children strongly preferred the upright display (where the scene could be easily recognized), 2-year-olds with ASD did not show this pattern, but were unique in detecting an audiovisual synchrony in the presented information.

In non-autistic toddlers, gaze disengagement from a face toward an unpredicted cue is longer than disengagement from a nonsocial stimulus. Chawarska, Volkmar, and Klin (2010) did not find this bias in 32–month-old toddlers with ASD and concluded that "toddlers with ASD [were] not captivated by faces to the same extent as toddlers without ASD" (p. 178). They further suggest (p. 182) that the bias might be due to "deeper obligatory processing triggered by faces" (e.g., semantic, emotional) in non-autistics.

SOCIAL STIMULI AS ATTENTIONAL CUES

In typically developing children, the social environment constitutes an important target of attention, but some social stimuli (e.g., gaze direction and pointing gestures) also direct attention toward other parts of the environment. A diminished tendency to spontaneously follow another person's eye gaze or pointing is one of the symptoms of autism and has been reported in various situations. Retrospective studies examining the home movies of infants prior to ASD diagnosis (Osterling & Dawson, 1994; Osterling et al., 2002; Werner et al., 2000) showed a diminished tendency to follow pointing. Older children with ASD were shown not to shift their attention toward novel objects when cued socially by a head turn and gaze shift (Leekam, Lopez, & Moore, 2000). Moreover, while viewing movie sequences, adults with autism and average IQ tend not to follow pointing (Klin et al., 2002b). However, older and more able autistic children do not have difficulty understanding what a person is looking at (Baron-Cohen, Campbell, Karmiloff-Smith, Grant, & Walker, 1995; Leekam, Baron-Cohen, Brown, Perrett, & Milders, 1997). Social orienting has also been tested by using variants of Posner's (1980) spatial cueing paradigm with social (eye gaze) and nonsocial (arrow) cues. The main automatic facilitation effect usually observed in non-autistics seems preserved in people with ASD. In most of these studies (see Nation & Penny, 2008, for a comprehensive review; see Ristic et al., 2005, for an exception), adults and children show faster responses to targets occurring in the cued location than in the noncued location, even when the direction of the cue (gaze) does not predict the location of the upcoming target (Senju, Tojo, Dairoku, & Hasegawa, 2004; Swettenham, Condie, Campbell, Milne, & Coleman, 2003).

Direct studies of neural correlates of social orienting in ASD are sparse. A recent fMRI study (Greene et al., 2011) examined the neural correlates of social orienting in children and adolescents with ASD and in a matched sample of typically developing controls while they performed a spatial cueing task with social (eye gaze) and nonsocial (arrow) cues. When attention was directed by social cues as opposed to nonsocial cues, despite similar performance, the typically developing group showed increased activity in frontoparietal attention networks, visual processing regions, and the striatum, whereas the group with ASD only showed increased activity in the superior parietal lobule. In previous studies, many differences of neural activation in social perception areas may be considered consequences of orientation. For example, Pelphrey, Morris, and McCarthy (2005) found that individuals with ASD showed typical activation of the superior temporal sulcus (STS) when viewing gaze shifts. However, STS activity varied depending on the intentions conveyed by the gaze shift in control participants, while no such difference was found in the group with ASD.

Social Cognition: Distinguishing Chickens from Eggs

The current data support different summary claims on social cognition in autism. Some people with ASD have a lower comprehension of their social environment; their perception of stimuli involved in social interactions, such as faces, is both atypical and less efficient in allowing an appropriate interpretation of social content; they display atypical patterns of attention to their social environment, including diminished fixations to typical targets of attention (actions, eyes), as well as increased attention to perceptual aspects of their surrounding.

Regardless of how they relate to non-domain-specific aspects of cognition, these atypicalities are usually considered as distinct deficits (Schultz & Robins, 2005) in social comprehension, social perception, and social orientation (we abbreviate these hereafter as SC, SP, and SO, respectively). How can alterations in these three domains be nevertheless causally related? Both SC and SP atypicalities may arise from lack of expertise, due to low SO. Whether lower SO is related to low-level perception of dimensions intrinsic to social material (e.g., orienting toward perceptive features of biological motion) or to more abstract representations (e.g., monitoring goals) is still uncertain. Whether preferential orientation toward sensory features of the environment may be a cause or a consequence of diminished SO is discussed in a later section of this chapter (see "Perception and Social Cognition," below).

An "SP-first" hypothesis might easily account for subsequent deficits in SO and SC. Indeed, an increased prevalence of autism in blind children (Hobson & Bishop, 2003) has been reported, although this report was based on fragmentary data and a possible limited phenotype similarity. However, a majority of children with sensory loss do not develop social features of autism, and a majority of people with ASD do not have any sensory loss; rather, they have perceptual enhancement. Moreover, although a moderately delayed performance in tasks tapping SC has been reported in children with congenital blindness

(Brambring & Asbrock, 2010), neural networks implied in SC in non-autistics with congenital blindness are strikingly similar to those of seeing people (Bedny, Pascual-Leone, & Saxe, 2009), suggesting limited influence of restricted perceptual input on the development of SC. Low-level perceptual abnormalities (e.g., differences in response to various spatial frequencies) instead suggest an "SP-first" mechanism, grounded on an atypical feedforward flow of information by basic, non-domain-specific aspects of perception. Alternatively, difficulties in labeling facial expressions are in favor of an "SC-first" mechanism—that is, a limited conceptual apprehension of emotions and intentions. Indeed, SC relies on a set of very abstract concepts from a very early age: Typically developing infants as young as 7 months seems to process *beliefs* (Kovacs, Teglas, & Endress, 2010). Differences in the use or mastering of such concepts may lead to a modified balance of attention toward social and nonsocial aspects of the environment, leading in turn to a diminished social expertise.

Social Features in a Pervasive Condition

Autism is defined as a combination of social and nonsocial features. At a cognitive level, nonsocial aspects of cognition, such as perception and executive functions, have attracted growing interest in the last decade. The question of whether social features may explain nonsocial features or vice versa has arisen. Before exploring such possible links, we examine to what extent the social and nonsocial features of autism form a unitary construct or a mere constellation of independent characteristics. In the former case, parsimony suggests that different accounts either share a common etiology or are causally related to each other. Conversely, if autism is better conceived as a fractionable entity (Happé & Ronald, 2008), different cognitive characteristics may account for distinct aspects of autistic symptoms and are not necessarily related.

Fractionability of Autistic Symptoms

To what extent might the different domains of alteration that define autism be caused by a single factor? Genetic and population studies, factor analysis of symptoms, and brain imaging studies can all provide insight into this question. Aggregation of autistic traits in non-autistic populations provides information about whether autism-related behaviors tend to cluster in general populations. The first large population-based study aimed at addressing this question (Ronald, Happé, & Plomin, 2005) was based on parental and teacher reports from 3000 twin pairs assessed between the ages of 7 and 9 years. The authors reported modest to low correlations (.20–.40) between autistic-like behavioral traits in the three core areas. When only children with extreme scores were considered, degree of social difficulty, communicative impairment, and RIRB were still only modestly related. However, parental reports for autism-like traits are of limited reliability, and these traits do not map straightforwardly onto autistic behaviors (Barbeau, Mendrek, & Mottron, 2009). A follow-up study by Dworzynski et al. (2007), based on cross-twin cross-trait correlations and restricting analysis to children diagnosed as having ASD, reported only small genetic effects between RIRB and other symptoms.

The factorial structure of autistic behaviors—that is, whether they represent unique or multiple dimensions—has been explored through more than 20 studies. Factor-analytic studies have produced contrasting results: In some (Constantino et al., 2004; Szatmari et al., 2002; Volkmar, Cohen, Hoshino, Rende, & Paul, 1988; Wadden, Bryson, & Rodger, 1991), a large proportion of variance is explained by a single principal component, which provides evidence that the so-called "autistic triad" of impairments represents facets of a unique factor. Other studies (Berument, Rutter, Lord, Pickles, & Bailey, 1999; DiLalla & Rogers, 1994; Lecavalier, Aman, Hammer, Stoica, & Mathews, 2004; Miranda-Linne & Melin, 2002; Stella, Mundy, & Tuchman, 1999; Tadevosyan-Leyfer et al., 2003; van Lang et al., 2006) report three- to six-factor solutions. According to a review by Mandy and Skuse (2008), after the studies that did not meet criteria for quality were excluded, evidence for multiple factors underlying autistic behaviors was supported by seven studies and challenged by one. Research on the fractionability of the autistic triad has thus reached a quasi-consensus.

The high heritability of autistic traits is an established fact. Moreover, twin and family studies both show that the broader autistic phenotype is more heritable than strictly defined autism is. However, whether single genetic factors underlie the triad or whether different elements are separately inherited remains debated. On the one hand, linkage studies have identified distinct loci for endophenotypes related to different core domains. Candidate gene studies have begun to explore the possibility of symptom-specific genetic effects in autism. A good example of such research is recent work on the serotonin transporter gene promoter (SLC6A4) and autism. The short version of the polymorphism (5-HTTLPR, S/L or S/S genotypes) may be associated with higher scores on the ADI-R item "failure to use nonverbal communication to regulate social interaction" (Brune et al., 2006; Tordjman et al., 2001), whereas the long version (L/L genotype) and other variants (Mulder et al., 2005; Sutcliffe et al., 2005) may be associated with higher scores on the item "stereotyped and repetitive motor mannerisms" and on an aggression measure.

Contrasting with this view of very selective effects of moderate-impact variants, statistical analyses of ASD family data suggest that a significant proportion of ASD cases may be the result of dominantly acting de novo mutations, indicating that autism can result from single genetic alterations. The list of mutations, each of which causes ASD in a substantial proportion of affected individuals, includes fragile X, tuberous sclerosis, and duplication on chromosome 15q11–13 (Freitag, Staal, Klauck, Duketis, & Waltes, 2010). However, it remains unclear whether this different form of syndromic autism is a valid model for other cases of autism.

Neuroimaging may also contribute to the debate on fractionability. Indeed, the definition of autism is based on a set of behavioral disturbances that more or less map onto specific functional systems of the brain. Although multiple areas are concerned, not all brain systems have been proven to be affected. It would thus be a mistake to consider autism as a general disorder of brain function (Schultz & Robins, 2005), despite its distributed nature (Muller, 2007). Many functions are spared or even enhanced in autism. However, the picture of multiple distinct, focal, co-occurring variations poorly constrains the number of possible etiological mechanisms.

In conclusion, since domains seem at least partly independent, there can be no universal explanation of one domain by one other. However, the observation that some cases seem caused by one unique (e.g., genetic) factor leads to exploration of possible links between domains of ASD at different levels of explanation. At the cognitive level, links among social cognition, executive functions, and perception may shed light on our comprehension of autism.

Executive Functions and Social Cognition

The association between executive dysfunction and ASD was widely explored in the late 1980s and 1990s, with recognition that the perseveration, planning, and set-shifting difficulties encountered in everyday life by those with ASD resembled problems found in patients with acquired frontal lesions. Several investigators suggested that frontal deficits might have developmental cascading effects, sufficient to cause the social and communication impairments seen in autism (Hill, 2004; Russell, 1998).

Hill (2004) divided the executive functions relevant to autism into planning, mental/cognitive flexibility (set shifting), inhibition, generativity, and self-monitoring. Her review of the available literature still did not support a unique deficit in one of these functions in autism. Alternatively, Russo et al. (2007), examining the mechanisms implicated in autistic perseverative errors in executive function tasks, concluded that impairments in cognitive flexibility are constant during adolescence/adulthood and may be noted at younger developmental levels as well. They also reported evidence of complex working memory impairments in ASD, affecting measures of working memory span but not measures of interference in later childhood and adolescence. In contrast, inhibition, strictly defined, appears to be intact in ASD after a developmental age of 6 years. In regard to the specificity of these various alterations of executive components in autism, it has been reported that young children with ASD differ from ability-matched children with intellectual disabilities in their executive function performance (Griffith, Pennington, Wehner, & Rogers, 1999). However, it remains unclear whether a specific profile of executive deficits distinguishes ASD from ADHD or other neurodevelopmental disorders (Geurts, Verte, Oosterlaan, Roeyers, & Sergeant, 2004; Happé, Booth, Charlton, & Hughes, 2006; Johnson et al., 2007; Ozonoff & Jensen, 1999).

Atypical executive functioning in autism is supported by fMRI studies. During executive function tasks, although a greater likelihood of activation was found in typically developed adults than in autistics in the dorsolateral prefrontal cortex (Brodmann's area [BA] 9–10) and lateral parietal cortex (e.g., supramarginal gyrus and inferior parietal lobule [BA 40]), a difference in the opposite direction was found for rostral anterior cingulate (Di Martino et al., 2009). However, atypical activation in subcomponents of the frontal lobes cannot be equated with deficits, and it is reported even in the presence of a typical level of performance (Samson et al., 2011).

Could an executive deficit be related to the social phenotype of autism, in either direction of causality? Theoretically, it has been suggested that development of social cognition may be crucial for executive functioning (Perner, 1995). To date, this hypothesis has received little empirical support. Executive function may be a prerequisite for ToM (Moses, 2001; Russell, 1998), since success on executive function tasks appears to be a necessary but not sufficient condition for success on the FBT, both in longitudinal designs (Pellicano, 2007) and in cross-sectional designs (Joseph, McGrath, & Tager-Flusberg, 2005). Performance on executive function tasks moderately predicts later performance on the FBT (Hughes, 1998; Pellicano, 2010), whereas the reverse has not been reported. Conversely, an executive deficit or limitation may be implicated in some autistic features and cognitive atypicalities, although an executive function deficit may exist without the social features of autism (e.g., in ADHD; see Pauli-Pott & Becker, 2011). Moreover, some autistic children pass the FBT, a complex social comprehension task, while failing executive function tasks (Ozonoff, Pennington, & Rogers, 1991). Conversely, able adults with severe ASD can perform perfectly well on executive function tests while still exhibiting dramatic difficulties in social reciprocity (Baron-Cohen, Wheelwright, Stone, & Rutherford, 1999). It thus remains unclear whether executive functions really play a role in socio-cognitive development as such, or only sustain performance on the FBT (Carlson & Moses, 2001; Leslie et al., 2004; Russell, 1998). In sum, the executive dysfunction hypothesis has multiple limitations. Not all autistic individuals show executive problems, those who do may have differing executive function profiles; and the relation of executive dysfunction to the social phenotype of autism is poorly established. Moreover, executive functions are multifaceted, hard to delineate conceptually, and to isolate empirically. We indicate in the next section that executive dysfunction is more convincingly involved through a different sharing of cognitive architecture in task performance, rather than as a mere "executive deficit."

Perception and Executive Function

Perception selects, organizes, provides a representation of, and interprets information coming from the senses; it is typically involved in bidirectional relationships with non-perceptual architecture, language, executive functions, and emotions. Perceptual processing is primarily composed of *low-level perception*—the extraction of the elementary features composing information. Visual or auditory elementary features are then grouped into patterns, at a level called *midlevel perception*. These patterns are matched with domain-specific memorized templates of objects, faces, language, and biological motion. An *expertise network*, located in the inferior temporal lobe of the FG, is involved in the categorization of highly similar social and nonsocial patterns. This expertise network has different foci according to domains of objects, such as the fusiform face area for faces and the lateral occipital cortex for objects. Although the cartography of perceptual functions in autism has not yet been accomplished, differences, mainly superiorities, have been found in a large array of low-level and midlevel domains. Auditory (pitch) and visual (luminance-defined information, high spatial frequencies, symmetry) low-level processes have lower discrimination thresholds in autism under various conditions. Superior performances are consistently reported in pattern detection and manipulation in the visual modality (for reviews, see Mottron, Dawson, & Soulieres, 2009; Mottron, Dawson, Soulieres, Hubert, & Burack, 2006); a striking example is mental rotation (Soulieres, Zeffiro, Girard, & Mottron, 2011). Besides enhanced performance, it is now also established that perception in autism relates to the rest of the cognitive architecture in a unique way: an enhanced role (superior feedforward) for, and superior autonomy of, perception in higher-level cognitive processes.

Reciprocal influences have been reported between perception and executive function. Mottron, Belleville, and Menard (1999) argued that limitations in the complexity of information that can be manipulated in short-term visual memory during graphic planning (i.e., executive load) might result in an improvement of performance by autistics in some global processing tasks (e.g., copying a drawing of a 3D-impossible object as a possible, easier 2D drawing). However, a more recent interpretation of this finding suggests an optional top-down influence, resulting in a possible access to raw perception. The notion of *local bias* may also be interpreted as an "attention trap" (or "glue"), which diminishes the ability of autistics to switch flexibly from a local to a global level (for examples, see Mann & Walker, 2003; Rinehart, Bradshaw, Moss, Brereton, & Tonge, 2000). A mere imbalance of activity between atypical perception and executive function, of the form "The more perception, the less executive function," is implausible: Using a limited set of both executive function and perceptual tasks, Pellicano (2010) found no correlation between executive function and perception either in cross-sectional or in longitudinal analysis, supporting the idea that the two domains are largely independent.

How then can we interpret the striking association between superior activity in the temporal and occipital regions involved in perceiving and recognizing patterns and objects, and diminished activity in more anterior frontal regions? This association was evident in a recent meta-analysis of fMRI studies involving visual presentation, encompassing a large variety of simple and complex tasks (Samson et al., 2011). The frontal areas include regions specialized for movement execution and planning, and for cognitive control: BA 4 (fine motor control), BA 6 and 8 (response selection and attention shifting), BA 9 (planning and monitoring of behavior according internal goals), BA 45 and 47 (decision making, response comparison, selection and inhibition based on stored representations), and BA 6 and 9 (cognitive control and adjustment). This finding has been robustly replicated across various tasks with different performance levels in many participant groups. Most of the tasks included in this meta-analysis were performed at a typical or superior level by autistics, and this performance was associated with hyperactivity in the visual expertise network, in a region specific to the category of stimuli presented. We therefore suggest that this atypical pattern represents a deep, plausibly early, and seemingly efficient reorganization of cortical architecture.

Perception and Social Cognition

General aspects of perception and socio-cognitive processing in ASD have also been more directly associated. Visual exploration of social scenes is determined by perceptual rather than social content in children with ASD (Klin et al., 2009). In an eye-tracking study, 2-year-olds with ASD did not show selective interest in a device depicting a social scene. Their preference seemed driven by audiovisual synchrony; that is, they mostly looked at screen regions where sound matched movement. Such percept-driven attention may be the consequence, rather than the cause, of a failure to treat the social content of the stimulus. Strong versions of the ToM deficit hypothesis endorse such a claim (Frith, Happé, & Siddons, 1994). However, the robustness of findings encompassing multiple intrinsic and extrinsic properties of autistic perception contrasts with the inconstancy of findings specific to social material. This has led some authors (Behrmann et al., 2006; Belmonte et al., 2004) to attribute (at least partially) the atypicalities evident in the processing of social material to domain-general perceptual alterations. The finding that neural response to facial expression is determined by details rather than configuration in 3- to 4-year-olds with ASD (Vlamings et al., 2010), as well as in older children with ASD (Deruelle et al., 2004), suggests that even the very social aspects of processing (e.g., emotion recognition) are modulated by a low-level perceptual feedforward flow of information. Atypical perception thus cannot be a mere consequence of a lack of social interpretation of the stimuli. Consistently, although longitudinal findings are limited to social comprehension, they support the idea of an influence of early detail-oriented perception on later social cognition (Pellicano, 2010).

Differences in the way low-level perceptual mechanisms (e.g., spatial frequency analyses) feed forward in face processing, possibly producing a comparatively local bias in comparison to typical face processing, have an equivalent in the auditory modality. There are strong indications of a relation (if not of a direction of causality) between atypical perception and speech acquisition, in the sense that superior pitch perception and visual–spatial peaks are highly associated with speech delay in autism (Bonnel et al., 2010; Jones et al., 2009).

In sum, according to recent data, perceptual functioning in people with ASD may influence their social phenotype in at least two ways. First, their percept-driven attention (i.e., optional top-down) may lead to different interpretations of their environment (the SO hypothesis), and developmentally may favor a superior expertise in perception at the expense of social experience. Second, their perceptual processing (i.e., detail-oriented) may result in (but not from) less efficient processing of features like emotions, identity, or gender (the SP hypothesis).

Conclusions

Whatever its place in the causal chain of autistic development, social cognition has long received the greatest emphasis in cognitive studies of ASD. This may be due to the massive impact of this very aspect of autistic cognition on adaptation to the current social environment, or possibly to the "normocentric" bias of non-autistic researchers. Although it cannot be ruled out that some autistics have a specific and/or primary deficit in socio-cognitive processes, current data do not support the view that the majority of the autistic behavioral and cognitive phenotype results from such a deficit. Indeed, no single cognitive deficit or superiority has the power either to *define* or to *generate* autism. The emerging view of autism is that of a highly heterogeneous group of conditions, sharing features in social as well as nonsocial aspects of cognition, but in which the different cognitive, behavioral, and biological characteristics are only partially dependent. Early influences among these features are plausible, although still poorly understood. Among those, general perceptive characteristics play an early role in social cognition, but the study of their developmental agenda and of their interrelation is still in its infancy. Understanding social cognition in autism remains a part of the puzzle, but it is crucial to focus flexibly on every piece and on the whole picture. Future research in social cognition should aim at unraveling the conceptual, perceptive, attentional, and expertise aspects of social cognition, and at further exploring their links with other cognitive domains.

References

Adolphs, R., Sears, L., & Piven, J. (2001). Abnormal processing of social information from faces in autism. *Journal of Cognitive Neuroscience, 13*(2), 232–240.

American Psychiatric Association. (2000). *Diagnostic and statistical manual of mental disorders* (4th ed., text rev.). Washington, DC: Author.

American Psychiatric Association. (2011). *Diagnostic and statistical manual of mental disorders* (5th ed., draft criteria). Washington, DC: Author.

Apperly, I. A. (2008). Beyond simulation–theory and theory–theory: Why social cognitive neuroscience should use its own concepts to study "theory of mind." *Cognition, 107*(1), 266–283.

Astington, J. W. (2003). Sometimes necessary, never sufficient: False belief understanding and social competence. In B. Repacholi & V. Slaughter (Eds.), *Individual differences in theory of mind: Implications for typical and atypical development* (pp. 13–38). New York: Psychology Press.

Atkinson, A. P. (2009). Impaired recognition of emotions from body movements is associated with elevated motion coherence thresholds in autism spectrum disorders. *Neuropsychologia, 47*(13), 3023–3029.

Back, E., Ropar, D., & Mitchell, P. (2007). Do the eyes have it?: Inferring mental states from animated faces in autism. *Child Development, 78*(2), 397–411.

Barbeau, E. B., Mendrek, A., & Mottron, L. (2009). Are autistic traits autistic? *British Journal of Psychology, 100*(Pt. 1), 23–28.

Baron-Cohen, S. (2000). Theory of mind and autism: A review. *International Review of Research in Mental Retardation, 23*, 169–184.

Baron-Cohen, S. (2009). Autism: The empathizing–systemizing (E-S) theory. *Annals of the New York Academy of Sciences, 1156*, 68–80.

Baron-Cohen, S., Campbell, A., Karmiloff-Smith, J., Grant, J., & Walker, J. (1995). Are children with autism blind to the mentalistic significance of the eyes? *British Journal of Developmental Psychology, 13*, 379–398.

Baron-Cohen, S., Leslie, A. M., & Frith, U. (1985). Does the autistic child have a "theory of mind"? *Cognition, 21*(1), 37–46.

Baron-Cohen, S., O'Riordan, M., Stone, V., Jones, R., & Plaisted, K. (1999). Recognition of faux pas by normally developing children and children with Asperger syndrome or high-functioning autism. *Journal of Autism and Developmental Disorders, 29*(5), 407–418.

Baron-Cohen, S., Ring, H. A., Wheelwright, S., Bullmore, E. T., Brammer, M. J., Simmons, A., et al. (1999). Social intelligence in the normal and autistic brain: An fMRI study. *European Journal of Neuroscience, 11*(6), 1891–1898.

Baron-Cohen, S., Wheelwright, S., Hill, J., Raste, Y., & Plumb, I. (2001). The "Reading the Mind in the Eyes" Test revised version: A study with normal adults, and adults with Asperger syndrome or high-functioning autism. *Journal of Child Psychology and Psychiatry, 42*(2), 241–251.

Baron-Cohen, S., Wheelwright, S., & Jolliffe, T. (1997). Is there a "language of the eyes"?: Evidence from normal adults and adults with autism or Asperger syndrome. *Visual Cognition, 4*(3), 311–331.

Baron-Cohen, S., Wheelwright, S., Stone, V. E., & Rutherford, M. (1999). A mathematician, a physicist, and a computer scientist with Asperger syndrome: Performance on folk psychology and folk physics tests. *Neurocase, 5*(6), 475–483.

Barton, J. J., Cherkasova, M. V., Hefter, R., Cox, T. A., O'Connor, M., & Manoach, D. S. (2004). Are patients with social developmental disorders prosopagnosic?: Perceptual heterogeneity in the Asperger and socio-emotional processing disorders. *Brain, 127*, 1706–1716.

Bedny, M., Pascual-Leone, A., & Saxe, R. R. (2009). Growing up blind does not change the neural bases of theory of mind. *Proceedings of the National Academy of Sciences USA, 106*(27), 11312–11317.

Behrmann, M., Thomas, C., & Humphreys, K. (2006). Seeing it differently: Visual processing in autism. *Trends in Cognitive Sciences, 10*(6), 258–264.

Belmonte, M. K., Cook, E. H., Jr., Anderson, G. M., Rubenstein, J. L., Greenough, W. T., Beckel-Mitchener, A., et al. (2004). Autism as a disorder of neural information processing: Directions for research and targets for therapy. *Molecular Psychiatry, 9*(7), 646–663.

Berument, S. K., Rutter, M., Lord, C., Pickles, A., & Bailey, A. (1999). Autism Screening Questionnaire: Diagnostic validity. *British Journal of Psychiatry, 175*, 444–451.

Bloom, P., & German, T. P. (2000). Two reasons to abandon the false belief task as a test of theory of mind. *Cognition, 77*(1), B25–B31.

Boddaert, N., Zilbovicius, M., Philipe, A., Robel, L., Bourgeois, M., Barthelemy, C., et al. (2009). MRI findings in 77 children with non-syndromic autistic disorder. *PLoS One, 4*(2), e4415.

Bonnel, A., McAdams, S., Smith, B., Berthiaume, C., Bertone, A., Ciocca, V., et al. (2010). Enhanced pure-tone pitch discrimination among persons with autism but not Asperger syndrome. *Neuropsychologia, 48*(9), 2465–2475.

Boucher, J., & Lewis, V. (1992). Unfamiliar face recognition in relatively able autistic children. *Journal of Child Psychology and Psychiatry, 33*(5), 843–859.

Boucher, J., Lewis, V., & Collis, G. (1998). Familiar face and voice matching and recognition in children with autism. *Journal of Child Psychology and Psychiatry, 39*, 171–181.

Bowler, D. (2007). *Autism spectrum disorders: Psychological theory and research.* Hoboken, NJ: Wiley.

Brambring, M., & Asbrock, D. (2010). Validity of false belief tasks in blind children. *Journal of Autism and Developmental Disorders, 40*(12), 1471–1484.

Braverman, M., Fein, D., Lucci, D., & Waterhouse, L. (1989). Affect comprehension in children with pervasive developmental disorders. *Journal of Autism and Developmental Disorders, 19*(2), 301–316.

Brune, C. W., Kim, S. J., Salt, J., Leventhal, B. L., Lord, C., & Cook, E. H., Jr. (2006). 5-HTTLPR genotype-specific phenotype in children and adolescents with autism. *American Journal of Psychiatry, 163*(12), 2148–2156.

Carlson, S. M., & Moses, L. J. (2001). Individual differences in inhibitory control and children's theory of mind. *Child Development, 72*(4), 1032–1053.

Castelli, F. (2005). Understanding emotions from standardized facial expressions in autism and normal development. *Autism, 9*(4), 428–449.

Castelli, F., Frith, C., Happé, F., & Frith, U. (2002). Autism, Asperger syndrome and brain mechanisms for the attribution of mental states to animated shapes. *Brain, 125*(Pt. 8), 1839–1849.

Celani, G., Battacchi, M. W., & Arcidiacono, L. (1999). The understanding of the emotional meaning of facial expressions in people with autism. *Journal of Autism and Developmental Disorders, 29*(1), 57–66.

Chawarska, K., & Shic, F. (2009). Looking but not seeing: Atypical visual scanning and recognition of faces in 2- and 4-year-old children with autism spectrum disorder. *Journal of Autism and Developmental Disorders, 39*(12), 1663–1672.

Chawarska, K., Volkmar, F., & Klin, A. (2010). Limited attentional bias for faces in toddlers with autism spectrum disorders. *Archives of General Psychiatry, 67*(2), 178–185.

Congiu, S., Schlottmann, A., & Ray, E. (2009). Unimpaired perception of social and physical causality, but impaired perception of animacy in high functioning children with autism. *Journal of Autism and Developmental Disorders, 40*(1), 39–53.

Constantino, J. N., Gruber, C. P., Davis, S., Hayes, S., Passanante, N., & Przybeck, T. (2004). The factor structure of autistic traits. *Journal of Child Psychology and Psychiatry, 45*(4), 719–726.

Critchley, H. D., Daly, E. M., Bullmore, E. T., Williams, S. C., Van Amelsvoort, T., Robertson, D. M., et al. (2000). The functional neuroanatomy of social behaviour: Changes in cerebral blood flow when people with autistic disorder process facial expressions. *Brain, 123*(Pt. 11), 2203–2212.

Dalton, K. M., Nacewicz, B. M., Johnstone, T., Schaefer, H. S., Gernsbacher, M. A., Goldsmith, H. H., et al. (2005). Gaze fixation and the neural circuitry of face processing in autism. *Nature Neuroscience, 8*(4), 519–526.

Decety, J., & Lamm, C. (2007). The role of the right temporoparietal junction in social interaction: How low-level computational processes contribute to meta-cognition. *Neuroscientist, 13*(6), 580–593.

Deruelle, C., Rondan, C., Gepner, B., & Tardif, C. (2004). Spatial frequency and face processing in children with autism and Asperger syndrome. *Journal of Autism and Developmental Disorders, 34*(2), 199–210.

DiLalla, D. L., & Rogers, S. J. (1994). Domains of the Childhood Autism Rating Scale: Relevance for diagnosis and treatment. *Journal of Autism and Developmental Disorders, 24*(2), 115–128.

Di Martino, A., Ross, K., Uddin, L. Q., Sklar, A. B., Castellanos, F. X., & Milham, M. P. (2009). Functional brain correlates of social and nonsocial processes in autism spectrum disorders: An activation likelihood estimation meta-analysis. *Biological Psychiatry, 65*(1), 63–74.

Dworzynski, K., Ronald, A., Hayiou-Thomas, M., Rijsdijk, F., Happé, F., Bolton, P., et al. (2007). Aetiological relationship between language performance and autistic-like traits in childhood: A twin study. *International Journal of Language and Communication Disorders, 42*(3), 273–292.

El Abd, S., Patton, M. A., Turk, J., Hoey, H., & Howlin, P. (1999). Social, communicational, and behavioral deficits associated with ring X Turner syndrome. *American Journal of Medical Genetics, 88*(5), 510–516.

Forgeot d'Arc, B., & Ramus, F. (2011). Belief attribution despite verbal interference. *Quarterly Journal of Experimantal Psychology (Colchester) 64*(5), 975–990.

Freitag, C. M., Konrad, C., Haberlen, M., Kleser, C., von Gontard, A., Reith, W., et al. (2008). Perception of biological motion in autism spectrum disorders. *Neuropsychologia, 46*(5), 1480–1494.

Freitag, C. M., Staal, W., Klauck, S. M., Duketis, E., & Waltes, R. (2010). Genetics of autistic disorders: Review and clinical implications. *European Child Adolescent Psychiatry, 19*(3), 169–178.

Frith, C. D. (1992). *The cognitive neuropsychology of schizophrenia.* Hillsdale, NJ: Erlbaum.

Frith, U., Happé, F., & Siddons, F. (1994). Autism and theory of mind in everyday life. *Social Development, 3*, 108–124.

Gepner, B., de Gelder, B., & de Schonen, S. (1996). Face processing in autistics: Evidence for a generalized deficit? *Child Neuropsychology, 2*, 123–139.

Gepner, B., Deruelle, C., & Grynfeltt, S. (2001). Motion and emotion: A novel approach to the study of face processing by young autistic children. *Journal of Autism and Developmental Disorders, 31*(1), 37–45.

Gervais, H., Belin, P., Boddaert, N., Leboyer, M., Coez, A., Sfaello, I., et al. (2004). Abnormal cortical voice processing in autism. *Nature Neuroscience, 7*(8), 801–802.

Geurts, H. M., Verte, S., Oosterlaan, J., Roeyers, H., & Sergeant, J. A. (2004). How specific are executive functioning deficits in attention deficit hyperactivity disorder and autism? *Journal of Child Psychology and Psychiatry, 45*(4), 836–854.

Gray, K., Jenkins, A. C., Heberlein, A. S., & Wegner, D. M. (2011). Distortions of mind perception in psychopathology. *Proceedings of the National Academy of Sciences USA, 108*(2), 477–479.

Greene, D. J., Colich, N., Iacoboni, M., Zaidel, E., Bookheimer, S. Y., & Dapretto, M. (2011). Atypical neural networks for social orienting in autism spectrum disorders. *NeuroImage, 56*(1), 354–362.

Griffith, E. M., Pennington, B. F., Wehner, E. A., & Rogers, S. J. (1999). Executive functions in young children with autism. *Child Development, 70*(4), 817–832.

Gross, T. F. (2004). The perception of four basic emotions in human and nonhuman faces by children with autism and other developmental disabilities. *Journal of Abnormal Child Psychology, 32*(5), 469–480.

Grossman, J., Klin, A., Carter, A. S., & Volkmar, F. (2000). Verbal bias in recognition of facial emotions in children with Asperger syndrome. *Journal of Child Psychology and Psychiatry, 41*(3), 369–379.

Hadjikhani, N., Joseph, R. M., Snyder, J., Chabris, C. F., Clark, J., Steele, S., et al. (2004).

Activation of the fusiform gyrus when individuals with autism spectrum disorder view faces. *NeuroImage, 22*(3), 1141–1150.

Happé, F. (1995). The role of age and verbal ability in the theory of mind task performance of subjects with autism. *Child Development, 66*(3), 843–855.

Happé, F., Booth, R., Charlton, R., & Hughes, C. (2006). Executive function deficits in autism spectrum disorders and attention-deficit/hyperactivity disorder: Examining profiles across domains and ages. *Brain and Cognition, 61*(1), 25–39.

Happé, F., & Ronald, A. (2008). The 'fractionable autism triad': A review of evidence from behavioural, genetic, cognitive and neural research. *Neuropsychology Review, 18*(4), 287–304.

Hauck, M., Fein, D., Maltby, N., Waterhouse, L., & Feinstein, C. (1998). Memory for faces in children with autism. *Child Neuropsychology, 4*, 187–198.

Hill, E. L. (2004). Executive dysfunction in autism. *Trends in Cognitive Sciences, 8*(1), 26–32.

Hobson, R. P. (1986). The autistic child's appraisal of expressions of emotion. *Journal of Child Psychology and Psychiatry, 27*(3), 321–342.

Hobson, R. P., & Bishop, M. (2003). The pathogenesis of autism: Insights from congenital blindness. *Philosophical Transactions of the Royal Society of London: Series B. Biological Sciences, 358*(1430), 335–344.

Hughes, C. (1998). Finding your marbles: Does preschoolers' strategic behavior predict later understanding of mind? *Developmental Psychology, 34*(6), 1326–1339.

Hyde, K. L., Samson, F., Evans, A. C., & Mottron, L. (2010). Neuroanatomical differences in brain areas implicated in perceptual and other core features of autism revealed by cortical thickness analysis and voxel-based morphometry. *Human Brain Mapping, 31*(4), 556–566.

Jemel, B., Mimeault, D., Saint-Amour, D., Hosein, A., & Mottron, L. (2010). VEP contrast sensitivity responses reveal reduced functional segregation of mid and high filters of visual channels in autism. *Journal of Vision, 10*(6), 13.

Jemel, B., Mottron, L., & Dawson, M. (2006). Impaired face processing in autism: Fact or artifact? *Journal of Autism and Developmental Disorders, 36*(1), 91–106.

Johnson, K. A., Robertson, I. H., Kelly, S. P., Silk, T. J., Barry, E., Daibhis, A., et al. (2007). Dissociation in performance of children with ADHD and high-functioning autism on a task of sustained attention. *Neuropsychologia, 45*(10), 2234–2245.

Jones, C. R., Happé, F., Baird, G., Simonoff, E., Marsden, A. J. S., Tregay, J., et al. (2009). Auditory discrimination and auditory sensory behaviours in autism spectrum disorders. *Neuropsychologia, 47*(13), 2850–2858.

Jones, C. R., Pickles, A., Falcaro, M., Marsden, A. J., Happé, F., Scott, S. K., et al. (2011). A multimodal approach to emotion recognition ability in autism spectrum disorders. *Journal of Child Psychology and Psychiatry, 52*(3), 275–285.

Joseph, R. M., McGrath, L. M., & Tager-Flusberg, H. (2005). Executive dysfunction and its relation to language ability in verbal school-age children with autism. *Developmental Neuropsycholology, 27*(3), 361–378.

Just, M. A., Cherkassky, V. L., Keller, T. A., & Minshew, N. J. (2004). Cortical activation and synchronization during sentence comprehension in high-functioning autism: Evidence of underconnectivity. *Brain, 127*(Pt. 8), 1811–1821.

Kana, R. K., Keller, T. A., Cherkassky, V. L., Minshew, N. J., & Just, M. A. (2009). Atypical frontal–posterior synchronization of theory of mind regions in autism during mental state attribution. *Social Neuroscience, 4*(2), 135–152.

Kanner, L. (1943). Autistic disturbances of affective contact. *Nervous Child, 2*, 217–250.

Kleinhans, N. M., Richards, T., Johnson, L. C., Weaver, K. E., Greenson, J., Dawson, G., et al. (2010). fMRI evidence of neural abnormalities in the subcortical face processing system in ASD. *NeuroImage, 54*(1), 697–704.

Klin, A., Jones, W., Schultz, R., Volkmar, F., & Cohen, D. (2002a). Defining and quantifying the social phenotype in autism. *American Journal of Psychiatry, 159*(6), 895–908.

Klin, A., Jones, W., Schultz, R., Volkmar, F., & Cohen, D. (2002b). Visual fixation patterns during viewing of naturalistic social situations as predictors of social competence in individuals with autism. *Archives of General Psychiatry, 59*(9), 809–816.

Klin, A., Lin, D. J., Gorrindo, P., Ramsay, G., & Jones, W. (2009). Two-year-olds with autism orient to non-social contingencies rather than biological motion. *Nature, 459*(7244), 257–261.

Kovacs, A. M., Teglas, E., & Endress, A. D. (2010). The social sense: Susceptibility to others' beliefs in human infants and adults. *Science, 330*(6012), 1830–1834.

Lahaie, A., Mottron, L., Arguin, M., Berthiaume, C., Jemel, B., & Saumier, D. (2006). Face perception in high-functioning autistic adults: Evidence for superior processing of face parts, not for a configural face-processing deficit. *Neuropsychology, 20*(1), 30–41.

Langdell, T. (1978). Recognition of faces: An approach to the study of autism. *Journal of Child Psychology and Psychiatry, 19*(3), 255–268.

Lecavalier, L., Aman, M. G., Hammer, D., Stoica, W., & Mathews, G. L. (2004). Factor analysis of the Nisonger Child Behavior Rating Form in children with autism spectrum disorders. *Journal of Autism and Developmental Disorders, 34*(6), 709–721.

Leekam, S., Baron-Cohen, S., Brown, S., Perrett, D., & Milders, M. (1997). Eye-direction detection: A dissociation between geometric and joint-attention skills in autism. *British Journal of Developmental Psychology, 15*, 77–95.

Leekam, S., Lopez, B., & Moore, C. (2000). Attention and joint attention in preschool children with autism. *Developmental Psychology, 36*(2), 261–273.

Leslie, A. M., Friedman, O., & German, T. P. (2004). Core mechanisms in "theory of mind." *Trends in Cognitive Sciences, 8*(12), 528–533.

Leveille, C., Barbeau, E. B., Bolduc, C., Limoges, E., Berthiaume, C., Chevrier, E., et al. (2010). Enhanced connectivity between visual cortex and other regions of the brain in autism: A REM sleep EEG coherence study. *Autism Research, 3*(5), 280–285.

Lopez, B., Donnelly, N., & Hadwin, J. A. (2004). Face processing in high-functioning adolescents with autism: Evidence for weak central coherence. *Visual Cognition, 11*, 673–688.

Lord, C., Pickles, A., McLennan, J., Rutter, M., Bregman, J., Folstein, S., et al. (1997). Diagnosing autism: Analyses of data from the Autism Diagnostic Interview. *Journal of Autism and Developmental Disorders, 27*(5), 501–517.

Lord, C., Risi, S., Lambrecht, L., Cook, E. H., Jr., Leventhal, B. L., DiLavore, P. C., et al. (2000). The Autism Diagnostic Observation Schedule—Generic: A standard measure of social and communication deficits associated with the spectrum of autism. *Journal of Autism and Developmental Disorders, 30*(3), 205–223.

Mandy, W. P., & Skuse, D. H. (2008). Research review: What is the association between the social-communication element of autism and repetitive interests, behaviours and activities? *Journal of Child Psychology and Psychiatry, 49*(8), 795–808.

Mann, T. A., & Walker, P. (2003). Autism and a deficit in broadening the spread of visual attention. *Journal of Child Psychology and Psychiatry, 44*(2), 274–284.

Miranda-Linne, F. M., & Melin, L. (2002). A factor analytic study of the Autism Behavior Checklist. *Journal of Autism and Developmental Disorders, 32*(3), 181–188.

Moran, J. M., Young, L. L., Saxe, R., Lee, S. M., O'Young, D., Mavros, P. L., et al. (2011). Impaired theory of mind for moral judgment in high-functioning autism. *Proceedings of the National Academy of Sciences USA, 108*(7), 2688–2692.

Moses, L. J. (2001). Executive accounts of theory-of-mind development. *Child Development, 72*(3), 688–690.

Mottron, L., Belleville, S., & Menard, E. (1999). Local bias in autistic subjects as evidenced by graphic tasks: Perceptual hierarchization or working memory deficit? *Journal of Child Psychology and Psychiatry, 40*(5), 743–755.

Mottron, L., Dawson, M., & Soulieres, I. (2009). Enhanced perception in savant syndrome:

Patterns, structure and creativity. *Philosophical Transactions of the Royal Society of London: Series B. Biological Sciences, 364*(1522), 1385–1391.

Mottron, L., Dawson, M., Soulieres, I., Hubert, B., & Burack, J. (2006). Enhanced perceptual functioning in autism: An update, and eight principles of autistic perception. *Journal of Autism and Developmental Disorders, 36*(1), 27–43.

Mulder, E. J., Anderson, G. M., Kema, I. P., Brugman, A. M., Ketelaars, C. E., de Bildt, A., et al. (2005). Serotonin transporter intron 2 polymorphism associated with rigid–compulsive behaviors in Dutch individuals with pervasive developmental disorder. *American Journal of Medical Genetics, 133B*(1), 93–96.

Muller, R. A. (2007). The study of autism as a distributed disorder. *Mental Retardation and Developmental Disabilities Research Review, 13*(1), 85–95.

Murphy, P., Brady, N., Fitzgerald, M., & Troje, N. F. (2009). No evidence for impaired perception of biological motion in adults with autistic spectrum disorders. *Neuropsychologia, 47*(14), 3225–3235.

Nation, K., & Penny, S. (2008). Sensitivity to eye gaze in autism: Is it normal? Is it automatic? Is it social? *Developmental Psychopathology, 20*(1), 79–97.

Onishi, K. H., & Baillargeon, R. (2005). Do 15-month-old infants understand false beliefs? *Science, 308*(5719), 255–258.

Ornitz, E. M., & Ritvo, E. R. (1968). Neurophysiologic mechanisms underlying perceptual inconstancy in autistic and schizophrenic children. *Archives of General Psychiatry, 19*(1), 22–27.

Osterling, J., & Dawson, G. (1994). Early recognition of children with autism: A study of first birthday home videotapes. *Journal of Autism and Developmental Disorders, 24*(3), 247–257.

Osterling, J. A., Dawson, G., & Munson, J. A. (2002). Early recognition of 1-year-old infants with autism spectrum disorder versus mental retardation. *Developmental Psychopathology, 14*(2), 239–251.

Ozonoff, S., & Jensen, J. (1999). Brief report: Specific executive function profiles in three neurodevelopmental disorders. *Journal of Autism and Developmental Disorders, 29*(2), 171–177.

Ozonoff, S., Pennington, B. F., & Rogers, S. J. (1990). Are there emotion perception deficits in young autistic children? *Journal of Child Psychology and Psychiatry, 31*(3), 343–361.

Ozonoff, S., Pennington, B. F., & Rogers, S. J. (1991). Executive function deficits in high-functioning autistic individuals: Relationship to theory of mind. *Journal of Child Psychology and Psychiatry, 32*(7), 1081–1105.

Pauli-Pott, U., & Becker, K. (2011). Neuropsychological basic deficits in preschoolers at risk for ADHD: A meta analysis. *Clinical Psychology Reviews, 31*(4), 626–637.

Pellicano, E. (2007). Links between theory of mind and executive function in young children with autism: Clues to developmental primacy. *Developmental Psychology, 43*(4), 974–990.

Pellicano, E. (2010). Individual differences in executive function and central coherence predict developmental changes in theory of mind in autism. *Developmental Psychology, 46*(2), 530–544.

Pelphrey, K. A., Morris, J. P., & McCarthy, G. (2005). Neural basis of eye gaze processing deficits in autism. *Brain, 128*(Pt. 5), 1038–1048.

Pelphrey, K. A., Sasson, N. J., Reznick, J. S., Paul, G., Goldman, B. D., & Piven, J. (2002). Visual scanning of faces in autism. *Journal of Autism and Developmental Disorders, 32*(4), 249–261.

Perner, J. (1995). *Understanding the representational mind.* Cambridge, MA: MIT Press.

Pierce, K., Haist, F., Sedaghat, F., & Courchesne, E. (2004). The brain response to personally familiar faces in autism: Findings of fusiform activity and beyond. *Brain, 127*(Pt. 12), 2703–2716.

Pierce, K., Muller, R. A., Ambrose, J., Allen, G., & Courchesne, E. (2001). Face processing occurs outside the fusiform 'face area' in autism: Evidence from functional MRI. *Brain, 124*(Pt. 10), 2059–2073.

Piggot, J., Kwon, H., Mobbs, D., Blasey, C., Lotspeich, L., Menon, V., et al. (2004). Emotional attribution in high-functioning individuals with autistic spectrum disorder: A functional imaging study. *Journal of the American Academy of Child and Adolescent Psychiatry, 43*(4), 473–480.

Posner, M. I. (1980). Orienting of attention. *Quarterly Journal of Experimental Psychology (Colchester), 32*(1), 3–25.

Premack, D., & Woodruff, G. (1978). Does the chimpanzee have a theory of mind? *Behavioral and Brain Sciences, 1*(4), 515–526.

Radua, J., Via, E., Catani, M., & Mataix-Cols, D. (2011). Voxel-based meta-analysis of regional white-matter volume differences in autism spectrum disorder versus healthy controls. *Psychological Medicine, 41*(7), 1539–1550.

Rinehart, N. J., Bradshaw, J. L., Moss, S. A., Brereton, A. V., & Tonge, B. J. (2000). Atypical interference of local detail on global processing in high-functioning autism and Asperger's disorder. *Journal of Child Psychology and Psychiatry, 41*(6), 769–778.

Ristic, J., Mottron, L., Friesen, C. K., Iarocci, G., Burack, J. A., & Kingstone, A. (2005). Eyes are special but not for everyone: the case of autism. *Brain Research: Cognitive Brain Research, 24*(3), 715–718.

Ronald, A., Happé, F., & Plomin, R. (2005). The genetic relationship between individual differences in social and nonsocial behaviours characteristic of autism. *Developmental Science, 8*(5), 444–458.

Rouse, H., Donnelly, N., Hadwin, J. A., & Brown, T. (2004). Do children with autism perceive second-order relational features?: The case of the Thatcher illusion. *Journal of Child Psychology and Psychiatry, 45*(7), 1246–1257.

Russell, J. (1998). How an executive disorder can bring about an inadequate 'theory of mind.' In J. Russell (Ed.), *Autism as an executive disorder* (pp. 256–304). Oxford, UK: Oxford University Press.

Russo, N., Flanagan, T., Iarocci, G., Berringer, D., Zelazo, P. D., & Burack, J. A. (2007). Deconstructing executive deficits among persons with autism: Implications for cognitive neuroscience. *Brain and Cognition, 65*(1), 77–86.

Rutter, M. (1968). Concepts of autism: A review of research. *Journal of Child Psychology and Psychiatry, 9*(1), 1–25.

Samson, F., Mottron, L., Soulieres, I., & Zeffiro, T. A. (2011, April 4). Enhanced visual functioning in autism: An ALE meta-analysis. *Human Brain Mapping.* [Epub ahead of print]

Saygin, A. P., Cook, J., & Blakemore, S. J. (2010). Unaffected perceptual thresholds for biological and non-biological form-from-motion perception in autism spectrum conditions. *PLoS One, 5*(10), e13491.

Schultz, R. T., Gauthier, I., Klin, A., Fulbright, R. K., Anderson, A. W., Volkmar, F., et al. (2000). Abnormal ventral temporal cortical activity during face discrimination among individuals with autism and Asperger syndrome. *Archives of General Psychiatry, 57*(4), 331–340.

Schultz, R. T., & Robins, D. (2005). Functional neuroimaging studies of autism spectrum disorders. In F. Volkmar, A. Klin, & R. Paul (Eds.), *Handbook of autism and pervasive developmental disorders* (3rd ed., pp. 515–533). Hoboken, NJ: Wiley.

Senju, A., Southgate, V., White, S., & Frith, U. (2009). Mindblind eyes: An absence of spontaneous theory of mind in Asperger syndrome. *Science, 325*(5942), 883–885.

Senju, A., Tojo, Y., Dairoku, H., & Hasegawa, T. (2004). Reflexive orienting in response to eye gaze and an arrow in children with and without autism. *Journal of Child Psychology and Psychiatry, 45*(3), 445–458.

Shic, F., Bradshaw, J., Klin, A., Scassellati, B., & Chawarska, K. (2010). Limited activity monitoring in toddlers with autism spectrum disorder. *Brain Research, 1380*, 246–254.

Simmons, D. R., Robertson, A. E., McKay, L. S., Toal, E., McAleer, P., & Pollick, F. E. (2009). Vision in autism spectrum disorders. *Vision Research, 49*(22), 2705–2739.

Singer, T. (2006). The neuronal basis and ontogeny of empathy and mind reading: Review of literature and implications for future research. *Neuroscience and Biobehavioral Reviews, 30*(6), 855–863.

Soulieres, I., Zeffiro, T. A., Girard, M. L., & Mottron, L. (2011). Enhanced mental image mapping in autism. *Neuropsychologia, 49*(5), 848–857.

Stella, J., Mundy, P., & Tuchman, R. (1999). Social and nonsocial factors in the Childhood Autism Rating Scale. *Journal of Autism and Developmental Disorders, 29*(4), 307–317.

Sutcliffe, J. S., Delahanty, R. J., Prasad, H. C., McCauley, J. L., Han, Q., Jiang, L., et al. (2005). Allelic heterogeneity at the serotonin transporter locus (SLC6A4) confers susceptibility to autism and rigid-compulsive behaviors. *American Journal of Human Genetics, 77*(2), 265–279.

Swettenham, J., Condie, S., Campbell, R., Milne, E., & Coleman, M. (2003). Does the perception of moving eyes trigger reflexive visual orienting in autism? *Philosophical Transactions of the Royal Society of London: Series B. Biological Sciences, 358*(1430), 325–334.

Szatmari, P., Merette, C., Bryson, S. E., Thivierge, J., Roy, M. A., Cayer, M., et al. (2002). Quantifying dimensions in autism: A factor-analytic study. *Journal of the American Academy of Child and Adolescent Psychiatry, 41*(4), 467–474.

Tadevosyan-Leyfer, O., Dowd, M., Mankoski, R., Winklosky, B., Putnam, S., McGrath, L., et al. (2003). A principal components analysis of the Autism Diagnostic Interview—Revised. *Journal of the American Academy of Child and Adolescent Psychiatry, 42*(7), 864–872.

Tantam, D., Monaghan, L., Nicholson, H., & Stirling, J. (1989). Autistic children's ability to interpret faces: A research note. *Journal of Child Psychology and Psychiatry, 30*(4), 623–630.

Tordjman, S., Gutknecht, L., Carlier, M., Spitz, E., Antoine, C., Slama, F., et al. (2001). Role of the serotonin transporter gene in the behavioral expression of autism. *Molecular Psychiatry, 6*(4), 434–439.

van Lang, N. D., Boomsma, A., Sytema, S., de Bildt, A. A., Kraijer, D. W., Ketelaars, C., et al. (2006). Structural equation analysis of a hypothesised symptom model in the autism spectrum. *Journal of Child Psychology and Psychiatry, 47*(1), 37–44.

Van Overwalle, F. (2011). A dissociation between social mentalizing and general reasoning. *NeuroImage, 54*(2), 1589–1599.

Vlamings, P. H., Jonkman, L. M., van Daalen, E., van der Gaag, R. J., & Kemner, C. (2010). Basic abnormalities in visual processing affect face processing at an early age in autism spectrum disorder. *Biological Psychiatry, 68*(12), 1107–1113.

Volkmar, F. R., Cohen, D. J., Hoshino, Y., Rende, R. D., & Paul, R. (1988). Phenomenology and classification of the childhood psychoses. *Psychological Medicine, 18*(1), 191–201.

Volkmar, F. R., Sparrow, S. S., Rende, R. D., & Cohen, D. J. (1989). Facial perception in autism. *Journal of Child Psychology and Psychiatry, 30*(4), 591–598.

Wadden, N. P., Bryson, S. E., & Rodger, R. S. (1991). A closer look at the Autism Behavior Checklist: Discriminant validity and factor structure. *Journal of Autism and Developmental Disorders, 21*(4), 529–541.

Wang, A. T., Dapretto, M., Hariri, A. R., Sigman, M., & Bookheimer, S. Y. (2004). Neural correlates of facial affect processing in children and adolescents with autism spectrum disorder. *Journal of the American Academy of Child and Adolescent Psychiatry, 43*(4), 481–490.

Wellman, H. M., & Cross, D. (2001). Theory of mind and conceptual change. *Child Development, 72*(3), 702–707.

Weng, S. J., Carrasco, M., Swartz, J. R., Wiggins, J. L., Kurapati, N., Liberzon, I., et al. (2011). Neural activation to emotional faces in adolescents with autism spectrum disorders. *Journal of Child Psychology and Psychiatry, 52*(3), 296–305.

Werner, E., Dawson, G., Osterling, J., & Dinno, N. (2000). Brief report: Recognition of autism spectrum disorder before one year of age: A retrospective study based on home videotapes. *Journal of Autism and Developmental Disorders, 30*(2), 157–162.

Wicker, B., Fonlupt, P., Hubert, B., Tardif, C., Gepner, B., & Deruelle, C. (2008). Abnormal cerebral effective connectivity during explicit emotional processing in adults with autism spectrum disorder. *Social Cognitive and Affective Neuroscience, 3*(2), 135–143.

Yoshida, W., Dziobek, I., Kliemann, D., Heekeren, H. R., Friston, K. J., & Dolan, R. J. (2010). Cooperation and heterogeneity of the autistic mind. *Journal of Neuroscience, 30*(26), 8815–8818.

Yu, K. K., Cheung, C., Chua, S. E., & McAlonan, G. M. (2011). Can Asperger syndrome be distinguished from autism?: An anatomic likelihood meta-analysis of MRI studies. *Journal of Psychiatry and Neuroscience, 36*(2), 100–138.

Zhao, X., Leotta, A., Kustanovich, V., Lajonchere, C., Geschwind, D. H., Law, K., et al. (2007). A unified genetic theory for sporadic and inherited autism. *Proceedings of the National Academy of Sciences USA, 104*(31), 12831–12836.

Zucker, N. L., Losh, M., Bulik, C. M., LaBar, K. S., Piven, J., & Pelphrey, K. A. (2007). Anorexia nervosa and autism spectrum disorders: Guided investigation of social cognitive endophenotypes. *Psychological Bulletin, 133*(6), 976–1006.

PART V
SOCIAL INTERVENTIONS

CHAPTER 16

Pragmatic Language Impairment after Brain Injury

Social Implications and Treatment Models

Skye McDonald, Lyn S. Turkstra, and Leanne Togher

P*ragmatic communication ability*, or simply *pragmatics,* is the ability to use language in context—beyond understanding and expressing basic word meanings (*semantics*) in the correct grammatical forms (*syntax*). In other words, pragmatics is concerned with the way language is used rather than the form it takes (Levinson, 1983). Pragmatics is generally thought to involve three major types of communication skills: (1) adhering to social conventions, such as looking at your communicative partner and maintaining an appropriate distance during conversational exchanges; (2) using language for different functions, such as requesting and greeting; and (3) adjusting language to meet the expectations and needs of the listener, such as talking differently to an adult versus a peer or in a professional versus a personal context (Turkstra, Burgess, Clark, Hengst, & Paul, 2011).

It is important to keep in mind that *pragmatics* is not synonymous with *social communication*, although these terms are often used interchangeably. Social communication requires more than just pragmatics—notably phonology, morphology, semantics, and syntax. It also requires awareness of social-linguistic factors (influence of culture, gender, and languages spoken), psychological attributes (e.g., psychosocial adjustment), and interpersonal skills (Adams, 2005), as discussed elsewhere in this text. Successful social communication also requires the opportunity to demonstrate and practice skills. Thus treatment of pragmatic communication skill deficits in children with neurological disorders must go beyond discrete pragmatic skill training to consider the "bigger picture" of social adjustment and social opportunities (Sohlberg & Turkstra, 2011). The majority of contemporary research has focused on (1) characterizing the nature of pragmatic impairment within the context of typical development, and determining its likely impact upon social function; (2) discovering the neuropsychological mechanisms underpinning impaired pragmatic competence; and (3) examining social opportunities and the ways in which these interact with communication skills. We review each of these endeavors below, and then we consider treatment approaches.

Characterizing the Nature of Pragmatic Impairment

Research Findings to Date

The particular communication difficulties experienced by children and adolescents with brain injuries were first documented on the basis of relatives' and teachers' reports (Ylvisaker, 1989). Reports of persistent aphasia are relatively uncommon following traumatic brain injury (TBI) (Heilman, Safran, & Geschwind, 1971), even in children with focal left-hemisphere lesions (Vicari et al., 2000). Nevertheless, many relatives reported that their family members with TBI had such communication problems as slow, hesitant speech; lack of initiative (e.g., failing to start conversations); rigid use of repeated expressions; or, conversely, overtalkative, tangential, and inappropriate conversation (Thomsen, 1975). These behaviors have been quantified in a large body of research studies since that time, which have shown various deficits in pragmatic communication skills: maintaining and organizing topic sequences in a narrative, turn taking, and ideational fluency (Morse et al., 1999); using complex language and organizing story elements in a logical and informative manner (Brookshire, Chapman, Song, & Levin, 2000); understanding linguistic inferences in text (Dennis & Barnes, 1990); differentiating truth from deception (Dennis, Purvis, Barnes, Wilkinson, & Winner, 2001); understanding written or spoken sarcasm (Turkstra, McDonald, & DePompei, 2001; Turkstra, McDonald, & Kaufmann, 1996); and comprehending figurative language, such as metaphors and idioms (Towne & Entwisle, 1993).

Although the effects of pragmatic communication impairments on quality of life are well documented in adults with TBI (Struchen et al., 2008), less is known about their effects in children. A critical consideration for children is that poor pragmatic communication skills in childhood will influence later development—not only because ongoing brain development may be permanently interrupted (Ciccia, Meulenbroek, & Turkstra, 2009), but also because children and adolescents learn social and intimacy skills by interacting with others (Dawson, 2001), and those who are socially isolated may miss this opportunity. To our knowledge, only one study has directly tested the relation of pragmatic communication skills to social outcome in children (Yeates et al., 2004). In that study, comprehension of figurative language made a significant and unique contribution to scores on a test of social problem solving and also to parent reports of social problems, 4 years after injury. This preliminary evidence that pragmatic communication is connected to social outcome in TBI is supported by a large body of research in children, adolescents, and young adults with developmental disorders, which has linked pragmatic communication problems to such negative outcomes as rejection by peers (Place & Becker, 1991), a reduction in the frequency of dating and making new friends (Asher & Hymel, 1986), and limitations related to employment (DeGroot & Motowidlo, 1999). The connection between pragmatic communication and social outcomes is also supported by studies focusing on social behavior more broadly (e.g., Yeates et al., 2004), which are likely to capture elements of pragmatic communication.

Whereas the evidence suggests that pragmatic communication impairments acquired in childhood may have profound effects on social outcomes, outcomes for adolescents with brain injury are less clear. Although the studies described above would predict negative social outcomes for adolescents with pragmatic communication disorders after TBI, Turkstra, Dixon, and Baker (2004) found that adolescents with moderate to severe TBI reported levels of social acceptance similar to those of their peers, despite differences in pragmatic communication skills. Thus it is not clear that impairments in communication ability or limited social opportunities necessarily translate into poorer quality of life at this particular stage. Our clinical experience and interviews with young adults with TBI suggest that the full effects of pragmatic communication problems may not be experienced until later in life, when social life becomes less structured, so that self-reports of adolescents with TBI may underestimate the potential for future problems.

A Theoretical Approach to Pragmatics: Grice and Speech Act Theory

Pragmatic communication impairments in children and adolescents have historically been described in terms of a theoretical model of pragmatic language that was first proposed by Grice (1978). Grice, a linguist, argued that all communication proceeds according to an implicit agreement to cooperate between speakers (the *cooperative principle*); that is, both parties can assume that the communication between them is designed to inform, persuade, or otherwise influence the other party. Furthermore, he argued that speakers adhere to four maxims: the maxims of *quality* (truthfulness), *quantity* (saying what is sufficient), *manner* (orderliness), and *relation* (relevance). Speakers often deliberately flout such maxims—for example, by saying something that is only partly sufficient or patently untrue, in order to achieve particular communication effects (diplomacy and sarcasm, respectively). Although such planned deviations can be effective, haphazard failures to adhere to these maxims will result in chaotic, socially unacceptable, or ineffective language.

Descriptions of discourse problems following TBI, such as lack of initiative, reliance on set expressions, overtalkativeness, tangentiality, and inappropriateness, do suggest that TBI leads to problems in adhering to implicit conversational maxims as outlined by Grice. Checklists based upon Grice's maxims have been developed for use in clinical evaluation. The oldest of these is the Pragmatic Protocol (Prutting & Kirchner, 1987), developed for the study of pragmatic communication ability in children. The Pragmatic Protocol provides a descriptive taxonomy of 30 pragmatic aspects of communication (e.g., topic selection and maintenance, turn taking), and each pragmatic behavior is rated as "appropriate" or "not appropriate" by a clinician. The protocol is based on Grice's maxims, as well as elements of *speech act theory* (Searle, 1976), which describes how a child uses language for different functions (e.g., requesting, clarifying, entertaining). It also encompasses research on interpersonal behaviors (such as physical proximity and body posture) and analysis of paralinguistic behaviors (such as prosody, vocal loudness, gestures, and facial expression).

Another popular pragmatic assessment tool is the Clinical Discourse Analysis (CDA; Damico, 1991), which has been used to evaluate children and adolescents with TBI both clinically and in research contexts (Morse et al., 1999). The CDA categorizes problem behaviors into four types of errors, which correspond to Grice's maxims: amount and efficiency of information provided (quantity); accuracy of information (quality); cohesion (e.g., matching pronouns to their referents) and topic management (relation); and aspects of conversational fluency at the word level (e.g., repeating words or phrases), discourse level (e.g., connecting story elements in a logical fashion), and conversation level (e.g., taking turns appropriately) (manner).

Derivatives of the Pragmatic Protocol and CDA are used in many settings today, and these two instruments are also reflected in many informal "in-house" pragmatic communication assessment tools used in pediatric TBI rehabilitation and school settings. The tradition of examining speech acts and interpersonal communication behaviors along the lines of Grice's maxims is also evident in tools designed for adults that are now being applied to children, such as the La Trobe Communication Questionnaire (LCQ; Douglas, Bracy, & Snow, 2007). The LCQ is sensitive to pragmatic impairments not only in adults with TBI, but also in adolescents (Douglas, 2010). The LCQ has two forms: a self-report form that is completed by the person with TBI, and a significant-other form that is completed by a family member or carer of the person with TBI. In both cases, the person is asked to reflect on the communication behavior of the person with TBI over the preceding few months. It consists of 30 questions about specific pragmatic communication behaviors (e.g., "When talking to others do you/does your relative have difficulty getting the conversation started?"), which are answered on a 4-point scale from "never" to "usually." This contrasts with the Pragmatic Protocol, which is completed by a speech–language pathologist after watching at least 15 minutes of videotaped conversation, and the CDA, which is typically completed by a speech–language pathologist using a transcription of a language sample. These time- and labor-intensive processes are often impractical in clinical settings.

A few standardized tests have items related to pragmatic communication ability, such as the Pragmatic Judgment test of the Comprehensive Assessment of Spoken Language (Carrow-Woolfolk, 1999) and the Test of Language Competence (Wiig & Secord, 1989). Many parent and caregiver behavior questionnaires also include items related to pragmatic communication skills, such as the Child Behavior Checklist (Achenbach & Rescorla, 2001), the Vineland Adaptive Behavior Scales, Second Edition (Sparrow, Cicchetti, & Balla, 2005), and the Behavior Rating Inventory of Executive Function (Gioia, Isquith, Guy, & Kenworthy, 2000). These instruments should be viewed with the caveat that "standardized" assessment of pragmatic communication has inherent limitations, given (1) the context dependence of pragmatic behaviors; (2) the range of skills among the typical population; and (3) the many variables that contribute to judgments of "appropriateness" in a given social, ethnic, or cultural group (Turkstra et al., 2011).

The use of Grice's model of pragmatic language is also useful for considering specific interpersonal and social functions of language, and for determining how inappropriate, rigid, or tangential conversation may disrupt social relationships. All speech acts carry with them underlying implications about the relationship between the speakers. It is extremely common to be indirect when speaking (Gibbs, 1986), and such indirectness tends to increase as the distance between speakers (power, familiarity, cultural imposition of the topic) increases (Brown & Levinson, 1978). Consequently, a speaker's language choices send clear messages to a listener about how the listener is viewed. Failure to adjust language to meet changing social constraints will lead to clumsy and ineffective language use and will impair social negotiations. Once again, this type of approach has proven sensitive for people with brain injury. For example, it has been demonstrated that some adolescents with TBI cannot hint effectively or address potential obstacles to compliance on the part of the addressee (Turkstra, McDonald, & Kaufmann, 1996).

These problems affect not just production, but also comprehension. Many children and adults with severe TBI have great difficulty comprehending indirect remarks meant to be taken sarcastically (e.g., "What a genius" said to mean the reverse) (McDonald, Flanagan, Rollins, & Kinch, 2003; Turkstra et al., 1996), and such deficits correspond with poor social skills. For example, we have established that adolescents with TBI who fail to understand such types of pragmatic language as sarcasm, indirect requests, and bragging also use fewer "thought" and "feeling" words in extemporaneous conversations (Stronach & Turkstra, 2008), putting them at risk for alienating peers. In adults who have sustained severe TBI, we have also found that those who experienced difficulty understanding white lies and irony were also those rated by independent observers as lacking appropriate humor and as being insensitive and egocentric when relating to another (McDonald, Flanagan, Martin, & Saunders, 2004).

Neuropsychological Mechanisms Contributing to Pragmatic Competence

Deconstructing pragmatic communication into its neuropsychological constituents has been a major endeavor of research over the past three decades or more—not only in speech–language pathology and neuropsychology, but also in cognitive psychology and cognitive neuroscience. A summary of neuropsychological processes hypothesized to be involved in pragmatic communication is shown in Figure 16.1.

Models such as this have been the basis for a growing body of research testing links among brain lesions, cognitive functions, and communication behaviors. Sensory–motor deficits, language impairment, and poor memory will certainly affect the capacity to demonstrate pragmatic competence. Even so, it can be argued that disorders in these domains will not necessarily lead to pragmatic deficits. People with aphasia, for example, can demonstrate relatively preserved pragmatic understanding (Chapman, Highley, & Thompson, 1998). On the other hand, evidence is accruing to suggest that many pragmatic elements and metapragmatic skills depend specifically upon executive functions (EFs) and social cognition.

Executive Dysfunction

Executive dysfunction can be characterized in a number of ways, but it essentially reflects the loss of executive control over other cognitive processes and behavior. Poor working memory (attention) and impaired processing speed will also contribute to poor executive control. Executive dysfunction is highly prevalent following brain damage in children and adolescents, especially TBI due to motor vehicle accidents, falls, or intentional injury (Chapman et al., 1998). In these events, blunt forces and rapid acceleration–deceleration lead to diffuse white matter injury and to contusions and bleeding, with a preponderance of damage to the ventrolateral surfaces of the frontal and temporal lobes (Bigler, 2001). Slowed information processing is thought to arise from diffuse axonal injury, which results in shearing of interconnections between networks (Felmingham, Baguley,

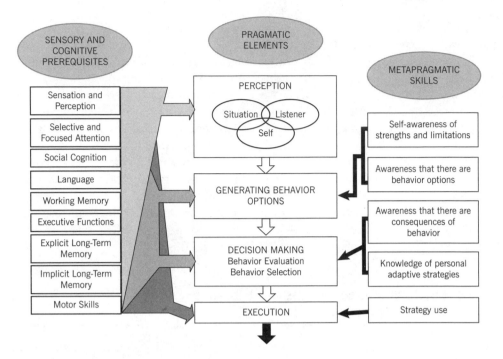

FIGURE 16.1. A heuristic for conceptualizing relations among the sensory, perceptual, cognitive, metacognitive, and motor processes involved in pragmatic communication. *Sensory, perceptual,* and *cognitive* prerequisites are expected to influence all pragmatic elements, although "earlier" elements may be more influenced by sensory and perceptual processes, whereas "later" elements may be more influenced by higher-order cognitive functions (e.g., executive functions). *Language skills* include phonology, morphology, syntax, and semantics. *Executive functions* include such processes as shifting between mental sets, updating and monitoring working memory, and inhibiting prepotent responses (Miyake et al., 2000). *Implicit memory* processes include emotional associations and social skills/routines. Missing from this figure are pragmatic functions related to execution of behaviors over time, such as acquisition of skills during development; self-monitoring and self-correcting behavior based on feedback and consequences of communication behaviors; and changing context factors, such as opportunity for and access to communication opportunities, as well as motivation to communicate with others.

& Green, 2004). Executive dysfunction is though to arise as a result of damage to frontal systems and their connections, although in both children (see review by Ciccia et al., 2009) and adults there is evidence that other factors—such as decreased integrity of white matter throughout the brain—also play a role. In general, damage to the dorsolateral frontal convexities appears to impair the coordination of cognitive systems in goal-directed activity, while lesions in dorsomedial and ventral frontal regions have an impact on self-regulation, including motivational states (Eslinger, 2008).

In studies of children with TBI, EF impairments typically include deficits both in basic EF processes (such as working memory and control of attention, behavior, and emotions) and in more complex EF-dependent processes (including concept formation and problem solving) (Ganesalingam et al., 2011; Levin & Hanten, 2005). These deficits are more apparent when evaluated in everyday life than on standardized tests (Ganesalingam et al., 2011; Mangeot, Armstrong, Colvin, Yeates, & Taylor, 2002), as the structure of the latter may act as a support for EFs, particularly for young children. A critical consideration in children is that EFs have a protracted developmental trajectory, so that early injury may not only disrupt previously developed skills, but also slow or even prevent the development of future skills. This disruption may occur in a nonlinear fashion (Jacobs, Harvey, & Anderson, 2007), so that injury at certain developmental stages (e.g., preadolescence) may have more severe effects than at others. EF development is also influenced by such factors as social stimulation and affective states, particularly in adolescence (Crone, 2009), and these may be altered by an injury in complex ways. Another critical consideration in children and adolescence is the presence of premorbid conditions that affect EF development and may interact with acquired injury, such as attention-deficit/hyperactivity disorder (Slomine et al., 2005).

Descriptions of communication patterns following TBI in childhood and adolescence directly suggest the impact of reduced information processing efficiency and impaired EFs affecting both drive and control. For example, language production has been described as impoverished in both amount and variety, and the complexity of language used is negatively correlated with impairments in working memory and other EFs (Chapman et al., 1992). In other cases, the effects of impaired control and inhibition are apparent. The language of adolescents has been described as overtalkative, inefficient, and tangential (Turkstra, 2000). Some adolescents with TBI may show reduced conversational fluency and inability to juggle multiple demands of conversation (Douglas, 2010).

Impairments in executive control, working memory, and impaired processing speed will also decrease comprehension by significantly reducing a child's ability to interpret conversation rapidly and process information flexibly, in order to draw pragmatic inferences and appreciate implied meanings (McDonald et al., 2006). This has been demonstrated empirically, with both children and adolescents experiencing impairments in comprehension of linguistic inferences such as implied meaning in text (Moran & Gillon, 2005) and figurative language (Towne & Entwisle, 1993), as well as social inferences including irony, sarcasm, and humor (Turkstra et al., 1996).

Finally, self-awareness, commonly viewed as an important facet of EF, is frequently affected as a result of childhood brain injury (Dennis, Barnes, Donnelly, Wilkinson, & Humphreys, 1996). Metacognition is a major concern for children and adolescents, as they are in the process of developing the ability to monitor, understand, and control their own thinking and performance (Hanten et al., 2004). These self-reflective abilities are integrally associated with metapragmatic skills as delineated in Figure 16.1. They can also be a major factor in a child's willingness to consider strategies for managing pragmatic communication problems in everyday life.

Social Cognition

Social cognition refers to the ability to identify and interpret social signals. It includes, at a minimum, such processes as the ability to recognize others' emotions from affective and vocal displays; the ability to demonstrate empathy; and *theory of mind* (ToM), defined as the understanding that others have thoughts separate from one's own and these thoughts can influence behaviors (Premack, 1976). As indicated in Figure 16.1, social-cognitive skills are part of the basic cognitive architecture of pragmatic communication, and failure to recognize or understand the feelings and intentions of others has an immediate deleterious impact on interpersonal behavior (Morrison & Bellack, 1981).

Thus it is unsurprising that social cognition has increasingly become a focus for research on disorders of pragmatics and communication.

As in other aspects of cognitive function, children and adolescents are particularly vulnerable to impairments in social cognition, because it is still developing well into the adolescent years (see review in Choudhoury, Blakemore, & Charman, 2006). Social-cognitive skills begin to develop very early in childhood: By 3–4 months of age, infants gaze at adults for longer periods of time and with more positive affect during interactions in which the adults share positive affect and attention (Striano & Stahl, 2005). Infants also show gaze following (Striano & Stahl, 2005), a precursor to joint attention, which in turn plays a critical role in social development (Tomasello & Farrar, 1986). By the preschool years, most children have established ToM (Wimmer & Perner, 1983) and begin to shift away from egocentric thinking to being able to consider the perspective of another person. With increasing age, children develop progressively sophisticated social-cognitive skills until, by adolescence, they are well able to take the perspective of others (Choudhoury et al., 2006), can quickly recognize complex emotions such as disgust (Herba & Phillips, 2004), and can make moral decisions requiring empathy (Eslinger et al., 2009). There is some evidence that developments in social cognition are not linear, but occur in waves in middle childhood and then again in early adolescence (Kolb, Wilson, & Taylor, 1992). Thus there may be times at which the developing brain is particularly vulnerable to injury or disease.

Emotion

In the 1980s, the first two reports were published that attested to the fact that people with TBI have deficits in recognizing emotional expressions in faces (Jackson & Moffat, 1987; Prigatano & Pribam, 1982). This is now a well-established finding in adults, who commonly have problems with recognizing emotional expressions in static facial expressions, dynamic visual portrayals, emotionally charged voices, and audiovisual displays (McDonald & Flanagan, 2004; Milders, Fuchs, & Crawford, 2003). Similar deficits have been reported in children and adolescents with brain injury, beginning with a study by Pettersen (1991), who found that children with TBI were impaired both in recognizing basic emotions from affective displays and in interpreting emotional tone from vignettes. These findings have subsequently been confirmed in larger samples of adolescents as well as children, and for both facial affect recognition and recognition of emotional prosody (Schmidt, Hanten, Li, Orsten, & Levin, 2010; Snodgrass & Knott, 2006; Tonks, Williams, Frampton, Yates, & Slater, 2007a, 2007b, 2007c; Turkstra et al., 1996). These deficits have been attributed to damage to frontal networks thought to be involved in processing emotions and vulnerable to damage in TBI (Schmidt et al., 2010). To date, no study has linked emotion recognition impairments to pragmatic language problems in children, though one study has demonstrated that adults with TBI who are poor at recognizing emotions are also perceived to lack humor in their social interactions (McDonald et al., 2004).

Theory of Mind

There is growing evidence that children with moderate to severe TBI have impairments in ToM (Dennis, Agostino, Roncadin, & Levin, 2009; Dennis, Barnes, Wilkinson, &

Humphreys, 1998; Dennis et al., 2001; Snodgrass & Knott, 2006; Turkstra et al., 2001). Deficits have been observed on formal ToM tests, as well as on tasks requiring inferences about others' thoughts and feelings in written stories, spoken stories, pictures, and video vignettes. In most studies, impairments are evident only on tasks designed to evaluate "higher-level" ToM tasks, such as understanding deception and faux pas.

Formal links between ToM and pragmatic communication have not been established to date (in part because such terms as *social cognition* and *ToM* did not appear in the pediatric TBI literature until fairly recently), but a review of previous studies reveals evidence that many of the pragmatic problems described could be attributed at least in part to underlying deficits in ToM. In an early model of discourse performance in children with TBI, Dennis (1991) included prerequisite skills in "intentional representations," which referred to knowledge about one's own and others' beliefs—clearly, an aspect of social cognition. In other work, Dennis and colleagues have described "semantic–pragmatic functions" in children with TBI, which include such functions as irony comprehension and detection of deception (Dennis & Barnes, 2001), as well as comprehension of emotions in written narratives (Dennis, Barnes, Wilkinson, & Humphreys, 1998)—again, features clearly linked to social cognition. Similarly, Bara, Bosco, and Bucciarelli (1999) included irony comprehension in their definition of pragmatic communication skills in children with TBI. Turkstra et al. (1996) included such functions as comprehension of hints and sarcasm, and adapting language to meet perceived needs of the listener (i.e., the listener's mental state).

From a theoretical point of view, it is clear that ToM is fundamental to many pragmatic language skills. For language expression, the speaker must be aware of the listener and his or her perspective in order to plan language appropriately. For comprehension, particularly of indirect speech or implicature, a listener is often required to make judgments about a speaker's intention. For example, if a speaker says something untrue (transgressing the maxim of truthfulness), nonverbal cues and context provide additional information to guide interpretation. The statement "Dinner was delicious, Daddy" (when the facts indicate otherwise) may be intended as a comforting white lie or a contemptuous joke, and can only be interpreted accurately by using contextual cues to determine what the speaker believes and intends by the remark. Indeed, it has been found that people with TBI who experience difficulties understanding second-order ToM inferences (i.e., what one person wants another person to believe) are also those most likely to experience problems understanding the meaning of sarcastic exchanges (Channon, Pellijeff, & Rule, 2005; McDonald & Flanagan, 2004).

Another aspect of pragmatic communication that directly reflects ToM is the use of *mental state terms* (MSTs), which reflect an appreciation of one's own and others' thoughts and feelings. An example of an MST is a *cognitive state predicate*—a verb or verb phrase that asserts something about the cognitive state of the child (e.g., "He thinks he will go" or "I made her do it"). Use of a cognitive state predicate provides evidence of ToM abilities, because an understanding of the underlying cognitive state is implied in the use of the term. As the development of ToM is in part dependent on talk about mental states (Nestoriuc, Martin, Rief, & Andrasik, 2008), there has been some interest in studying MSTs in children with language disorders (e.g., Karavidas et al., 2007), to further explore the interaction of language and social cognition in development.

Stronach and Turkstra (2008) examined conversations of adolescents with TBI for evidence of MSTs. Participants were 16 adolescents with TBI, divided into two groups based on scores on a test of ToM and compared to 8 typically developing adolescents matched for age, sex, and race. Each participant completed a 3-minute conversation with a peer or researcher partner, and conversations were analyzed to determine the number of MSTs relative to total words produced. The group with TBI and low scores on the ToM test expressed significantly fewer MSTs than either of the other two groups, although the three groups produced an equivalent number of words and diverse vocabulary (i.e., the results were not due to a language impairment in the low-scoring group). There was no significant difference between the typically developing group and the group with high ToM scores and TBI. These findings suggested that deficits in social cognition were reflected in specific aspects of language use in adolescents with TBI. As lack of MST use might convey the impression that a speaker is not considering others' thoughts and feelings, these specific linguistic behaviors may ultimately influence adolescents' acceptability as social communication partners.

The neural basis for ToM impairments in acquired brain injury is unclear. On one hand, the lack of correlation between measures of EFs and ToM in adults with focal frontal lesions has led some to argue that ToM is a specialized module of cognitive processing (Havet-Thomassin, Allain, Etcharry-Bouyx, & Le Gall, 2006). ToM has also been reported to be more severely affected following ventromedial lesions (Shamay-Tsoory, Tomer, & Aharon-Peretz, 2005) and is not associated with dorsolateral frontal lesions, which commonly give rise to EF impairment. On the other hand, problems with general inference making secondary to EF deficits in TBI have been shown to affect both non-social inferences and ToM (Martin & McDonald, 2005). Furthermore, young children's performances on false-belief tasks are influenced by working memory capacity (Hughes, 1998), although the evidence suggests that this is not the sole contributing factor (Tager-Flusberg, Sullivan, & Boshart, 1997). In acquired brain injury, poor performance on tasks reliant upon working memory is associated with poor ToM performance (Bibby & McDonald, 2005), although such deficits do not totally explain the difficulties seen (Havet-Thomassin et al., 2006). In all, it appears that there are likely to be common processes for social and nonsocial tasks, depending upon the medium and response requirements (spoken, written, etc.) as well as unique ToM judgments.

Empathy

Empathy plays a pivotal role in successful interpersonal communication by providing speakers with the capacity to be aware of and respond to the emotional experiences of those with whom they interact (Decety & Jackson, 2004). Egocentrism, alienation, and conflict are consequences of low empathy (Eslinger, 2002). Individuals who have sustained severe TBI are frequently reported to be egocentric, self-centered, and insensitive to another person's needs (Elsass & Kinsella, 1987) and low empathy on self-report questionnaires is seen in a large proportion of individuals with TBI (de Sousa et al., 2011; Williams & Wood, 2009). Disorders of empathy are well characterized in several childhood disorders, such as autism and conduct disorder; however, research in children and adolescents with brain injury is only beginning to emerge. Studies to date have focused primarily on ToM and emotion recognition (i.e., impaired performance on empathy tasks, due to failure to recognize and appreciate the thoughts and feelings of others), rather than on the subjective experience of emotion in response to another's emotions, referred to as *emotional* empathy (Mehrabian & Epstein, 1972). More than 60% of adults with TBI self-report a loss of emotional empathy (de Sousa et al., 2011; Williams & Wood, 2009). A significant proportion also describes postinjury blunting of emotional experience more generally (Hornak, Rolls, & Wade, 1996). Adults with TBI also often show reduced facial mimicry (McDonald et al., 2011) and arousal responses to emotion, indexed by skin conductance (Hopkins, Dywan, & Segalowitz, 2002), especially when they are viewing angry facial expressions (de Sousa et al., 2011).

To date, the only study of emotional empathy in children with brain injury was a case report of two adolescents who had sustained frontal lobe damage in early childhood— one as the result of being run over by a motor vehicle, and the other of brain tumor resection (Anderson, Bechara, Damasio, Tranel, & Damasio, 1999). Both had deficits in all aspects of empathy noted above in relation to adults: lack of emotional empathy, blunting of emotional experiences, and lack of autonomic responses to arousing stimuli. The authors noted three important differences, however, between these adolescents and adults with similar injuries: They had severe behavior problems that persisted from childhood into adolescence; their behavior problems were more severe than those seen in adults; and they lacked declarative social and moral knowledge that they could use in decision making. These adolescents appeared to have failed to develop moral and social reasoning, leading the authors to describe them as showing a syndrome that could be described as "acquired sociopathy," except that their behaviors appeared to be "impulsive rather than goal-directed" (p. 1036). This finding has major implications for recovery from brain injury in early childhood, before the foundations of empathy and moral reasoning have been established, and it suggests that empathy should be considered in evaluating and treating children with TBI.

The role of poor empathy in communication skills following TBI has never been examined empirically in either adults or children. However, it is obvious that failure to empathize will render social exchanges more difficult for persons with TBI to negotiate, as they are unable to respond appropriately to another person's emotional state.

Speaking in Context: Social Opportunities Available to People with TBI

It is clear that people with TBI experience difficulties with their communication during everyday social interactions. However, social opportunities are limited not only by deficits in these persons' communication skills, but also by the way others interact with them. The behavior of their conversational partners is critically important, facilitating or diminishing opportunities for the individuals with brain injury to continue conversations in a successful manner. For example, in a study of telephone conversations where participants with TBI requested information from a range of communication partners, they were asked for and were given less information than matched control participants were (Togher, Hand, & Code, 1996, 1997). Therapists and mothers never asked people with TBI questions to which they did not already know the answers. In addition, the participants with TBI were more frequently questioned regarding the accuracy of their contributions, and their contributions were followed up less often than those of matched control participants. Communication partners used a patronizing tone and slowed speech production when talking to the people with TBI. This was in contrast to the control interactions, where participants were asked for unknown information, were encouraged to elaborate, did not have their contributions checked frequently, and had their contributions followed up.

The role of communication partners is particularly important for children, who develop pragmatic communication skills in large part from the models of parents and other important people in their lives. Parental stress and burden are well documented after pediatric TBI (Donders, 1992; Wade, Carey, & Wolfe, 2006) and are likely to influence the extent to which parents can engage in positive, meaningful communication interactions with their children. Parents also vary in the extent to which they have the skills and capacity to provide communication supports for children with TBI (Bedell, Cohn, & Dumas, 2005); thus some children may have richer communication opportunities than others. These findings, along with evidence that parent training can improve social communication skills in typical children (McCollum, 1984), support the need for research on communication interactions among children and adolescents with TBI and their partners.

One method for analyzing interactions that has shown promise in adults is based upon the theory of *systemic functional linguistics* (SFL). SFL has been used extensively to evaluate the discourse of people with TBI (Jorgensen & Togher, 2009; Kilov, Togher, & Grant, 2008; Togher, Taylor, Aird, & Grant, 2006); it is designed to examine interactions across different genres (e.g., shopping encounters, storytelling, letter writing). Various novel genres have now been developed that are sensitive to discourse difficulties of people with TBI. For example, a problem-solving discourse task, where two people are asked to determine the name and function of an unknown object, provides an opportunity to observe their jointly produced discourse as they work together to establish the name and function of the object before them (Kilov et al., 2008). Analysis of such discourse, using an SFL framework known as *generic structure potential analysis*, has provided a number of sensitive measures—including the participants' provision of a thesis about the possible identity of the object, arguments to support this thesis, possible solutions to the task, challenges to the other person's suggestions, and personal comments.

Another task, extending the substantial body of research into narrative elicitation, has been developed to foster naturalistic communicative opportunities for the person with TBI. In this task, an everyday communication partner (e.g., a friend or relative) is invited to retell a story jointly with the person, and this has proven to be facilitatory for the person with TBI. An example of this type of task is asking a person with TBI and a friend to retell a story they have watched on TV to a third person who has not seen the TV show. In one study where 10 people with TBI jointly produced this type of narrative with a friend, they showed similar productivity scores and similar levels of communication exchange as matched control dyads (Jorgensen & Togher, 2009). Research into the role of conversational partners in facilitating (or limiting) communication opportunities for people with TBI has, to date, been restricted to work with adults. However, this approach has clear implications for children and adolescents. Friends, parents, and teachers play a critical role in providing adequate support and opportunities for young people with TBI to communicate effectively. Focus upon such communication partners has the potential to provide interesting and fruitful avenues for treatment, as we discuss below.

Treatment Approaches

Several years ago, Teichner, Golden, and Giannaris (1999) made the following statement about intervention for TBI:

> Despite the increasing number of head-injury survivors faced with the difficulties of resuming a normal life, very little research has been aimed at finding reliable treatment methods suited to their specific needs. . . . As a result, a common approach to treatment for children and adolescents with acquired brain injuries, especially those who exhibit aggressive or difficult-to-manage behavior, has been to apply techniques that have been successful in other populations. (p. 207)

Nowhere is this truer than in pragmatic communication disorders. Although our knowledge about the basis of pragmatic communication disorders in children and adolescents with acquired brain injury has evolved considerably over recent years, this progress has not translated into changes in clinical practice. Two evidence reviews have shown the limitations of our knowledge about best practices in the area of pragmatic communication disorders in children and adolescents with TBI. Coelho, Ylvisaker, and Turkstra (2005) reviewed evidence in support of nonstandardized assessment methods for children and adults with TBI, primarily focusing on analysis of spoken discourse. This review revealed strong evidence that impairments in pragmatic communication are most evident in unstructured communication, and can be revealed by using standard analysis techniques used in this area (e.g., examination of discourse cohesion), which support the analysis of genres like conversation rather than story retelling (although see also Jorgenson & Togher, 2009, cited above). Turkstra et al. (2005) reviewed standardized tests of communication skills for children and adults with TBI, and found no pragmatics test that met internationally accepted standards for reliability and validity.

The literature on intervention for children and adolescents with pragmatic communication disorders is similarly sparse. A review of current treatment materials suggests that therapists primarily use models of social skills training (SST) that have been developed for other populations, such as children with developmental disabilities. These traditional models based in behavioral principles do appear to have some efficacy, but this has been mainly demonstrated for adults with TBI (Carney et al., 1999). SST normally occurs in a group format, using group problem solving and practice, role plays, homework, feedback, and positive reinforcement to shape more socially skilled behavior. Two recent randomized controlled trials (RCTs) examined the effectiveness of a 12-week SST program for adults with chronic moderate to severe TBI (Dahlberg et al., 2007; McDonald et al., 2008). The treatment programs differed, but used many of these common elements to tackle interpersonal social-interactional skills. In the former study, improvements were noted on independent assessors' rating of the social style of participants. More specifically, improvements were noted in their ability to link comments to preceding comments and in the general cohesiveness, relevance and relatedness of ideas expressed, along with improved quantity and clarity. In the latter study, participants improved on one measure of social behavior, the Partner-Directed Behavior Scale of the Behaviorally Referenced Rating System of Intermediary Social Skills—Revised (Wallander, Conger, & Conger, 1985). This scale focuses on the ability to adapt to the social requirements of others. The Self-Centered Behavior and Partner Involvement Behavior subscales in particular showed improvement, indicating that participants with SST were less inclined to talk about themselves and more inclined to encourage their conversational partners to contribute to the conversation.

Despite the limited success of these approaches in adults with TBI, traditional SST makes several assumptions about participants that may not be true for adolescents with TBI (Ylvisaker, Turkstra, & Coelho, 2005); therefore, careful consideration is required before applying these approaches with this population. First, traditional SST assumes that participants lack knowledge about social rules and behaviors. There is evidence that this is not the case for adolescents (or adults) with TBI (Turkstra et al., 2004), and it also makes sense that an adolescent with typical development before his or her injury would have prior social knowledge. So traditional SST requires modification to have relevance for this population. Second, traditional SST approaches assume that the participants are motivated to change their behavior, which is often not the case in TBI. Careful clinical interventions and the use of group formats with peer feedback may ameliorate this issue to some extent for some participants.

Another major concern for pragmatic remediation approaches, including SST, is that there is an assumption that gains in treatment settings will generalize to other contexts. Yet there is little evidence for this. Although both RCTs reported above demonstrated posttreatment gains on specific measures, improvements in everyday function as rated by significant others were more difficult to demonstrate. Indeed, recent evidence-based reviews have revealed that generalization is the exception in therapy for individuals with TBI. This is true in multiple domains, including attention training (Sohlberg et al., 2003), the use of memory aids (Sohlberg et al., 2007), and social and behavioral intervention (Ylvisaker et al., 2007). This problem is especially salient in the field of pragmatic communication, given that such behaviors by and large are learned implicitly; didactic (explicit) training is therefore unlikely to produce analogous results, particularly in a group at high risk for declarative memory impairments. Failure to generalize characterized the only study to date that has directly tested the effectiveness of intervention for pragmatic communication disorders in children or adolescents with TBI. In this study, Wiseman-Hakes, Stewart, Wasserman, and Schuller (1998) administered a structured pragmatic communication skills intervention program to six adolescents with severe TBI, ages 14–17 years (three males, three females), and 3 months to 9 years postinjury. All had been functioning at grade level premorbidly, but at the time of the study had evidence of difficulty in social situations, based on self-reports and parent reports of social problems and on behavioral ratings by a trained social worker. Therapy consisted of five hierarchical modules, including initiation, topic maintenance, turn taking, and active listening. Generalization was encouraged by practice with peers, staff, and parents, and peer feedback was incorporated into treatment. Participants received therapy four times each week for 6 weeks, for 1 hour per session. The authors reported significant improvements on two pragmatic rating scales, both administered by the social worker who delivered the therapy, and these gains were maintained at 6 months posttreatment. Unfortunately, consistent with general trends in remediation research, there were no significant changes in parent reports of social behavior on the original Vineland Adaptive Behavior Scales (Sparrow, Balla, & Cicchetti, 1984).

Overall, a review of research on social skills interventions for children and adults with TBI found little evidence that discrete SST in children leads to improvements in everyday communication functioning (Ylvisaker, Turkstra, & Coelho, 2005). This problem is exacerbated by the general paucity of research and poor quality of existing evidence in the field. Turkstra and Burgess (2007) reviewed the literature on social skills intervention for adolescents with TBI, of which the study by Wiseman-Hakes et al. (1998), cited above, was the only one that addressed pragmatic communication directly. These reviews revealed major limitations in the existing literature: Studies lacked controls for such confounds as experimenter bias and maturation; treatment that focused on discrete symptoms did not generalize beyond the therapy setting; and training methods were underspecified. The most promising programs were those in which family members were trained to support positive social behaviors (Bedell et al., 2005; Braga, Da Paz, & Ylvisaker, 2005), but none of these focused specifically on pragmatic communication functions. This is not to say that assessment and intervention for pragmatic communication disorders is ineffective, but rather that evidence is lacking. There is a critical need for studies specifically addressing pragmatic communication competence in children and adolescents with TBI, particularly when the injury affects EFs and declarative learning.

Special Considerations for Adolescents

In regard to acquisition of pragmatic communication disorders, adolescents with TBI should be considered as a unique group. Although there have been few studies in this population, the available data suggest that they have long-term intervention needs in the social and behavioral domain (Burke, Wesolowski, Buyer, & Zawlocki, 1990). In our experience, a loss of friends and social networks is a common complaint of adolescents with moderate to severe postinjury cognitive impairments. For the first year or so after the injury, friends may be supportive and sympathetic, but over time the challenges in dealing with such an adolescent's atypical social behaviors may outweigh the benefits of staying in touch. Long-term social isolation is a common concern of parents, who often worry about their children's future independent living and social networks. Thus there is a strong motivation for intervention to improve social outcomes, particularly in the domain of communication.

Providing Contextual Support

Treating communication difficulties of people with TBI can be additionally complicated by confounding problems of impaired sensory and cognitive prerequisites, as outlined in Figure 16.1. These problems often limit the degree to which such a person can learn new skills. An alternative approach to remediation is to examine the communicative environment within which interactions occur and modify these to enhance the opportunities for the person with brain injury to talk. One way of modifying the environment is to think about how different types of speaker roles may provide different types of opportunities. For example, asking people with TBI to speak in a powerful information-giving role, such as being a guest speaker talking about the experience of having a serious injury, enabled them to speak in a similar way to matched control participants (who had a spinal injury) (Togher, 2000). People with TBI also had better levels of participation (compared to pre-intervention and postintervention baseline periods) when they were placed in an environment with a trained mentor who provided prompts, modeling, and structured activities (Bellon & Rees, 2006). Similarly, in interactions between staff members and people with TBI, there was a trend toward increased compliance, attention, and participation of the persons with TBI when the staff members used more positive communication strategies (Shelton & Shryock, 2007).

These results suggest that greater opportunities and increased conversational competence can be created for a person with TBI within a facilitative context. Furthermore, by adopting a broader perspective that regards communication as an interaction between at least two speakers, it is possible to broaden our scope for training to include not just persons with TBI, but their communication partners, and several studies have addressed this. The first RCT of training communication partners was conducted with police officers, who were trained to manage specific service encounters with people with TBI whom they had not previously met (Togher, McDonald, Code, & Grant, 2004). The speakers with TBI called the police to ask their advice both before and after the police had been trained. Training resulted in more efficient, focused interactions. In other words, this study confirmed that training communication partners improved the competence of people with TBI.

More recently, a program entitled TBI Express has been evaluated in a three-arm clinical trial where everyday communication partners were trained (Togher, McDonald, Tate, Power, & Rietdijk, 2009). Based on communication partner availability, participants were allocated to one of three groups: a TBI solo group (where only the persons with TBI were trained), a joint group (both the everyday communication partners and the person with TBI were trained together), or a control delayed-treatment condition. The TBI solo and joint groups received individual and group training in strategies to maximize communicative effectiveness; behavioral approaches including role plays, cues to assist self-monitoring, and positive reinforcement were used. Treatment included concepts based on social-linguistic theories of communication and principles of Vygotskian learning theory, with a focus on everyday discourse.

There were several possible benefits to this kind of training. First, the training program that included everyday communication partners should substantially increase the amount of practice completed by the joint group at home. Second, it should increases engagement of the communication partners with the acknowledgment that they could make a significant contribution to the way their relatives with TBI communicated. Third, the focus on communication as a collaborative and elaborative process (Ylvisaker & Szekeres, 1998), along with the application of the concept originally espoused by Kagan et al. (2004) that training an everyday communication partner can reveal competence in a disabled speaker, should make this kind of approach qualitatively different from conventional methods. Most of the communication partners in the Togher et al. (2004) study were wives and mothers who had changed their communication styles following their husbands' or sons' injury, and these changes in some cases had been detrimental to successful everyday interactions. Sensitively targeting the behaviors of the everyday communication partners, such as their use of test questions and speaking on behalf of the persons with TBI, appeared to lead to a significant change in everyday interactions.

Treating Social Perception

As an alternative approach to improving pragmatic skills, treatment can be targeted at those deficits in social cognition that are thought to underpin aspects of pragmatic language competence. Two obvious target areas are emotion perception and mentalizing (or ToM). Despite this, virtually no work has been conducted to evaluate treatments for difficulties with social perception following child or adolescent TBI, and only a few studies with adults (Bornhofen & McDonald, 2008a) have focused on emotion perception specifically. Two small RCTs have been reported that examined targeted remediation of emotion perception in adults with severe TBI, and both of these demonstrated modest improvements on direct measures of emotion perception (Bornhofen & McDonald, 2008b, 2008c). Techniques were adapted from the cognitive remediation literature, including *errorless learning* (Wilson, Baddeley, Evans, & Shiel, 1994) and *self-instructional training*. Errorless learning involves repeated rehearsal of information (e.g., repeatedly practicing the identification of facial patterns associated with particular expressions, with the explicit instruction not to guess if unsure). This technique boosts encoding and retention of knowledge, and is especially useful for participants with TBI who have severe and/or widespread cognitive impairment. Alternatively, self-instructional training (Meichenbaum & Cameron, 1973) involves the guided use of verbalization of procedural steps by participants engaged in complex tasks. This approach is more appropriate for novel emotion perception tasks that require information to be integrated from multiple sources. To date, there has been no work examining remediation of ToM deficits in children and adolescents with TBI, although there has been some work in children with autism (Ozonoff & Miller, 1995).

Conclusions

In conclusion, research into communication disorders after TBI has advanced significantly from the early observational studies based on relative reports in the 1970s to more recent empirical studies of discourse patterns. The use of theoretical approaches such as pragmatics and SFL has provided sensitive means to elucidate these disorders of communication. Whereas primary disorders of language are relatively infrequent after brain injury, it is apparent that for many children, adolescents, and adults with severe TBI, the ability to communicate with others in a way that is clear, effective, and sensitive to their conversational partners' needs is often compromised.

The approaches to assessment discussed in this chapter emphasize the social nature of communication in everyday settings. As such, they represent an important advance over the "context-free" assessments of standard language batteries of the past. Social theories of communication expand notions of communicative competence to encompass the ability to use language in context. In doing so, they have been shown to be sensitive to subtle but pervasive disorders of communication, wherein basic language abilities remain intact but the ability to apply these sensitively and adaptively in everyday life is impaired.

An analysis of communication using pragmatic and social-linguistic frameworks also provides us with the means to examine how cognitive factors contribute to communication competence, especially disorders of EFs, information processing, and attention. New advances in our understanding of social cognition and of how this becomes disordered following TBI have added a critical facet to models of pragmatic competence. Better elucidation of the way in which pragmatic communication occurs, combined with better understanding of the cognitive and social processes that are involved in communication, should provide us with exciting new directions for both assessment and remediation. Using social-linguistic frameworks, it has also been possible to reveal how important conversational partners are in providing speakers with TBI with adequate and/or typical opportunities to demonstrate communicative competence.

Much of the research into assessment and remediation of pragmatic communication following brain injury stems from work on adults with TBI. Many of the deficits are equally relevant for children and adolescents. But there are also unique factors to take into account when we are considering the developmental and social consequences of impaired communication skills in younger people. Disorders in the basic building blocks needed for developing social skills and communicative competence both impede the acquisition of interpersonal skills and reduce social opportunities for skill development. Consequently, accurate assessment, effective remediation, and contextual support of these skills at age-appropriate stages of development is critical. It is here that we need to concentrate our research efforts.

References

Achenbach, T. M., & Rescorla, L. A. (2001). *Manual for the ASEBA School-Age Forms and Profiles*. Burlington: University of Vermont, Research Center for Children, Youth, and Families.

Adams, C. (2005). Social communication intervention for school-age children: Rationale and description. *Seminars in Speech and Language, 26*(3), 181–188.

Anderson, S. W., Bechara, A., Damasio, H., Tranel, D., & Damasio, A. R. (1999). Impairment of social and moral behavior related to early damage in human prefrontal cortex. *Nature Neuroscience, 2*(11), 1032–1037.

Asher, S. R., & Hymel, S. (1986). Coaching in social skills for children who lack friends in school. *Social Work in Education, 8*(4), 205–218.

Bara, B. G., Bosco, F. M., & Bucciarelli, M. (1999). Developmental pragmatics in normal and abnormal children. *Brain and Language, 68*(3), 507–528.

Bedell, G. M., Cohn, E. S., & Dumas, H. M. (2005). Exploring parents' use of strategies to promote social participation of school-age children with acquired brain injuries. *American Journal of Occupational Therapy, 59*(3), 273–284.

Bellon, M. L., & Rees, R. J. (2006). The effect of context on communication: A study of the language and communication skills of adults with acquired brain injury. *Brain Injury, 20*(10), 1069–1078.

Bibby, H., & McDonald, S. (2005). Theory of mind after traumatic brain injury. *Neuropsychologia, 43*(1), 99–114.

Bigler, E. D. (2001). The lesion(s) in traumatic brain injury: Implications for clinical neuropsychology. *Archives of Clinical Neuropsychology, 16*(2), 95–131.

Bornhofen, C., & McDonald, S. (2008a). Emotion perception deficits following traumatic brain injury: A review of the evidence and rationale for intervention. *Journal of the International Neuropsychological Society, 15*, 511–525.

Bornhofen, C., & McDonald, S. (2008b). Comparing strategies for treating emotion perception deficits in traumatic brain injury. *Journal of Head Trauma Rehabilitation, 23*, 103–115.

Bornhofen, C., & McDonald, S. (2008c). Treating emotion perception deficits following traumatic brain injury. *Neuropsychological Rehabilitation, 18*, 22–24.

Braga, L. W., Da Paz, A. C., & Ylvisaker, M. (2005). Direct clinician-delivered versus indirect family-supported rehabilitation of children with traumatic brain injury: A randomized controlled trial. *Brain Injury, 19*(10), 819–831.

Brookshire, B. L., Chapman, S. B., Song, J., & Levin, H. S. (2000). Cognitive and linguistic correlates of children's discourse after closed head injury: A three-year follow-up. *Journal of the International Neuropsychological Society, 6*(7), 741–751.

Brown, P., & Levinson, S. C. (1978). Universals in language usage: Politeness phenomena. In E. N. Goody (Ed.), *Questions and politeness: Strategies in social interaction* (pp. 56–311). Cambridge, UK: Cambridge University Press.

Burke, W., Wesolowski, M., Buyer, D., & Zawlocki, R. (1990). The rehabilitation of adolescents with traumatic brain injury: Outcome and follow-up. *Brain Injury, 4*(4), 371–378.

Carney, N., Chesnut, R., Maynard, H., Mann, N., Patterson, P., & Helfand, M. (1999). Effect of cognitive rehabilitation on outcomes for persons with traumatic brain injury: A systematic review. *Journal of Head Trauma Rehabilitation, 14*(3), 277–307.

Carrow-Woolfolk, E. (1999). *Comprehensive Assessment of Spoken Language*. Circle Pines, MN: American Guidance Service.

Channon, S., Pellijeff, A., & Rule, A. (2005). Social cognition after head injury: Sarcasm and theory of mind. *Brain and Language, 93*, 123–134.

Chapman, S. B., Culhane, K. A., Levine, H. S., H., H., Mendelsohn, D., Ewing-Cobbs, L., et al. (1992). Narrative discourse after closed head injury in children and adolescents. *Brain and Language, 43*, 42–65.

Chapman, S. B., Highley, A. P., & Thompson, J. L. (1998). Discourse in fluent aphasia and Alzheimer's disease: Linguistic and pragmatic considerations. *Journal of Neurolinguistics, 11*(1–2), 55–78.

Choudhury, S., Blakemore, S.-J., & Charman, T. (2006). Social cognitive development during adolescence. *Social Cognitive and Affective Neuroscience, 1*(3), 165–174.

Ciccia, A.H., Meulenbroek, P., & Turkstra, L.S. (2009). Adolescent brain and cognitive developments: Implications for clinical assessment in traumatic brain injury. *Topics in Language Disorders, 29*(3), 249–265.

Coelho, C., Ylvisaker, M., & Turkstra, L. S. (2005). Nonstandardized assessment approaches for individuals with traumatic brain injuries. *Seminars in Speech and Language, 26*(4), 223–241.

Crone, E. A. (2009). Executive functions in adolescence: Inferences from brain and behavior. *Developmental Science, 12*(6), 825–830.

Dahlberg, C. A., Cusick, C. P., Hawley, L. A., Newman, J. K., Morey, C. E., Harrison-Felix, C. L., et al. (2007). Treatment efficacy of social communication skills training after traumatic brain injury: A randomized treatment and deferred treatment controlled trial. *Archives of Physical Medicine and Rehabilitation, 88*(12), 1561–1573.

Damico, J. S. (1991). Clinical Discourse Analysis: A functional approach to language assessment. In C. S. Simon (Ed.), *Communication skills and classroom success: Assessment and therapy methodologies for language and learning disabled students* (pp. 125–148.). Eau Claire, WI: Thinking Publications.

Dawson, G. (Chair). (2001, April). *Neural bases of social behavior: Insights from autism*. Symposium conducted at the meeting of the Society for Research in Child Development, Minneapolis, MN.

Decety, J., & Jackson, P. L. (2004). The functional architecture of human empathy. *Behavioral and Cognitive Neuroscience Reviews, 3*, 71–100.

DeGroot, T., & Motowidlo, S. J. (1999). Why visual and vocal interview cues can affect interviewers' judgments and predict job performance. *Journal of Applied Psychology, 84*(6), 986–993.

Dennis, M. (1991). Frontal lobe function in childhood and adolescence: A heuristic for assessing attention regulation, executive control, and the intentional states important for social discourse. *Developmental Neuropsychology, 7*(3), 327–358.

Dennis, M., Agostino, A., Roncadin, C., & Levin, H. (2009). Theory of mind depends on domain-general executive functions of working memory and cognitive inhibition in children with traumatic brain injury. *Journal of Clinical and Experimental Neuropsychology, 31*(7), 835–847.

Dennis, M., & Barnes, M. A. (1990). Knowing the meaning, getting the point, bridging the gap, and carrying the message: Aspects of discourse following closed head injury in childhood and adolescence. *Brain and Language, 39*, 428–446.

Dennis, M., & Barnes, M. A. (2001). Comparison of literal, inferential, and intentional text comprehension in children with mild or severe closed head injury. *Journal of Head Trauma Rehabilitation, 16*(5), 456–468.

Dennis, M., Barnes, M. A., Donnelly, R. E., Wilkinson, M., & Humphreys, R. P. (1996). Appraising and managing knowledge: Metacognitive skills after childhood head injury. *Developmental Neuropsychology, 12*, 77–103.

Dennis, M., Barnes, M. A., Wilkinson, M., & Humphreys, R. P. (1998). How children with head injury represent real and deceptive emotion in short narratives. *Brain and Language, 61*(3), 450–483.

Dennis, M., Purvis, K., Barnes, M. A., Wilkinson, M., & Winner, E. (2001). Understanding of literal truth, ironic criticism, and deceptive praise following childhood head injury. *Brain and Language, 78*(1), 1–16.

de Sousa, A., McDonald, S., Rushby, J., Li, S., Dimoska, A., & James, C. (2011). Understanding deficits in empathy after traumatic brain injury: The role of affective responsivity. *Cortex, 47*, 526–535.

Donders, J. (1992). Premorbid behavioral and psychosocial adjustment of children with traumatic brain injury. *Journal of Abnormal Child Psychology, 20*(3), 233–246.

Douglas, J. M. (2010). Using the La Trobe Communication Questionnaire to measure perceived social communication ability in adolescents with traumatic brain injury. *Brain Impairment, 11*(2), 171–182.

Douglas, J. M., Bracy, C. A., & Snow, P. C. (2007). Exploring the factor structure of the La Trobe Communication Questionnaire: Insights into the nature of communication deficits following traumatic brain injury. *Aphasiology, 21*(12), 1181–1194.

Elsass, L., & Kinsella, G. (1987). Social interaction following severe closed head injury. *Psychological Medicine, 17*(1), 67–78.

Eslinger, P. J. (2002). *Neuropsychological interventions: Clinical research and practice.* New York: Guilford Press.

Eslinger, P. J. (2008). The frontal lobes: Executive, emotional and neurological functions. In P. Marien & J. Abutalebi (Eds.), *Neuropsychological research: A review* (pp. 379–408). New York: Psychology Press.

Eslinger, P. J., Robinson-Long, M., Realmuto, J., Moll, J., deOliveira-Souza, R., Tovar-Moll, F., et al. (2009). Developmental frontal lobe imaging in moral judgment: Arthur Benton's enduring influence 60 years later. *Journal of Clinical and Experimental Neuropsychology, 31*(2), 158–169.

Felmingham, K. L., Baguley, I. J., & Green, A. M. (2004). Effects of diffuse axonal injury on speed of information processing following severe traumatic brain injury. *Neuropsychology, 18*(3), 564–571.

Ganesalingam, K., Yeates, K. O., Taylor, H. G., Walz, N. C., Stancin, T., & Wade, S. (2011). Executive functions and social competence in young children 6 months following traumatic brain injury. *Neuropsychology, 25*(4), 466–476.

Gibbs, R. W. (1986). What makes some indirect speech acts conventional? *Journal of Memory and Language, 25*(2), 181–196.

Gioia, G. A., Isquith, P. K., Guy, S. C., & Kenworthy, L. (2000). *Behavior Rating Inventory of Executive Function.* Odessa, FL: Psychological Assessment Resources.

Grice, H. P. (1978). Further notes on logic and conversation. In P. Cole (Ed.), *Syntax and semantics: Vol. 9. Pragmatics* (pp. 113–127). New York: Academic Press.

Hanten, G., Dennis, M., Zhang, L., Barnes, M., Roberson, G., & Archibald, J. (2004). Childhood head injury and metacognitive processes in language and memory. *Developmental Neuropsychology, 25*(1–2), 85–106.

Havet-Thomassin, V., Allain, P., Etcharry-Bouyx, F., & Le Gall, D. (2006). What about theory of mind after severe brain injury? *Brain Injury, 20*(1), 83–91.

Heilman, K. M., Safran, A., & Geschwind, N. (1971). Closed head trauma and aphasia. *Journal of Neurology, Neurosurgery and Psychiatry, 34*, 265–269.

Herba, C., & Phillips, M. (2004). Annotation: Development of facial expression recognition from childhood to adolescence: Behavioural and neurological perspectives. *Journal of Child Psychology and Psychiatry, 45*(7), 1185–1198.

Hopkins, M. J., Dywan, J., & Segalowitz, S. J. (2002). Altered electrodermal response to facial expression after closed head injury. *Brain Injury, 16*, 245–257.

Hornak, J., Rolls, E., & Wade, D. (1996). Face and voice expression identification in patients with emotional and behavioural changes following ventral frontal lobe damage. *Neuropsychologia, 34*(4), 247–261.

Hughes, C. (1998). Executive function in preschoolers: Links with theory of mind and verbal ability. *British Journal of Developmental Psychology, 16*, 233–253.

Jackson, H. F., & Moffat, N. J. (1987). Impaired emotional recognition following severe head injury. *Cortex, 23*, 293–300.

Jacobs, R., Harvey, A. S., & Anderson, V. (2007). Executive function following focal frontal lobe lesions: Impact of timing of lesion on outcome. *Cortex, 43*(6), 792–805.

Jorgensen, M., & Togher, L. (2009). Narrative after traumatic brain injury: A comparison of monologic and jointly-produced discourse. *Brain Injury, 23*(9), 727–740.

Kagan, A., Winckel, J., Black, S., Duchan, J. F., Simmons-Mackie, N., & Square, P. (2004). A set of observational measures for rating support and participation in conversation between adults with aphasia and their conversation partners. *Topics in Stroke Rehabilitation, 11*(1), 67–83.

Karavidas, M. K., Lehrer, P. M., Vaschillo, E., Vaschillo, B., Marin, H., Buyske, S., et al. (2007). Preliminary results of an open label study of heart rate variability biofeedback for the treatment of major depression. *Applied Psychophysiology and Biofeedback, 32*(1), 19–30.

Kilov, A. M., Togher, L., & Grant, S. (2008). Problem solving with friends: Discourse participation and performance of individuals with and without traumatic brain injury. *Aphasiology, 23*(5), 584–605.

Kolb, B., Wilson, B., & Taylor, L. (1992). Developmental changes in the recognition and comprehension of facial expression: Implications for frontal lobe function. *Brain and Cognition, 20*(1), 74–84.

Levin, H. S., & Hanten, G. (2005). Executive functions after traumatic brain injury in children. *Pediatric Neurology, 33*(2), 79–93.

Levinson, S. (1983). *Pragmatics.* Cambridge, UK: Cambridge University Press.

Mangeot, S., Armstrong, K., Colvin, A. N., Yeates, K. O., & Taylor, H. G. (2002). Long-term executive function deficits in children with traumatic brain injuries: Assessment using the Behavior Rating Inventory of Executive Function (BRIEF). *Child Neuropsychology, 8*(4), 271–284.

Martin, I., & McDonald, S. (2005). Exploring the causes of pragmatic language deficits following traumatic brain injury. *Aphasiology, 19*, 712–730.

McCollum, J. A. (1984). Social interaction between parents and babies: Validation of an intervention procedure. *Child: Care, Health and Development, 10*(5), 301–315.

McDonald, S., Bornhofen, C., Shum, D., Long, E., Saunders, C., & Neulinger, K. (2006). Reliability and validity of 'The Awareness of Social Inference Test' (TASIT): A clinical test of social perception. *Disability and Rehabilitation, 28*, 1529–1542.

McDonald, S., & Flanagan, S. (2004). Social perception deficits after traumatic brain injury:

The interaction between emotion recognition, mentalising ability and social communication. *Neuropsychology, 18,* 572–579.

McDonald, S., Flanagan, S., Martin, I., & Saunders, C. (2004). The ecological validity of TASIT: A test of social perception. *Neuropsychological Rehabilitation, 14,* 285–302.

McDonald, S., Flanagan, S., Rollins, J., & Kinch, J. (2003). TASIT: A new clinical tool for assessing social perception after traumatic brain injury. *Journal of Head Trauma Rehabilitation, 18,* 219–238.

McDonald, S., Li, S., de Sousa, A., Rushby, J., James, C., & Tate, R. L. (2011). Impaired mimicry response to angry faces following severe traumatic brain injury. *Journal of Clinical and Experimental Neuropsychology, 33,* 17–29.

McDonald, S., Tate, R., Togher, L., Bornhofen, C., Long, E., Gertler, P., et al. (2008). Social skills treatment for people with severe, chronic acquired brain injuries: A multicenter trial. *Archives of Physical Medicine and Rehabilitation, 89*(9), 1648–1659.

Mehrabian, A., & Epstein, N. (1972). A measure of emotional empathy. *Journal of Personality and Social Psychology, 40*(4), 525–543.

Meichenbaum, D., & Cameron, R. (1973). Training schizophrenics to talk to themselves: A means of developing attentional controls. *Behavior Therapy, 4*(4), 515–534.

Milders, M., Fuchs, S., & Crawford, J. R. (2003). Neuropsychological impairments and changes in emotional and social behaviour following severe traumatic brain injury. *Journal of Clinical and Experimental Neuropsychology, 25*(2), 157–172.

Miyake, A., Friedman, N. P., Emerson, M. J., Witzki, A. H., Howerter, A., & Wager, T. D. (2000). The unity and diversity of executive functions and their contributions to complex "frontal lobe" tasks: A latent variable analysis. *Cognitive Psychology, 41,* 49–100.

Moran, C., & Gillon, G. (2005). Inference comprehension of adolescents with traumatic brain injury: A working memory hypothesis. *Brain Injury, 19*(10), 743–751.

Morrison, R. L., & Bellack, A. S. (1981). The role of social perception in social skill. *Behavior Therapy, 12,* 69–79.

Morse, S., Haritou, F., Ong, K., Anderson, V., Catroppa, C., & Rosenfeld, J. (1999). Early effects of traumatic brain injury on young children's language performance: A preliminary linguistic analysis. *Pediatric Rehabilitation, 3*(4), 139–148.

Nestoriuc, Y., Martin, A., Rief, W., & Andrasik, F. (2008). Biofeedback treatment for headache disorders: A comprehensive efficacy review. *Applied Psychophysiology and Biofeedback, 33*(3), 125–140.

Ozonoff, S., & Miller, J. N. (1995). Teaching theory of mind: A new approach to social skills training for individuals with autism. *Journal of Autism and Developmental Disorders, 25,* 415–437.

Pettersen, L. (1991). Sensitivity to emotional cues and social behavior in children and adolescents after head injury. *Perceptual and Motor Skills, 73,* 1139–1150.

Place, K. S., & Becker, J. A. (1991). The influence of pragmatic competence on the likability of grade-school children. *Discourse Processes, 14,* 227–241.

Premack, D. (1976). *Intelligence in ape and man.* Hillsdale, NJ: Erlbaum.

Prigatano, G. P., & Pribam, K. H. (1982). Perception and memory of facial affect following brain injury. *Perceptual and Motor Skills, 54*(3), 859–869.

Prutting, C. A., & Kirchner, D. M. (1987). A clinical appraisal of the pragmatic aspects of language. *Journal of Speech and Hearing Disorders, 52,* 105–119.

Schmidt, A. T., Hanten, G. R., Li, X., Orsten, K. D., & Levin, H. S. (2010). Emotion recognition following pediatric traumatic brain injury: Longitudinal analysis of emotional prosody and facial emotion recognition. *Neuropsychologia, 48*(10), 2869–2877.

Searle, J. (1976). A classification of illocutionary acts. *Language in Society, 5*(1), 1–23.

Shamay-Tsoory, S. G., Tomer, R., & Aharon-Peretz, J. (2005). The neuroanatomical basis of understanding sarcasm and its relationship to social cognition. *Neuropsychology, 19,* 288–300.

Shelton, C., & Shryock, M. (2007). Effectiveness of communication/interaction strategies with patients who have neurological injuries in a rehabilitation setting. *Brain Injury, 21*(12), 1259–1266.

Slomine, B. S., Salorio, C. F., Grados, M. A., Vasa, R. A., Christensen, J. R., & Gerring, J. P. (2005). Differences in attention, executive functioning, and memory in children with and without ADHD after severe traumatic brain injury. *Journal of the International Neuropsychological Society, 11*(5), 645–653.

Snodgrass, C., & Knott, F. (2006). Theory of mind in children with traumatic brain injury. *Brain Injury, 20*(8), 825–833.

Sohlberg, M. M., Avery, J., Kennedy, M. R. T., Coehlo, C., Ylvisaker, M., Turkstra, L. S., et al. (2003). Practice guidelines for direct attention training. *Journal of Medical Speech–Language Pathology, 11*(3), xix–xxxix.

Sohlberg, M. M., Kennedy, M., Avery, J., Coehlo, C., Turkstra, L. S., Ylvisaker, M., et al. (2007). Evidence-based practice for the use of external memory aids as a memory compensation technique. *Journal of Medical Speech–Language Pathology, 15*(1), xv–li.

Sohlberg, M. M., & Turkstra, L. S. (2011). *Optimizing cognitive rehabilitation: Effective instructional methods.* New York: Guilford Press.

Sparrow, S., Balla, D., & Cicchetti, D. (1984). *Vineland Adaptive Behavior Scales.* Circle Pines, MN: American Guidance Service.

Sparrow, S., Cicchetti, D. V., & Balla, D. A. (2005). *Vineland Adaptive Behavior Scales, Second Edition.* Circle Pines, MN: American Guidance Service.

Striano, T., & Stahl, D. (2005). Sensitivity to triadic attention in early infancy. *Developmental Science, 8*(4), 333–343.

Stronach, S. T., & Turkstra, L. S. (2008). Theory of mind and use of cognitive state terms by adolescents with traumatic brain injury. *Aphasiology, 22*(10), 1054–1070.

Struchen, M. A., Clark, A. N., Sander, A. M., Mills, M. R., Evans, G., & Kurtz, D. (2008). Relation of executive functioning and social communication measures to functional outcomes following traumatic brain injury. *NeuroRehabilitation, 23*(2), 185–198.

Tager-Flusberg, H., Sullivan, K., & Boshart, J. (1997). Executive functions and performance on false belief tasks. *Developmental Neuropsychology, 13*, 487–493.

Teichner, G., Golden, C. J., & Giannaris, W. J. (1999). A multimodal approach to treatment of aggression in a severely brain-injured adolescent. *Rehabilitation Nursing, 24*(5), 207–211.

Thomsen, I. V. (1975). Evaluation and outcome of aphasia in patients with severe closed head trauma. *Journal of Neurology, Neurosurgery and Psychiatry, 38*, 713–718.

Togher, L. (2000). Giving information: The importance of context on communicative opportunity for people with traumatic brain injury. *Aphasiology, 14*(4), 365–390.

Togher, L., Hand, L., & Code, C. (1996). A new perspective on the relationship between communication impairment and disempowerment following head injury in information exchanges. *Disability and Rehabilitation, 18*(11), 559–566.

Togher, L., Hand, L., & Code, C. (1997). Analysing discourse in the traumatic brain injury population: Telephone interactions with different communication partners. *Brain Injury, 11*(3), 169–189.

Togher, L., McDonald, S., Code, C., & Grant, S. (2004). Training communication partners of people with traumatic brain injury: A randomised controlled trial. *Aphasiology, 18*(4), 313–335.

Togher, L., McDonald, S., Tate, R. L., Power, E., & Rietdijk, R. (2009). Training communication partners of people with traumatic brain injury: Reporting the protocol for a clinical trial. *Brain Impairment, 10*, 118–204.

Togher, L., Taylor, C., Aird, V., & Grant, S. (2006). The impact of varied speaker role and communication partner on the communicative interactions of a person with traumatic brain injury: A single case study using systemic functional linguistics. *Brain Impairment, 7*(3), 190–201.

Tomasello, M., & Farrar, M. J. (1986). Joint attention and early language. *Child Development, 57*(6), 1454–1463.

Tonks, J., Williams, W. H., Frampton, I., Yates, P., & Slater, A. (2007a). Assessing emotion recognition in 9–15-years olds: preliminary analysis of abilities in reading emotion from faces, voices and eyes. *Brain Injury, 21*(6), 623–629.

Tonks, J., Williams, W. H., Frampton, I., Yates, P., & Slater, A. (2007b). Reading emotions after child brain injury: A comparison between children with brain injury and non-injured controls. *Brain Injury, 21*(7), 731–739.

Tonks, J., Williams, W. H., Frampton, I., Yates, P. J., & Slater, A. M. (2007c). The neurological bases of emotional dysregulation arising from brain injury in childhood: A "when and where" heuristic. *Brain Impairment, 8*, 143–153.

Towne, R. L., & Entwisle, L. M. (1993). Metaphoric comprehension in adolescents with traumatic brain injury and in adolescents with language learning disability. *Language, Speech and Hearing Services in Schools, 24*, 100–107.

Turkstra, L. S. (2000). Should my shirt be tucked in or left out?: The communication context of adolescence. *Aphasiology, 14*(4), 349–364.

Turkstra, L. S., & Burgess, S. (2007). Social skills intervention for adolescents with TBI. *Perspectives on Neurophysiology and Neurogenic Speech and Language Disorders, 17*(3), 15–20.

Turkstra, L. S., Burgess, S., Clark, A., Hengst, J., & Paul, D. (2011). *Assessment of pragmatic communication ability: A review for the neuropsychologist.* Manuscript submitted for publication.

Turkstra, L. S., Coelho, C., Ylvisaker, M., Kennedy, M., Sohlberg, M. M., Avery, J., et al. (2005). Practice guidelines for standardized assessment for persons with traumatic brain injury. *Journal of Medical Speech–Language Pathology, 13*(2), ix–xxviii.

Turkstra, L. S., Dixon, T. M., & Baker, K. K. (2004). Theory of mind and social beliefs in adolescents with traumatic brain injury. *NeuroRehabilitation, 19*(3), 245–256.

Turkstra, L. S., McDonald, S., & DePompei, R. (2001). Social information processing in adolescents: Data from normally developing adolescents and preliminary data from their peers with traumatic brain injury. *Journal of Head Trauma Rehabilitation, 16*(5), 469–483.

Turkstra, L. S., McDonald, S., & Kaufmann, P. M. (1996). Assessment of pragmatic communication skills in adolescents after traumatic brain injury. *Brain Injury, 10*(5), 329–345.

Vicari, S., Albertoni, A., Chilosi, A. M., Cipriani, P., Cioni, G., & Bates, E. (2000). Plasticity and reorganization during language development in children with early brain injury. *Cortex, 36*(1), 31–46.

Wallander, J. L., Conger, A. J., & Conger, J. C. (1985). Development and evaluation of a behaviorally referenced rating system for heterosocial skills. *Behavioral Assessment, 7*(2), 137–153.

Wade, S. L., Carey, J., & Wolfe, C. R. (2006). An online family intervention to reduce parental distress following pediatric brain injury. *Journal of Consulting and Clinical Psychology, 74*(3), 445–454.

Wiig, E., & Secord, W. (1989). *Test of Language Competence (Expanded ed.).* San Antonio, TX: Psychological Corporation.

Williams, C., & Wood, R. L. (2009). Alexithymia and emotional empathy following traumatic brain injury. *Journal of Clinical and Experimental Neuropsychology, 22*, 1–11.

Wilson, B. A., Baddeley, A. D., Evans, J. J., & Shiel, A. (1994). Errorless learning in the rehabilitation of memory impaired people. *Neuropsychological Rehabilitation, 4*, 307–326.

Wimmer, H., & Perner, J. (1983). Beliefs about beliefs: Representation and constraining function of wrong beliefs in young children's understanding of deception. *Cognition, 13*, 103–128.

Wiseman-Hakes, C., Stewart, M. L., Wasserman, R., & Schuller, R. (1998). Peer group training of pragmatic skills in adolescents with acquired brain injury. *Journal of Head Trauma Rehabilitation, 13*(6), 23–38.

Yeates, K. O., Swift, E., Taylor, H. G., Wade, S. L., Drotar, D., Stancin, T., et al. (2004). Short- and long-term social outcomes following pediatric traumatic brain injury. *Journal of the International Neuropsychological Society, 10*(3), 412–426.

Ylvisaker, M. (1989). Cognitive and psychosocial outcome following head injury in children. In J. T. Hoff, T. E. Anderson, & T. M. Cole (Eds.), *Mild to moderate head injury* (pp. 203–216). Oxford: Blackwell Scientific.

Ylvisaker, M., & Szekeres, S. F. (1998). A framework for cognitive rehabilitation. In M. Ylvisaker (Ed.), *Traumatic brain injury rehabilitation: Children and adolescents* (2nd ed., pp. 125–158). Boston: Butterworth-Heinemann.

Ylvisaker, M., Turkstra, L. S., & Coelho, C. (2005). Behavioral and social interventions for individuals with traumatic brain injury: A summary of the research with clinical implications. *Seminars in Speech and Language, 26*(4), 256–267.

Ylvisaker, M., Turkstra, L., Coehlo, C., Yorkston, K., Kennedy, M., Sohlberg, M. M., et al. (2007). Behavioural interventions for children and adults with behaviour disorders after TBI: A systematic review of the evidence. *Brain Injury, 21*(8), 769–805.

CHAPTER 17

Family-Centered and Parent-Based Models for Treating Socio-Behavioral Problems in Children with Acquired Brain Injury

Damith T. Woods, Cathy Catroppa, and Vicki Anderson

S ocio-behavioral problems have been identified as the most pervasive and enduring challenges experienced by children with acquired brain injury (ABI), their families, the professionals treating them, and the community at large (Braine, 2011; Brenner et al., 2007; Catellani, Lombardi, Brianti, & Mazzucchi, 1998; Slomine et al., 2006). They are particularly common, with 30–70% of these children experiencing difficulties that did not predate the injury (Fletcher, Ewing-Cobbs, Miner, Levin, & Eisenberg, 1990; Schwartz et al., 2003; Max, Koele, et al., 1998; Ylvisaker et al., 2007). The manifestation of these problems is usually not a simple cause-and-effect relationship, because diverse child-related, injury-related, family- and parent-related, and environmental influences interact to determine outcomes (Rivara et al., 1992; Max et al., 1997; Taylor et al., 2001).

After a child has suffered a brain injury and is medically stabilized, the child will return home to his or her family. Rehabilitation specialists agree that interventions for childhood ABI should be focused within the family environment (Cavallo, Kay, & Ezrachi, 2005; Ylvisaker, 1998; Wade, 2006). Indeed, with investigations highlighting that the family environment can moderate the impact of childhood brain injury (Anderson, Catroppa, Morse, Haritou, & Rosenfeld, 2005; Hooper, Williams, Wall, & Chua, 2007; Rivara et al., 1993, 1994; Taylor et al., 2002; Wade, 2006), it seems logical for interventional strategies to be family-oriented and to consider parenting factors in the rehabilitation process.

This chapter begins by describing common socio-behavioral difficulties following childhood ABI, as well as their relationship with family/parental problems. The existing research evidence for family-centered and parent-based models to manage and prevent socio-behavioral problems in children with ABI is then examined. The chapter concludes by exploring some of the limitations on research into effective socio-behavioral interventions in childhood populations with ABI, and by suggesting directions for future research.

Socio-Behavioral Problems in Children after ABI

The most frequently reported challenges and difficulties for families and parents of children with ABI include inattention to task (Anderson, Godber, Smibert, Weiskop, & Ekert, 2004; Bruce, Selznick-Gurdin, & Savage, 2004); aggressive and antisocial behavior (Andrews, Rose, & Johnson, 1998; Cole et al., 2008); maladaptive behavior (Asarnow, Satz, Light, Lewis, & Neumann, 1991); personality change (Max et al., 2000); disinhibition (Brown, Chadwick, Shaffer, Rutter, & Traub, 1981; Schachar, Levin, Max, Purvis, & Chen, 2004); hyperactivity (Max et al., 2004); emotional instability and irritability (Fuemmeler, Elkin, & Mullins, 2002); self-dysregulation (Ganesalingam, Sanson, Anderson, & Yeates, 2006); poor social functioning or outright impropriety (Janusz, Kirkwood, Yeates, & Taylor, 2002; Muscara, Catroppa, & Anderson, 2008; Yeates et al., 2003); and noncompliance (Bruce et al., 2004). Understanding the evolution of these problems requires a multideterministic view, as they can result from a number of different pathways (Taylor et al., 2001). One pathway may represent a level of socio-behavioral disturbance that antedates the injury and has even been suggested to contribute to the risk for incurring the ABI (Light et al., 1998). This pathway may also represent a preinjury level of behavior disturbance that is exacerbated by an ABI (Schwartz et al., 2003). A second pathway is the direct effect of neuropathology on behavior. This can include a neurophysiological response (e.g., headache, lethargy, sensitivity to various stimuli) (Ayr, Yeates, Taylor, & Browne, 2009; Necajauskaite, Endziniene, & Jureniene, 2005; Yeates et al., 1999) and/or neurocognitive disturbance (e.g., deficits in memory, attention, planning, problem solving, self-regulation) (Anderson et al., 2004, 2005; Catroppa & Anderson, 2006; Ganesalingam et al., 2006), underpinned by damage to brain regions implicated in affective regulation (Drevets, 2003; Max, Levin, Schachar, & Landis, 2006). A third pathway may embody a child's own psychopathological response to injury and its aftermath, such as posttraumatic stress or adjustment disorder (Fuemmeler et al., 2002; Levi, Drotar, Yeates, & Taylor, 1999; Poggi et al., 2005). A fourth pathway may be a child's response to long-term cognitive deficits or other impairments (e.g., physical); this response may take the form of frustration and/or conduct problems (Max, Lindgren, et al., 1998). In a final pathway, pre- and postinjury family/parental dysfunction and environmental factors may influence socio-behavioral outcomes via interactional processes, which in turn may interact with any of the other pathways (Anderson et al., 2001; Kinsella, Ong, Murtagh, Prior, & Sawyer, 1999; Rivara et al., 1993, 1996; Taylor et al., 2001; Wade, 2006).

Bidirectional Influences Affecting Child and Family/Parental Outcomes

Increased levels of parental stress, anxiety, depression, and other psychopathology (Fuemmeler, Mullins, & Marx, 2001; Hawley, Ward, Magnay, & Long, 2003; Prigatano & Gray, 2007), as well as family burden and dysfunction, have been well documented in the literature (Foley, Barakat, Herman-Liu, Radcliffe, & Molloy, 2000; Ganesalingam et al., 2008; Wade, 2006; Wade, Taylor, et al., 2005). Interestingly, clinical research in samples of children without brain injury but with significant socio-behavioral problems have often found these family/parental factors to be significantly correlated with ineffectual parenting practices, such as a lack of support, coercive exchanges, and harsh and ineffective punishment (Amato & Fowler, 2002; Ge, Conger, Cadoret, & Neiderhiser, 1996; Patterson, 1982).

Recent research highlights some of the effects childhood ABI can have on parents' abilities to manage the socio-behavioral challenges of their children. Hooper et al. (2007) assessed the impact of childhood ABI on parenting practices, using the Disciplinary Strategy Questionnaire (Jelalian, Stark, & Miller, 1997). They found that a majority of parents made use of proactive disciplinary strategies, such as positive reinforcement, logic, and rationalization. Conversely, parents also endorsed (but to a lesser extent) the use of more negative strategies, such as punishment, force, and giving in to child behavior. Child executive deficits associated with affective and behavioral dysregulation were significantly correlated with parents' use of punishment as a disciplinary strategy. Furthermore, parents who reported greater psychological distress (e.g., anxiety and depression) were less likely to use positive disciplinary approaches when managing socio-behavioral problems. We (Woods, Catroppa, Barnett, & Anderson, 2011) conducted similar research examining the relationship between parental disciplinary use, family adversities, and parental distress on the one hand, and socio-behavioral sequelae in children with mild, moderate, and severe ABI on the other. We found that overreactive and lax parental disciplinary approaches had significant and positive associations with children's externalizing socio-behavioral difficulties, regardless of injury severity. We also found that for parents who reported clinically dysfunctional levels of parenting in disciplinary situations, a clear profile emerged: These families were more distressed, had higher rates of family dysfunction and social adversity, and rated their children as having significantly more socio-behavioral problems.

Other researchers have attempted to explore particular dimensions of the parent–child relationship following childhood brain injury. Wade, Taylor, et al. (2008) attempted to describe parent–child dyads and their interactions following either traumatic brain injury (TBI) or orthopedic injury (OI). Under experimental conditions, ratings were made of videotaped interactions between parents and children approximately 40 days after injury on levels of parental warmth and responsiveness. The authors found that parents in the TBI group exhibited less warm responsiveness to their children than parents in the OI group did during separate free-play and teaching tasks. Also, after socio-demographic factors were controlled for, a child's own inability to regulate behavior helped explain a significant portion of the variance in parents' responsiveness to the child. Although this study did not directly investigate parents' abilities to enact control over behavior, the findings demonstrate the complex reciprocal influences children and parents may have on each other following brain injury—influences that can lead to an unstable pattern of future interactions. A parent's inability to manage a child's socio-behavioral challenges following ABI is likely to contribute to an emergence of maladaptive parent–child interactions, which may become well established in the absence of any formal help (Rivara et al., 1996; Taylor et al., 2001, 2002). It is evident from this and the other studies described here that intervention strategies following childhood ABI need to incorporate an emphasis on family interactions, family-centeredness, and parenting skills.

Family-Centeredness and Parent Management

Family-centered intervention draws on the theoretical framework of family-centered care (FCC), which espouses these guiding principles: (1) the idea that parents and families should work collaboratively with health care professionals to maximize rehabilitation outcomes for children with ABI; and (2) heightened recognition that parents and family members are the experts about their children and should be involved in making the decisions about their care (Hostler, 1999). Research suggests that FCC is associated with reduced parental distress, higher parental satisfaction with services, and better socio-behavioral outcomes for children with disabilities (Law et al., 2003; Rosenbaum, King, Law, King, & Evans, 1998). The accession of FCC as a front-line treatment for socio-behavioral problems in children following brain injury has also been prompted by the analysis of social learning models of parent–child interactions in typically developing children (e.g., Patterson, 1982). These models identify mechanisms that can maintain a level of parent–child interaction maladaptive, dysfunctional in nature and predicts future antisocial behavior in children (Patterson, 1982). This subsequently guided the development of family-centered socio-behavioral programs to focus on parenting strategies in order to manage child behavior more effectively (Patterson, Chamberlain, & Reid, 1982; Sanders, 1999; Sanders & Dadds, 1993). Current, efficacious family-centered interventions for children with significant socio-behavioral disturbance have utilized a parent management training (PMT) approach to program delivery (Patterson et al., 1982; Dangel & Polster, 1984).

PMT is a psychosocial treatment in which parents are taught skills and strategies that *empower them* to manage their children's challenging socio-behavioral difficulties. Practitioners of the PMT approach have demonstrated that, through altering parent behavior, children can be assisted in developing adaptive prosocial behaviors and thus reducing challenging behavior (Patterson et al., 1982; Sanders, 1999; Sanders & Dadds, 1993; Sanders, Dadds, & Bor, 1989). This approach is deeply rooted in social learning theory (Skinner, 1953), positive behavior support (PBS; Carr et al., 2002), and applied behavior analysis (ABA; Sulzer-Azaroff & Mayer, 1991). PMT has been validated as an effective treatment approach for a wide variety of childhood socio-behavioral problems, including noncompliance (Sanders & Dadds, 1993), aggressiveness (Patterson et al., 1982), oppositional defiant or conduct disorder (McNeil, Eyberg, Eisenstadt, Newcomb, & Funderburk, 1991; Sanders et al., 1989; Webster-Stratton, 1984), hyperactivity (Dubey, O'Leary, & Kaufmann, 1983), poor impulse control, and attention-deficit hyperactivity disorder (Anastopoulos, Shelton, DuPaul, & Guevremont, 1993; Johnston, 1992). This body of research is particularly pertinent, given that such challenging social behaviors are often seen in children following ABI (Dooley, Anderson, Hemphill, & Ohan, 2008; Max et al., 1997, 2000, 2004, 2006; Max, Koele, et al., 1998; Max, Lindgren, et al., 1998; Schwartz et al., 2003). PMT has also been applied within differing age groups (from preschool children to adolescents) with equal efficacy and success (e.g., Sanders, 1999; Graziano & Diament, 1992).

How to Treat: Theoretical Underpinnings

Prominent specialists within the field of childhood ABI contend that there are three theoretical intervention paradigms for the treatment and prevention of socio-behavioral problems in children with brain injury (Ylvisaker et al., 2005, 2007): contingency management procedures, most often associated with ABA; antecedent management procedures, most often associated with PBS; or a balanced combination of both.

Contingency Management Procedures

ABA is based on the assumption that behaviors can increase or decrease in frequency as a result of the consequences assigned to them. ABA focuses on a three-term contingency (stimulus–response–consequence) that aims to shed light on events that precede behavior, but even more so on the events that directly follow the behavior, which increase the probability of its future occurrence (Chance, 1998; Miltenberger, 1997). Within the ABA framework, the specific systematic assessment of environmental contingencies on behavior is termed *functional analysis* (Iwata, Dorsey, Slifer, Bauman, & Richman, 1982). ABA, together with functional analysis, has assisted in the development of a number of educational methods (such as shaping, fading, and contingent reinforcement) for reducing socio-behavioral problems in children (Sulzer-Azaroff & Mayer, 1991).

Antecedent Management Procedures

On the other hand, Carr et al. (2002) have described PBS as "an applied science that uses educational and systems change methods *[environmental redesign and manipulation]* in an attempt to enhance the individuals quality of life by way of minimizing problem behavior" (p. 4). In the field of ABI socio-behavioral rehabilitation, PBS specifically emphasizes the manipulation of antecedents—that is, events that occur prior to behavior. PBS integrates the technology and philosophy of ABA, but with a person-centered approach (Carr et al., 2002; Johnston, Foxx, Jacobson, Green, & Mulick, 2006). Person-centered therapy embraces the idea that the specific needs and goals of the individual should drive the creation of services; as such, intervention is carefully tailored to address the unique characteristics of the individual. This is a pertinent concept for the childhood ABI population, considering the heterogeneous nature of brain injury and the varied clinical presentations that can be seen in clinical practice. Clinical research also reveals a neuropsychological basis for using PBS with children who display challenging behaviors following ABI. Due to the various mechanisms of injury and sites of brain pathology, reinforcement learning is often reduced, and this reduction in turn diminishes the capacity to learn from consequences and inhibit behavior accordingly (Rolls, 2000; Schlund, 2002). With this knowledge, growing numbers of clinical researchers are investigating the utility of teaching parents and other family members (e.g., siblings) skills and strategies for the socio-behavioral management of their children/siblings with ABI.

Current Research in Populations of Children with ABI

In this section, the existing evidence for interventions with children who have ABI and display socio-behavioral difficulties is reviewed and summarized.

Wade, Wolfe, Brown, and Pestian (2005) set out to examine the feasibility of a web-based family problem-solving (FPS) intervention for families, siblings, and children with ABI. This FPS paradigm ("Aim, Brainstorm, Choose, Do it, and Evaluate skills") represented Wade and colleagues' adaptation of the original five-step FPS intervention process: problem orientation, problem definition and formulation, generation of alternative solutions, decision making, and solution implantation/verification (D'Zurilla & Nezu, 1999). Participants in the initial study consisted of six families, comprising eight parents, five siblings, and six children with moderate and severe ABI (mean age of these children = 10.5 years). All children with ABI were between 5 and 16 years of age and had been injured on average 18 months prior to participation. The FPS intervention was delivered over a 5-month period. After participants' consent was obtained, computers, printers, web cameras, and high-speed Internet connection were installed in all families' homes except one. Families were trained to operate the computers, to access the requisite web pages, and to navigate therein. The FPS web-based intervention consisted of eight core sessions and an additional four sessions that addressed family/parental stressors and burdens. The outline and key features of session content are provided in Table 17.1.

With one exception, all families were randomly allocated to one of two intervention groups for videoconferencing. Both intervention groups had access to two-way audio and video, as well as application sharing; however, during two-way videoconferencing with a trained ABI behavioral therapist, picture size, video- and sound-enhancing features, and cost all differed. A primary purpose of this study was to determine whether a larger picture size and higher-quality video images would be significantly more beneficial to a family's experience with the intervention, and thus would justify the additional expense.

TABLE 17.1. Outline and Key Features of Core Sessions for Wade, Wolfe, et al. (2005) FPS Design

Session	Content	Key feature
1–3	• Identifying goals wanting to be changed (not limited to injury-related issues). • Describing steps in problem-solving framework. • Implementing problem-solving process with one goal.	• Learning about family and overviewing FPS. Working as a team to achieve goals.
4–5	• Didactic information regarding cognitive and behavioral sequelae of TBI. • Instruction on use of antecedent behavior management.	• Creating positive behavior impetus.
6	• Addressing communication skills.	• Providing families with skills to discuss problems without conflict.
7	• Crisis management; problem solving, communication, behavior management strategies reviewed. • Addressing unresolved stressors.	• Reviewing progress and concerns with intervention.
8–11	• Individualized sessions (managing behavior, sibling issues, communication with spouse/family, working with school, stress management).	• Assigning content based on family need.
12	• Planning for the future.	• Developing plans for addressing future challenges.

Families participated in the self-guided web sessions at any time (day or night), and each contained a number of didactic portions. Video clips were also part of the didactic content, modeling for parents the skills that they were to learn. All participants' workbook exercises and other information were stored and reviewed by a therapist. Once each individual online session and materials were completed, a 45- to 60-minute videoconference session was scheduled with the therapist to review exercises and worksheets, and to discuss any issues the families had.

Families participated in an average of 10.67 web sessions. Participants ranked the website content as moderately easy or easy to use overall. Siblings rated the videoconferences as easier to use than did the children with ABI. However, children with ABI who used the smaller and less expensive videoconferencing rated it as significantly more difficult to use than the children with ABI who used the larger videoconferencing platform rated that platform. The authors suggested that the smaller screen might have placed too many demands on the injured children to attend. In spite of this, families' perceptions of the intervention's helpfulness were high, particularly for the brain-injury-related content. Parents rated the behavior management strategies as helpful, and all participants (parents, injured children, and siblings) reported satisfaction levels between 70% and 96% (Wade, Wolfe, & Pestian, 2004). Some of the specific problems addressed by the families in the study centered on issues involving academic performance, homework completion, chore completion, anger control, and maternal stress. Although the sample population was small, the authors reported a statistically significant reduction in children's antisocial behaviors, as well as improvements in planning and organization (Wade, Wolfe, et al., 2005). Children also reported significant reductions in conflict with their parents regarding school issues. As the research was primarily a feasibility study, no control group was employed. The authors did not report a measure of parental distress, despite parents' identifying this as a problem area. Of the 19 participants, all but 1 (an injured child) indicated that they would recommend the program to others.

In a subsequent study utilizing the same online FPS paradigm, Wade, Carey, and Wolfe (2006) reported that parents receiving the Internet-delivered FPS intervention ($n = 20$) had significantly less anxiety, depression, and psychiatric symptoms than parents in an Internet resources control group ($n = 20$) had, after the authors controlled for baseline symptoms. Families in the control group had access to all the same Internet resources as the FPS group, except that they did not participate in the FPS intervention and did not have access to any FPS material. The FPS group also reported better child self-management and compliance at follow-up than did the control group. Interestingly, the researchers also found that children's age and socioeconomic status moderated treatment effects, with older children and those of lower socioeconomic status who received the FPS intervention showing greater improvements in self-management and socio-behavioral problems.

Wade, Michaud, and Brown (2006) then reported on the delivery of an adapted FPS intervention in a randomized controlled trial (RCT) involving 32 families of children with moderate and severe ABI. Families were randomly allocated either to an FPS group or to a control group that received standard care. The study design was slightly different, however, in that the intervention was delivered in the more customary face-to-face method. Within the study, there was also an emphasis on how to deal with the specific cognitive and behavioral sequelae of ABI. The intervention was delivered over a 6-month period. Families completed an average of 8.31 treatment sessions, and 78.9% completed seven core sessions (see Table 17.1). Eighty percent of parents in the FPS group reported that they had reached the problem-solving goals that they had targeted at the beginning of the program. Families in the FPS group also reported significant reductions in child internalizing behaviors (anxiousness, depression, withdrawal). Didactic teaching of skills and strategies helped parents to report that they understood their children's injury better and were aware of strategies for improving their children's behavior. No significant changes were reported in parental distress, child externalizing behaviors, or parent–child conflict. No measure of family functioning was incorporated within the study. For those parents who provided satisfaction data, all indicated that they would recommend the program to others.

Most recently, Wade, Walz, Carey, and Williams (2008) briefly reported on the feasibility and satisfaction ratings for nine adolescents and their families using (once again) an FPS web-based intervention. Both teens with moderate and severe ABI and their parents reported high ratings for helpfulness and ease of use for the online problem-solving paradigm (the same "Aim, Brainstorm, Choose, Do it, and Evaluate skills" paradigm as in the Wade, Wolfe, et al. [2005] study). The researchers concluded that this intervention may prove beneficial for teens with moderate and severe ABI in the future.

Using an alternative approach, Braga, Da Paz, and Ylvisaker (2005) conducted a large RCT investigating whether a contextualized family-centered cognitive intervention would be superior to one delivered by trained clinicians. Their research compared a direct clinician-delivered to an indirect family-supported (IFS) intervention for children with chronic impairment in cognitive and/or physical domains following ABI. Children were between 5 and 12 years of age. The sample consisted of 87 families and children with moderate and severe ABI who were randomly allocated to the two different treatment approaches. The indirect intervention incorporated an integrated program created by a team of rehabilitation specialists, including physical therapists, speech pathologists, and psychologists. Individualized treatment programs were developed to meet each child's needs, and each program consisted of approximately 14 pages of graphic illustrations aimed to guide families in specific rehabilitation objectives. (A graphically organized manual was chosen to help parents who had reading difficulties to learn the program procedures.) After an intensive 2-week training regimen, the members of each family took "ownership" of their manualized program and began implementing it into their homes and typical daily routines. Programs were regularly assessed by trained professionals and modified to meet the evolving needs and goals of the children and families. Alternatively, children in the direct-intervention group received specialist rehabilitation programs for 2 hours a day, 5 days a week, without their parents' being present. Parents received information about their children's programs, but they were not trained to implement any facet of the intervention. Each rehabilitation specialist communicated with the others about the treatments, but no attempt was made to coordinate services. During the 2-week assessment period parents in the direct-intervention group attended information and support sessions, but again did not participate in any aspect of the rehabilitation with their children.

Parents in the indirect-intervention group were evaluated for their learning of program procedures every 3 months. The authors reported that these parents consistently demonstrated learning percentages in the high 90s, regardless of prior educational levels. On measures of cognitive, physical, and functional change, both groups revealed that they had benefitted from the intervention; however, only families receiving the indirect intervention showed statistically significant change following 1 year of treatment. In addition, children with more severe injuries showed greater gains on physical and functional measures than moderately injured children did. No measure of family functioning, challenging child behaviors, or parental distress was reported in the study methodology. Nevertheless, the RCT did show that the use of indirect family support coupled with professional integration was superior to conventional multidisciplinary, clinic-based direct intervention.

Further Development of a Family-Centered, Parent-Based Model

We (Woods, Catroppa, Anderson, Matthews, et al., 2008) noted that socio-behavioral family-centered interventions for childhood ABI up to that time had been somewhat circumscribed with regard to ameliorating parental distress, improving family functioning, lowering levels of associated family burden, and reducing and/or preventing challenging social behaviors in children. In light of this, we proposed an adapted version of the Signposts for Building Better Behavior program (Hudson & Cameron, 2001). Signposts is a flexible family-centered socio-behavioral intervention drawing on much published scientific literature in disability research (Dunlap, Johnson, & Robbins, 1990; Lutzker & Campbell, 1994); Koegel, Koegel, & Dunlap, 1996; O'Neill et al., 1997). Signposts utilizes a PMT design, incorporating ABA and PBS strategies, and was originally developed to provide support to parents and families of children with intellectual disabilities and/or developmental delay (Hudson et al., 2003). In a small controlled pilot treatment study consisting of three families of children with ABI and professionals (including neuropsychologists, clinical psychologists, a speech pathologist, an educational psychologist, special educators, and an occupational therapist), the Signposts program was rated highly feasible for use with an ABI population by parents and professionals alike (Giallo, Matthews, & Anderson, 2005).

The main aims of the adapted Signposts program are (1) to help parents of children with ABI, by way of various educational methods, to identify the purposes of their children's challenging socio-behavioral problems; (2) to teach parents effective ways of managing these problems; and (3) to learn strategies in order to prevent/reduce these challenging behaviors in the future (Woods, Catroppa, Anderson, Matthews, et al., 2008). Within the program, families focus on antecedents, behaviors, and consequences in order to alter both child and parent behaviors (Hudson et al., 2003). Examples of *antecedent events* that are used to promote and directly facilitate behavioral change include verbal instruction, physical aids (prompts), and parental modeling. In regard to *behaviors*, despite the fact that the primary goal of the intervention is to eliminate unwanted behaviors, the program focuses on developing prosocial, positive behaviors that will reduce the likelihood of challenging behaviors' occurring (Hudson et al., 2003). *Consequences* are the events that parents use to reinforce their children's appropriate behaviors. Through instruction from trained ABI behavior therapists, parents are assisted in reducing unnecessary commands, increasing the clarity of the limits they do set, and increasing the consistency with which they follow through on their limits. They are helped to assess and evaluate the precipitants of socio-behavioral problems in their children, to monitor their children effectively, and to engage in effective conflict resolution strategies with them (Hudson & Cameron, 2001; Woods, Catroppa, Anderson, Matthews, et al., 2008).

Resources that are available to increase the effectiveness of the Signposts program for an ABI population include a facilitator manual for clinicians who deliver the program and provide the structure, content, and flow of treatment sessions (Woods, Catroppa, & Anderson, 2008a), as well as information modules and a workbook for the parents. DVDs also assist in illustrating key points from the program for parents (Hudson et al., 2003), and an ABI Information Booklet helps parents identify family, parental, and child adjustment difficulties that can uniquely arise following childhood ABI (Woods, Catroppa, & Anderson, 2008b). See Table 17.2 for a list of all Signposts modules and material covered by the ABI Information Booklet. Note that the booklet is a stand-alone resource and is used as an adjunct to the Signposts program (Woods et al., 2008b).

Using the Signposts program as modified for ABI, we (Woods, Catroppa, Anderson, Matthews, et al. (2008; Woods et al., 2009) sought (1) to investigate whether the manualized program could reduce challenging behaviors in children with ABI and improve family/parental well-being and functioning in the short and long term; and (2) to ascertain the clinical feasibility of delivering the program in group support and telephone support modes. A telephone support delivery mode was investigated in this research because many children with ABI and their families live in rural areas, far from metropolitan rehabilitation services. Although such families can manage the necessary acute care in major cities or health care facilities, they eventually must return home to more isolated regions and find it difficult to access follow-up services because of physical, financial, and/or geographical constraints (Wade & Wolfe, 2005).

The study recruited 61 parents of children between 3 and 12 years of age with mild (n = 20), moderate (n = 18), or severe (n = 10) ABI. Parents received the Signposts program in either the group support (n = 23) or the telephone support (n = 25) format. Families were invited to choose whichever of the two intervention conditions was better suited to their family commitments. Parents were asked to provide demographic information, ratings of child behavior, family functioning, parental distress, and parenting practices at time of recruitment (baseline), approximately 4 months prior to intervention delivery. Parents then partook of the Signposts intervention, which lasted 5 months for both groups. Data collection procedures were then repeated at 1 month postintervention (approximately 10 months postbaseline), and again at 6- and 12-month follow-ups. Data from all participants were successfully collected at the 1-month postintervention point. At 6 months following the delivery of the intervention, 42 of the original 48 parent–child

---

TABLE 17.2. Signposts Modules and Material Covered by the ABI Information Booklet

- Introduction
- Dealing with a head injury in the family (ABI Information Booklet)
- Module 1—Measuring your child's behavior
- Module 2—Systematic use of daily interactions
- Module 3—Replacing difficult behavior with useful behavior
- Module 4—Planning for better behavior
- Module 5—Developing more skills in your child
- Module 6—Dealing with stress in the family
- Your family as a team

dyads provided follow-up data on family/parental and child socio-behavioral functioning (retention rate of 87.5%). At the 12-month follow-up, 17 parent–child dyads could not be located, did not want to participate further in the study, or had incomplete data sets. Thirty-one parent–child combinations (64.5% of the original sample) participated in the 12-month follow-up. Analyses were employed to represent the two different modes of intervention support (group vs. telephone) and two levels of child injury severity (mild vs. moderate/severe). The moderate and severe groups were collapsed into one representative group due to the relatively small numbers of such children recruited into the study (Woods, Catroppa, Anderson, Matthews, et al., 2008; Woods et al., 2009).

With respect to consumer satisfaction (measured via standard questionnaire), all parents receiving the Signposts intervention approved of the skills taught, and a majority felt the materials helped them in managing socio-behavioral difficulties as well as in learning new skills (Woods, Catroppa, Giallo, Matthews, & Anderson, 2012). Parents also reported feeling significantly less stressed; this is of clinical significance because parental stress has been shown to be a significant predictor of ineffective parenting practices, which act as a mediator for poor socio-behavioral outcomes in children (Amato & Fowler, 2002; Brenner & Fox, 1998). We (Woods, Catroppa, Anderson, Matthews, et al., 2008) discovered that parents who had endorsed maladaptive levels of parental disciplinary behaviors prior to intervention found that after the intervention, these dropped below levels deemed to be dysfunctional in nature.

In both of its two service delivery modes, the program produced significant effect size changes in externalizing and total number of challenging behaviors for both injury severity groups. This is an important finding for the group with mild ABI, as children with mild head injuries constitute approximately 90% of all children with TBI (Kraus, Rock, & Heymari, 1990; McKinlay et al., 2008) and have often been a neglected and unreported group (Evans, 1992; McKinlay et al., 2008; McKinlay, Grace, Horwood, Fergusson, & MacFarlane, 2009). Those children whose socio-behavioral scores did not fall into the clinically significant range on the Child Behavior Checklist (CBCL; Achenbach, 1991) before intervention displayed no elevation and in fact a small decrease in scores after intervention (see Figures 17.1 and 17.2). This result suggests that the program has a prophylactic capacity both to restrict socio-behavioral problems and to prevent them from developing. In addition, a majority of parents who had rated their children's problem behaviors above the clinically at-risk range prior to intervention (CBCL T-score ≥ 60) reported that it had fallen to within typical limits after the intervention, thus presenting evidence of clinically meaningful as well as statistically significant change. At both the 6- and 18-month follow-ups, the socio-behavioral scores for the mildly injured children remained well below the clinically at-risk range (Woods et al., 2009). Children who had sustained moderate and severe brain injuries showed maintenance of treatment effects across the two follow-up periods, with no significant elevations in socio-behavioral difficulties. Overall, the results demonstrated that children's levels of socio-behavioral functioning at the 6- and 12-month follow-ups were relatively comparable to their immediate postintervention scores. Parental disciplinary practices also remained at levels below the clinically dysfunctional range at the two follow-up time points for all parents. At later follow-ups, parental distress and family burden revealed no significant increases for either group and remained within typical limits.

Despite these promising findings, there are some limitations to our investigation that are common in child ABI interventions more generally and should be brought to light.

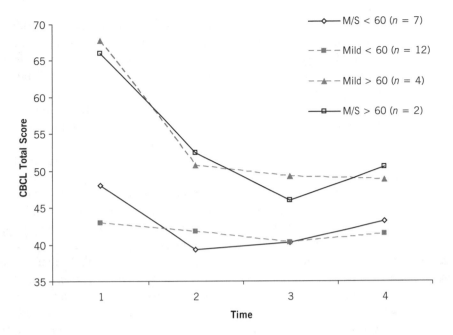

FIGURE 17.1. Child behavior outcomes for the telephone support intervention across injury severity groups (M/S, moderate/severe ABI; Mild, mild ABI) and behavioral groups (< 60, typical behavior; > 60, behavioral problem).

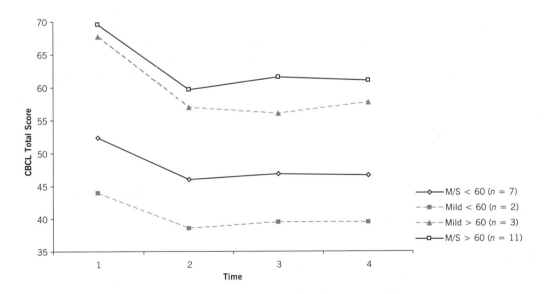

FIGURE 17.2. Child behavior outcomes for the group support intervention across injury severity groups (M/S, moderate/severe ABI; Mild, mild ABI) and behavioral groups (< 60, typical behavior; > 60, behavioral problem).

The first centers on the absence of random allocation. We (Woods et al., 2009) have acknowledged that the inherent strength of an RCT is its ability to control for unknown confounding variables. However, there are both ethical and practical limitations on the use of RCTs in ABI rehabilitation research. This population of children and families is not representative of the general population, but is already vulnerable due to reported elevations in rates of preinjury child and family/parental dysfunction (Kinsella et al., 1999; Rivara et al., 1993, 1994). Random allocation runs the risk of minimizing adherence, and thus biasing findings and leading to significant attrition (Hudson et al., 2003). Consequently, in the Signposts study the distribution of parents to the various intervention groups was determined by the families, based on their personal needs and circumstances at the time of recruitment. This resulted in a disproportionate number of parents of mildly injured children choosing the less burdensome telephone support delivery, and in more parents of moderately or severely injured children choosing the group support delivery. Also, the measures in this study were reliant on parents' subjective appraisal of either their own, their children's or their families' levels of functioning. Such subjective appraisal can lead to biases, imprecisions, sensitivities, and exaggerations. Finally, it is difficult to discern whether the amount of observed postintervention change was due to aspects of the Signposts program alone or to other factors unrelated to treatment. Nonetheless, it is evident from the literature that where no treatment or socio-behavioral support is provided, challenging behaviors are persistent, and we can reasonably infer that they would not have improved without intervention.

We have further acknowledged that one of the challenges in the childhood ABI socio-behavioral intervention domain is moving from *feasibility* and *efficacy* research (demonstration that a treatment can work under well-controlled circumstances) to *effectiveness* research (demonstration of the treatment's applicability in typical clinical settings). However, we contend that providing families with options rather than random allocation replicates real-world conditions, and further serves the purpose of placing control back into the hands of the parents. For the ABI socio-behavioral rehabilitation field, *external validity* means that research results need to be generalized to other families of children with ABI, not to the general population. Because of the limitations on randomizing treatment and/or withholding therapy, socio-behavioral intervention strategies have needed to take a different approach from the "gold standard" of RCTs. The lack of a wait-list control in some of the above-described studies relates directly to a therapist's duty to provide the best possible care—in particular, the finding that early intervention results in the best outcomes (Kolb, Gibb, & Gorny, 2000). Thus it may be argued that comparing two intervention approaches is a more acceptable design than withholding therapy for an extended period of time.

Conclusions

Much research now confirms that interventions targeting the environment can result in changes in brain circuitry (Kolb et al., 2000; Chen, Cohen, & Hallett, 2000). Indeed, Kolb et al. (2000) have shown that rats receiving tactile stimulation soon after sham-operated cerebral trauma showed an attenuation of the behavioral deficits of injury and performed nearly as well on a skilled reaching task as control rats did. Thus, in order to promote adaptive neuronal connections that subserve advantageous social and behavioral functioning, it is suggested that early intervention is best. In effect, early intervention helps children with ABI to "practice" the correct way to behave at home and in social situations, which in turn helps to prevent aberrant and maladaptive neuronal connections.

A goal of future investigations will be to evaluate multicomponent interventions for enhancing treatment effects and promoting greater rehabilitation in children with ABI. This may involve combining socio-behavioral family-centered interventions with other empirically supported treatments. For example, children in the chronic stage of head injury exhibit discourse and language deficits and show disrupted information structure in their communication (Chapman, Culhane, Levin, Harward, & Mendelsohn, 1992). Research has demonstrated that training parents how to read to children in an interactive manner results in increases in the children's vocabulary development (Arnold, Lonigan, Whitehurst, & Epstein, 1994). Combining these strategies with socio-behavioral family-centered interventions may result in even greater benefits for children.

Following discharge from hospitals, most children with ABI go home to their families. Some families will experience significant stress, burden, and dysfunction. Children with severe socio-behavioral issues following ABI can be particularly difficult to care for, and it is here that parents and families can play a critically important role for their children. It is widely accepted that the most salient feature of interventions for a child with ABI is the premise that socio-behavioral support procedures should be implemented by natural support agents, the parents. It is in this context that parent training and family-centered interventions enable parents, as natural agents, to teach skills to their children through the use of natural everyday routines. Research on this approach is demonstrating that embedding intervention strategies within the family context offers the greatest promise of success in working with parents and families of children with ABI.

References

Achenbach, T. M. (1991). *Manual for the Child Behavior Checklist*. Burlington, VT: University Associates in Psychiatry.

Amato, P. R., & Fowler, F. (2002). Parenting practices, child adjustment, family diversity. *Journal of Marriage and Family, 64*(3), 703–716.

Anastopoulos, A. D., Shelton, T. L., DuPaul, G. J., & Guevremont, D. C. (1993). Parent training for attention deficit hyperactivity disorder: Its impact on parent functioning. *Journal of Abnormal Child Psychology, 21*(5), 581–596.

Anderson, V. A., Catroppa, C., Haritou, F., Morse, S., Pentland, L., Rosenfeld, J., et al. (2001). Predictors of acute child and family outcome following traumatic brain injury in children. *Pediatric Neurosurgery, 34*, 138–148.

Anderson, V. A., Catroppa, C., Morse, S., Haritou, F., & Rosenfeld, J. (2005). Attentional and processing skills following traumatic brain injury in early childhood. *Brain Injury, 19*, 699–710.

Anderson, V. A., Godber, T., Smibert, E., Weiskop, S., & Ekert, H. (2004). Impairments of attention following treatment with cranial irradiation and chemotherapy in children. *Journal of Clinical and Experimental Neuropsychology, 26*, 684–697.

Andrews, T. K., Rose, F. D., & Johnson, D. A. (1998). Social and behavioral effects of traumatic brain injury in children. *Brain Injury, 12*, 133–138.

Arnold, D. H., Lonigan, C. J., Whitehurst, G. J., & Epstein, J. N. (1994). Accelerating language development through picture book reading: Replication and extension to a videotape training format. *Journal of Educational Psychology, 86*, 235–243.

Asarnow, R. F., Satz, P., Light, R., Lewis, R., & Neumann, E. (1991). Behavior problems and adaptive functioning in children with mild and severe closed head injury. *Journal of Pediatric Psychology, 16,* 543–555.

Ayr, L. K., Yeates, K. O., Taylor, H. G., & Browne, M. (2009). Dimensions of postconcussive symptoms in children with mild traumatic brain injuries. *Journal of the International Neuropsychological Society, 15,* 19–30.

Braga, L. W., Da Paz, A. C., & Ylvisaker, M. (2005). Direct clinician-delivered versus indirect family-supported rehabilitation of children with traumatic brain injury: A randomized control trial. *Brain Injury, 19,* 819–831.

Braine, M. (2011). The experience of living with a family member with challenging behavior post acquired brain injury. *Journal of Neuroscience Nursing, 43,* 156–164.

Brenner, L. A., Dise-Lewis, J. E., Bartles, S. K., O'Brien, S. E., Godleski, M., & Selinger, M. (2007). The long-term impact and rehabilitation of pediatric traumatic brain injury: A 50–year follow-up case study. *Journal of Head Trauma Rehabilitation. 22,* 56–64.

Brenner, V., & Fox, R. A. (1998). Parental discipline and behavior problems in young children. *Journal of Genetic Psychology, 159*(2), 251–256.

Brown, G., Chadwick, O., Shaffer, P., Rutter, M., & Traub, M. (1981). A prospective study of children with head injuries: III. Psychiatric sequelae. *Psychological Medicine, 11,* 63–78.

Bruce, S., Selznick-Gurdin, L., & Savage, R. (2004). *Strategies for managing challenging behaviors of students with brain injuries.* Wake Forest, NC: Lash and Associates Publishing and Training.

Carr, E. G., Dunlap, G., Horner, R. H., Koegel, R. L., Turnball, A. P., Sailor, W., et al. (2002). Positive behaviour support: Evolution of an applied science. *Journal of Positive Behavior Intervention, 4,* 4–20.

Catroppa, C., & Anderson, V. (2006). Planning, problem-solving and organizational abilities in children following traumatic brain injury: Intervention techniques. *Pediatric Rehabilitation, 9*(2), 89–97.

Cattelani, R., Lombardi, F., Brianti, R., & Mazzucchi, A. (1998). Traumatic brain injury in childhood: Intellectual, behavioural and social outcome into adulthood. *Brain Injury, 12,* 283–296.

Cavallo, M. M., Kay, T., & Ezrachi, O. (2005). Problems and changes after traumatic brain injury: Differing perceptions within and between families. *Brain Injury, 6,* 327–335.

Chance, P. (1998). *First course in applied behavior analysis.* Pacific Grove, CA: Brooks/Cole.

Chapman, S. B., Culhane, K. A., Levin, H. S., Harward, H., & Mendelsohn, D. (1992). Narrative discourse after closed head injury in children and adolescents. *Brain and Language, 43,* 42–65.

Chen, R., Cohen, L. G., & Hallett, M. (2002). Nervous system reorganization following injury. *Neuroscience, 111,* 761–773.

Cole, W. R., Gerring, J. P., Gray, R. M., Vasa, R. A., Salorio, C. F., Grados, M., et al. (2008). Prevalence of aggressive behaviour after severe paediatric traumatic brain injury. *Brain Injury, 22*(12), 932–939.

Dangel, R. F., & Polster, R. A. (Eds.). (1984). *Parent training: Foundations of research and practice.* New York: Guilford Press.

Dooley, J. J., Anderson, V., Hemphill, S., & Ohan, J. (2008). Aggression after paediatric traumatic brain injury: A theoretical approach. *Brain Injury, 22,* 836–846.

Drevets, W. C. (2003). Neuroimaging abnormalities in the amygdala in mood disorders. *Annals of the New York Academy of Sciences, 985,* 420–444.

Dubey, D. R., O'Leary, S. G., & Kaufmann, K. F. (1983). Training parents of hyperactive children in child management: A comparative outcome study. *Journal of Abnormal Child Psychology, 11,* 229–246.

Dunlap, G., Johnson, L. F., & Robbins, F. R. (1990). Prevention of serious behavior problems through skill development and early interventions. In A. C. Repp & N. N. Singh (Eds.), *Perspectives on the use of nonaversive and aversive interventions for persons with developmental disabilities* (pp. 273–286). Sycamore, IL: Sycamore.

D'Zurilla, T. J., & Nezu, A. M. (1999). *Problem-solving therapy: A social competence approach to clinical intervention.* New York: Springer.

Evans, R. W. (1992). The postconcussion syndrome and the sequelae of mild head injury. *Neurologic Clinics, 10,* 815–847.

Fletcher, J. M., Ewing-Cobbs, L., Miner, M. E., Levin, H. S., & Eisenberg, H. M. (1990). Behavioral changes after closed head injury in children. *Journal of Consulting and Clinical Psychology, 58,* 93–98.

Foley, B., Barakat, L. P., Herman-Liu, A., Radcliffe, J., & Molloy, P. (2000). The impact of childhood hypothalamic/chiasmatic brain tumors on child adjustment and family functioning. *Children's Health Care, 29*(3), 209–223.

Fuemmeler, B. F., Elkin, D. T., & Mullins, L. L. (2002). Survivors of childhood brain tumors: Behavioral, emotional, and social adjustment. *Clinical Psychology Review, 22,* 547–585.

Fuemmeler, B. F., Mullins, L. L., & Marx, B. P. (2001). Posttraumatic stress and general distress among parents of children surviving brain tumor. *Children's Health Care, 30*(3), 169–182.

Ganesalingam, K., Sanson, A., Anderson, V., & Yeates, K. O. (2006). Self-regulation and social and behavioral functioning following childhood traumatic brain injury. *Journal of the International Neuropsychology Society, 12,* 609–621.

Ganesalingam, K., Yeates, K. O., Ginn, M. S., Taylor, H. G., Dietrich, A., Nuss, K., et al. (2008). Family burden and parental distress following mild traumatic brain injury in children and its relationship to post-concussive symptoms. *Journal of Pediatric Psychology, 33* (6), 621–629.

Ge, X., Conger, R. D., Cadoret, R. J., & Neiderhiser, J. M. (1996). The developmental interface between nature and nurture: A mutual influence model of child antisocial behavior and parent behaviors. *Developmental Psychology, 32,* 574–589.

Giallo, R., Matthews, J., & Anderson, V. (2005). *A pilot investigation of the applicability of the Signposts program for families with children with ABI.* Paper presented at the 6th World Congress on Brain Injury, Melbourne.

Graziano, A. M., & Diament, D. M. (1992). Parent behavioral training: An examination of the paradigm. *Behavior Modification, 16,* 3–38.

Hawley, C., Ward, B., Magnay, A., & Long, J. (2003). Parental stress and burden following traumatic brain injury amongst children and adolescents. *Brain Injury, 17,* 1–23.

Hooper, L., Williams, H., Wall, S. E., & Chua, K. C. (2007). Caregiver distress, coping and parenting styles in cases of childhood encephalitis. *Neuropsychological Rehabilitation, 17,* 621–637.

Hostler, S. L. (1999). Pediatric family-centered rehabilitation. *Journal of Head Trauma Rehabilitation, 14,* 384–393.

Hudson, A., & Cameron, C. (2001). *Signposts for Building Better Behaviour: Materials for parents of children with an intellectual disability and challenging behaviour.* Paper presented at the Helping Families Change Conference, Brisbane, Queensland, Australia.

Hudson, A., Matthews, J., Gavidia-Payne, S., Cameron, C., Mildon, R., & Radler, G. (2003). Evaluation of an intervention system for parents of children with intellectual disability and challenging behaviour. *Journal of Intellectual Disability Research, 47,* 238–249.

Iwata, B. A., Dorsey, M. F., Slifer, K. J., Bauman, K. E., & Richman, G. S. (1982). Toward a functional analysis of self-injury. *Analysis and Intervention in Developmental Disabilities, 2,* 3–20.

Janusz, J. A., Kirkwood, M. W., Yeates, K. O., & Taylor, H. G. (2002). Social problem-solving skills in children with traumatic brain injury: Long-term outcomes and prediction of social competence. *Child Neuropsychology, 8,* 179–194.

Jelalian, E., Stark, L. J., & Miller, D. (1997). Maternal attitudes towards discipline: A comparison of children with cancer and non-chronically ill peers. *Children's Health Care, 26,* 169–182.

Johnston, C. (1992). Parent characteristics and parent–child interactions in families of nonproblem children and ADHD children with higher and lower levels of opposition-defiant behavior. *Journal of Abnormal Child Psychology, 24*(1), 85–104.

Johnston, J. M., Foxx, R. M., Jacobson, J. W., Green, G., & Mulick, J. A. (2006). Positive behavior support and applied behavior analysis. *The Behavior Analyst, 29,* 51–74.

Kinsella, G., Ong, B., Murtagh, D., Prior, M., & Sawyer, M. (1999). The role of the family for behavioral outcome in children and adolescents following traumatic brain injury. *Journal of Consulting and Clinical Psychology, 67,* 116–123.

Koegel, L. K., Koegel, R. L., & Dunlap, G. (1996). *Positive behavioral support: Including people with difficult behavior in the community.* Baltimore: Brookes.

Kolb, B., Gibb, R., & Gorny, G. (2000). Cortical plasticity and the development of behavior after early frontal cortical injury. *Developmental Neuropsychology, 18,* 423–444.

Kraus, J. F., Rock, A., & Hemyari, P. (1990). Brain injuries among infants, children, adolescents, and young adults. *American Journal of Diseases of Children, 144,* 684–691.

Law, M., Hanna, S., King, G., Hurley, P., King, S., Kertoy, M., et al. (2003). Factors affecting family-centred service delivery for children with disabilities. *Child: Care, Health and Development, 29*(5), 357–366.

Levi, R. B., Drotar, D., Yeates, K. O., & Taylor, H. G. (1999). Posttraumtic stress symptoms in children following orthopedic or traumatic brain injury. *Journal of Clinical Child Psychology, 28,* 232–243.

Light, R., Asarnow, R., Satz, P., Zaucha, K., McCleary, C., & Lewis, R. (1998). Mild closed-head injury in children and adolescents: Behavior problems and academic outcomes. *Journal of Consulting and Clinical Psychology, 66,* 1023–1029.

Lutzker, J. R., & Campbell, R. (1994). *Ecobehavioral family interventions in developmental disabilities.* Belmont, CA: Brooks/Cole.

Max, J. E., Koele, S. L., Castillo, C., Lindgren, S. D., Arndt, S., Bokura, H., et al. (2000). Personality change disorder in children and adolescents following traumatic brain injury. *Journal of the International Neuropsychological Society, 6,* 279–289.

Max, J. E., Koele, S. L., Smith, W. L., Sato, Y., Lindgren, S. D., Robin, S. D., et al. (1998). Psychiatric disorders in children and adolescents after severe traumatic brain injury: A controlled study. *Journal of the American Academy of Child and Adolescent Psychiatry, 37,* 832–840.

Max, J. E., Lansing, A. E., Koele, S. L., Castillo, C. S., Bokura, H., & Schachar, R. (2004). Attention deficit hyperactivity disorder in children and adolescents following traumatic brain injury. *Developmental Neuropsychology, 25,* 159–177.

Max, J. E., Levin, H. S., Schachar, R. J., & Landis, J. (2006). Predictors of personality change due to traumatic brain injury in children and adolescents six to twenty-four months after injury. *Journal of Neuropsychiatry and Clinical Neurosciences, 18*(1), 21–32.

Max, J. E., Lindgren, S. D., Knutson, C., Pearson, C. S., Ihrig, D., & Welborn, A. (1998). Child and adolescent traumatic brain injury: Correlates of disruptive behaviour disorders. *Brain Injury, 12*(1), 41–52.

Max, J. E., Robin, D. A., Lindgren, S. D., Smith, W. L., Jr., Sato, Y., Mattheis, P. J., et al. (1997). Traumatic brain injury in children and adolescents: Psychiatric problems at two years. *Journal of the American Academy of Child and Adolescent Psychiatry, 36,* 1278–1285.

McKinlay, A., Grace, R. C., Horwood, L. J., Fergusson, D. M., & MacFarlane, M. R. (2009). Long-term behavioural outcomes of pre-school mild traumatic brain injury. *Child: Care, Health and Development, 36*(1), 22–30.

McKinlay, A., Grace, R. C., Horwood, L. J., Fergusson, D. M., Ridder, E. M., & MacFarlane, M. R. (2008). Prevalence of traumatic brain injury among children, adolescents and young adults: Prospective evidence from a birth cohort. *Brain Injury, 22,* 175–181.

McNeil, C. B., Eyberg, S., Eisensdtadt, T. H., Newcomb, K., & Funderburk, B. (1991). Parent–child interaction therapy with behaviour problem children: Generalization of treatment effects to the school setting. *Journal of Clinical Child Psychology, 20*, 140–151.

Miltenberger, R. (1997). *Behavior modification: Principles and procedures*. Pacific Grove, CA: Brooks/Cole.

Muscara, F., Catroppa, C., & Anderson, V. (2008). Social problem-solving skills as a mediator between executive function and long-term social outcome following pediatric traumatic brain injury. *Journal of Neuropsychology, 2*, 445–461.

Necajauskaite, O., Endziniene, M., & Jureniene, K. (2005). The prevalence, course and clinical features of post-concussion syndrome in children. *Medicina (Kaunas), 41*, 457–464.

O'Neill, R. E., Horner, R. H., Albin, R. W., Sprague, J. R., Storey, K., & Newton, J. S. (1997). *Functional assessment and program development for problem behavior: A practical handbook* (2nd ed.). Pacific Grove, CA: Brooks/Cole.

Patterson, G. R. (1982). *Coercive family processes*. Eugene, OR: Castalia.

Patterson, G. R., Chamberlain, P., & Reid, J. B. (1982). A comparative evaluation of a parent-training program. *Behavior Therapy, 13*, 638–650.

Poggi, G., Liscio, M., Adduci, A., Galbiati, S., Massimino, M., Sommovigo, M., et al. (2005). Psychological and adjustment problems due to acquired brain lesions in childhood: A comparison between post-traumatic patients and brain tumour survivors. *Brain Injury, 19*, 777–785.

Prigatano, G. P., & Gray, J. A. (2007). Parental concerns and distress after paediatric traumatic brain injury: A qualitative study. *Brain Injury, 21*, 721–729.

Rivara, J. B., Fay, G. C., Jaffe, K. M., Polissar, N. L., Shurtleff, H. A., & Martin, K. M. (1992). Predictors of family functioning one year following traumatic brain injury in children. *Archives of Physical Medicine and Rehabilitation, 73*, 899–910.

Rivara, J. B., Jaffe, K. M., Fay, G. C., Polissar, N. L., Martin, K. M., Shurtleff, H., et al. (1993). Family functioning and injury severity as predictors of child functioning one year following traumatic brain injury. *Archives of Physical Medicine and Rehabilitation, 74*, 1047–1055.

Rivara, J. B., Jaffe, K. M., Polissar, N. L., Fay, G. C., Liao, S. L., & Martin, K. M. (1996). Predictors of family functioning and change 3 years after traumatic brain injury in children. *Archives of Physical Medicine and Rehabilitation, 77*, 754–764.

Rivara, J. B., Jaffe, K. M., Polissar, N. L., Fay, G. C., Martin, K. M., Shurtleff, H. A., et al. (1994). Family functioning and children's academic-performance and behavior problems in the year following traumatic brain injury. *Archives of Physical Medicine and Rehabilitation, 75*, 369–379.

Rolls, E. T. (2000). The orbitofrontal cortex and reward. *Cerebral Cortex, 10*(3), 284–294.

Rosenbaum, P., King, S., Law, M., King, G., & Evans, J. (1998) Family-centred service: A conceptual framework and research review. *Physical and Occupational Therapy in Pediatrics, 18*, 1–20.

Sanders, M. R. (1999). Triple P Positive Parenting Program: Towards an empirically validated multilevel parenting and family support strategy for the prevention of behaviour and emotional problems in children. *Clinical Child and Family Psychology Review, 2*, 71–90.

Sanders, M. R., & Dadds, M. R. (1993). *Behavioral family intervention*. Needham Heights, MA: Allyn & Bacon.

Sanders, M. R., Dadds, M. R., & Bor, W. (1989). Contextual analysis of child oppositional and maternal aversive behaviors in families of conduct-disordered and nonproblem children. *Journal of Clinical Child Psychology, 18*, 72–83.

Schachar, R., Levin, H. S., Max, J. E., Purvis, K., & Chen, K. (2004). Attention deficit hyperactivity disorder symptoms and response inhibition after closed head injury in children: Do preinjury behavior and injury severity predict outcome? *Developmental Neuropsychology, 25*, 179–198.

Schlund, M. W. (2002). Effects of acquired brain injury on adaptive choice and the role of reduced sensitivity to contingencies. *Brain Injury, 16,* 527–535.

Schwartz, L., Taylor, G., Drotar, D., Yeates, K., Wade, S., & Stancin, T. (2003). Long-term behavior problems following pediatric traumatic brain injury: Prevalence, predictors, and correlates. *Journal of Pediatric Psychology, 28,* 251–263.

Skinner, B. F. (1953). *Science and human behavior.* New York: Macmillan.

Slomine, B. S., McCarthy, M. L., Ding, R., MacKenzie, E. J., Jaffe, K. M., Aitken, M. E., et al. (2006). Health care utilization and needs after pediatric traumatic brain injury. *Pediatrics, 117,* e663–e674.

Sulzer-Azaroff, B., & Mayer, G. R. (1991). *Behavior analysis for lasting change.* Fort Worth, TX: Holt, Rinehart & Winston.

Taylor, H. G., Yeates, K. O., Wade, S. L., Drotar, D., Stancin, T., & Burant, C. (2001). Bidirectional child–family influences on outcomes of traumatic brain injury in children. *Journal of the International Neuropsychological Society, 7,* 755–767.

Taylor, H. G., Yeates, K. O., Wade, S. L., Drotar, D., Stancin, T., & Minich, M. (2002). A prospective study of short and long-term outcomes after traumatic brain injury in children: Behaviour and achievement. *Neuropsychology, 16,* 15–27.

Wade, S. L. (2006). Interventions to support families of children with traumatic brain injuries. In J. E. Farmer, J. Donders, & S. Warschausky (Eds.), *Treating neurodevelopmental disabilities: Clinical research and practice* (pp. 170–185). New York: Guilford Press.

Wade, S. L., Carey, H., & Wolfe, C. R. (2006). An online family intervention to reduce parental distress following pediatric brain injury. *Journal of Consulting and Clinical Psychology, 74* (3), 445–454.

Wade, S. L., Michaud, L., & Brown, T. M. (2006). Putting the pieces together: preliminary efficacy of a family-problem solving intervention for children with traumatic brain injury. *Journal of Head Trauma Rehabilitation, 21,* 57–67.

Wade, S. L., Taylor, H. G., Chertkoff, N., Salisbury, S., Stancin, T., Bernard, L. A., et al. (2008). Parent–child interactions during the initial weeks following brain injury in young children. *Rehabilitation Psychology, 53,* 180–190.

Wade, S. L., Taylor, H. G., Yeates, K. O., Drotar, D., Stancin, T., Minich, N. M., et al. (2005). Long-term parental and family adaptation following pediatric brain injury. *Journal of Pediatric Psychology, 31*(10), 1072–1083.

Wade, S. L., Walz, N. C., Carey, J. C., & Williams, K. M. (2008). Brief report: Description of feasibility and satisfaction findings from an innovative online family problem-solving intervention for adolescents following traumatic brain injury. *Journal of Pediatric Psychology, 34*(5), 517–522.

Wade, S. L., & Wolfe, C. R. (2005). Telehealth interventions in rehabilitation psychology: Postcards from the edge. *Rehabilitation Psychology, 50*(4), 323–324.

Wade, S. L., Wolfe, C. R., Brown, T. M., & Pestian, J. P. (2005). Can a web-based family problem-solving intervention work for children with traumatic brain injury? *Rehabilitation Psychology, 50,* 337–345.

Wade, S. L., Wolfe, C. R., & Pestian, J. P. (2004). A web-based family problem-solving intervention for families of children with traumatic brain injury. *Behavioral Research Methods, Instruction, and Computing, 36,* 261–269.

Webster-Stratton, C. (1984). Randomized trial of two parent training programs for families with conduct-disordered children. *Journal of Consulting and Clinical Psychology, 52,* 666–678.

Woods, D., Catroppa, C., & Anderson, V. (2008a). *Dealing with a head injury in the family: Facilitator manual.* Collingwood, Victoria, Australia: Gill Miller Press.

Woods, D., Catroppa, C., & Anderson, V. (2008b). *Dealing with a head injury in the family: ABI booklet.* Collingwood, Victoria, Australia: Gill Miller Press.

Woods, D., Catroppa, C., Anderson, V., Matthews, J., Giallo, R., & Barnett, P. (2008). Preliminary efficacy for a family-centred intervention for parents of children with acquired brain injury. *Australian Journal of Psychology, S60*, 209–226.

Woods, D., Catroppa, C., Anderson, V., Matthews, J., Giallo, R., & Barnett, P. (2009). *Efficacy for a family-centred intervention for parents of children with acquired brain injury.* Paper presented at the 6th Satellite Symposium on NeuroRehabilitation, Tallin, Estonia.

Woods, D., Catroppa, C., Barnett, P., & Anderson, V. (2011). Parental disciplinary practices following acquired brain injury in children. *Developmental Neurorehabilitation, 14*(5), 274–282.

Woods, D., Catroppa, C., Giallo, R., Matthews, J., & Anderson, V. (2012). Feasibility and consumer satisfaction ratings following group and telephone delivered intervention for parents who have a child with ABI. *NeuroRehabilitation, 30*, 1–10.

Yeates, K. O., Luria, J., Bartkowski, H., Rusin, J., Martin, L., & Bigler, E. (1999). Postconcussive symptoms in children with mild closed head injuries. *Journal of Head Trauma Rehabilitation, 14*, 337–350.

Yeates, K. O., Swift, E., Taylor, H. G., Wade, S., Drotar, D. & Stancin, T. (2003). Short- and long-term social outcomes following pediatric traumatic brain injury. *Journal of the International Neuropsychological Society, 10*, 412–426.

Ylvisaker, M. (1998). Traumatic brain injury in children and adolescents: Introduction. In M. Ylvisaker (Ed.), *Traumatic brain injury rehabilitation: Children and adolescents* (2nd ed., pp. 1–10). Boston: Butterworth-Heinemann.

Ylvisaker, M., Adelson, D., Braga, L. W., Burnett, S. M., Glang, A., Feeney, T., et al. (2005). Rehabilitation and ongoing support after pediatric TBI twenty years of progress. *Journal of Head Trauma Rehabilitation, 20*(1), 95–109.

Ylvisaker, M., Turkstra, L., Coehlo, C., Yorkston, K., Kennedy, M., Sohlberg, M. M., et al. (2007). Behavioural interventions for children and adults with behaviour disorders: A systematic review of the evidence. *Brain Injury, 21*(8), 769–805.

CHAPTER 18

Social Anxiety and Its Treatment in Children and Adolescents with Acquired Brain Injury

Cheryl Soo, Robyn L. Tate, and Ronald M. Rapee

Acquired brain injury (ABI) includes a number of nonprogressive neurological conditions, including traumatic brain injury (TBI), stroke, hypoxia, tumors, and brain infections. It is the leading cause of death and permanent disability in children and young people (Australian Bureau of Statistics, 2006; Sharples, 1998). TBI, caused by an external mechanical force to the head (e.g., force resulting from a road traffic crash or fall), represents the majority of injuries in this group. The incidence of hospital admissions for TBI is estimated at 75 per 100,000 per year in children ages 0–15 years (Mitra, Cameron, & Butt, 2007), with the majority of this group sustaining mild injuries. The impact of childhood ABI is seen across all areas of the framework of the *International Classification of Functioning, Disability and Health* (ICF; World Health Organization, 2001). Injury-related consequences may be seen at the level of *body function and/or structure,* including impairments in physical (Wallen, McKay, Duff, McCartney, & O'Flaherty, 2001), cognitive-behavioral (Anderson & Catroppa, 2006), and emotional (Bloom et al., 2001) functioning. Difficulties may also be seen in a child's level of *activity* (i.e., the execution of a task or action) and *participation* (i.e., involvement in life situations). Such difficulties may affect his or her ability, for example, to get dressed; catch public transport; and engage in recreational, school, and social activities (Bedell & Dumas, 2004).

The occurrence of anxiety following ABI in children and adolescents in particular has been well documented (Grados et al., 2008; Vasa et al., 2002). Indeed, anxiety disorders are the most common mental disorders experienced by children and adolescents in the general population (Fergusson, Horwood, & Lynskey, 1993). Several conditions are classified under anxiety disorders (American Psychiatric Association, 2000), including social phobia (also known as social anxiety disorder), generalized anxiety disorder, specific phobia, panic disorder (with or without agoraphobia), obsessive–compulsive disorder, posttraumatic stress disorder, and acute stress disorder. Past studies have found that childhood brain injury increases the risk of subsequent psychiatric conditions, including anxiety and depression (Luis & Mittenberg, 2002; Vasa et al., 2002). Luis and Mittenberg (2002) compared postinjury anxiety and mood disorders in children ages 6–15 years with mild TBI (*n* = 42), moderate/severe TBI (*n* = 19), or orthopedic injury (the control group; *n* = 35). They found that the percentage of participants with new-onset anxiety disorders in the group with moderate/severe TBI was 63.2%, compared to 35.7% in the group with mild TBI and only 11.4% in the control group.

In this chapter, we outline the literature on social anxiety in children and adolescents with reference to ABI. We present a model proposed by Kendall and Terry (1996), which we use to highlight the complex and interactive factors that may contribute to psychosocial outcome (including the development of social anxiety) in this population. We also present a case study illustrating the treatment of social anxiety with a cognitive-behavioral therapy (CBT) program that was adapted to suit the needs of people with ABI.

Social Anxiety in Children and Adolesecnts with ABI

To date, few studies have specifically examined social anxiety in children and adolescents with ABI. At the extreme end of the continuum, social phobia is characterized by an intense fear of social or performance situations in which embarrassment may occur (American Psychiatric Association, 2000). Social phobia is one of the most common anxiety disorders, with a lifetime prevalence of 7–13% in Western countries (Furmark, 2002). Among younger populations, social phobia affects between 0.3% and 1.5% of children and adolescents within a 6- to 12-month period (Rapee, Schniering, & Hudson, 2009). In regard to the occurrence of social anxiety in childhood ABI, limited data are available. Although no studies have yet examined the prevalence of social anxiety symptomatology, Luis and Mittenberg (2002) reported that 10.5% of their sample of children between 6 and 15 years of age with moderate/severe TBI (*n* = 19) met criteria for social phobia.

A Model of Psychosocial Adjustment Following ABI

The etiology of social anxiety following ABI is complex and may depend on a number of factors. A biopsychosocial model of adjustment following TBI proposed by Kendall and Terry (1996) provides a useful heuristic to illustrate the multiple factors that may contribute to psychosocial outcome following ABI, including the development of anxiety conditions. We use this model to explain social anxiety in particular, but the model also refers to other psychosocial areas more broadly, including interpersonal/family relationships and social behavior. Kendall and Terry (1996) based their framework on a model of psychosocial adjustment proposed by Lazarus and colleagues (see Lazarus, 1993) for populations without ABI. In the Lazarus model, appraisal and coping are influenced by a number of antecedent variables—including personal (e.g., beliefs or traits) and environmental (e.g., social and financial support) resources, as well as situational factors. Individuals rely on these antecedents when they make appraisals of their situation and when they select and utilize coping strategies, the combination of which contributes to psychosocial outcome.

Figure 18.1 displays the psychosocial adjustment framework proposed by Kendall and Terry (1996). They suggest that psychosocial outcome can be predicted by direct (cognitive and neurological/medical) and indirect (personal, environmental, and situational) factors; mediating factors that include cognitive appraisals and coping, as well as the individual's adjustment styles, also contribute. Factors influencing primary appraisal of the situation include perceived uncertainty of the situation, perceived stigma as a result of the situation, and perceived injury severity. Secondary appraisals include perception of self-efficacy and control of the situation. Consistent with this framework, cognitive theories of social phobia also emphasize the importance of these appraisals in the development and maintenance of social anxiety (Clark & Wells, 1995; Rapee & Heimberg, 1997). These appraisals are often in the form of maladaptive beliefs and thought processes; people with high levels of social anxiety attach fundamental and often excessive importance to being positively appraised by others (Rapee & Heimberg, 1997). In addition, coping strategies will have an impact on psychosocial well-being—that is, whether emotion-focused or problem-focused coping strategies are used in the context of the situation's being controllable or uncontrollable.

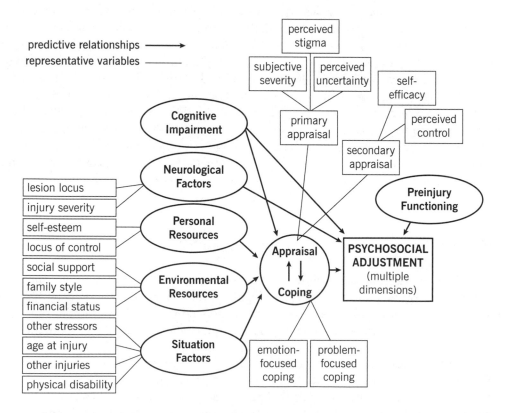

FIGURE 18.1. Kendall and Terry's (1996) model of psychosocial outcome following brain injury. From "Psychosocial Adjustment Following Closed Head Injury: A Model for Understanding Individual Differences and Predicting Outcome" by E. Kendall and D. J. Terry, 1996, *Neuropsychological Rehabilitation, 6,* 101–132. Copyright 1996 by Lawrence Erlbaum Associated Ltd. Reprinted by permission of Taylor & Francis.

According to Kendall and Terry's (1996) model, neurological and cognitive factors also play a part in psychosocial outcome, including locus of the lesion, severity of the injury, and degree of cognitive impairment. Importantly, cognitive impairment can directly influence a person's ability to perform psychosocial tasks (e.g., capacity for effective social interaction); it can also indirectly influence outcome through reducing the ability to appraise the situation and to select, initiate, and maintain coping strategies.

Indirect components of the model comprise personal resources (such as self-esteem and locus of control) and environmental resources (including social/family support, family style, and family financial status). Other indirect factors in the model are situational variables, including, for example, the timing of injury. Childhood ABI outcome research has long debated the relationship between age of injury/onset and outcome. There is growing support for the early-vulnerability model, suggesting that children sustaining severe TBI in early childhood are particularly vulnerable to significant residual impairment (Anderson, Catroppa, Morse, Haritou, & Rosenfeld, 2005). When the Kendall and Terry (1996) model is applied to young people with social anxiety, developmental periods such as adolescence play a major timing role. Adolescence is an important developmental stage, in which the need for belonging and concerns about peer acceptance are pronounced (Petersen & Leffert, 1995). For adolescents with ABI, social anxiety is likely to be compounded by ABI-related difficulties and by developmental considerations relating to adolescence. In addition, the occurrence of the ABI either immediately after or in conjunction with other major stressors may increase the likelihood of poorer outcome by depleting the level of resources a person has available for dealing with the event (Terry, 1991). Other situational variables contributing to psychosocial outcome outlined by the model include existence of multiple physical injuries and degree of residual physical disability.

Kendall and Terry (1996) also emphasize that postinjury psychosocial functioning may be simply a continuation or worsening of circumstances that existed before the injury, including poor family cohesion and dynamics prior to injury. Finally, to further illustrate the complexity of outcome following ABI, the model highlights that we need to consider more than the individual's direct response to the single stressful event of the ABI. It is likely that emotional adjustment following ABI will be affected by the chronic strains and changes (e.g., disabilities) that accompany the ABI, as well as the event of the ABI itself.

Conceptual Issues Relating to Social Anxiety in ABI

As a group, children and adolescents with ABI display a number of disabilities that have specific implications for the presentation of social anxiety in this population. For example, increasing research is appearing on the impact of childhood brain injury on social function (e.g., Hanten et al., 2008; Turkstra, McDonald, & DePompei, 2001; Yeates et al., 2007). People with ABI are particularly vulnerable to social difficulties, given that cerebral regions involved in ABI may include those associated with social function (Hanten et al., 2008; see also Thomas & Tranel, Chapter 4, this volume), and that the high rate of functional disability (e.g., motor, speech) experienced by this group hinders social confidence and self-esteem. In addition, cognitive problems associated with ABI may result in difficulties in social cognition and communication, and hence may impede participation (and practice) in social activities. Factors relating to the injury (e.g., hospitalization, recovery, fatigue) may also disrupt opportunities to develop and maintain social networks (Bohnert, Parker, & Warschausky, 2007). According to models of social phobia, it is the *belief* that one lacks skills and abilities, more than the actual lack of skills, that leads to subsequent anxiety (Rapee & Heimberg, 1997). Among populations without ABI, empirical evidence has generally shown that people high in social anxiety do not typically lack adequate social skills, but rather have a biased perception of inadequate skill (Rapee & Lim, 1992). The relationship between social anxiety and social skills in a population typically considered to have social difficulties (such as individuals with ABI) remains unclear. However, Rapee and Heimberg (1997) propose that where true social skills deficits exist and are recognized by the individual, this will result in heightened anxiety.

Assessment of Social Anxiety in ABI

A number of measures have been developed to assess social anxiety in children and adolescents, including the Social Phobia and Anxiety Inventory (SPAI; Turner, Beidel, Dancu, & Stanley, 1989), the SPAI for Children (SPAI-C; Beidel, Turner, & Morris, 1998), and the Social Anxiety Scales for Children and Adolescents (SAS-C and SAS-A; La Greca, 1999). Other, more general measures of anxiety in children also contain social anxiety subscales (e.g., Birmaher et al., 1997; Spence, 1998). A key limitation on use of these measures for the assessment of social anxiety symptoms in populations with ABI, however, is that none of these scales have been developed with such populations in mind. For example, the SPAI, which can be used for adolescents, contains somatic items that include sweating, shaking, and heart palpitations in social situations. It is possible that some of these physiological signs accompany the medical symptoms relating to ABI, and hence scores on these scales may be inflated. In addition, these scales do not take into account the neuropsychological impairments related to ABI (e.g., the response format may be too complicated), and hence care is required when clinicians are interpreting scores on these measures for children and adolescents with ABI.

Psychological Treatment of Social Anxiety in ABI

Clinicians treating anxiety in children and adolescents with ABI need to consider a number of factors. Pharmacological interventions for this group may not be ideal because of the increased risk of side effects (Williams, Evans, & Fleminger, 2003), such as possible exacerbating effects on cognitive difficulties associated with ABI (Perna, Bordini, & Newman, 2001). Therefore, much of the attention in the field has turned to psychological therapies. Clinicians and researchers alike now believe that CBT is the psychotherapeutic treatment of choice for children and adolescents from the general population who experience anxiety or other internalizing disorders (James, Soler, & Weatherall, 2005; Rapee et al., 2009). There are numerous studies supporting the efficacy of CBT for treating anxiety; a Cochrane Database Systematic Review of 13 studies has found positive support for the efficacy of these approaches for managing anxiety disorders in children and adolescents (James et al., 2005). CBT for anxiety comprises a variety of psychological techniques: primarily exposure (involving the individual's real or imagined confrontation with a feared stimulus), as well as cognitive restructuring using various activities to promote realistic thinking (Rapee et al., 2009). Most of the components of CBT programs are consistent with recommendations of the National Institutes of Health (1999), which indicate that the most successful interventions for individuals with ABI are structured, systematic, and goal-directed, and allow for learning, practice, and social contact in a relevant context. Indeed, there is evidence supporting the use of CBT for managing anxiety in adults with ABI (for a review, see Soo & Tate, 2007). A number of CBT interventions for children are available and have been discussed in the literature, including the Coping Cat (Kendall & Hedtke, 2006), the Cool Kids Child and Adolescent Anxiety Program (Rapee et al., 2006), FRIENDS for Life (Barrett, 2004), and Social Effectiveness Therapy for Children (Beidel, Turner, & Morris, 2004). For a detailed comparison of these programs, the reader is referred to Schoenfield and Morris (2009).

Only one previous study has examined the efficacy of CBT for treatment of social anxiety following brain injury (Hodgson, McDonald, Tate, & Gertler, 2005). In this study, which focused on adults, participants with ABI were randomly allocated to either CBT (*n* = 6) or a wait-list control group (*n* = 6). Those in the treatment group received between 9 and 14 hours of individual sessions of CBT. Compared to the wait-list controls, those in the treatment group showed significant improvements in general anxiety and depression, as well as on a transient mood measure of tension and anxiety, with these gains maintained at a 1-month follow-up. For social anxiety, although moderate to large effect sizes were found between pre- and posttreatment, there was no time × group interaction. It is possible that the non-significant findings for social anxiety reflect the lack of sufficient numbers in the study to detect treatment effects, and thus the study may have been underpowered in this respect. It is also possible, as noted by the authors, that the lack of social anxiety treatment effect was due to the absence of a specific social skills component in the program. Inclusion of such a component (focusing on both the development and self-awareness of social skills) would directly target the social skills difficulties that those with ABI have been reported to display.

An Adapted CBT Program for Social Anxiety in Adolescents with ABI

As noted above, the Cool Kids Child and Adolescent Anxiety Program is one of a number of anxiety management programs available for children and adolescents. It is a structured CBT-based program that has been empirically validated for people without brain injury (see Rapee et al., 2009, for a review of the evidence). This program is used in several countries for the treatment of childhood/adolescent anxiety. To date, there are no data available on the efficacy of Cool Kids as a program for adolescents with ABI. Below, we describe a case study in which Cool Kids was used for the treatment of social anxiety following childhood TBI.

Adapting the Program to the Needs of People with ABI

A number of adaptations to the Cool Kids program were required in order to tailor it to the specific needs of people with ABI. In the adapted program, the emphasis on gradual exposure is retained, remaining the essential core of the program. Several key components distinguish the adapted program from the original version. First, activities for promoting realistic thinking from the original program have been simplified, to make it more likely that adolescents with ABI will be able to follow these concepts. Second, where appropriate, emphasis has been placed on concrete behavioral learning strategies (e.g., within the graded exposure activities). Third, more emphasis has been placed on social and communication skills training. This focus on skills building is considered essential, given the evidence for such difficulties in people with ABI. Finally, a number of strategies have been emphasized in order to minimize the impact of neuropsychological and other difficulties that people with ABI are likely to display, such as problems with memory, executive function, concentration, and fatigue (Ponsford, Sloan, & Snow, 1995). These strategies include the use of written aids (handouts), provision of activities in a structured form (including breaks within sessions), frequent repetition, and tailored timing of practice tasks to account of ABI difficulties (e.g., scheduling tasks for the afternoon if fatigue occurs earlier in the day).

Outline of the Adapted Cool Kids Therapy Program

The Cool Kids program as adapted for ABI comprises 11 sessions. Sessions 1 and 2 are psychoeducation-based, during which ABI and anxiety concepts are outlined. The cognitive, behavioral, and emotional effects of brain injury are discussed, as are possible strategies for overcoming them. The treatment rationale and model of how anxiety develops are also reviewed and discussed. Session 3 (Realistic Thinking) focuses on concepts of realistic thinking and rational responses. In this session, simplified techniques appropriate for people with cognitive difficulties are used, including discussion of examples and role playing, with emphasis on written handouts. Session 4 (Introduction to Exposure and Parenting Anxiety) targets parental management of anxiety-related behaviors and introduces the concept of graded exposure for adolescents. In Session 5 (Exposure), principles of graded exposure and development of relevant hierarchies or "stepladders" are covered. These "stepladders" allow the individual to confront the feared social situation gradually. Graded exposure continues as an active component of the program in subsequent sessions. Session 6 (Social Skills and Assertiveness) focuses on social and communication skills, emphasizing the use of role plays and between-session tasks involving the adolescent's family and peer group. Emphasis on the enhancement of these skills continues where appropriate throughout subsequent sessions of the program. Identifying and addressing issues affecting the implementation of the "stepladders" constitute the focus of Session 7 (Managing Emotions), and strategies for managing emotions are also introduced. Next, Session 8 (Behavioral Experiments) focuses on use of behavioral strategies to determine whether "stepladder" expectations are realistic. Session 9 (Teasing) continues the work on gradual exposure and also covers strategies for dealing with being teased and other social difficulties with peers. Graded exposure activities are again covered in Session 10 (Coping Strategies), with coping skills such as techniques for managing stress also incorporated into this session. The final session (The Future and Maintaining Gains) concentrates on review of the program, relapse prevention, and planning for the future.

Throughout the program, emphasis is placed on encouraging the adolescent to be independent; however, parents are also involved as appropriate to the family's situation. Sessions are provided on a weekly basis and last for approximately 1–1.5 hours, with breaks included as required. Although the adaptations to the program described above were made in order to tailor it to the needs of people with ABI, it should be highlighted that participation in the adapted program still requires an adolescent to have a certain level of ability, such as sufficient cognitive functioning and self-awareness of his or her deficits (e.g., self-reported anxiety). We have recently pilot-tested the adapted Cool Kids program to investigate the effectiveness for use in individuals with a history of ABI.

Below, we outline a case study in which the adapted version of the Cook Kids program was used to treat social anxiety in a 15-year old boy who had sustained a TBI 5 years previously. In this case presentation, the program was delivered via the telephone, because the family resided a long distance from the hospital and face-to-face weekly meetings were not feasible. Such novel delivery modes are important to maximize access to treatment, as many young people with ABI live at a distance from tertiary health care facilities and are reluctant to commit the significant time required to participate in face-to-face behavioral treatments. Previous research including a systematic review on telerehabilitation has found growing support for this mode of therapy for a range of physical and neurological conditions, including TBI (Kairy, Lehoux, Vincent, & Visintin, 2009). In one randomized controlled trial of adults with TBI 1 year after injury, scheduled telephone counseling and education resulted in improved overall outcome, particularly functional status and quality of life, when compared to usual outpatient care (Bell et al., 2005). In addition, studies have shown the benefit of supplementing bibliotherapy (where a program is implemented solely through the use of written or computerized materials, with little or no therapist contact) with telephone sessions in the treatment of anxiety disorders in children without brain injury (Lyneham & Rapee, 2006).

Case Presentation

This case study aimed to examine the effectiveness of the adapted Cool Kids program for managing social anxiety in a 15-year-old who had sustained a TBI. The success of the program was evaluated in terms of social anxiety (primary outcome measure), as well as secondary effects on other anxiety, depression, self-efficacy, participation in everyday activities (generalization effects on the self), and parental stress ("flow-on" effects on the family). Levels of fatigue were also tracked during treatment in order to implement strategies to maximize the effectiveness of the program for the adolescent.

History

Sean (a pseudonym)[1] had sustained a mild complicated TBI from a fall 5 years previously, when he was 10 years of age. He lost consciousness at the scene of the accident and was taken to a hospital. At hospital admission, his Glasgow Coma Score was 13. A computerized tomography scan revealed a small right frontal contusion with minimal mass effect and midline shift. He also sustained a right parietal skull fracture and extradural hematoma. Posttraumatic amnesia information was unavailable; however, Sean was able to recall day-to-day events occurring approximately 4–5 days after trauma (suggesting a duration of posttraumatic amnesia of at least 3 days). He was discharged home 7 days after the injury.

In regard to premorbid functioning, Sean's mother reported no difficulties during his birth. He experienced no delays in achieving developmental milestones. Prior to the

---

[1] Sean and his parents furnished consent to participate in this study and for the results to be written up for publication.

injury, he was an above-average student, and he did not report any preinjury anxiety difficulties.

At the time of assessment, Sean was 15 years old. He was performing at grade level (Year 10 in the Australian educational system) at school and was living with his parents and older brother (20 years old). Sean was referred to the research study by his parents, who expressed concern regarding his high social anxiety levels. They reported that he worried about participating in sporting and leisure activities, and that he refused to go to crowded places. They reported no obvious family stressors in their lives. Sean was generally able to complete activities of daily living without difficulty, although at times he was disorganized. Sean's mother also expressed concerns about Sean's elevated level of fatigue, which she believed was related to his TBI, and which she believed was having an impact on his ability to participate in everyday activities. Sean reported that he still had a few close friends but avoided large groups of people. He described himself as quiet and not able to engage in conversation easily, especially with unfamiliar people. He indicated feeling nervous and worrying a lot, especially about his peers and what they might think of him. He also reported frequent stomachaches and headaches, which he associated with his anxiety.

Initial Assessment

Sean's neuropsychological profile indicated that his skills were consistently within the average range. His verbal and nonverbal intelligence scores were in the average range on the Wechsler Abbreviated Scale of Intelligence (Wechsler, 1999; Verbal IQ = 88, Performance IQ = 110, Full Scale IQ = 100). His learning and memory functions were also in the average range on the Rey Auditory Verbal Learning Test (Rey, 1964; Acquisition z-score = 0.7), as were his attentional abilities on the Trail Making Test (Delis, Kaplan, & Kramer, 2001; Combined Number and Letter Sequencing subtest, standard score = 8; Number–Letter Switching subtest, standard score = 10). In addition, Sean did not display difficulties on a task requiring planning and organization, the Zoo Map Version 1 and 2 (Emslie, Wilson, Burden, Nimmo-Smith, & Wilson, 2006; 79.8–87.8 and 43.4–56.6 percentile bands, respectively). He also performed at an age-appropriate level on a measure of social skills and communication, the Pragmatics Profile of the Clinical Evaluation of Language Fundamentals (Semel, Wiig, & Secord, 2003; criterion score = 147).

Sean's mean score on the Fatigue Severity Scale (Krupp, LaRocca, Muir-Nash, & Steinberg, 1989) was 5.2, indicating significant fatigue (cutoff score of 4 or higher on a 7-point scale where higher scores indicate greater fatigue). When his fatigue levels were tracked over a weekly period early in therapy (Week 3), the pattern indicated that Sean's fatigue level was generally highest in the morning (see Figure 18.2). Self-awareness (the ability to appraise one's current strengths and weaknesses) was measured on the Patient Competency Rating Scale (Prigatano et al., 1986; we adapted this scale for use with adolescents). Sean's mean score was 4.0, and in overall terms was similar to that of his parents' mean score rating of 3.7, suggesting fairly intact level of self-awareness.

Intervention Design and Evaluation Instruments

Given that the Cool Kids program had not been previously evaluated in adolescents with ABI, a case study approach was used to provide preliminary data on the effectiveness of the program.

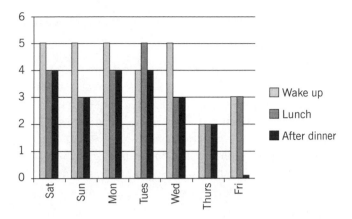

FIGURE 18.2. Sean's fatigue ratings (0, not at all, 10, very much) during Week 3.

The following measures were administered at pretreatment and immediately post-treatment, to evaluate the impact of the program on both social anxiety and more general well-being.

Primary Outcome Measures

SOCIAL ANXIETY

The Social Anxiety Scale for Adolescents (SAS-A; La Greca, 1999) is a 22-item self-report measure divided into three subscales: (1) Fear of Negative Evaluation, (2) Social Avoidance and Distress—new situations, and (3) Social Avoidance and Distress—general social inhibition and distress. Test–retest reliability coefficients of the SAS-A range from .54 to .78 over 2-month and 6-month intervals, and evidence of concurrent validity has been reported (see La Greca, 1999).

Secondary Outcome Measures

OTHER ANXIETY

The Screen for Child Anxiety Related Disorders (SCARED; Birmaher, et al., 1997) is a 41-item self-report questionnaire that assesses the severity of anxiety symptoms broadly in line with DSM-IV dimensions of anxiety disorders. The coefficient alpha value for the total score is reported as .90. The scale has demonstrated discriminant validity both among anxiety, depressive, and disruptive disorders and within anxiety disorders (Birmaher et al., 1999).

DEPRESSION

The Children's Depression Inventory—Short Form (CDI-S; Kovacs, 2003) is a 10-item self-report questionnaire that focuses on cognitive and somatic symptoms of depression. Internal-consistency alpha coefficients of the CDI ranging from .71 to .89 are reported in the manual, along with evidence for discriminant validity: CDI scores have been shown to distinguish persons diagnosed with depression from nondepressed individuals (Kovacs, 2003).

SELF-EFFICACY

The Self-Perception Profile for Adolescents (SPPA; Harter, 1988) is a 36-item self-report questionnaire of perceived competence, consisting of nine subscales covering such domains as Athletic Competence, Scholastic Competence, Social Acceptance, and Global Self-Worth. Internal-consistency alpha coefficients range from .55 to .93 (Harter, 1988). The Social Acceptance and Global Self-Worth subscales were used in this study.

PARTICIPATION IN EVERYDAY ACTIVITIES

The Child and Adolescent Scale of Participation (CASP; Bedell, 2004) consists of 20 items focusing on home, neighborhood/community, and school participation. High internal consistency has been reported, with Cronbach alpha coefficients ranging from .95 to .98. Correlation coefficients between the CASP and the Pediatric Evaluation of Disability Inventory range from .51 to .72, supporting the scale's concurrent validity (Bedell, 2004).

PARENTAL STRESS

The Stress Index for Parents of Adolescents (SIPA; Sheras, Abidin, & Konold, 1998) is a 90-item scale that measures parental stress. It contains a total score and three domain scores: Adolescent Domain, Parent Domain, and Adolescent–Parent Relationship Domain. All subscale Cronbach alpha coefficients exceed .80, with the majority ranging from the high .80s to .90 (Sheras et al., 1998).

The Cool Kids program adapted for ABI as described previously was delivered by a psychologist (Cheryl Soo). Sean completed 11 weekly sessions, each approximately 1–1.5 hours in duration, with breaks provided as needed. Because Sean and his family lived a long distance from the hospital, treatment was provided via the telephone. Handouts and other materials were sent to Sean each week so that he could follow the session with structured visual aids. He was asked to place the handouts from each session in a three-ring binder. Sean had these materials in front of him while speaking on the telephone with the therapist in all of the sessions. Sean's mother was also involved in therapy, as guided by the Cool Kids program. Following the session with Sean, the therapist discussed the concepts and practice tasks outlined in the week's session with Sean's mother as appropriate. Throughout treatment, Sean was actively engaged in the therapy sessions. He was able to follow the handouts easily while the therapist covered the material on the telephone. Pre- and posttreatment assessments were completed by a research assistant who was not involved in the delivery of the therapy.

Evaluation Approach

In order to evaluate the effectiveness of the program, the following statistical approach was employed. First, effect sizes were calculated for pre- to posttreatment changes. The reliable change index (RCI) was used to determine reliable change, according to the method recommended by Jacobson and Truax (1991). The test–retest reliability coefficients were obtained from test manuals and papers evaluating the psychometric properties of the outcome variables. Significant change in test scores were evaluated by using z-scores derived from the RCI, with nonsignificant levels of change being indicated by z-scores between –1.96 and 1.96. A *clinically significant* change was defined as a test score that was statistically significant *and* shifted from the impaired to average range (Jacobson, Follette, & Revenstorf, 1984), or as the individual's making a functional change in his or her everyday life (Perdices, 2005).

Treatment Progress

Sean's overall goal in therapy was to increase his level of participation in social and recreational activities, particularly activities involving groups/crowds of people. In particular, Sean and the therapist worked together to construct a graded exposure hierarchy focusing on going to the local shops, and leading toward purchasing an item on his own in a crowded shop. Sean rated this graded exposure activity as very important to him, as success at the task would contribute to his sense of independence. He progressed well throughout the program, although he displayed a considerable level of inertia; that is, he found it difficult to generate information relating to a number of activities in the program, often stating that he was "not sure." Adjustments were made to assist Sean in generating examples from his own life that would be relevant to the activities of the program. For example, when Sean was asked what rewards he would like for achieving his graded exposure subgoals, the therapist provided a number of reward categories for him to choose from, allowing him to select his rewards more easily.

Outcome Measurement

Primary Outcome

As shown in Table 18.1, Sean's pretreatment social anxiety scores as measured by the SAS-A were in the clinical range (>1.5 standard deviations below the mean). Although Sean's scores on all subscales improved at posttreatment, these scores did not show statistically significant change and, with one exception (Social Avoidance and Distress in new situations), remained in the clinical range. In contrast, on the broader measure of anxiety, the SCARED, Sean's score on the Social Anxiety subscale moved from clinical levels at pretreatment to nonclinical levels at posttreatment. Improvement on this subscale was statistically significant, according to the RCI.

Secondary Outcomes

Sean's scores on the SCARED subscales for other anxiety difficulties, also improved from pre- to posttreatment (with the exception of the Somatic/Panic and Generalized Anxiety subscales). Notably, the School Avoidance subscale moved from clinical levels at pretreatment to nonclinical levels at posttreatment. However, the changes on these SCARED subscales did not reach statistical significance. Sean's score on the brief depression measure,

TABLE 18.1. Sean's Pre- and Posttreatment Scores

Measure	Pretreatment	Posttreatment
Primary outcome measure		
SAS-A (self-report)	ZS	ZS
Fear of Negative Evaluation	−2.6	−2.1
Social Avoidance and Distress—new situations (SAD-new)	−1.7	−0.7
Social Avoidance and Distress—general	−2.6	−2.2
Secondary outcome measures		
SCARED (self-report)	RS	RS
Somatic/Panic	3	4
Generalized Anxiety	6	8
Separation Anxiety	3	1
Social Anxiety	12[a]	7
School Avoidance	4[a]	2
CDI-S (self-report)	TS	TS
Total score	65	54
SPPA (self-report)	ZS	ZS
Social Acceptance	−1.05	−0.70
Global Self-Worth	0.00	−0.62
CASP (parent report)	CS	CS
Home Participation	88 (average)	95 (average)
Neighborhood/Community Participation	75 (average)	69 (borderline)
School Participation	95 (average)	100 (average)
Home/Community Living Activities	85 (average)	70 (borderline)
SIPA (parent report)	%ile	%ile
Adolescent Domain	70	71
Parent Domain	47	58
Adolescent–Parent Relationship Domain	42	52

Note. SAS-A, Social Anxiety Scale for Adolescents; SCARED, Screen for Childhood Anxiety Related Disorders; CDI-S, Children's Depression Inventory—Short Form; SPPA, Self-Perception Profile for Adolescents; CASP, Child and Adolescent Scale of Participation; SIPA, Stress Index for Parents of Adolescents; ZS, z-score; RS, raw score; TS, *T*-score; CS, converted score, according to the test author's scoring instructions; %ile, percentile.
[a]Clinical range.

the CDI-S, improved from clinical to nonclinical levels; however, again, this improvement was not statistically significant.

In regard to self-concept, Sean's pretreatment score on the Social Acceptance subscale of the SPPA was one standard deviation below the mean, and his score on the Global Self-Worth subscale was in the average range. At posttreatment, his scores on both scales were in the average range. However, the changes in scores on the SPPA subscales did not reach statistical significance, according to the RCI. Participation (CASP) and parental stress (SIPA) measures were in the average range at pretreatment and generally remained unchanged (borderline/average scores) at posttreatment.

Summary of Evaluation

The findings from this case study indicate that although Sean's scores on the primary outcome measure of social anxiety (SAS-A) showed numerical improvement following the completion of the Cool Kids program, the change was not statistically significant. In contrast, scores on the Social Anxiety subscale of the SCARED not only showed statistically significant change, but moved from the clinical to nonclinical range, indicating *clinically significant* change. It is not clear why the results from the two social anxiety measures were conflicting. It may be because the two scales contain items emphasizing different aspects of social anxiety. For example, compared to the SCARED, the SAS-A contains considerably more items (and a subscale) representing the construct of "fear of negative evaluation," and Sean obtained high scores on this subscale at both pre- and posttreatment. Results from secondary outcome measures of other anxiety, depression, self-concept, participation in everyday activities, and parental stress showed variable results. Notably, school avoidance, depression, and self-concept (social acceptance) scores moved from clinical to nonclinical levels. On the other hand, scores on participation and parental stress scales were in the average range at pretreatment and generally showed no change at posttreatment.

Although improvements across test scores were mixed, functional improvements in Sean's life should be highlighted, as suggested by Perdices (2005). In particular, the Cool Kids program was also evaluated in regard to attainment of everyday goals Sean identified prior to therapy. Both Sean and his mother reported a number of positive behavioral changes at posttreatment that were directly related to his goal of increasing his participation in social and recreational activities. Sean's mother reported that he became more confident and noticed that he was more willing to try different activities posttreatment. For example, following treatment Sean was able to go to a crowded café to buy a soft drink—a task he refused to do before treatment. In addition, during Week 10 of the program, Sean reported that he attended a job interview for a cashier's position at the local convenience store and stated that he "felt pretty proud of himself" afterwards.

It is noteworthy that Sean's score on the Fear of Negative Evaluation subscale of the SAS-A showed little improvement following treatment and remained in the clinical range. This is actually consistent with the way the Cool Kids program is structured, because much of the program directly targets the behavioral avoidance aspects of anxiety (primarily graded exposure activities). Less therapeutic work was focused on Sean's belief relating to fear of what people would think of him (i.e., negative evaluation), and thus it could be expected that his score on this construct might not show substantial improvement. Theoretically, it might be expected that Sean's fear of negative evaluation might change more gradually as long as he continued to engage with social activities. On the other hand, the lack of change on this cognitive variable might point to an increased risk for relapse. Interestingly, Sean's scores showed high levels of social anxiety, despite his being reported as having age-appropriate social and communication skills on a parent-rated questionnaire. This is consistent with cognitive models of social anxiety (e.g., Clark & Wells, 1995; Rapee & Heimberg, 1997), which have proposed that those with social anxiety do not necessarily lack adequate social skills. It is also in line with the concept that individuals' negative appraisal of their social skills could undermine their performance and confidence in social situations (Rapee & Heimberg, 1997), and it can be explained within the appraisal components of the Kendall and Terry (1996) model.

Generalization of Findings to Treatment of Adolescents with ABI

A number of factors should be considered in generalizing findings from this case study to treatment of adolescents with ABI. It is noteworthy that on testing Sean did not display cognitive impairments. Thus, when engaging in the cognitive components of the program (e.g., the realistic thinking exercises), he generally had no difficulties in following the program materials. In addition, Sean's level of self-awareness (particularly regarding his anxiety and fatigue levels) was fairly intact, and this was essential for his ability to engage in the program. Sufficient cognitive ability and self-awareness have been identified in previous research as key influences on treatment outcome (Prigatano et al., 1984). Nevertheless, this case study shows the importance of providing scaffolding to help those with ABI to engage in the activities provided in therapy (Khan-Bourne & Browne, 2003). A number of clinical observations during therapy indicated that Sean benefited from structured therapy sessions, and particularly from the modifications made to accommodate the typical consequences of ABI. Sean readily followed and completed the therapy worksheets, which provided visual representations of concepts. The use of a binder with dividers to organize the therapy worksheets also allowed him to easily locate the materials from earlier sessions when needed. For each session, the therapist began by reviewing the previous session's materials and practice tasks, which provided Sean with reminders to assist him to learn the materials.

The modest gains demonstrated in Sean's case may be due to a number of factors, including the administration of the therapy by telephone. Although the provision of handouts helped Sean to follow the materials, it is likely that gains would have been greater if the therapy had been provided in a face-to-face format—where, for example, nonverbal cues between Sean and the therapist could have been utilized. Pragmatically, treatments for young people with ABI need to account for difficulties in accessing therapy via traditional routes. Although the medium of the telephone may not be optimal, advances in technology provide a number of alternative opportunities, such as web-based applications (e.g., Skype). Sean's fatigue and inertia levels may have also affected his therapy outcome. This possibility is in keeping with clinical reports and highlights the importance of considering the effects of fatigue on social participation in young people with ABI. A number of strategies were undertaken to minimize these factors, including scheduling therapy sessions and practice tasks in the afternoon (i.e., times of less fatigue for Sean). On a number of occasions, however, Sean's fatigue and lack of ability to initiate may have reduced his ability to complete the practice tasks.

It is important to acknowledge that the use of a case study design is less methodologically rigorous and is subject to such limitations as bias in the way treatment outcome is measured. The next step in our research will be to reevaluate the modified Cool Kids program by using a single-case experimental design trial in which cause-and-effect relationships can be established (e.g., a multiple-baseline design).

Conclusions

The findings from this case study suggest that CBT tailored to each individual's needs may be helpful for young people with ABI, particularly in the context of sufficient levels of cognitive ability and self-awareness. To our knowledge, this case study represents the first published illustration of the use of CBT for managing social anxiety in childhood ABI; however, the findings presented here should be considered preliminary. Further controlled trials, including single-case experimental design and randomized controlled trials, are needed to support the efficacy of CBT in this population. Investigating the therapeutic value of specific components of these programs in the context of the unique characteristics of childhood brain injury (e.g., neurological, cognitive, situational factors) will assist us in our understanding of the developmental social neuroscience relating to the treatment of social anxiety in children and adolescents with ABI.

Acknowledgments

This chapter was written with the support of a research fellowship from the Victorian Neurotrauma Initiative awarded to Cheryl Soo. We are grateful to Dr. Rowena Conroy for her helpful comments on the case study detailed in this chapter. We also acknowledge Dr. Conroy, Associate Professor Cathy Catroppa, and Dr. Heidi Lyneham for their involvement in the adaptation of the therapy program presented in this chapter. Finally, we would like to thank Sean and his parents for their involvement in this research.

References

American Psychiatric Association (2000). *Diagnostic and statistical manual of mental disorders* (4th ed., text rev.). Washington, DC: Author.

Anderson, V., & Catroppa, C. (2006). Advances in post-acute rehabilitation after childhood brain injury: A focus on cognitive, behavioral and social domains. *American Journal of Physical and Medical Rehabilitation, 85,* 767–778.

Anderson, V., Catroppa, C., Morse, S., Haritou, F., & Rosenfeld, J. (2005). Functional plasticity or vulnerability after early brain injury? *Pediatrics, 116*(6), 1374–1382.

Australian Bureau of Statistics. (2006). *Health of children in Australia: A snapshot 2004–2005.* Canberra: Author.

Barrett, P. M. (2004). *FRIENDS for Life group leaders' manual for children.* Bowen Hills, Queensland: Australian Academic Press.

Bedell, G. M. (2004). Developing a follow-up survey focused on participation of children and youth with acquired brain injuries after discharge from inpatient rehabilitation. *NeuroRehabilitation, 19,* 191–205.

Bedell, G., & Dumas, H. M. (2004). Social partvicipation of children and youth with acquired brain injury discharged from inpatient rehabilitation: A follow up study. *Brain Injury, 18*(1), 65–82.

Beidel, D. C., Turner, S. M., & Morris, T. L. (1998). *Social Phobia and Anxiety Inventory for Children (SPAI-C) manual.* North Tonawanda, NY: Multi-Health Systems.

Beidel, D. C., Turner, S. M., & Morris, T. L. (2004). *Social Effectiveness Therapy for Children: A treatment manual.* North Tonawanda, NY: Multi-Health Systems.

Bell, K. R., Temkin, N. R., Esselman, P. C., Doctor, J. N., Bombardier, C. H., Fraser, R. T., et al. (2005). The effect of a scheduled telephone intervention on outcome after moderate to severe traumatic brain injury: A randomized trial. *Archives of Physical Medicine and Rehabilitation, 86,* 851–856.

Birmaher, B., Brent, D. A., Chiappeta, L., Bridge, J., Monga, S., & Baugher, M. (1999). Psychometric properties of the Screen for Child Anxiety Related Emotional Disorders (SCARED): A replication study. *Journal of the American Academy of Child and Adolescent Psychiatry, 38,* 1230–1236.

Birmaher, B., Khetarpal, S., Brent, D., Cully, M., Balach, L., Kaufman, J., et al. (1997). The Screen for Child Anxiety Related Emotional Disorders (SCARED): Scale construction and psychometric characteristics. *Journal of the American Academy of Child and Adolescent Psychiatry, 36,* 545–553.

Bloom, D. R., Levin, H. S., Ewing-Cobbs, L., Saunders, A. E., Song, J., Fletcher, J. M., et al. (2001). Lifetime and novel psychiatric disorders after pediatric traumatic brain injury. *Journal of the American Academy of Child and Adolescent Psychiatry, 40,* 572–579.

Bohnert, A. M., Parker, J. G., & Warschausky, S. A. (1997). Friendship and social adjustment of children following a traumatic brain injury: An exploratory investigation. *Developmental Neuropsychology, 13,* 477–486.

Clark, D. M., & Wells, A. (1995). A cognitive model of social phobia. In R. G. Heimberg, M. R. Liebowitz, D. A. Hope, & F. R. Schneier (Eds.), *Social phobia: Diagnosis, assessment, and treatment* (pp. 69–93). New York: Guilford Press.

Delis, D. C., Kaplan, E., & Kramer, J. H. (2001). *Delis–Kaplan Executive Function System.* San Antonio, TX: Psychological Corporation.

Emslie, H., Wilson, F. C., Burden, V., Nimmo-Smith, I., & Wilson, B. A. (2006). *Behavioural Assessment of the Dysexecutive Syndrome in Children (BADS-C) manual.* London: Psychological Corporation.

Fergusson, D. M., Horwood, L. J., & Lynskey, M. T. (1993). Prevalence and comorbidity of DSM-III-R diagnoses in a birth cohort of 15 year olds. *Journal of the American Academy of Child and Adolescent Psychiatry, 32,* 1127–1134.

Furmark, T. (2002). Social phobia: Overview of community surveys. *Acta Psychiatrica Scandinavica, 105,* 84–93.

Grados, M. A., Vasa, R., Riddle, M. A., Slomine, B. S., Salorio, C., Christensen, J., et al. (2008). New onset obsessive–compulsive symptoms in children and adolescents with severe traumatic brain injury. *Depression and Anxiety, 25,* 398–407.

Hanten, G., Wilde, E. A., Deleene, M., Li, X., Lane, S., Vasquez, S., et al. (2008). Correlates of social problems solving during the first year after traumatic brain injury in children. *Neuropsychology, 22,* 357–370.

Harter, S. (1988). *Manual for Self-Perception Profile for Adolescents.* Denver, CO: University of Denver.

Hodgson, J., McDonald, S., Tate, R., & Gertler, P. (2005). A randomised controlled trial of a cognitive-behavioural therapy program for managing social anxiety after acquired brain injury. *Brain Impairment, 6,* 169–180.

Jacobson, N. S., Follette, W. C., & Revenstorf, D. (1984). Psychotherapy outcome research: Methods for reporting variability and evaluating clinical significance. *Behavior Therapy, 15,* 336–352.

Jacobson, N. S., & Truax, P. (1991). Clinical significance: A statistical approach to defining meaningful change in psychotherapy research. *Journal of Consulting and Clinical Psychology, 59*(1), 12–19.

James, A., Soler, A., & Weatherall, R. (2005). Cognitive behavioural therapy for anxiety disorders in children and adolescents. *Cochrane Database Systematic Reviews,* Issue 4. Retrieved from *onlinelibrary.wiley.com/doi/10.1002/14651858.CD004690.pub2/pdf*

Kairy, D., Lehoux, P., Vincent, C., & Visintin, M. (2009). A systematic review of clinical outcomes, clinical process, healthcare utilization and costs associated with telerehabilitation. *Disability and Rehabilitation, 31*(6), 427–447.

Kendall, E., & Terry, D. J. (1996). Psychosocial adjustment following closed head injury: A model for understanding individual differences and predicting outcome. *Neuropsychological Rehabilitation, 6,* 101–132.

Kendall, P. C., & Hedtke, K. A. (2006). *Cognitive-behavioral therapy for anxious children: Therapist manual.* Ardmore, PA: Workbook Publications.

Khan-Bourne, N., & Browne, R. G. (2003). Cognitive behaviour therapy for the treatment of depression in individuals with brain injury. *Neuropsychological Rehabilitation, 13,* 89–107.

Kovacs, M. (2003). *Children's Depression Inventory.* North Tonawanda, NY: Multi-Health Systems.

Krupp, L. B., LaRocca, N. G., Muir-Nash, J., & Steinberg, A. D. (1989). The Fatigue Severity Scale: Application to patients with multiple sclerosis and systemic lupus erythematosus. *Archives of Neurology, 46*(10), 1121–1123.

La Greca, A. M. (1999). *Manual for Social Anxiety Scales for Children and Adolescents.* Miami, FL: University of Miami.

Lazarus, R. (1993). Coping theory and research: Past, present and future. *Psychosomatic Medicine, 55,* 234–247.

Luis, C. A., & Mittenberg, W. (2002). Mood and anxiety disorders following pediatric traumatic brain injury: A prospective study. *Journal of Clinical and Experimental Neuropsychology, 24,* 270–279.

Lyneham, H. J., & Rapee, R. M. (2006). Evaluation of therapist-supported parent-implemented CBT for anxiety disorders in rural children. *Behaviour Research and Therapy, 44*(9), 1287–1300.

Mitra, B., Cameron, P., & Butt, W. (2007). Population-based study of pediatric head injury. *Journal of Paediatrics and Child Health, 43*(3), 154–159.

National Institutes of Health. (1999). NIH consensus development panel on rehabilitation of persons with traumatic brain injury. *Journal of the American Medical Association, 282,* 974–983.

Perdices, M. (2005). How do you know whether your patient is getting better (or worse)?: A user's guide. *Brain Impairment, 6*(3), 219–226.

Perna, R. B., Bordini, E. J., & Newman, S. A. (2001). Pharmacological treatments: Considerations in brain injury. *Journal of Cognitive Rehabilitation, 19,* 4–7.

Petersen, A. C., & Leffert, N. (1995). What is special about adolescence? In M. Rutter (Ed.), *Psychosocial disturbances in young people: Challenges for prevention* (pp. 3–37). Cambridge, UK: Cambridge University Press.

Ponsford, J., Sloan, S., & Snow, P. (1995). *Traumatic brain injury: Rehabilitation for everyday adaptive living.* Hove, UK: Psychology Press.

Prigatano, G. P., Fordyce, D. J., Zeiner, H. K., Roueche, J. R., Pepping, M., & Wood, B. C. (1984). Neuropsychological rehabilitation after closed head injury in young adults. *Journal of Neurology, Neurosurgery and Psychiatry, 47,* 505–513.

Prigatano, G. P., Fordyce, D. J., Zeiner, H. K., Roueche, J. R., Pepping, M., & Wood, B. C. (1986). *Neuropsychological rehabilitation after brain injury.* Baltimore: Johns Hopkins University Press.

Rapee, R. M., & Heimberg, R. G. (1997). A cognitive behavioural model of anxiety in social phobia. *Behaviour Research and Therapy, 35*(8), 741–756.

Rapee, R. M., & Lim, L. (1992). Discrepancy between self-and observer ratings of performance in social phobia. *Journal of Abnormal Psychology, 19*(4), 728–731.

Rapee, R. M., Lyneham, H. J., Schniering, C. A., Wuthrich, V., Abbott, M. J., Hudson, J. L., et al. (2006). *The Cool Kids Child and Adolescent Anxiety Program therapist manual.* Sydney: Centre for Emotional Health, Macquarie University.

Rapee, R. M., Schniering, C. A., & Hudson, J. L. (2009). Anxiety disorders during childhood and adolescence: Origins and treatment. *Annual Review of Clinical Psychology, 5,* 311–341.

Rey, A. (1964). *L'examen clinique en psychologie*. Paris: Presses Universitaires de France.

Schoenfield, G., & Morris, R. (2009). Cognitive-behavioral treatment for childhood anxiety disorders: Exemplary programs. In M. Mayer, R. Van Acker, J. E. Lochman, & F. Gresham (Eds.), *Cognitive-behavioral interventions for emotional and behavioral disorders: School-based practice* (pp. 204–232). New York: Guilford Press.

Semel, E., Wiig, E. H., & Secord, W. A. (2003). *Clinical Evaluation of Language Fundamentals— Fourth Edition*. San Antonio, TX: Psychological Corporation.

Sharples, P. M. (1998). Head injury in children. In M. P. Ward Platt & R. A. Little (Eds.), *Injury in the young* (pp. 263–299). Cambridge, UK: Cambridge University Press.

Sheras P. L., Abidin, R. R., & Konold, T. R. (1998). *SIPA: Stress Index for Parents of Adolescents*. Odessa, FL: Psychological Assessment Resources.

Soo, C., & Tate, R. L. (2007). Psychological treatment for anxiety in people with traumatic brain injury. *Cochrane Database Systematic Reviews,* Issue 3. Retrieved from *onlinelibrary.wiley. com/doi/10.1002/14651858.CD005239.pub2/pdf*

Spence, S. H. (1998). A measure of anxiety symptoms among children. *Behaviour Research and Therapy, 36*(5), 545–566.

Terry, D. J. (1991). Predictors of subjective stress in a sample of new parents. *Australian Journal of Psychology, 43*, 29–36.

Turkstra, L., McDonald, S., & DePompei, R. (2001). Social information processing in adolescents: Data from normally-developing adolescents and preliminary data from their peers with traumatic brain injury. *Journal of Head Trauma Rehabilitation, 16*, 469–483.

Turner, S. M., Beidel, D. C., Dancu, C. V., & Stanley, M. A. (1989). An empirically derived inventory to measure social fears and anxiety: The Social Phobia and Anxiety Inventory. *Psychological Assessment, 1*, 35–40.

Vasa, R., Gerring, J., Grados, M., Slomine, B., Rising, W., Christensen, J., et al. (2002). Anxiety after severe pediatric closed head injury. *American Academy of Child and Adolescent Psychiatry, 41*, 148–156.

Wallen, M., McKay, S., Duff, S., McCartney, L., & O'Flaherty, S. (2001). Upper-limb function in Australian children with traumatic brain injury: A controlled, prospective study. *Archives of Physical Medicine and Rehabilitation, 82*(5), 642–649.

Wechsler, D. (1999). *Wechsler Abbreviated Scale of Intelligence*. San Antonio, TX: Psychological Corporation.

Williams, W. H., Evans, J. J., & Fleminger, S. (2003). Neurorehabilitation and cognitive behavioural therapy of anxiety disorders after brain injury: An overview and a case illustration of obsessive compulsive disorder. *Neuropsychological Rehabilitation, 13*, 133–148.

World Health Organization. (2001). *International classification of functioning, disability and health*. Geneva: Author.

Yeates, K. O., Bigler, E. D., Dennis, M., Gerhardt, C. A., Rubin, K., Stancin, T., et al. (2007). Social outcomes in childhood brain disorder: A heuristic integration of social neuroscience and developmental psychology. *Psychological Bulletin, 133*(3), 535–556.

Index